TECHNOLOGY ASSESSMENT IN A DYNAMIC ENVIRONMENT

Technology Assessment in a Dynamic Environment

Edited by
Marvin J Cetron
and
Bodo Bartocha

GORDON AND BREACH SCIENCE PUBLISHERS
LONDON · NEW YORK · PARIS

Gordon and Breach, Science Publishers, Inc.
440 Park Avenue South
New York, N.Y. 10016

Editorial office for the United Kingdom

Gordon and Breach, Science Publishers Ltd.
42 William IV Street
London W.C.2.

Editorial office for France

Gordon & Breach
7-9 rue Emile Dubois
Paris 14e

ISBN 0 677 13150 X
Library of Congress catalog card number 72-75869.

Printed in the United Kingdom

TABLE OF CONTENTS

FOREWORD

"Slow down, you're going too fast" runs the refrain of a popular song of recent memory. To a generation born in the age of the rush to technology that followed the Second World War, these words express more than an existential philosophy, but perhaps the fear that our technological society is out of control. Our great cities decay, our transportation, communication and service systems deteriorate, the natural environment is degraded, and a dehumanization seems to permeat ever more deeply into our personal lives.

Too many of us, though we know that correlation is not cause, feel that these ills are an inescapable concomitant of technological progress — and question whether technology is progress at all. Yet at the same time the decision makers of western society are often wedded to the concept that technology is a panacea for all social-economic-problems.

But does not the truth, as often, lie somewhere in the vast middle ground? The ill effects complained of are not inevitable. In the rapid thrust of the Western world in the last thirty years, there have undoubtedly been far too many examples of technology overhastily introduced, of insufficient planning, of unnecessary failures of vision.

Technology is politically, ideologically and philosophically neutral; many engineers, scientists and technologists are the same. Yet it is unwise, even perilous, for either technologists or policymakers to assume that this neutrality means that its social (and political, ideological and philosophic) effects can be ignored. Technology is not asocial. Technological activity is one of the most fundamentally human pursuits, and innovation is introduced into society *because* it affects it. And technology need not be a tangible product; it is also system, process, people working with people. The uses to which technology are put are social decisions. That so often they are taken with insufficient knowledge of their consequences is an indictment of bad planning, not bad technology.

This important new book compiled and edited by Drs. Cetron and Bartocha represents a milestone in technology and government. The words, technology assessment, will from now on be a significant part of the language of government and technology. Technology Assessment is a term applied to a class of research activities intended to provide essential background facts to aid public policymakers in the Executive and Legislative branches of government. Technology Assessment may be defined as a systematic forecasting process that yields options, costs, and benefits of a new technology with special focus on the secondary and higher order effects of that technology.

When science and technology were young, and the population density small, the second order and long term effects of a new technology received scant attention. Surely, little thought was given to the long term influence of the crude vehicle first marketed by Henry Ford. It is hard to believe that Ford and his backers, or any responsible scientists, engineers or government officials, could have imagined the full and ultimate impact of the automobile on society. The automobile example illustrates clearly the challenges and difficulties of making a credible technology assessment. In its fullest form, it should forecast and evaluate the impact on all sectors of society.

The world's leading authorities on Technology Assessment have contributed to the present book. The state of the art in the United States and elsewhere is discussed together with the methods available and proposed for conducting Technology Assessments. Representative case studies are presented with analyses of their strengths and weaknesses. The authors suggest further research in the field and discuss priority areas for this research. Finally, the outlook for Technology Assessment is discussed by several authors.

The pioneering character and timeliness of the present book can be judged by the attention which is being given by governments throughout the world to Technology Assessment. One illustration of this timeliness was the special attention given Technology Assessment at the Fall 1971 meeting of the Organization for Economic Cooperation & Development (OECD) in Paris, Another was the House of Representatives approval on 8 February 1972 of a bill encompassing Congressman Emilio Q. Daddario's 1967 plan for an Office of Technology Assessment (OTA) for the Congress of the United States. The preamble to the bill explains the importance of Technology Assessment to the Congress of the United States and indeed to all governments of industrial nations.

"Emergent national problems, physical, biological and social, are of such a nature and are developing at such an unprecedented rate as to constitute a major threat to the security and general welfare of the United States...

"The present mechanisms of the Congress do not provide the legislative branch with adequate independent and timely information concerning the potential application or impact of such technology, particularly in those instances where the Federal Government may be called on to consider support, management, or regulation of technological applications."

"It is therefore imperative that the Congress equip itself with new and effective means for securing competent, unbiased information concerning the effects, physical, economic, social and political, of the applications of technology, and that such information be utilized whenever appropriate as one element in the legislative assessment of matters pending before the Congress."

Because technology encompasses the full spectrum from hardware to human factors, the introduction of technology into society must be accompanied with a clearer view of what ends will be accomplished by the act than heretofore. The clearer view can be provided by Technology Assessment. And whether Technology Assessment is a technological art or a social science is a philosophical question in the main. Suffice it to say that it lies at the interface between the science of society and science *for* society. In Technology Assessment lies the hope for technological advance that is an advance on all fronts for mankind, and thereby for a society with both the material wealth technology can bring and the humane qualities that will make it worth the effort.

Washington, D.C.

Raymond L. Bisplinghoff
Deputy Director, National
Science Foundation

PREFACE

Technology Assessment, like motherhood, is hard to oppose, but there seems to be a plethora of views on what exactly it is.

Many ingenious lawmakers state they are looking forward to the "objective" and "impartial" study that Technology Assessment will produce on such complex matters as the supersonic transport, the advanced ballistic missile, etc.: but Congressman John W. Davis, the Chairman of the House Subcommittee on Sçience and Atronautics, has recently released a report which will lead one to believe that impartiality in Technology Assessment is impossible. In fact, we also feel Technology Assessment cannot exclude all biases.

Bias lurks in the basic assumptions explicit or implicit in every study. It is enhanced in the omissions and neglected challenges. In fact, no study can be unbiased because someone must select the factual information which must be considered and since everything cannot be considered, naturally, bias creeps in. Therefore, the author(s) of any Technology Assessment should not claim that his assessment is the last word.

The planning of "Technology Assessment in a Dynamic Environment" began originally in 1969. During the summer of 1971, after much deliberation, the editors decided that technology assessment had been under sufficient discussion to warrant a "state-of-the-art" synthesis, the results of which make up this book.

In selecting the contents, the editors reviewed the works of approximately 100 authors, who have written on this subject both in the United States and on a worldwide basis, and asked some 40 of those authors to write specific chapters for this volume.

During subsequent reviewing, it was resolved that certain redundancies or overlaps in contents would be permitted, in order for each separate chapter to be internally consistent and coherent to the reader. This approach was also used in accepting articles from abroad, so that the reader would have a more realistic picture regarding the state of the art of technology assessment in each country represented.

Many viewpoints, including opposing ones, are presented in this collection. The editors felt strongly that technology assessment had not matured sufficiently for it to adopt its own particular rules and rigorous methodologies. Case studies were also used to illustrate the various approaches currently employed in technology assessment.

This book is divided into four general areas: Part I — An Overview of Technology Assessment; Part II — Methodological Considerations; Part III — Case Studies; and, Part IV — The Future of Technology Assessment.

The major goal of the editors in producing "Technology Assessment in a Dynamic Environment" was to present the many disputed and varied concepts of technology assessment in a way that can be readily utilized by a wide range of managers and decision makers who will become the future users of technology assessment.

A special note of thanks is due to Joe Coates, Program Manager in the Exploratory Research and Assessment Office of the National Science Foundation in Washington, D.C., who served with us as a reviewer of these chapters and to Liz Pattison who acted as our editorial assistant. She, with the able secretarial assistance of Mary Jo Hardesty, helped combine the chapters and improve the continuity of what we hope will be a book for an interdisciplinary audience.

Marvin J. Cetron
Bodo Bartocha

Arlington, Virginia

PART I

TECHNOLOGY ASSESSMENT IN VARIOUS PARTS OF THE WORLD

TECHNOLOGY ASSESSMENT: A LAYMAN'S OVERVIEW*

DAVID M. KIEFER

Senior Editor
Chemical and Engineering News, Washington, D.C., U.S.A.

Technology assessment looks like a great idea. The argument goes something like this: Science and technology have vastly benefited man and contributed to his well-being. At the same time, however, they all too often have done him and his environment much harm. Man has used his technology to dominate nature. Now, however, man may be dominated, in turn, by technology. Now, an overriding need is to devise a way of protecting man from his own technological creativity.

Technology assessment is an attempt - still halting and uncertain - to establish an early-warning system to control, direct, and, if necessary, restrain technological development so as to maximize the public good while minimizing the public risks. It is, no less, a new approach to allocating scientific resources, setting technological priorities, and seeking more benign alternatives to the technology already at hand.

It is not so much technology itself that is the monster, of course, but the way in which it is employed to bring about economic or social change. Generally, however, whenever technology becomes available, the urge to use it is overwhelming if it offers profits or political or military

---

* Reprint from *Chemical and Engineering News*, Volume 48, October, 1970, with permission from the copyright owner, The American Chemical Society.

advantages. This is "progress." Efforts thus far to promote and accelerate innovation far exceed any attempts to limit it. Little or no concern has been shown for indirect or by-product effects of new developments. Yet second-order and third-order or even more remote consequences, in a society as complex and interlocking as ours, are no less significant than first-order effects. In the long run second-order consequences may have even greater impact.

In an urban community, for instance, transportation, communications, housing, retailing, education, medical care, public services, and industrial activities are all interdependent. A change in one system will be reflected through all and thereby affect the quality of life in the entire community - even though the decision to make the initial change may be reached in one small part of the community remote from the usual centers of political power or control.

The costs or hazards stemming from indirect effects, moreover, frequently fall not upon those who benefit from new technology but upon uninvolved, or even unidentifiable, bystanders, or upon the public at large, or even upon generations unborn. The benefits may be local whereas the costs are far-ranging or even global. Ill effects, too, may appear as a large number of incremental changes, each in itself small and perhaps hardly perceptible, that build up slowly into what in sum is a danger of considerable dimensions. Often the damage may well be irreversible, as well; we find ourselves painted into an uncomfortable corner from which we cannot readily escape.

As a result, a new development that is acclaimed as a technological and economic success in the short run and on a microscale may lead unwittingly to a long-run sociological macrofailure.

Our society tends, moreover, to be crisis oriented. Yet decisions made merely in reaction to crises become less and less effective. Policy does not anticipate future problems; rather it merely deals with those of the past and present.

Technology assessment, then, is a mechanism for focusing on second- and higher-order consequences for balancing these consequences against first-order benefits, and for asking new questions instead of merely trying to come up with better answers to old ones.

Speaking to a meeting of the National Academy of Sciences seven years ago, President John F. Kennedy mused that "every time you scientists make a major invention, we politicians have to invent a new institution to cope with it." Technology assessment is an attempt to put the process of coping in parallel with the process of invention, rather than after the fact.

As a concept, therefore, technology assessment is difficult to fault. Looked at on a hard, practical, how-to-do-it basis, though, it poses more fundamental questions - of the traditional who, where, when, what, and why variety - than now can be answered. It calls for decisions concerning a variety of ill-understood and intractable matters relating to where technology is taking us, our basic values and objectives, and the ties between technology and our environment and social welfare. Nobody really is sure how to go about doing it, even though obviously so much is at stake that if it is going to be done at all it must be done well.

Congress and several federal agencies are now grappling with this concept and these questions. The issue is still in doubt. What comes out of current debate on Capitol Hill and moves afoot within the executive agencies could well result in a radical reordering of the relationships between technology and society at large. Or the idea may only fade away into bureaucratic limbo - another aborted, paper-producing attempt to manipulate the unmanipulatable. In any event, the odds favor the erection soon of some sort of formal structure to tackle the job. The foundations of that structure already are being laid.

A CRY FOR ACTION

"Technology assessment is almost inevitable," says Prof. Louis H. Mayo, head of the program of policy studies in science and technology at George Washington University, which has been exploring the concept since 1966.

"There is no doubt in my mind that we should do it and that eventually we will have to do it," adds Gabor Strasser, of Battelle Memorial Institute, who developed several pilot studies for technology assessment while with the Office of Science and Technology.

"It is highly likely that some formal structure for technology assessment will be established," the first (and perhaps last) report of the President's National Goals Research Staff, issued in July 1970, states flatly.

"Unless we very soon develop better measures for technology assessment, it is likely that ill-considered political reaction against technical progress will produce a crisis in our society which will make the environmental crisis look tame by comparison," says Dr. Harvey Brooks, dean of engineering and applied physics at Harvard. "Thus I regard technology assessment as essential to continued progress."

"Technology assessment will be the salvation of our democratic, free-enterprise system," asserts Richard A. Carpenter of the Library of Congress' Congressional Research Service, a man who has been closely involved with early steps in Congress to explore the idea. "Only by this means will we be able to convince the public that our economic and political institutions can bring technology under control."

Thus it seems likely that much as laissez-faire economics became socially and politically untenable in the latter half of the 19th century, so laissez-faire technology may become outmoded during the final decades of the 20th century. It won't happen overnight or in a single sweeping stroke, to be sure, any more than government regulation of business practices took place in one step.

The aim, moreover, may not be to eliminate all untoward side effects of technology. In many cases, this would not be feasible, just as it is not possible to cure all maladies with medicines that are completely devoid of risky side reactions. Even where society must learn to put up with such second-order consequences, however, the consequences should at least be made as tolerable as possible.

Although no one is really talking much as yet about "science assessment," it follows, as well, that science, too, is likely to come under greater guidance from outside the scientific community. If technology is to be regulated, the science from which it stems is hardly likely to maintain for long its traditional neutral and apolitical stance, independent of economic, social, or political pressures.

In a narrow and rudimentary fashion, of course, the appraisal of technology probably is as old as technology itself. The Federal Government has long promoted and thereby to a considerable extent controlled the development of inland waterways, railroads, aviation, highways, agriculture, forestry, irrigation, mining, and other technological activities. Nearly 150 years ago, Congress took steps to investigate the cause of boiler explosions. More recently, atomic energy, water desalinization, space exploration, and other technological feats have been fostered by government efforts. The stimulus has been found in the federal mandate to promote the general welfare and provide for the common defense. The aim was more to support the introduction of new technology than to weigh its social implications, however. Society tended to support technological development without questioning what the return on its investment might be.

The allocation of time, effort, and resources to the commercialization of new inventions and the nurturing of innovation has always hinged on general acceptance that they will prove beneficial or fill some need. As business and government poured more and more energy and capital into research, development, and the application of technology, they have sharpened their tools for assessment. Product testing, market research, commercial development, budgeting, and planning all involve evaluations of technology to some degree, although largely in economic terms.

Technology assessment, however, goes beyond all this by focusing on the interactions, side effects, spillovers, and trade-offs among several technologies or between technology and other aspects of living. A full-fledged technology assessment would not just look at transportation as a means for moving people and things about. Rather, it would examine the effects of transportation developments on housing and urban development, for example, or at the ways new methods of communication might affect the need for transportation. Consequently, even the most sophisticated techniques of modern-day planning, such as operations research and systems analysis, may prove inadequate or inappropriate for the job. The undertaking seems to call for the development of quite new tools.

Concern about the unexpected and untoward effects of wayward technology likewise is nothing new. Opposition, often bitter, to industrialization sprang up concurrently with the industrial revolution. Initially, opposition was largely confined to small special-interest groups, selfishly fearful of the effects of new machines or new ways of doing things on their own private interests. As industrialization and new technology spread to remake the western world, concern about deleterious by-products increased rapidly. Writers and social critics as varied as Karl Marx and John Ruskin, Aldous Huxley and Lewis Mumford, George Orwell and Herbert Marcuse, Rachel Carson and Ralph Nader have sounded alarms about the growing tendency for technology to benefit the few at the expense of the many or the environment of all.

Where attempts have been made to assess technology they have been piecemeal and sporadic, for the most part.

Benefits have been evaluated primarily as they accrue to the innovator (or his employer). Overall effects on the public at large or on the total earthbound biological system generally have been neglected - except to the extent that what is good (or profitable) for innovation and the innovator presumably carries good for society as a whole.

The first atom bombs were dropped without any consideration on the part of the decision makers of the dangers of fallout or the genetic effects of radiation. At the time synthetic detergents were introduced - in the late 1940's - no advance studies had been undertaken to determine the impact their resistance to degradation or their phosphate content might have on water resources.

A NEW CONCEPT

What is new, then, is not just the label on the concept but the idea that decisions regarding the exploitation of technological developments must rest on more than simple, self-interested analysis. The stakes are too high for innovation to be dictated merely by affirmative answers to the traditional questions: Is it technologically feasible? Is it economically profitable? The profit motive and the market mechanism no longer suffice, at least in the classic manner.

Several developments suggest that the time is now ripe for a quantum jump in efforts to assess and control technology. Technological mistakes and malfunctions are potentially more disruptive. The way in which technology is cultivated and functions and its long-run implications, too, are better understood. It now at least appears to be more manageable, especially as its development depends more and more on government support. Coupled with this is a growing awareness that intermittent attempts to do the job are increasingly ineffective at the same time that the potential for far-ranging injury is expanding rapidly.

Meanwhile, many people within and without government are concluding that the tools needed for assessment are within grasp, at least in some roughhewn form. They believe, too, that the capability will only be achieved by actually putting full-scale assessments to a practical test. Some experience can already be drawn upon. The efforts of the Food and Drug Administration to evaluate new pharmaceuticals and food additives, the varied assessments that have been attempted of the supersonic transport program, and the many recent studies dealing with pollution and its causes, DDT, and the uses of nuclear energy - these and similar activities represent attempts to prejudge the potential effects of technology in terms of overall public welfare rather than by narrow economic considerations alone.

Probably most important, however, is the spreading public recognition - and outcry - that the job needs doing now. Without public demand or support, any major assessment effort would be little more than an academic exercise. Perhaps some assessment mechanism should have been functioning a generation ago. Many current social and ecological crises might then have been avoided. Even five or 10 years ago, the urgency was largely unrecognized.

The current thrust for establishing a federal assessment mechanism can be largely credited to a single individua former Congressman Emilio Q. Dadderio (D.-Conn.). Mr. Daddario headed the Subcommittee on Science, Research and Development after it was set up by its parent House Committee on Science and Astronautics in 1963. His interest awakened by suggestions from Charles A. Lindbergh, Dr. Jerome W. Wiesner, and others, Mr. Daddario, together with several members of his subcommittee staff, came to the conclusion about 1965 that the Government as a whole, and Congress especially, needed some type of early-warning system to help ward off the unfavorable effects of burgeoning technology.

"We have reached a point where the tendency to use technology just because it is here must be limited or controlled," he has stated. "Our goal is a legislative capability for policy determination in applied science and technology which will be anticipatory and adaptive rather than reactionary and symptomatic."

The term "technology assessment" itself is generally credited to Philip Yeager, counsel for the House Committee on Science and Astronautics, who used it in a progress report issued by the Daddario subcommittee in 1966. That report called for an apparatus for "keeping tabs on the potential dangers, as well as the benefits inherent in new technology."

Congressman Daddario introduced a bill to establish a Technology Assessment Board early in 1967, more as a vehicle to stimulate discussion of the overall idea than to enact legislation. In the fall of that year, his subcommittee sponsored a seminar to explore the subject. The proposed bill focused on the need for "identifying the undesirable by-products and side effects of such applied research and technology in advance of their crystallization and informing the public of their potential in order that appropriate steps be taken to eliminate or minimize them."

In 1969, the subcommittee published three groundbreaking reports based on work done at its request by the Legislative Reference Service of the Library of Congress, the Committee on Science and Public Policy of the National Academy of Sciences, and the Committee on Public Engineering Policy of the National Academy of Engineering. The LRS report, "Technical Information for Congress," is a carefully detailed discussion of how Congress has obtained technological information and handled technological issues since World War II. Although it makes no specific recommendations, it does suggest that Congress needs an improved early-warning system for obtaining and interpreting in making decision on technological matters.

The NAS report, "Technology: Processes of Assessment and Choice," prepared by an *ad hoc* panel of academy members, addresses itself to the organization of a technology assessment mechanism within the Government and evaluates some of the problems and pitfalls inherent in practical implementation of such an apparatus. The NAS panel recommends "the creation of a constellation of organizations, with components located strategically within both political branches, that can create a focus and a forum for responsible technology-assessment activities throughout government and the private sector." This, it adds, should be established soon. Although the NAS panel does not spell out in detail how a government mechanism would operate, its report probably is the cornerstone study of the theory and practice of assessment.

The NAE report, also the work of an *ad hoc* panel, is essentially a study of the methodology, feasibility, and potential value of assessment, as developed from three pilot exercises in assessing electronic teaching aids, subsonic aircraft noise, and multiphasic health screening. Although the experiments themselves are by no means complete assessments, the NAE report concludes that the concept is indeed feasible and would provide useful information to Congress, particularly as an alerting device that could flag emerging problem areas. The report contains a wealth of how-to-do-it information.

A fourth study for the Daddario subcommittee was completed in the summer of 1970 by the National Academy of Public Administration. It is concerned primarily with the role of technology assessment in the executive branch of Government. The NAPA report recommends that assessment should be a continuing responsibility integrated into the routine operations of each department or agency, with overall coordination and review vested in an enlarged Council of Environmental Quality.

Working for eventual passage of some type of legislation, certainly, is the growing public concern about the impact of technology. Whatever legislation emerges, the resulting assessment mechanism is sure to move cautiously at first. "We have no idea yet how well a Congressional technology assessment component will work," Phil Yeager concedes. "It might take three or four years of trial and error to find out, and we are sure to make a lot of false starts and experience a lot of growing pains in the process."

Meanwhile, increasing activity is under way in the executive branch of Government. The National Science Foundation, through its relatively new Research Applied to National Needs program, has made several grants for studies with a technology assessment flavor.

The Office of Science and Technology likewise has sponsored studies aimed at developing assessment methodologies.

With the scent of new government money in the air, interest also is spreading to several universities.

One thing is clear. The idea has an appealing ring of relevancy about it that can arouse the interest of both academic scientists and politicians.

DIFFICULT QUESTIONS

Despite all the recent activity, proponents clearly are still advancing hesitantly into what is largely unexplored terrain. They recognize that any attempt at assessment made now should be viewed clearly as an experiment - one that no doubt needs to be undertaken, but only with a considerable sense of caution and skepticism. Doubters abound, too, who question whether any formally structured assessment mechanism can really function safely in practice.

A full-scale technology assessment must not only ask but also in some way answer many difficult questions: How will a new development or innovation be used? What consequences, direct or indirect, for good or ill, will these applications have on any or all sectors of society or the environment? What responses can be expected in other areas of technology? How do the desirable results balance against those that are undesirable or uncertain? Are the effects reversible in the short term or the long term? What alternative technologies might achieve the same results?

These questions should be answered as quickly as possible, for the longer a technology is in use the more difficult and costly it is to redirect or restrict. All this obviously is a very large order indeed. An early-warning system, for one thing, requires not only the ability to evaluate present technology but to envision where emerging technological trends are leading. Technological forecasting, an art still in its infancy, remains highly imprecise and uncertain. Recognizing this problem, the NAE report recommends, in fact, that, most assessments "be concentrated on the near future and supported at a relatively modest funding level." The reliability of forecasts, it points out, "declines rapidly with extension into the future and with the number of sequential events predicted."

Technological discoveries have sometimes been heralded by precursor events, to be sure. The possibility of harnessing the energy of the atom, for example, was recognized by many physicists (as well as science-fiction writers) long before any feasible way of unleashing that energy was devised. The advent of both the transistor and the laser was foreseeable on the basis of earlier scientific work. Speculation on the use of hormones for contraception preceded by more than two decades the introduction of the contraceptive pill. When scientists have had inklings of the future course (or consequences) of new technology, however, their voice has usually gone unheeded. They may have lacked a platform from which they could be heard, their predictions may have seemed incredible, or perhaps they were too uncertain of their speculations to press them on a doubting world.

More often, though, forecasting the course and rate of technological change has largely been beyond the reach of the imagination, except perhaps for the short run or in rather narrow areas. Forecasting technology's positive and negative consequences has been even more murky. Who could have perceived, at the time it was being developed, how the transistor would result in a proliferation of computer applications or spur the economies of several nations of the Far East? Who, at the turn of the century, could foresee that the automobile would become the major source of urban air pollution or that it would so drastically affect where Americans live, how they spend their spare time, their shopping habits, the way in which they court, or countless other aspects of their lives?

NEED FOR GOALS

If technological forecasting is an uncertain link in the technology assessment machine, methods for appraising the ways in which technology interacts with society are even more ambiguous. What is needed is some means for measuring social change and some standard by which to judge social well-being - a human welfare or "happiness" index, as it were. Yet no good set of social indicators exists today that permits a determination in any but the crudest manner of whether society as a whole is changing for better or for worse. "Until some such indicators come into being," says Dr. Lee A. DuBridge, until recently director of the President's Office of Science and Technology, "technological progress will always be subject to criticism of what it should have done but didn't, based on hindsight, because adequate standards or goals were not available at the beginning."

Meaningful assessment, moreover, would best be made within a context of broad national goals - or at least some identification of what the nation wants and what it should become. Here, too, there is lack of general agreement. The recent report of the President's National Goals Research Staff, "Toward Balanced Growth: Quantity with Quality," provides little more than a framework around which the decision-making machinery needed to establish national goals may be erected.

Without agreement on such goals, assessments can only be presented in terms of alternatives, from which a choice would have to be made by traditional political processes. Perhaps this is the way it should be done, to be sure, in a democratic society. The task of sorting out alternatives may well be beyond the scope of meaningful assessment. On the other hand, if an assessment does nothing more than set forth a confusingly broad array of alternatives, its basic purpose - rational guidance for arriving at decision - might only be lost. "The problem is: How do we introduce rationality into the process?" argues Franklin P. Huddle, who wrote the Legislative Reference Service's report to the Daddario subcommittee. "The assessment of technology is a demonstration of man's rational capability. It is his use of the scientific method in evaluating the works of science."

MANY PITFALLS

It should not be surprising that even advocates of technology assessment can point to many pitfalls that await even the most cautious would-be assessor. Both the NAS and the NAE reports cite several: inadequate information; lack of understanding or of imagination in attempting to visualize future implications of new technology; inability to look beyond current self-interest, however enlightened it might be; and difficulties in providing to all affected segments of society fair representation in the assessment process, among others.

"How to do it effectively, and even more important, how to do it 'properly' so that the deleterious side effects of the assessment process itself are eliminated or at least minimized, is the real question," says Gabor Strasser. "Institutionalized technology assessment must not discourage innovation, stifle progress, or adversely affect the productivity of the nation as a whole. What is new is that the problems that are today in need of some form of technology assessment have become more numerous, more severe, and more complex. Hence the methodologies or tools that are required to address such problems are likely to be also more complex than what we have been using in the past. Such tools must not only be useful but credible to the decision maker. They are anything but in abundance today."

Mr. Strasser points out that although technology assessment is a natural follow-up to systems engineering techniques, there are important differences: objectives are more diffuse because they are not oriented primarily toward economics or the development of technical hardware, more disciplines are involved, greater (perhaps even undue) emphasis is placed on deleterious side effects.

"It's a social and technological experiment that should not be delayed", says former Du Pont vice president Samuel Lenher. "But it should only be set up as an experiment at this stage, rather than as a rigidly structured new arm of government that might be very difficult to halt if the experiment failed to work as expected."

"It would be ironic indeed if in developing a mechanism for technology assessment we were to stifle technological innovation because of failure to take into consideration the broad range of measures, social, economic, and political as well as technological, that can guide technological developments along beneficial lines," Dr. DuBridge told the Daddario subcommittee in 1969. "We must be sure we are designing a technology assessment, rather than a technology arrestment, mechanism. We must develop a proper context for technology assessment, including an appreciation of its limitations as well as its potentials."

Testifying during the subcommittee's hearings in 1970, former NSF director William D. McElroy warned that technology assessment "can be done today only in a limited fashion, in particular instances, and primarily with respect to the direct or first-order effects of technology." He noted that economic, social, and political effects are particularly uncertain. "We have not yet reached a point where technology assessment has available to it, except in limited areas, the quantitative techniques necessary to establish standards, to optimize the choice of alternative technologies, or to anticipate the effects of innovation," he added.

It would be a mistake, however, to emphasize too highly the need for attaching hard, quantitative values to everything assessed. A formal economic analysis of the cost-benefit type is fine, whenever it is possible. "But we must not throw up our hands when faced with the problem of being unable to quantify the unquantifiable," Congressional Research Service's Dick Carpenter stresses. "You can't put a dollar sign on everything that you want to - the worth of a clean river, for example, or an uncluttered landscape, or a live osprey."

It is clear that even the staunchest supporters are still groping for a methodology. What is needed is both a technology for technology assessment and a means of assessing the assessors. Obviously, any assessment on which a major policy decision hinges will have to be flexible, so that it bends in response to new information, new uncertainties, or new developments.

DO IT NOW

Despite the caveats and concerns, the case is good for undertaking assessments now. "It is not unreasonable to ask if we have all the basic concepts and tools necessary to do a creditable job," GWU professor Louis Mayo concedes. "But we do have available in our present techniques for systems analysis, operations research, model building, and handling large masses of data electronically at least some of the fundamental resources from which such tools may be fashioned. So lack of tools no longer is a reasonable argument for shunning technology assessment."

"There is a need to do it, without question", adds Charles Falk, director of planning for NSF's division of science resources and policy studies. "And the only way we are going to learn how is by doing it now, with whatever tools are available and as best we can. Only then will we be able to sharpen those tools."

WHERE TO DO IT

Tools are useful, of course, only insofar as they are used, and the question of where the responsibility for technology assessments should be focused is still controversial. An extensive, if ill-defined and uncoordinated, apparatus for technology assessment is already routinely at work in most of the regulatory and science-oriented agencies of the Government, such as FDA, AEC, and FCC. A framework for the task exists, too, in most major corporations involved with technology.

Thus the need for better assessment could be met, presumably, by expanding and strengthening present capabilities. This, in fact, is the advice of many who have studied the concept.

There are many inherent drawbacks, though. No present organization has assessment as its primary mission. "Existing mechanisms," the NAS report concludes, "whether they involve government agencies, private industries, or professional groups, possess intrinsic limitations, some structural and others psychological, that leave serious gaps in the spectrum of processes that assess and direct the development of technology in our society." There are obvious questions of bias, self-interest, conflict of purpose, competitive pressures, limited objectives, inadequate incentives, narrow vision, and insufficient jurisdiction and power. Can AEC, for example, with its vested interest in promoting the use of nuclear power, be expected to weigh sufficiently both the economic benefits and the environmental drawbacks of installing new reactors for generating electricity? Or can FAA, with its interest in advancing supersonic aviation, judge objectively all the economic and social effects stemming from an SST?

Or, for that matter, would any group of scientists be able to look at the allocation of national resources for research and development with other than a science-biased tunnel vision? Or could producers of agricultural chemicals assess objectively the ecological damage inherent in pesticides?

These are not questions of callousness, dishonesty, or malevolence, really. "Any organization working on a new development has a built-in bias to move forward," Prof. Don Kash of the University of Oklahoma points out. "Everyone inherently tends to view his own component of a system rationally but finds it difficult to see the big overall picture."

The question of credibility also is significant. Even an assessment performed with utmost objectivity will be suspect if the recipients or the public at large assume, with or without reason, from the affiliation of the assessor that it might be self-serving.

Nevertheless, technology assessment is not a job for dilettantes or do-gooders. It requires expertise rather than superficiality. The problem of locating and enlisting experts who are qualified but whose objectivity is unimpeachable could be serious. Expertise is not necessarily matched with wisdom, moreover. So at the same time, somewhere within the assessment process the voice of all interested or affected parties, including the general public, must be heard. "The gap between the experts, who know, and the public, which is affected, must somehow be bridged," says NSF's Joel Snow. Dr. Snow sees an additional problem in manning the assessment mechanism. "Are there enough scientists and engineers capable of making meaningful assessments who are also interested in doing it?" he asks. "The task is not likely to seem to many to be as creative or intellectually stimulating an endeavor, in the traditional sense, as the laboratory research most of them have been trained to do."

EVOLVING MECHANISM

What at the present juncture seems to be emerging from the political debate and the studies of possible ways to tackle technology assessment is a multifaceted federal apparatus with two focal points, one acting for Congress and a second, entirely separate structure within the executive branch. The latter probably will function on several levels. Although such a complex system of multiple assessment may seem inefficient and prone to redundancy, it would at least assure that the work of the individual assessors would itself be assessed in turn.

Both Congress and the executive mechanisms are likely to look to outside sources for most assessments. Whether these would be *ad hoc* task forces set up temporarily to conduct a particular study or whether the assessment projects would best be farmed out under contract to existing entities such as universities, nonprofit research organizations, and industrial or research firms is still unclear. Certainly it seems unlikely that any single agency set up for assessment purposes within the Federal Government could permanently employ within itself sufficient talent spread across the wide spectrum of disciplines needed to conduct assessments. Thus whatever formal agencies are established within the Government probably will be responsible only for locating the authorities in the outside scientific community who would do the work and for overseeing and coordinating all the various assessment projects.

Much of this work might logically fall to universities, with their pool of presumably disinterested scholars. Few universities, if any, however, are really equipped to do the job as they are now organized. Technology assessment requires a broad range of technological, economic, social, and political inputs that can be provided only by a mission- or problem-oriented interdisciplinary approach. Universities customarily are structured relatively rigidly along disciplinary lines.

The technology assessment arm of Congress could be fitted into either the Library of Congress or the General Accounting Office, both of which already provide information and advice to the legislative branch. Either of these organizations would have to be considerably expanded and reoriented, however, to handle such a sweeping new responsibility. Neither has shown in the past the futuristic outlook that is implicit in technology assessment.

The executive branch has several potential candidates. Perhaps the most obvious are NSF, OST, and the relatively new Council of Environmental Quality. Both NSF and OST, at least, seem anxious to carve out an assessment niche for themselves. NSF, for example, regards itself as uniquely equipped for the job. It already has tentacles stretching throughout the scientific community and well-established channels for funneling information and funds between the Government and the technological world.

The NAS report from the National Academy of Public panded to provide the "natural focus for any new technology assessment activity to be established in the executive branch," either by forming a new technology assessment department within OST or by distributing assessment responsibility throughout the present organization. The panel also suggests that NSF set up a new technology assessment division, to work in cooperation with OST, that would administer a program of contracts and grants to nongovernmental organizations.

The report from the National Academy of Public Administration, on the other hand, argues that OST is too strongly oriented to technology alone to handle the job. "Effective technology assessment must also include social, economic, and political factors which OST has not been staffed to consider on a regular basis," it states. Other critics of a role for OST point out that as part of the Executive Office of the President it is too much a staff function. This could cast doubts on its freedom from political pressures or the exigencies of Presidential policy.

The NAPA panel recommends, instead, that the assessment function be distributed throughout the departments and agencies of the executive branch as a new part of their regular planning-programming-budgeting system. The Council of Environmental Quality, however, would be charged with monitoring and reviewing the efforts of all the executive agencies.

There is legislative precedent for this. The National Environmental Policy Act of 1969, which established the council, implicitly authorizes the assessment of technology by federal agencies, although only insofar as it has an impact on the environment. It also vests broad assessment responsibilities in CEQ.

Focusing responsibility for assessments in CEQ could mean, however, that each assessment would take on a strong cast of environmental protection and pollution control. Yet in many assessments, considerations of an environmental nature may be of secondary importance. Whether CEQ, whose primary function, as it now operates, lies in preventing damage to the environment, could be neutral rather than negatively disposed when it comes to assessing technologies that would affect ecology is debatable.

As Harvard's Harvey Brooks points out: "Technology assessment is not identical with the protection of the environment, although it overlaps with it strongly. If we go exclusively along the route of environment and ecology, rather than the more balanced and comprehensive approach of technology assessment, I think we are going to find ourselves faced with a great many bad side effects of overconcentration on environmental concerns just as we are now worrying about the side effects of technology."

The NAPA panel questions, in any event, the wisdom of setting up an entirely new agency within the executive branch of Government as a focal point for assessment. It comments that many people are convinced that such independent agencies already have proliferated beyond reason. Creation of an additional one would only reduce the President's flexibility to act and magnify his problems of management and coordination.

Similarly, the NAS panel dismissed as politically unrealistic the idea that what is needed is an organization entirely separated from any of the existing branches of Government. Such an entity would have advantages of freedom of action and isolation from political pressures, the panel conceded, but it also probably would lack access to political power and policy making as well as to a public platform.

Nevertheless, some observers foresee the need for a much more sweeping assessment mechanism than is found in any of these proposals - even to one that would demand a rewriting of the Constitution or the establishment of an entirely new branch of Government, a group of philosopher-scientists, as it were, watching over everyone. (Why the public would not view such an elite as just a political pressure group with a technocratic slant is unclear.)

Prof. Don Kash suggests that Congress should foster what he calls a "counterbureaucracy, functioning throughout all levels of government and within industry, the universities, and research institutions as well, that would be charged with evaluating the second- and third-order consequences of new technology." Such an organization would act as an early-warning system to "alert society to the deleterious effects of technology and propose alternative ways of attaining the first-order goals which have generated this new technology."

Edward Wenk, Jr., professor of engineering at the University of Washington, urges the formation of a full-time commission for social management of technology that would be independent of both branches of Government and would review federal and private programs involving technology in order to identify both neglected opportunities and inadvertent consequences or risks. The commission would publicize its findings and "if remedial action is not taken promptly . . . be empowered to take necessary legal action in the courts to protect the public interest."

Whether such an ambitious undertaking is practical or desirable any time in the near future is questioned by many, however, in light of the current state of the assessment art. Mr. Daddario notes, for example, that both the need and the capability for making assessments of technology probably will have to be demonstrated on a relatively modest scale before any massive effort will be accepted.

Prof. Harold P. Green of GWU's national law center has still another idea. He concludes that technology assessment should be the function of an agency (either entirely independent of government or set up as part of Congress) "which would act as a responsible devil's advocate or technological ombudsman and play the role of adversary in the Congressional and public forums." He argues that technology does not suffer from a lack of influential proponents, generally well endowed, pressing its benefits, but that the risks of technology frequently get insufficient hearing. He is distrustful of experts and insists that scientists "have a bias in favor of accomplishing what they think can be accomplished."

WHO ACTS

No matter where the work is concentrated, within Government or outside, some authority must dictate what should be assessed and initiate proposals. No mechanism conceivable at present could assess all new technology all the time.

The Daddario bill stated that activities undertaken by the proposed Office of Technology Assessment can be initiated by the chairman of any Congressional committee, by the board set up to oversee the office, or by the director of the office. As a practical matter, however, most assessments, at least initially, are likely to be done at the behest of Congress. The proposed bill's backers, of course, do not expect OTA to look into the consequences and implications of all new technology. The bill itself, in fact, singles out the need for assessments "particularly in those instances where the Federal Government may be called upon to consider support, management, or regulation of technological applications" (a rather widely encompassing grasp, to be sure).

This poses a possible problem: Those with their hands on the purse strings may request probes only of the most visible problems and most obvious crises - assessments of the squeaky-wheel-getting-the-grease type. Thus in practice limited funds may not be channeled toward assessments of potential threats that are not yet in the public spotlight. Yet it is just such newly emerging technology, whose hazards and benefits have not surfaced, that is perhaps most amenable to the assessment process. This type of technology, which has attracted few vested interests or little significant capital outlay, is most susceptible to control or redirection and least likely to prove irreversible.

Other questions can be raised regarding the actions that should stem from an assessment. Many advocates of the assessment concept are convinced that no matter what mechanism is established, the assessor should serve only in an advisory rather than a regulatory capacity if he is to maintain a neutral and unbiased stance and not lose his credibility. They see the function for the most part as an educational or fact-finding institution or as a channel for information transfer, setting forth unselectively all sides of a question and pointing out alternative courses without making a choice among them. "Once the problems have been identified, the interactions have been put together, and the alternatives laid out, the job of assessment ends," says NSF's Dr. Charles E. Falk. "An assessment is only a tool for the decision makers, not a decision-making function in itself."

This could prove a difficulty. Early warnings are not worth much if they go unheeded. Unless some follow-up action is taken, any assessment, no matter how thorough, will have served little useful purpose. Assessments used only to justify action already taken or, even worse, only as an excuse to stall and study a problem to death while avoiding any action whatsoever can only be counted as failures. If the assessment mechanism should become merely another means for shuffling masses of paper or a machine for grinding out exhaustive reports that few people read, it might be worse than a waste of money and high-caliber talent, for it would only lead to greater frustration and disappointment about the inability of government to govern or provide direction.

Dr. Falk thinks such fears are unfounded. "There is a greater danger in action taken in the absence of an appraisal of the technological effects than there is in the possibility that assessments will be inadequate or incomplete or an excuse for inaction."

Others, though, are not so sure. "Assessments made for people who don't care, or who don't follow through and continue to care, are almost sure to become routine and irresponsible," cautions Dr. John Pierce of Bell Telephone Laboratories.

"Unless an assessment makes specific suggestions or recommends specific actions, what is anyone going to do with it?" asks Eric Ward of OST. "Technology assessment must not be just another excuse for conducting more research and processing more data."

In the final analysis, of course, the proof of technology assessment will come only if it causes institutions - whether they be Congress, the executive agencies, or private enterprise - to act in a manner different from the way they otherwise would have.

It is not difficult to envision the political tumult, too, that is likely to greet any actual assessment that goes against some well-established and influential segment of society or some entrenched and powerful political or economic institution. Such a baptism under fire could well prove what - if any - basic political support the assessment concept really has. The NAS report speaks of the need to foster the creation of unrecognized constituencies whose protests against adverse technologies should be heard. As one astute advocate of assessment points out, though, the development of a constituency to help protect the embattled assessors from outraged pressure groups may be an even more pressing need.

"It's easy to agree with technology assessment as a generality," OST's Eric Ward notes. "But how do you implement it on a practical basis? We do things not because of technological opportunities that are available but for political and economic reasons. Consequently, technology assessment itself will not make the decision-making process more rational. If technological inputs are to be balanced with those of politics and economics, technology needs not just assessment but advocacy as well."

This underscores another problem. If assessments are to be made at an early stage of a development, much of the information about consequences and directions that might be desired is bound to be unobtainable. Questions, especially in the social sciences, will be easier to raise than resolve. The natural inclination of the assessors, then, will be to ask for more time and more research. This would only defeat the purpose of the entire activity. Both those who make assessments and those who use them will have to learn to live with results that contain a considerable content of unquantifiable speculation and are based more on probabilities than certainties.

"It is not possible to have a complete analysis and at the same time have it timely," cautions Dr. W.E. Hanford, vice president for research and development of Olin's chemical group. "Timeliness is more important than completeness; if it's complete from a technical standpoint, it is history."

ASSESS OR HARASS

The amount of money involved, although it is sure to be great, may be the least important cost of a formal structure for assessing technology within the Government. The second-order consequences of a more thorough search for second-order effects may themselves be considerable. The recent report of the National Goals Research Staff points out: "The issue at stake is whether or not we can establish wise criteria and appropriate choice mechanisms so that we can balance the gains in quality of life appropriately against the economic costs, since these costs represent reductions in our capacity to improve the quality of life in competing respects."

Fears that technology assessment could in fact lead to harassment or arrestment of industrial technology seem widespread. Dr. Morris Tanenbaum, general manager of the engineering division of Western Electric (and one of the two members of the NAS panel on technology assessment who are affiliated with an industrial firm), is convinced that some type of assessment mechanism is indeed needed, but he fears that unless it is applied very carefully and thoughtfully it might do more harm than good. In a separate statement appended to the NAS report he cautions that any assessment mechanisms, to the extent that they create new uncertainties, "could discourage private investment in areas of technology which are at the focus of the assessment activity. These may be the areas where innovation is most important and private participation most desirable." Dr. Tanenbaum points out that attempts to overcome this problem (for example, through government subsidies or increased prices) may in themselves have undesired second-order consequences. He warns that the assessment process must be contrived to operate "without irreparably damaging the systems of innovation that it is designed to stimulate and guide."

"It is a risky business," adds former Du Pont vice president Sam Lenher, "because so much is dependent upon the judgment of so few people. And if you look at the SST situation, you quickly realize that technology assessment must be done on a worldwide basis if it is not to undercut our nation's competitive position in world markets. You can't restrain technology in only one geographical area."

Industry's fears are understandable. It may be a real challenge to apply the consequences of technology assessment to business - by forcing businessmen to account internally for many of the costs and spillovers that they have long been accustomed to ignore - without fundamentally altering our capitalistic, free enterprise system in the process. Many industries, such as the drug makers and pesticide producers, already have experienced frequent and often abrasive confrontations with the relatively narrowly focused assessments prevalent in many federal agencies today. The far more searching investigations now proposed would probe effects far removed from those generally evaluated in laboratory testing and commercial development. Even when a technology assessment is not negative or does not result in restrictive action - and many, by their very nature, will do just that - the time required to conduct such studies will put an additional burden on the process of developing and marketing new products and ideas. Moreover, the ever present possibility that a new product or process will be subject to assessment, either before or after its commercial introduction, will only add another element of high risk into the already risk-strewn research and development equation.

Congressman Daddario and the staff of his House subcommittee insist that such fears are frequently overstated and that industry people in particular put too much emphasis on the negative aspects of technology assessment. They argue that the assessment function need not be looked upon as just another attempt at government empire building or regulation of industry. "Why should technology assessment only inhibit innovation?" Mr. Daddario asks. "Will it not, if it is done well, actually foster the use of unemployed technology and reduce the costs of innovation in the long run? And by providing Congress with better information at an early stage, will it not save the taxpayers many millions of dollars that might otherwise be wasted on poorly planned, unfeasible, or unproductive developments?"

NSF's Joel Snow agrees. "An important aspect of technology assessment that is often overlooked lies in guiding the development of beneficial technologies," he points out, "especially those that fall outside the normal market place economy because they are too costly or the potential market appears to be too small or too fragmented to warrant exploitation. Such technologies need to be identified and promoted if we are to get full advantage of our technological capabilities."

Although many people in industry remain skeptical about the need or the practicality of an institutionalized federal technology assessment function, the majority (even among those directly involved with directing research and development) appear to be either uninformed or unconcerned about recent proposals. The voice of industrial experts has been conspicuous largely by its absence in the deliberations that have taken place so far. Yet the stakes for any technologically oriented corporation obviously are great.

ROLE FOR INDUSTRY

What role, then, for industry? It certainly will be involved to some extent in supplying the experts needed in conducting assessments for government agencies. After all, it is in industry that many of the experts are to be found. As former NSF director Dr. McElroy points out: "Industry has characteristically been able to assemble multidisciplinary project teams to attack broad problems."

Olin's Dr. W.E. Hanford adds, "The success of this operation depends on getting top professionals to serve when they are needed. We should not worry about whether or not they have some possible conflict of interest. We technologists are used to wearing more than one hat. It is possible, in fact, for technical people to wear several hats and still maintain the highest standards of ethics. You must, in my opinion, select a high percentage of these experts from industry."

Nevertheless, it seems likely that questions of objectivity, self-interest, and credibility would be sure to plague any industry-dominated assessment team. Thus industry's role in government-sponsored assessments probably will not be a primary one.

If industry may not have a major involvement in asking the questions, it must in any event be ready with the answers. Perhaps its best defense, then, is a good offense: an internal assessment mechanism for its own self-policing. The company that doesn't assess its own technology and business goals, not only in the traditional terms of economic cost-benefit and market analysis but also in terms of social costs and environmental consequences as well, is only opening the door to government intervention in the process of innovation. It can be sure that what it does not assess itself will only be assessed by others. Then the losses in frustrated development efforts and aborted capital investments, as well as from a tarnished public image, could be catastrophic. Perhaps the greatest benefit to the nation from any assessment legislation, however, will stem not from a new federal appraisal apparatus but from a heightened self-policing effort on the part of industry itself.

"Industry needs its own technology assessment capability not only to keep itself out of trouble and answer criticism, but also to win credit for a job well done", says CRS's Dick Carpenter. "A better view of secondary consequences may remove assumed obstacles to certain markets or suggest additional selling points for a product."

The function of corporate assessor, to be sure, will not be an easy one to staff. What seems to be needed is a counterdevelopment structure at a position of influence within the corporate hierarchy, able and willing to raise objections, to ask embarrassing questions, and to throw up obstacles to pet projects. That's hardly the way to win friends or gain commendation in the executive suite. It is a function that, nonetheless, industry may soon find it cannot afford to be without.

In any event technological choices have to be made. Too often, though, we just don't know what the trade-offs are in the decision we make. We fail to recognize how an arbitrary choice made today may affect, or even preclude, the choices we may want to make tomorrow.

What the advocates of technology assessment are seeking are hardheaded, practical ways to set policies and priorities in a more rational manner based on a better informed viewpoint. Few of them would claim that they have found the means for tackling the job so far, but the seeds of the idea already have quietly flourished much more than most would have believed possible when they were initially planted six or seven years ago.

TECHNOLOGY ASSESSMENT IN GREAT BRITAIN*

C. SINCLAIR

The University of Sussex, Science Policy Research Unit, Great Britain

INTRODUCTION

The socio-political innovation of "Technology Assessment", as a distinct discipline, has had its origins and its greatest development, in the United States, as witness the sophistication of the U.S. contributions to this seminar. The readiness and ability to produce descriptive titles for wide ranging and complex socio-technical processes is matched, apparently, by the greater flexibility of political institutions to adapt to the novel administrative procedures required by these processes. The seminal descriptions of Technology Assessment [1, 2] have encouraged [3] wide discussion and rapid formulation of the rules for the application of the methods within Government Departments. In the context, however, of the U.K. administrative system it might be more true to say that Technology Assessment does not exist or even that the phrase is as yet unkown. The recommendation to vest central responsibility for "policy, coordination and review" of Technology Assessment, in the U.S., in the Council on Environmental Quality helps to demonstrate that the impulse from the political area was motivated by a regard for the increasing concern about environmental problems. Public debate in this area is probably a little

*Originally presented at OECD on 26 January, as TECHNICAL CHANGE AND RISK DISTRIBUTION: A METHODOLOGICAL NOTE FOR TECHNOLOGY ASSESSORS

more muted in the U.K., and in Europe at large for that matter, but the principal reason for the conspicuous lack of a rapid and concerted impetus to set up special institutions for Technology Assessment on the European scene is a relatively long historical background of procedures and measures, which are still extant, for safeguarding public versus private interests. Thus many authorities are involved in the defense of the general public good against possible harm from development, particularly those deriving from technology. Implicit in the above remarks is the emphasis on the idea of examining and weighing the costs rather more searchingly than promoting the measurement of the benefits of technological progress though it is, however, fully recognised that any Technology Assessment methodology will essentially be concerned with both aspects. Notwithstanding this, the study presently being described has concerned itself, too, with the examination of the controlling mechanisms and institutions and associated risk levels already existing in the U.K. rather than attempt a full balance. These arrangements may be fairly described as the embryo, or better, the skeleton, Technology Assessment system, on the control side, which it is hoped to promote. The argument and reasons for proceeding in this way are outlined in the following.

Rationale of the Method

Public Reaction and Technology. It is hoped that the results of the presently described method will be useful and informative in the context of a broad survey of Technology Assessment, particularly from the position of establishing the need for a more comprehensive approach. It may, however, be necessary to outline briefly the bases of making an approach in the manner described. To utilize the metaphor of the initial quotation, Technology Assessment, seen as a method of comprehensive cost-benefit-analysis, is concerned with the balancing of the levels of good and bad arising from technology and of ensuring that the former is always greater than the latter. If, however, such individual assessments are carried out without making some overall comparisons between the levels of the bads then a cumulative process of disbenefit may arise which is undesirable. Thus, in the report of the Secretary-General's Ad Hoc Group on New Concepts of Science Policy [4] the conclusion is reached that "*The side effects of economic growth are often of a highly non-linear character, appearing rapidly, and creating a new collective need for their abatement and control, often with almost a threshold effect. These thresholds are partly physical: for example, photochemical smog from motor-car exhausts appears only when the concentration of exhaust products reaches a critical point. Other thresholds are simply thresholds of public perception, which find political expression only when a certain segment of the population is affected, as in the case of jet-aircraft noise.*"

The task of a Technology Assessment system, on the risk or disbenefit side, is to keep these undesirable effects below the thresholds and, additionally, to attempt quantification of the many disparate disbenefits so that an intercomparison can be made. To quote further the above report, "*Such a situation engenders a general feeling of insecurity and instability, which may lie at the root of certain social behaviour; the sacrifice of the individual as producer to the individual as consumer, which is one of the consequences of the type and form of growth we know today, is reflected in the last analysis by a 'personality split', the character and effects of which have not yet been measured.*"

This theme too was the basis of the present author's section of a paper produced together with colleagues at the Science Policy Research Unit on "The Goals of R & D in the 1970s" [5] where it is suggested that the rationalisation of national R & D policies may be effectively assisted by the interconnection of the relative risks arising from alternative technologies with implicit numerical valuations.

Thus, to introduce immediately some numbers into the discussion, the Table 1 shows how the changes in output per man in various sectors of U.K. industry, assisted by the application of technology, has been coupled to changes in relative risk levels. The relative performance as demonstrated by these values are markedly different. Increases in social good as indicated solely by considerations of GNP have not been matched by correspondingly

TABLE 1. Comparison of Output per Worker and per Accident for Agriculture, Forestry and Fishing, Manufacturing and Pharmaceutical Industries

Year	Agriculture, Forestry and Fishing		Manufacturing		Pharmaceuticals	
	Output per worker	Output per accident	Output per worker	Output per accident	Output per worker	Output per accident
1958	85.6	94.2	92.3	113.4	73.4	100.6
1963	100.0	100.0	100.0	100.0	100.0	100.0
1967	129.7	116.1	110.7	89.6		
1968					169.6	120.3
% increase from 1958 to 1967	51.5	23.2	19.9	-21.0		
% increase from 1958 to 1968					131.1	19.6

good improvements in the occupational risk levels at least. Further examination of the changes of risk distribution with respect to different risk groups and consequent valuation changes must be a goal of assessment.

Research and Development directed towards Risk Reduction

The introduction in the above of the theme of R & D directed towards the abatement of hazard allows the second principal plank of the present method to be laid. The development of a Technology Assessment system which allows the demonstration of the paths to an optimisation of the level of risk or disamenity should enhance its acceptability as an instrument of science policy. Thus, while R & D may traditionally have been used to reduce the uncertainty and level of ignorance of the entrepreneur in searching and promoting the acceptance of a new product and hence to maximise living standards so that R & D too can be used to reduce and perhaps minimise the level of ignorance associated with the concomitant production of disamenities and risk [6]. Ideally, one would hope to allocate R & D resources to both functions for an optimal position to be reached. On the benefit side of any assessment the market mechanism will provide some measure of how effort might be distributed, but on the disbenefit side, since many of these are public "bads" the costing mechanism is less obviously optimised. The model proposed below allows a suboptimisation to be made for the allocation of anti-disamenity R & D, for example, in terms of implicit values arising from the existing or proposed control mechanisms. Some of the general conclusions arising from a pilot survey of such R & D spending in the U.K. are given below.

THEORETICAL FRAMEWORK

It is convenient to start the description of the theoretical framework in terms of cost benefit analysis, and to limit the methodological description, particularly one disbenefit, though this technique will be quickly modified in the following. The initial question, the basic Technology Assessment query, posed was how should decision be made as to the acceptable risk (health risk) to be taken when promoting new technologies? More specifically the inquiry was how could a risk level to inhabitants local to projected nuclear power stations be determined as acceptable or not? The concept of risk to an individual from such siting had begun to be discussed during the sixties in technical and administrative circles and several points of view were emerging which are fully discussed in the main report.

On closer examination of the wider picture it could be discerned that:-

(a) within existing technologies and industries risks had already a wide range of values and had been arrived at from pressures deriving from the existence of an enormously diverse set of factors — social, psychological, legal and administrative.

(b) other new technologies, in different ways were faced with problems of a similar kind which it might prove useful to incorporate into the same scheme of assessment. In this category obvious candidates were pharmaceuticals and aircraft. Considerations relating ultimately to the same problem of balancing risk and benefit were coming to the fore in the field of medical treatment where many technological possibilities were forcing decision makers into difficulties over the allocation of funds between competing projects. Again, in the ultimate this could be seen as an allocation between different life saving areas. The obvious and most visible examples are heart transplants, kidney machines and prophylactic examinations for cervical carcinoma.

(c) the consideration of assessment of risk in a comparative manner as suggested here brought to light many considerations not lying primarily in this field but strongly influenced by and influencing considerations of the quality of life.

Definition of "Risk" and "Accident"

It is necessary first to define risk and accident as used below, since in some studies the lack of precise definition has led to difficulties of interpretation. Both

Ryle and Dubos have categorised accidents as "social diseases" though difficulties arise here, too, because of the multiplicity and complexity of causation. Backett has in this context suggested that, in regard at least to domestic accidents, that is not too far fetched to regard domestic accidents as taking over where the threat of tuberculosis is declining since with these two "diseases" the so-called "poverty complex" associations are shared by both. Technological, here medical, advance has eradicated the latter threat but not yet the former. The imbalance of technical advance exemplified here is further demonstrated by the figures shown in Fig. 1, where the importance of accidental death here as a predominating cause up to around age 37 is illustrated.

The World Health Organisation have defined an accident as "an event, independent of the will of man, caused by a quickly acting extraneous force, and manifesting itself by an injury to body or mind". The addition of "quickly acting" concept to the original "unpremeditated event resulting in a recognisable injury" is unnecessarily complicating and in the present study erroneous in the sense that accidental poisoning may be far from instantaneous. The sudden rupture of a nuclear reactor pressure vessel fits the definition but not the subsequent induction of thyroid carcinoma in the affected group. Similarly, the use of "extraneous" could also lead to much subtle philosophy. The semantic discussion is largely made redundant by the uncertainties that presently exist in the recorded statistics. The view of accidents used here subsumes the definition used above but the method is concerned also with "undesirable" outcomes of activity which include accidents, pollution and the like.

The unit of exposure to these undesirable outcomes, as far as physical injury is concerned, is the unit of risk, that is the statistically derived average annual probability of damage in terms of individual injury or death. Occasionally other exposures may be used however this definition follows the majority of present day usage and is consistent with an influential earlier definition which utilised the term "man-year" of exposure to risk.

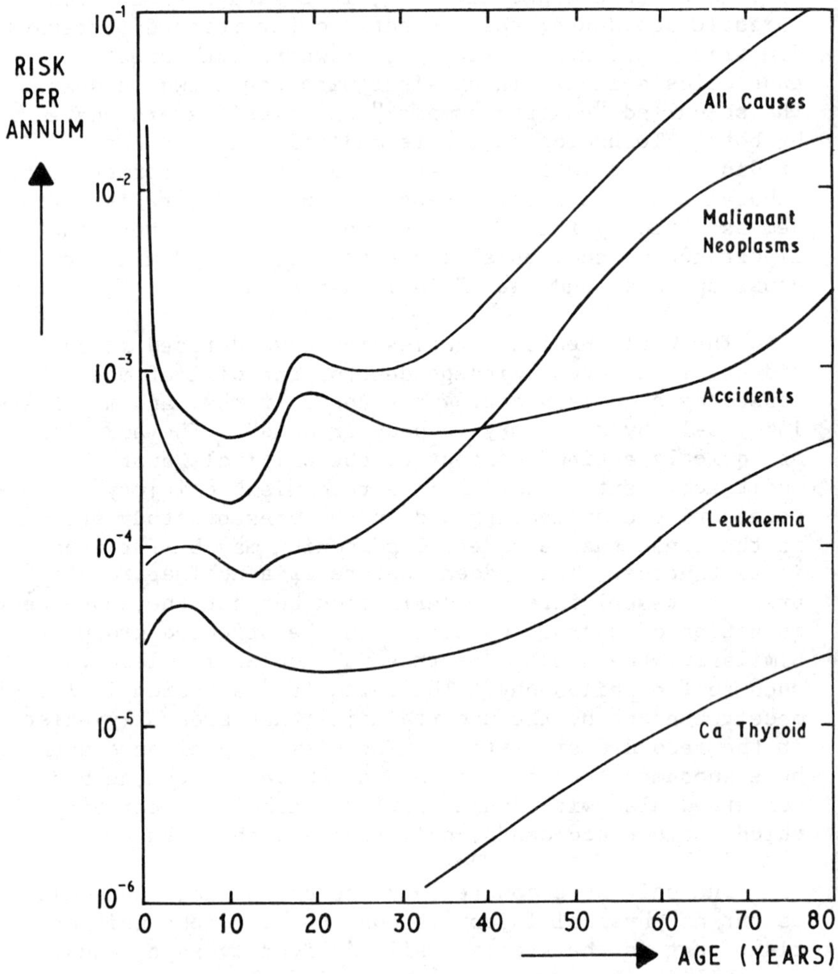

Fig. 1. Age distribution of risk of death from various causes in England and Wales 1965.

While it is not intended to treat domestic accidents in the study, some indicators of general use to the study can be mentioned here. Since the classification of accidents is presently done, in the main by the locus of the "unpremeditated event", e.g. domestic, those which occur in and around the home; the scheme used here of producer, user and third party risk will sometimes cut across these lines. Many domestic accidents may fall into the user or third party categories associated with a particular industry. The picture of changing risk patterns from technical change may be illustrated by examples drawn from the domestic field where new products were involved; for instance, asphyxiation from polythene bags has been reported, both for children and adults. Rising living standards have led to rising housing standards, though the demand for space heating has led to increased hazards from such items as paraffin heaters. When this hazard was removed by the increased use of oil fired central heating, however, the risk of damage is transferred from the home to the rivers. The rise of the oil pollution of rivers and consequent risk to human health and happiness from this source has been investigated.

Just as farming injuries are beginning to resemble more and more factory accident injuries, so in industrialised countries the home contains more and more technically advanced machinery. Backett lists the following as commonplace hazards in the home, even in 1965 — rotary cutters in the kitchen and garden, electric tin openers, and other electric household appliances, aerosols, air conditioning plant, dishwashing machines, do-it-yourself workshops, hypoxy resins and other glues. The responsibility for the control of risks from these sources often lies with the manufacturers in their design function, but the overall position is erratically organised and obscured by the methods of information gathering. One area which was examined, which lies within the usual scope of domestic accidents is that of drug safety as it affects third parties.

Cost Benefit Analyses

Starting then from a cost benefit approach one can describe this as "an analysis designed to compare two or more solutions to a given problem and to provide a framework in which comparisons can be usefully discussed". The essential aspect as far as the present problem is concerned is that the method allows decisions to be made in public expenditures (and private) on alternatives on a quantified basis. In the area of the present field of study the road transport authorities have been using cost benefit techniques to order priorities in road construction for some time. Death or injury risks were dealt with by calculating costs on the basis of material damage, administrative costs and economic losses and adding a value determined arbitrarily for the "value of life". The cost benefit analysis technique had a major exposure during the discussions about the Third London Airport, though here life risks were a very minor item (see below).

Official decisions about acceptable risks in other areas, where this has been done explicitly, have used the technique of calculating the economics of the proposed technology without regard to the social costs making economic comparisons with the competitor on this basis. Thus the development of nuclear power was based on the estimated condition of long-term supply and demands for electrical power and the known fuel resources. Considerations of risk were then dealt with by estimating roughly what order of risk of major accident was involved, taking into account possible safeguards and remedies and then estimating whether such a situation could be handled. The detailed course of the comparable process in other areas is dealt with in greater detail in the report. We are here concerned with the effective results in numerical terms. In aircraft, nuclear power and pharmaceutical development characteristic assessment methods developed. Where, as in road transport, calculations were made of a cost benefit type on safety or risk diminution matters, the monetary cost of improvement or safety device was set against the potential savings in damage and lives and these latter were determined separately in monetary terms by essentially *ad hoc* methods.

Method

The methodology proposed and tested in the following avoids many of the disadvantages arising from the cost benefit approach to such items as life valuation which usually depend on attributing a value from earnings loss, insurance, legal or other source and appears to be, in this respect, unique. It allows decisions to be made as to future action in terms of the existing levels of risk in a way which has only limited value judgments inherent in it. It is best described in terms of a very simple mathematical and geometrical model.

It is postulated that in any design it is possible to specify items that are allocated and present for reasons of safety, that is to reduce risk. By "design" is meant physical design of equipment, the operational design of the process and the pattern of social activity associated with the particular activity. When the level of risk is plotted against the extent in value of these design items and operational measures etc., as in Fig. 2, an inverse relation appears. The greater the outlay or effort towards risk reduction the lower the risk level. If the recorded accident level is used as a measure of risk and this can be turned into a monetary value and added to other monetary costs of accidents, i.e. from physical damage or administrative costs, then Fig. 3 can be drawn relating to expenditures "P", before the event, to "A" the total accident cost, and if we define the "cost" of risk or disamenity as the sum of the two then this will have a minimum value at some particular risk level. Thus an economic optimum can be determined. This is described more fully below. Similarly, by working not in total costs but in marginal increases in prevention cost and reductions in damage costs, a balance can be struck as indicated in Fig. 4. The development in mathematical terms of this model is given elsewhere.

Assuming now that costs and balances can be made for each activity or project for producer risk — i.e. conventional industrial safety, and for user risk and third party risk, then one can draw Fig. 5 on which along each of the three axes, relating to the three categories above, the risk level of the activity is marked l_1, l_2, l_3. A plane joining these points can be produced which represents the "risk plane". Changes in technology may alter the position of each or any of the axial points and thus

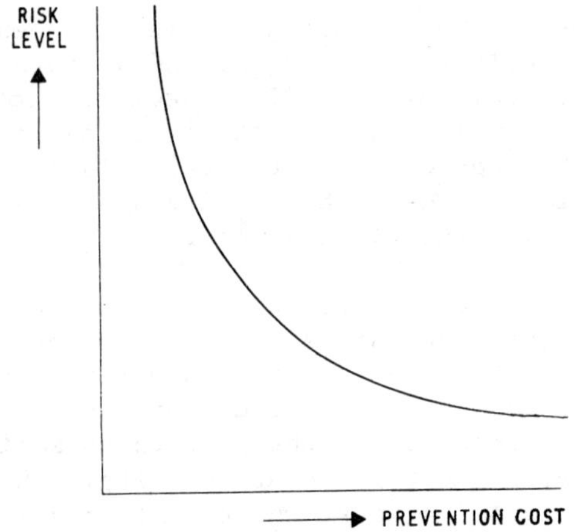

Fig. 2. Relationship between risk level and cos
prevention cost

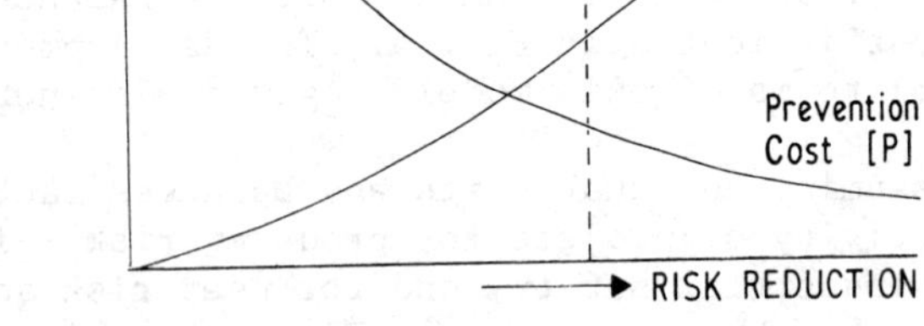

Fig. 3. Relationship between damage and prevention.
prevention cost

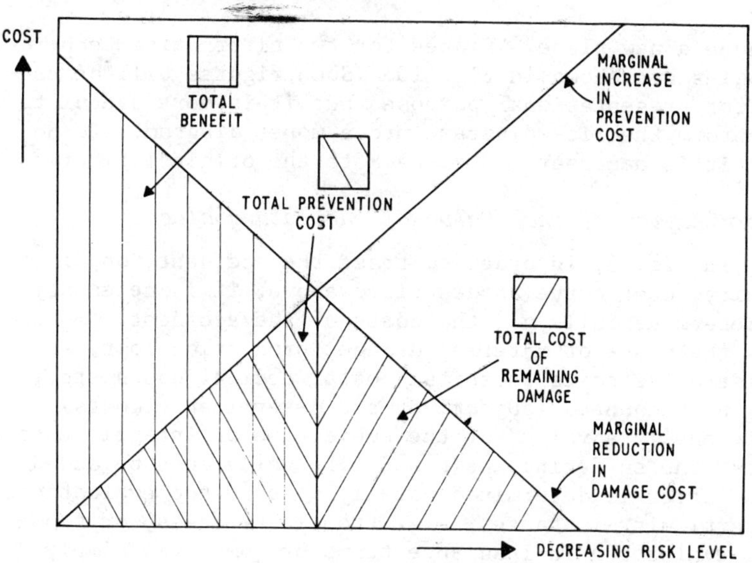

Fig. 4. Relationship between marginal changes in damage and prevention costs

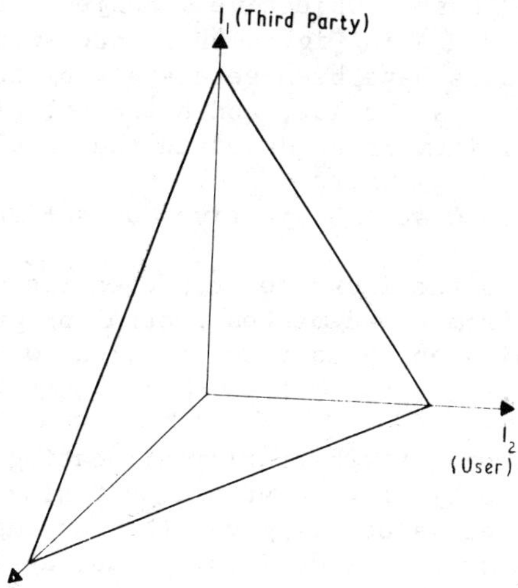

Fig. 5. Characteristic technology plane in amenity risk space

produce a new plane. Planes for two alternative technologies are shown in Fig. 10. Such figures will be useful for presentational purposes but it is more useful to transform the risk diagram into a money diagram. To do this it is necessary to go back to the original figures.

Determination of the "Value of the Intangibles

In Fig. 3, in order to trace the accident, or disamenity, cost curve as described above, it is necessary to determine fully *all* the costs of the accident. To do this the costs of material damage, production lost, administrative costs, hospital costs, medical costs etc., must be accounted (objective, and determinable costs). Additionally a value for the subjective or intangible cost costs, the suffering, pain and life value must be added. This latter is determined usually by post hoc evaluations of particular instances, e.g. from legal cases, or from self evaluation in insurance terms or sometimes simply from lifetime production or earnings suitably discounted and adjusted or not, as the case may be for consumption.

Thus "Disamenity" Cost = Material + Administrative Costs + Intangible Cost,

i.e. "Disamenity" Cost = Objective + Subjective Costs

We thus derive curve A in Fig. 6 where prevention costs and disamenity costs have been separately plotted against risk level. For a particular, and hitherto, given subjective cost a minimum is produced in the Total "Disamenity" Cost (T.D.C.) where

T.D.C. = Objective Cost + Subjective Costs + Prevention Costs.

This is so because the increase in prevention cost, P in Fig. 6, arising from an augmented control programme say, will cause the disamenity cost to fall, but with diminishing returns.

If now, instead of arbitrarily allocating a particular value to the subjective cost and so fixing the total cost curve a set of values is given then an implicit value for subjective cost can be derived. Thus, as in Fig. 7 where for clarity only two values have been given to subjective costs, a set of minima are produced at different risk levels. For the existing prevention expenditure

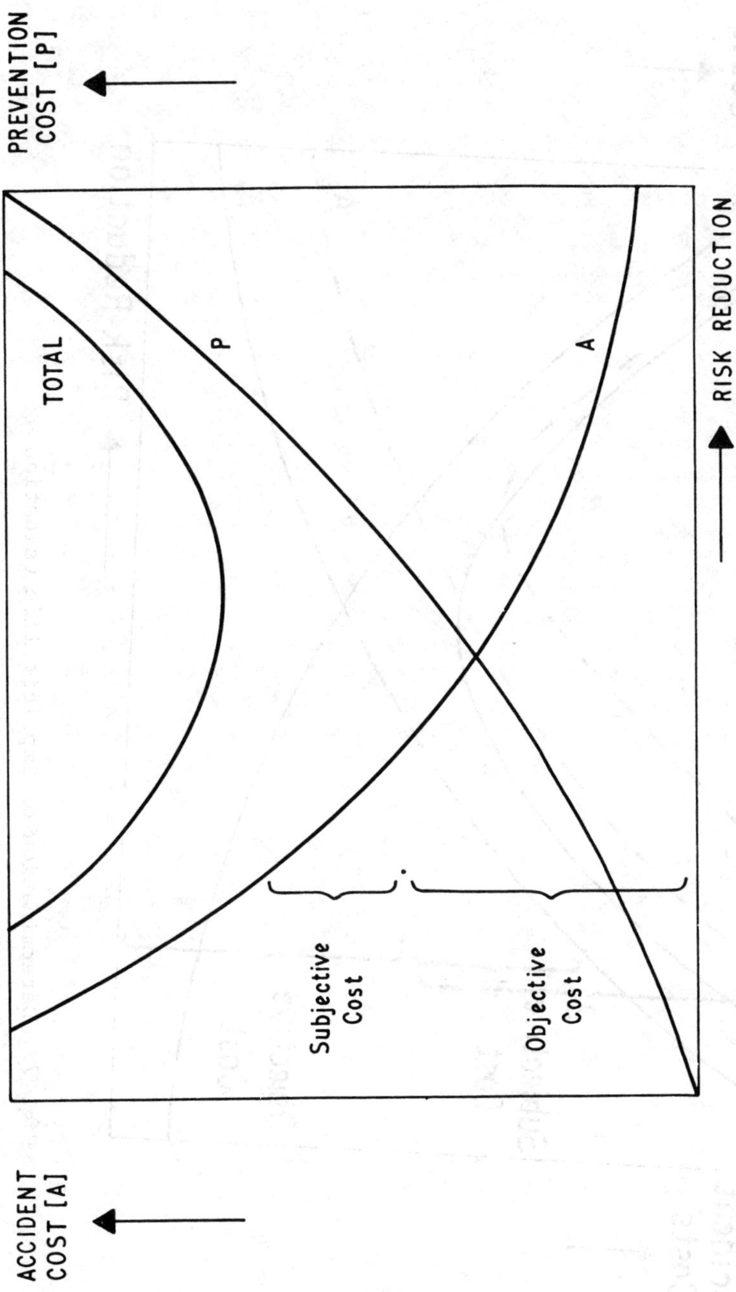

Fig. 6. Subjective and objective costs

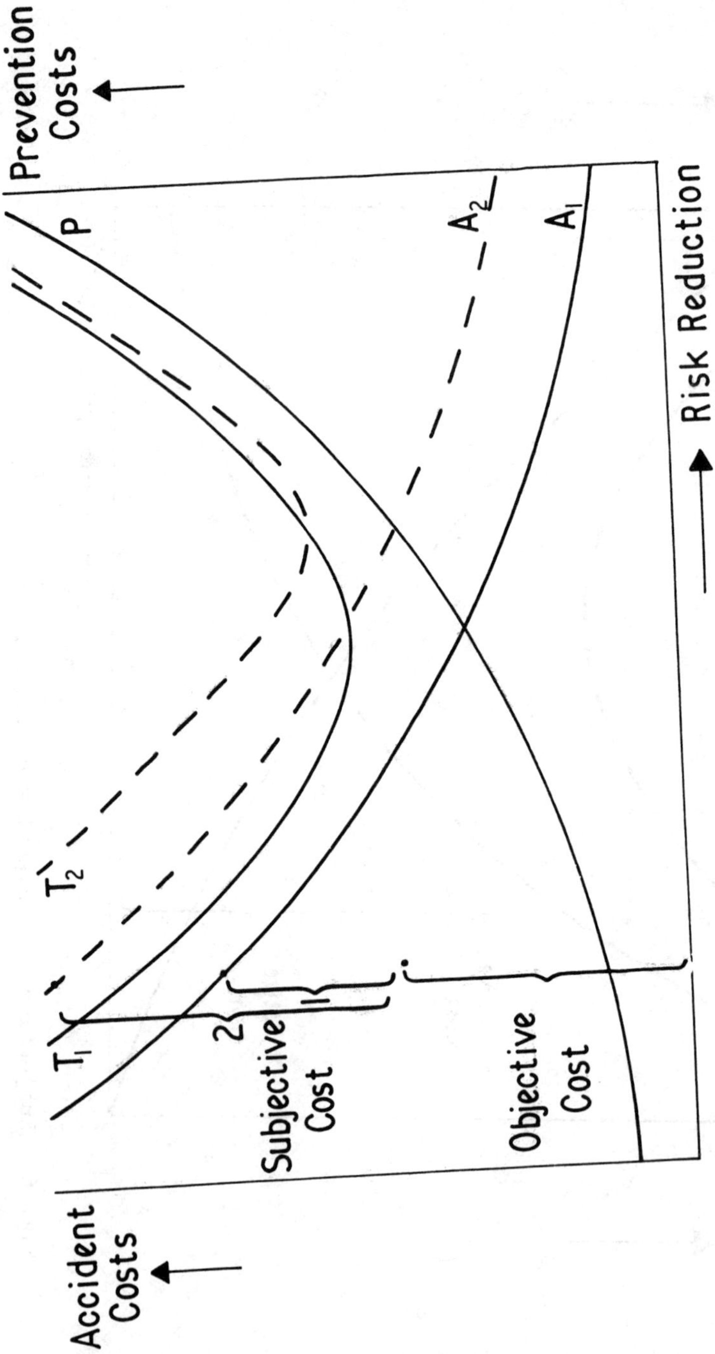

Fig. 7. Determination of implicit life valuation

level, only one value of subjective cost will produce a minimum. Thus determination of the existing prevention budget in all its aspects and the present objective costs, an implicit valuation is obtained for that activity, industry or technology. Similar analysis can be applied to other risk categories and other disamenities.

The present study was mainly interested in the implicit life valuation, for existing institutional and legal arrangements regulating certain activity in respect of accidental injury and death. The assumption is that present expenditures are judged to be at least near optimum. Implicit life valuations made in this way may be expected to show large differences between activities. If it is accepted that a movement towards greater equality of valuation is preferable then by reference to the values determined for marginal improvements a priority listing of areas for action will be obtained. When expressed in money terms, as in Fig. 8, the risk diagram may be seen to have axes which are not necessarily orthogal since many safety features protect both worker and user or worker and third party and a change in effort directed towards reduction of risk to one group may protect another group. Thus an interesting measurement is that of correlating risk levels imposed on the public by particular activities with industrial risk levels. Action towards the protection of plant where this is distinguishable and protection of the workers also makes an interesting comparison.

Implicit Valuations

If the attempt is to be made, as here, to develop a method of intercomparison of risk levels between projects and technologies as the basis for central government action and not, as hitherto, to balance cost and benefit within particular projects, then a further reason emerges for quantifying such items as human life in money terms. This is the fact that the "inherent" risk involved in different activities has a very wide range. Coal mining and trawler fishing (see below) are recognized as "extreme" occupations and are "more dangerous" than many manufacturing industries or clerking or banking. A relatively superficial attempt has been made to classify or order activities in terms of

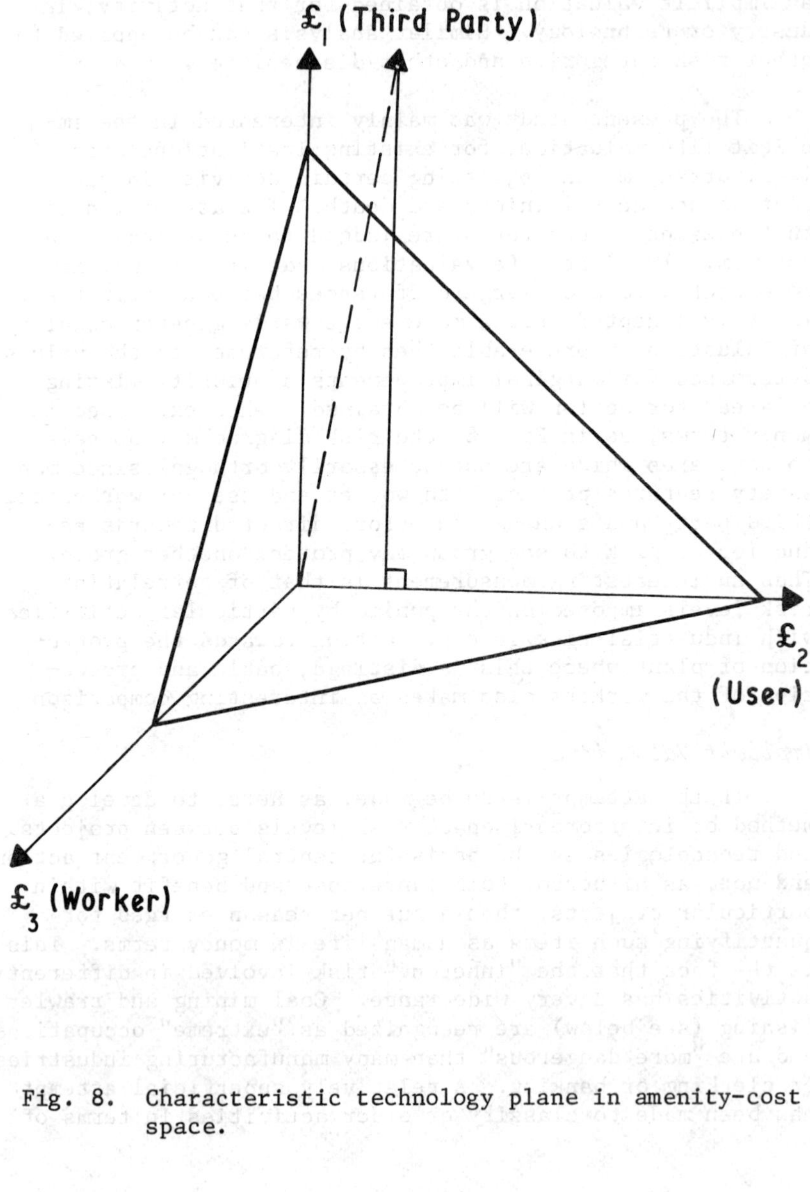

Fig. 8. Characteristic technology plane in amenity-cost space.

the energy exchanges involved in the activity. While this has some advantages — the most dangerous activities being those involved with the forces of nature in its most elemental state — fishing, farming and mining, it was abandoned as being unsuitable to encompass the range of risks considered here. The idea has associations with previous work done in the field of accident prevention, notably that of Gordon's work on the epidemiology of accidents which sought an analogue for the host-agent-environment model of disease. The realisation, in examining these lines of thought appeared that it would be necessary to seek for a method which compared not only risk levels but the costs of altering them. Thus the nature of the activity presents a range of hazard which is then altered to a great or less degree by man's manipulation through a range of prevention techniques with that level. Thus the claim is that annual prevention expenditures "E" are equated, in the optimum position, to the product of a risk level "P" of damage, e.g. loss of life and the implicit valuation of the damage "L". Thus "L" must represent the upper limit to a valuation of a human life and subtraction of the objective costs from this will lead implicit life valuations. Demonstration of the range of valuations allows efficient resource allocation.

Research and Development

Some corollaries to this approach can be mentioned here. The determination of the prevention curves beyond points representing current risk levels may be uncertain, giving rise to several possible positions of the curves, each of which will produce a different value of optimal risk. (Figure 9). The expected or estimated position of the two curves gives an indicator for the expenditure of R and D funds to reduce ignorance in this area and hence fix the optimum positions. The relative values of shifts in optimal positions between the various curves will depend on the relative slopes of the accident, or disamenity, prevention curves thus giving some cost effectiveness indications for the funding of R & D. This concept is of importance with reference to the many remarks about the need for social relevance indicators for R & D.

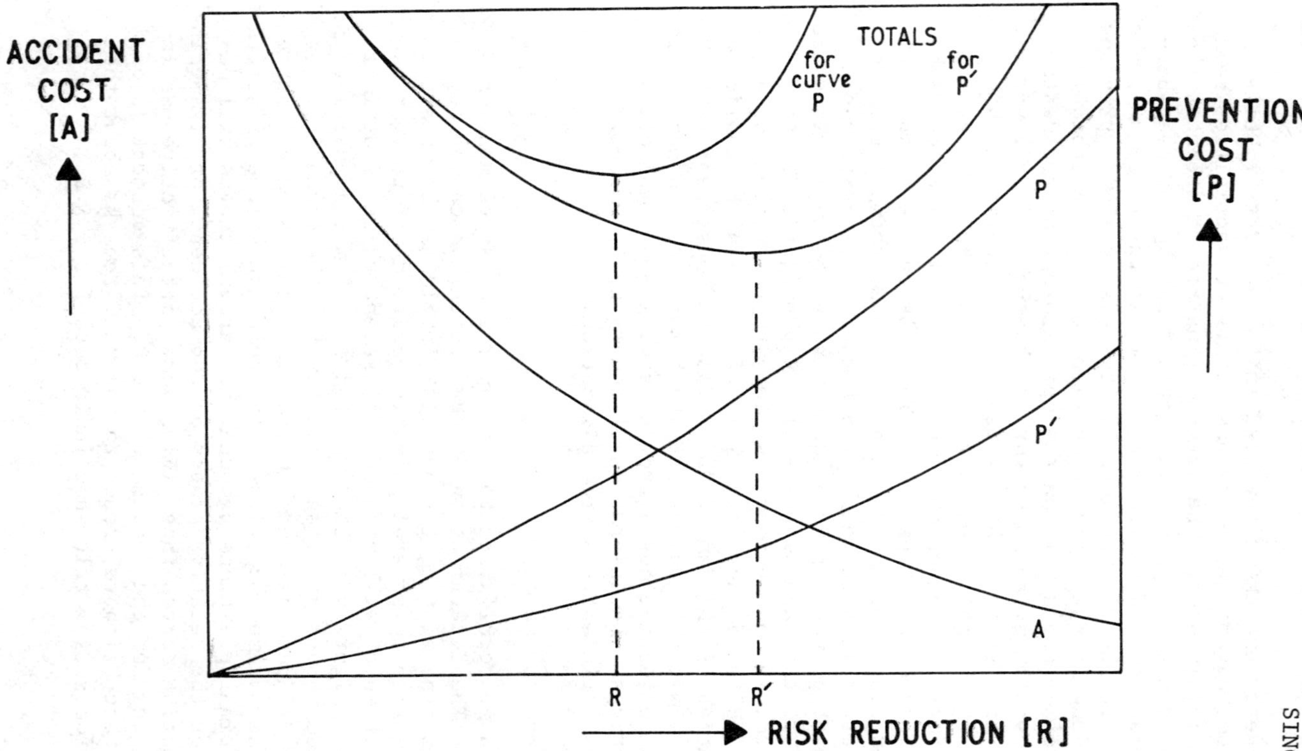

Fig. 9. Uncertain prevention function

It is argued here that added relevance would be given to such considerations when demonstrated against a background of the quantification of risk as is here developed.

When the modified Cobb-Douglas function was used to explain growth it appears that "technical progress" is a more weighty factor than capital stock though simplistic R & D investment correlations with natural economic growth have been unsuccessful. However, the use of R & D to increase social welfare through other than increase of GNP, the principal goal of much current R & D might be attempted in the field of amenity by the presently proposed methodology. Thus the determination of the appropriate cost curve for postulated or potential prevention costs will yield evidence for correct positioning of the optimal risk or pollutant level. Estimates can also be derived for each of the disamenities considered of the costs of undertaking R & D in that particular area with reference to the possible returns to be had from optimisation of the total cost of the disamenity.

By converting the total risk cost diagrams into individual valuation diagrams the intrinsic danger of the particular activity is taken into account. If, further, the cost of altering the individual risk levels as displayed in the diagrams is compared with total cost and effectiveness by taking account of the total numbers at risk then a further dimension is added to the set of data available to the decision maker (Fig. 10). This is essentially the position in the sphere of public health where overall risk to society and risk to the individual, from say vaccination, must both be taken into account. In siting nuclear power stations the same dilemma is present.

Suboptimisation

The foregoing assumes that activities cannot be judged comparatively in terms of an overall balance for each industry of benefit and risk but that it is possible to suboptimise within a given set of economically chosen projects and activities in such a way that disamenities arising from these activities are not disproportionate the one with the other. Thus, given that decisions are made on

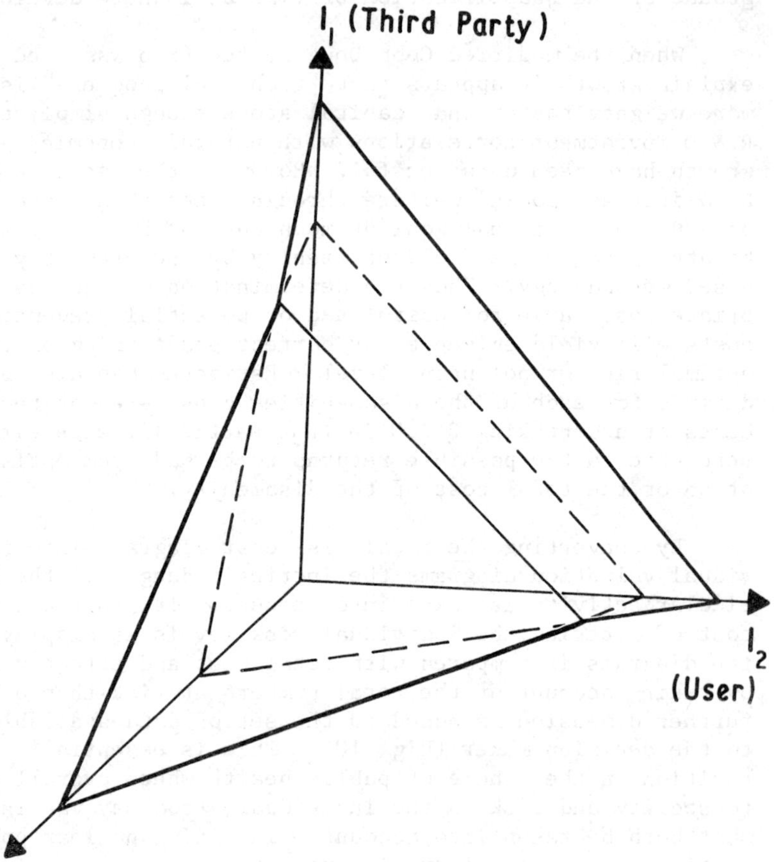

Fig. 10. Alternative technology planes in amenity-risk space

narrow economic or other grounds to adopt a particular technology it is possible, using the proposed method, to suboptimise in such a way that the spectrum of risks run by individuals in society is determined in an optimal way with regard to prevention costs.

Such a framework of assessment and investigation would, it is hoped, allow judgments to be framed in a way which the value judgments were apparent and minimal. The comparison of both existing risks and potential risks would thus be facilitated and they could be integrated into a single scheme. The underlying logic or otherwise of existing arrangements for risk assessment and risk distributions would be demonstrated and new initiatives developed for action based on numerical concepts.

Level of Precision

However, while a "systems analysis" approach to the problem of setting objectives for risk taking with respect to the health risk from pollution and accidents is being proposed, it is recognised that in the present state of development precise analyses are often not possible. For example, it has been argued that two specific weaknesses exist in the present application of this type of approach. These are, firstly, that the specification of objectives in the fields of health, safety, education, etc., compared with industrial or military objectives, are so broad as to vague and that consequently measurement of achievement is difficult. Secondly that consequent upon this first difficulty objectives are measured in terms of input rather than output and criteria developed from this aspect are unhelpful. The contention in the present study is that objectives in safety can be specified numerically in terms of the reduction or equalisation of overall risk relative to cost. Safety levels can be specified, at least in terms of the level of death risk, even if injury covers a wide spectrum of possibilities and while uncertainties exist in the statistics, it is possible to use them at least for a first analysis. With regard to the use of input measurements, in the present case prevention budgets, as criteria the present method combines these with risks to demonstrate that for existing control mechanisms implicit valuations of the so-called intangibles do exist. If these can be shown to have a sufficiently wide range then a basis for action exists, even taking account of the uncertainties.

Wider Impacts

Analysis of safety in terms of cost benefit analysis is often, indeed always doen in terms of the ceteris paribus beloved of economists. It is always assumed in calculating benefits, for instance, that increases in safety will not affect the extent to which the activity is prosecuted. In respect of private decisions about, for example, transport this may well be wrong. It seems likely that the use of particular modes of transport is becoming more dependent upon public knowledge of the risks involved. It is certainly wrong with respect to, for example, the doctor's decision to use chemotherapy or for that matter any form of therapy since an acute awareness of the offsetting danger deriving from the treatment relative to the possible benefit is present. Viewing incremental cost reductions (i.e. reduction in accident costs) simply as incremental benefits does not exhaust the analysis, and predictions must include comprehensive changes in the use and production of the good in question, arising from the reduction in risk.

Further, in using, as is often the case, incremental discounted earnings as the sole measure of the benefits from reduction in accident rates, a too simple view is being taken of the benefit, that is of the value of human life. This method has been criticised by Schelling as follows:

"There is no reason to suppose that what a man would pay to eliminate some specific probability, P, of his own death is more than, less than, or equal to, P times his discounted expected earnings. In fact, there is no reason to suppose that a man's future earnings, discounted in any pertinent fashion, bear any particular relation to what he would pay to reduce some likelihood of his own death."

The implicit measure advocated here removes some of the objections. This topic is further discussed in the main study where the full range of the presently used methods of obtaining "life valuations" are critically examined. It will be apparent firstly that the costing of many disamenities rest finally on a life valuation and secondly that other "intangibles" in the assessment balances can be substituted in the above analysis and thus "measured". This methodological excursion has been necessitated by the need to view the general comments made on Technology Assessment in the U.K. in the light of the author's standpoint. Thus no official body has utilised explicitly such an approach, although the nuclear power industry has moved perceptibly towards this rationalisation of central expenditures.

APPLICATION

An investigation of existing Technology Assessment-type mechanisms in U.K. practice was thus carried out using the foregoing theoretical framework. It ranged across industry from those typical, perhaps of the "laisser-faire" economics of previous times to the "laisser-innover" [7] technologies of the present day.

The forms of the administrative and financial controls used in industries of different age have been shown by Stinchcombe to be derived from the social forms dominant at the time of foundation. And further, in the development of the theory of "laisser-faire" economics. List has shown the permanent disadvantage to which latecomers to the technological and industrial revolution have been subject.

The advent of "Technology Assessment" must mark the end of any policies which attempt "laisser-innover". In its effectively step-by-step introduction in many economies, particularly those of Western Europe, the idea of the control of the effects of science and technology has left an imbalance which may be demonstrated by the wide range of implicit valuations which can be established. Following the analogy of the previous paragraph it can be shown that later technologies have, in general, carried a greater burden of control measures than some of the less developed. Thus a major gap in Technology Assessment in the U.K. as far as it is represented by the mechanisms controlling aircraft travel, pharmaceutical development and nuclear power hazards is the dependence on only partial assessments, limited particularly to the possible adverse effects and a lack of intercomparison on this side. This has led to different criteria, for example, of acceptable risk between the technologies, unrelated to cost. Further, the assessment in systematic analytical terms of older, and possible more fruitful in cost effectiveness terms of the more hazardous activities has not been attempted. This is not to say, of course, that forms of control like, for example, the uniquely British system of public inquiry, which represents a rudimentary control method, have not achieved certain successes. However, the form of the inquiry and its ultimate dependence on legal forms of definition have sometimes been overtaken by technological change. An example of this was the fact that when the ocean oil drilling rig Sea Gem suffered an accident in October 1967 it legally was neither ship, pipe line, mine or any other item covered by statute [8]. However, the very flexibility of the inquiry method makes it possible to investigate such cases in a discretionary or non-statutory manner. Land planning and electricity are two of the largest areas affording public inquiries and consequently safeguards against damage to the public welfare and here the siting of nuclear power or other stations is a case in point. However, such inquiries [9] have never struck anything like the form of a cost-benefit analysis or technology assessment. Those mentioned here tend to be *post hoc* examinations, for instance with regard to accidents, but in the planning area, where they are concerned with future development, has given rise to a body of research devoted to prediction. In passing, two retrospective inquiries should be mentioned which produced information of the type which would be utilised in a full-scale Technology Assessment to the Inquiry into the Collapse of the Ronan Point building in London in 1968 [10]. The report and subsequent discussion revealed much information both about the technical aspects of industrialised building systems and their potential failure modes and current civil engineering practice with respect to public safety. Estimates were also made of the risk levels of using gas for cooking in such buildings and arguments put forward of the relative national advantages. Together with the estimated cost of the Inquiry's recommended changes in building method, the risk level produces a valuation per life saved as in Table 2. The role of economics in setting safety margins in civil engineering was also exposed as being more important than is sometimes allowed.

The Hixon level crossing disaster [11] inquiry, also in 1968, not only investigated the details of the accident itself but invoked a formal general investigation into the safety of the new continental 'half-barriers' and the procedure for escorting heavy road vehicles carrying exceptional loads. The inquiry, conducted by a Queen's Counsel assisted by two technical assessors, lasted 41 days and produced recommendations of wide implication.

The possibly increasing frequency in the institution of such *post hoc* inquiries into technical failure and the fact that they frequently reveal the prior existence of relevant information is a further argument for the establishment of a formal Technology Assessment system.

Roskill Commission

The public inquiry, in this case non-statutory, which has most closely approached a comprehensive assessment of technology in any field is of course the Inquiry into the location of the Third London Airport [12]. Briefly, the chain of events, culminating in this largest example of the cost-benefit approach to Governmental decision making, arose from the report of an inter-departmental committee, in 1963, to the Minister of Aviation, published in 1964, on the subject of siting the Third London Airport at Stanstead. By ministerial decision at the Ministry of Housing and Local Government a public inquiry was held to establish local objections as to the suitability of Stanstead and was non-statutory, dealing as it did with Crown Land. The objectors were permitted to suggest alternative sites, though not to question the need for a third airport. This

TABLE 2. Specimen Implicit Life Valuations Derived from Present Assessment Techniques

Area	Risk Level p.a. (a) Death (b) Injury			Implicit Valuation £		
	Worker	User	Third Party	Worker	User	Third Party
Agriculture	(a) 0.4×10^{-3} (b) 25×10^{-3}	(b) 6×10^{-5} Food Poisoning	—	(a) 7.5×10^{3} (A)	(b) 15×10^{3} (M)	—
Steel Handling	(a) 0.22×10^{-3}	—	—	(a) 3.0×10^{4} (A)	—	—
Pharmaceuticals	(a) 0.02×10^{-3} (b) 30×10^{-3}	(a) 2×10^{-5} (b) 2×10^{-3} Adverse Drug Reactions	(a) 2×10^{-5}	(a) 7.5×10^{6} (A)	(a) (b) 0 or-ve	(a) (b) 0 or-ve
Nuclear Elec.	—	(a) 2×10^{-5} Electrocution	(a) 10^{-7}	1.0×10^{6} (M)	—	2×10^{6} (A)

Air Transport	(a) 1.4×10^{-3} (Aircrew civil)	—	—		(a) 17×10^{3} (M) (Life rafts)	—
Trawling	(a) 1.0×10^{-3} —	—	—	(a) -ve (current) (A) or 0.1×10^{6} recommendations (M)		
High Rise Flats	—	(a) 2×10^{-6} Explosion and Collapse	—	—	(a) 20.0×10^{6} (M)	

aspect has been a source of comment and criticism throughout the debate and represents a retreat from a full technology assessment. The inquiry, conducted by counsel, had a large amount of technical information available and some of the criticisms levelled pertained to this, particularly the question of accessibility and travel times to and from central London. After a year's review the original departmental decision was reaffirmed and a storm of protest ensued from local individuals through to the Council on Tribunals. A volte-face by the Government to suspend the decision and later to set up a major Inquiry was the origin of the Roskill Commission and report. The reasons for the change of decision were never completely stated. According to one commentator [13] "the catastrophic implications of a large airport had been under-rated by the Inter-Department Committee". This was primarily concerned with airport technology and only secondarily with agriculture and amenity and did not take into account long-term physical planning in the area. Thus, the original proposal was attacked in the "Governmental Responsibility for the Natural Environment" [14] on the grounds that "an informal inter-departmental understanding concerning the probable location of a third London Airport was allowed to acquire the status of a firm decision without ever being critically tested either within government or in public." The cost-benefit method applied to the value of building the Victoria Underground line in the early 1960s had not been applied. This was met by the appointment of an eight man Commission supported by a large research staff who, over 1968-1970 applied a thorough-going, if still criticized, cost-benefit analysis. It seems that such Commissions in future will be able to rely on the assistance of necessary researchers.

In turn, the outcome of this analysis — to site the airport at Cublington — met with renewed public outcry and final rejection. That this should occur is, of course, of political significance to the extent that it recognizes the difficulties of reconciling values derived from expert procedures with the values ascribed to by the public at large. The stimulation of public debate by the display of decision making factors is a valuable, if not prime, aspect of Technology Assessment.

Demonstrated Range of Values

To return to a description of the present results, the values displayed in Table 2 are of varying statistical robustness, but can be used to make certain statements about the varying levels of sophistication of current technical and other assessment and control methods that have determined them.

The assessment and control system that covers industrial worker risk is, of course, one of long and piecemeal application of legal, technical and social adaptation and this is reflected in the range of values shown. The original paper from which they are drawn examines the inadequacies and anomalies that have arisen from the uncoordinated approaches [15]. The rapid application of technology to agriculture, for example, has been coupled to a hazard control program, both with respect to worker safety and to special areas of wider concern such as the use of antibiotics as animal growth promoters and the use of chemical fertilizer. In respect of the former, while overall rates have fallen, serious injury and death rates have risen. In respect of the public hazard aspects, the investigation and ultimately the Swann Committee [16] recommendations on antibiotics represent again a partial Technology Assessment constrained by the lack of an adequate data base and a model procedural basis. The tenuous nature of the conclusions has been discussed by Smart [17] and the implicit valuation shown in the Table. The comparative complexity of present arrangements for the control of the use of antibiotics in the U.K. is indicated in Diagrams 1 and 2.

The relatively high valuations of workers in both the pharmaceutical and nuclear power fields are probably reflections of the stronger administrative control procedures applying in these areas in consideration of the wider hazards to users and third parties. Detailed consideration of the assessment methods in respect of these latter areas of public impact and, for example, similar concepts with respect to air travel, has shown that in each case, although the state insists upon a centralised assessment, this is almost exclusively concerned with hazards.

Diagram 1. The voluntary control of antibiotics for use in agriculture under the Veterinary Products Safety Precaution Scheme

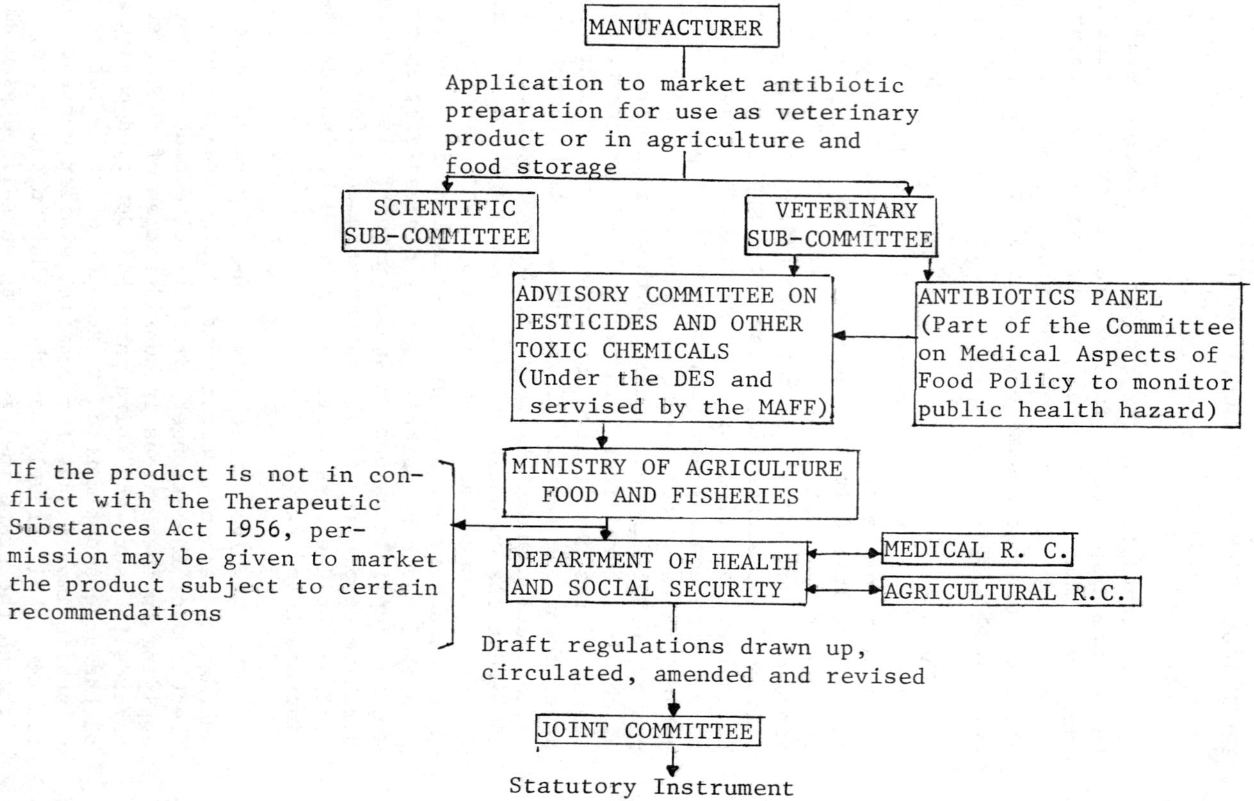

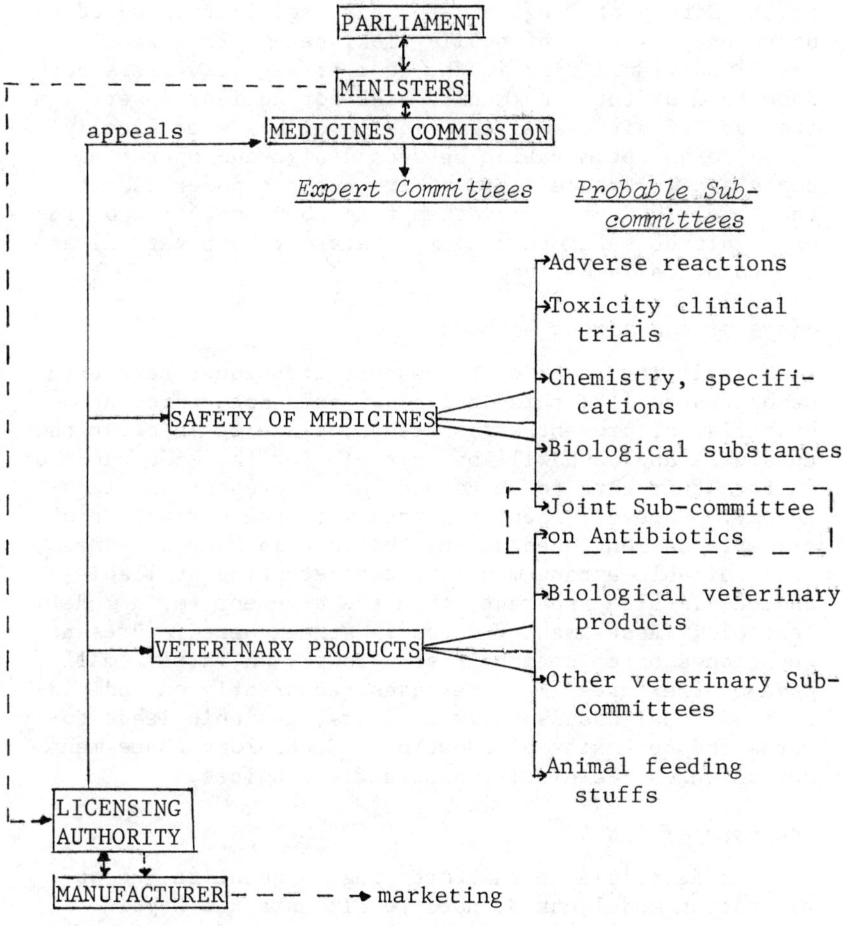

Diagram 2. A scheme for the statutory control of drugs under the Medicines Act of 1968. Adapted from a figure in the *British Medical Journal*, *3*, 761, 1970.

Further, they do not relate these to benefits or to the costs of reducing the risk. Thus the assessment techniques of the Inspectorate of Nuclear Installation, the Committee on the Safety of Drugs and the Air Registration Board use different criteria of public risk, being, for example, equal individual risk in the case of air travellers but some form of total community risk for nuclear power. In the case of air travel the risk level has specifically related to an optimisation between design and operating costs and insurance costs. With nuclear power essentially the level is decided by attempting to be one or two orders of magnitude safer than common hazards, both natural and technologically derived.

Range of Assessment Methods

While the range of assessment techniques have been demonstrated [18] this on its own does not represent a criticism of present arrangements, since it is clear that no single approach will be suitable for the wide range of factors that have to be considered in present day technology. However, when, together with the variable risk criteria as mentioned above, the lack in many instances, of a suitable arrangement for demonstrating available choices is also apparent, then the argument for a model Technology Assessment process is strengthened. Present techniques often seem to rest strongly on fairly small groups of experts, in some cases, apparently on individuals as final assessors or arbiters, and this leads towards the necessity of adopting a Technology Assessment method which can display alternative choices.

The Role of R & D

As described in the foregoing, a quantitative optimisation model can be used to allocate R & D priorities. The present investigation has indicated several difficulties in this area however. Firstly, the definition of what constitutes health, safety and anti-pollution R & D is ill-defined, industrial work in this field is often identified under another project title, and the results consequently not diffused as applying in the disamenity abatement field. Innovations arising from R & D which affect directly disamenity levels are more likely to arise consequently from R & D pursued for other reasons.

Examples of this which have been examined in the main report are those of the concrete pressure vessel for nuclear power reactors in the U.K., and the advent of the stern trawler. The latter, which has had, incidentally, a marked effect on accident rates, was introduced for reasons of strictly economic efficiency. The Inquiry into Trawler Safety arising from a disaster situation (see main report) made a thorough examination of a set of arrangements for improved amenity but a costing against the method proposed above suggests that the implementation of many of these will be delayed because of the high valuations implied with respect to the state of technological advance of the industry.

The concrete pressure vessel, as a replacement for steel arose from R & D performed for strictly economic ends and has resulted concomitantly in an improved safety assessment which has encouraged the Government risk assessors, the Inspectorate of Nuclear Installations to move towards less restricted siting of reactors. In this field it may be demonstrated by use of the above analysis and by examination of the course of events that the relatively stricter regulatory procedures have resulted in high implicit valuations.

Secondly the survey showed that while the range of percentage shares of particular R & D budgets falling to safety, pollution and the like was wide, perhaps 1-10% represented a majority of the respondents. Breakdowns were obtained of the proportions devoted to the various disbenefits but the total sample, two hundred firms and forty one Research Associations were not sufficiently large to make expenditures under the various sub headings relevant to the overall level of the disamenity. This, together with the difficulties of matching inputs and outputs, in terms of disbenefit abated puts the balancing of effort in this area, which may become a major theme of science policy, at an early stage. In total the expenditures were only a few per cent at most of the total costs attributed to the social "bads".

No convincing evidence was obtained, at least in this preliminary stage of an impetus towards disamenity R & D being stimulated by other than reaction to disaster situations. Thus formalised Technology Assessment methods might fruitfully be organised to produce recommendations for the performance of relevant R & D.

An area where policies for R & D have always been of prime importance for the growth of the industry is pharmaceuticals. Here a large percentage of the control expenditures against risk are accounted for within the R & D of private companies. The low, relatively, value for implicit valuation derived from the examination of such expenditures against the risk levels, is an indication that in this field the rationale of such research has been such that further expenditures (on the control side) cannot be scientifically justified. An improvement in risk levels, for the lack of adequate theory of drug action, must be sought elsewhere, possibly in better education of doctors and public. The controlling body in the U.K. for this risk is the Committee on the Safety of Drugs. This organisation does not attempt any assessment other than of the possible toxic effects of products and much of the detailed judgments in this field must be in the hands of the local doctor.

CONCLUSIONS

Since this paper is in part the description of a technique useful in Technology Assessment and in part a commentary on an examination of current assessments in the U.K. using this technique as a basis, it will be appropriate to gather the conclusions of this and the larger work together.

(a) Technology Assessment as such has no explicit existence as a tool of Government in the U.K., however, techniques of assessment arising from a large body of common and statute law, and custom have grown lately into forms similar to a formalised method.

(b) A particular method, unique to the U.K., is that of the Inquiry, often public, which may be (a) a retrospective examination of a technical failure or (b) in the form of a planning inquiry which may include aspects of a full Technology Assessment.

(c) Latterly two such planning inquiries have explicitly taken the form of full-scale cost benefit analysis with attempts at full social costing. The most recent of these was itself the outcome of unsuccessful attempts to arrive at a major decision, an airport siting, by less extensive and more traditional methods.

(d) Public reaction in this case was such as to cause the reversal of the recommendations of the appointed experts after much public debate.

(e) In the case of retrospective inquiries these have been almost invariably concerned with the cost or disamenity side of the equation.

(f) In respect of this cost, particularly social cost, a method is described for making intercomparisons between particular Technology Assessments.

(g) Application of this method has demonstrated a wide range of values for one particular variable relating to social cost. This arises in turn from the range of partial Technology Assessment methods currently employed in assessing, and setting levels in worker, user and third party risk. The method offers a way of optimising for risk or pollution level.

(h) Use of the technique has, when applied to the existing institutions, given statistical backing to the generalisation that the advanced technologies have greater burdens of risk control. This aspect requires further investigation. Also large areas of disbenefit, or risk, existing or potential, are unexplored by this or other techniques.

(i) Finally, the use of such a model of risk optimisation contains the possibilities of deriving priorities for welfare R & D budgets as far as the abatement of health risk, pollution and the like are concerned. The specification of such budgets from Technology Asessments awaits further definition of categories and measurement of the present distribution of both disbenefit and abatement effort.

REFERENCES

1. National Academy of Engineering, *A Study of Technology Assessment*, Committee on Science and Astronautics, U.S., G.P.O., July 1969.

2. National Academy of Sciences, *Technology: Processes of Assessment and Choice*, Committee on Science and Astronautics, U.S., G.P.O., July 1969.

3. National Academy of Public Administration, *A Technology Assessment System for the Executive Branch*, Committee on Science and Astronautics, U.S., G.P.O., July 1970.

4. Ad hoc Group on New Concepts of Science Policy, *Science Growth and Society*, O.E.C.D., Paris 1971.

5. FREEMAN, C. et al., "The Goals of R & D in the 1970s", *Science Studies, 1*, 357-406, 1971.

6. JONES, P.M.S., "Lessons from the Objective Appraisal of Programmes at the National Level", *Research Policy, 1*, 10-28, 1971.

7. FREEMAN, C., "Technology Assessment and its Social Context", *Stud. Gen., 24*, 1038-50, 1971.

8. Cmnd. 3409, "The Accident to the Drilling Rig, Sea Gem," October 1967, H.M.S.O., London, 1968.

9. SINCLAIR, C., *The Bases of the U.K. Energy Policy*, J.F. Kennedy Inst., Tilburg, 1972.

10. Ministry of Housing & Local Government, "Inquiry into the Collapse at Ronan Point", H.M.S.O., London, 1968.

11. Cmnd. 3706, "Inquiry into the Disaster at Hixon Level Crossing", H.M.S.O., London, 1968.

12. Ministry of Housing and Local Government, *Report of the Roskill Commission* of Enquiry into the Siting of the Third London Airport, H.M.S.O., London, 1971.

13. WRAITH, R.E., and LAMB, G.B., *Public Inquiries as an Instrument of Government*, Allen & Unwin, London,, 1971.

14. LORD HURCOMB, (ch.), "Governmental Responsibility for the Natural Enviornment", Rep. No. 17, H.M.S.O., London, 1971.

15. SINCLAIR, C. et al., *Human Life and Safety in Relation to Technical Change*, Chapter Three, Science Policy Research Unit, University of Sussex, 1972.

16. Cmnd. 4190, "Report of the Joint Committee on the use of Antibiotics in Animal Husbandry and Vetinary Medicine",H.M.S.O., London, .969.

17. SMART, C., "Antibiotic Technology in Agriculture an Aspect of Pollution". Unpublished thesis, Sussex, 1971.

18. SINCLAIR, C., *op cit.*, Chapters 3, 5, 6, 7 and 8.

TECHNOLOGY ASSESSMENT IN THE U.S.S.R. OR THE SCIENCE OF SCIENCE: A BASIS FOR THE EFFECTIVE MANAGEMENT OF SCIENTIFIC ACTIVITY*

G.M. DOBROV

Doctor of Economic Sciences, U.S.S.R. Academy of Sciences

INTRODUCTION

Science** is a mechanism for the acquisition of new knowledge. We must think of science as a systematic process; and, as such, a manageable activity. The role of science as an agent of economic growth and social change suggests that the public sector must play an important role in its planning and support. There were one million scientists in the U.S.S.R. in 1971 (and about four million persons engaged in some aspect of scientific work) with a budget of about thirteen billion rubles (about $14,430,000,000). This represents an 8.3% increase over

*Editor's Note: Although only a literal translation was available we have tried to be faithful to the author's original language and intent. Readers familiar with U.S. efforts in Assessment, Forecasting, Planning and Managing R & D will see much that is familiar. This article documents the current thinking in the U.S.S.R. and illustrates the universality of both the process and management of science and technology. (M.J.C.)

Adopted from translation of the original Russian version.

**The author uses science as a general term covering Technology, Research and Development activities.

the 1970 science budget concurrent with a 6.1% increase in the national income. In return for this investment we expect that about 3/4 of the increase in national production will be due to the use of the fruits of scientific research. To realize this potential for socially useful work the scientific process must be accompanied by: appropriate social conditions, a perfect organizational mechanism, and efficient management. Thus, the effective and efficient management of science is an integral part of the modern scientific and technological revolution, the quintessence of which is the transformation of the various disciplines of science into an independent productive and social force.

As science enters a new stage — a transition from extensive to intensive forms of organization — the conversion to more scientific methods of organizing and managing science is inevitable and mandatory. In recent years science has become more quantitative and more complex and its management must keep pace. For many years a rapid increase in scientific "yield" has been the result of an exponential increase in the number of scientists, funding levels, volume of technical facilities and rate of information flow. Now, however, this rate of increase has begun to level off and we must look to better organization and management for further improvements in the effectiveness and efficiency of scientific effort. The theoretical concepts and principles and their practical applications required to obtain this objective are being evolved today by a new discipline of science: THE SCIENCE OF SCIENCE.

THE SCIENCE OF SCIENCE

The Science of Science is a new scientific discipline whose objective is a systematic theoretical analysis of scientific effort and its results for purposes of forecasting technology and for determining needed changes in organization, management and policy as science itself changes. It offers more than can be achieved by merely putting together a mosaic from disassociated studies in economics, sociology, psychology, history and scientific logic. None of these approaches to the problem provides the overall breadth available from the Science of Science approach.

The Science of Science investigates the general principles underlying scientific and technological development, and the correlation between economic, social, historical, psychological, and structural-organizational aspects of scientific and technological development. Its objective is both the development of theoretical principles of organization and a system of measurement based on objective logic which will further and promote science and technology. In the end, the Science of Science approach will provide for optimum progress in scientific-technical development and an increase in the efficacy of scientific research and planning; thus becoming the theoretical basis for science policy formulation.

The Science of Science approaches the problem of increasing the effectiveness of scientific effort at three levels:

1. The immediate working environment of the researcher. Enhancing creativity by improving the quality of research methods, developing new and more original scientific concepts, and in adopting and applying productive and promising research techniques.

2. Improving the management of scientific activities and optimizing the organization of scientific activities. Creating better conditions for productive work for all classes of scientific workers within the system.

3. Improving and perfecting the social and economic mechanisms for the most rapid adoption of the fruits of scientific research by industry and the community at large.

PRINCIPLES AND MECHANISMS IN THE MANAGEMENT OF SCIENCE

Scientific activity is, in reality, a social system which requires the management of relationships among people, groups, and organizations who/which are in the process of achieving a specified set of goals and objectives. The essence of scientific management may be defined as the conscious and purposeful activation of the political, economic, organizational, informational, juridical, moral, and aesthetic relationships which arise in the process of scientific work. These relationships and ties, each in its own way, reflect the social-economic nature of society and the level of development reached by science, i.e., the degree of its conversion into a direct and independently productive social force. Science today is an enormous, complex, dynamic system facing the transition from extensive to intensive forms of progress. Concurrently, the management of science has begun to adopt a systems approach to the problem of solving the complex relationships which arise during the process of scientific work. What does this actually mean? One of the answers to this question was provided by D.M. Gvishiani, an outstanding Soviet specialist in social administration; "The scientific spirit in management, above all, means the rejection of a subjective attitude in evaluating production problems as well as the results obtained, a definite transition and conversion to an all inclusive evaluation of available resources and of the real possibilities of using them to an optimum. The scientific spirit in management means that the management is based on modern methods of quantitative measurements, on statistical-mathematical analyses, and on the mathematical modeling of the processes and of the industrial-economic complexes being managed. The modern system of management is inconceivable without overall mechanization and automation of the managerial function".

Let us now analyze the organizational mechanism and basic principles of the scientific management of science in greater detail. For this purpose we shall use the chart shown in Fig. 1.

It should be pointed out that the management of scientific activity is not something that is outside of or "foreign" to the system being managed. It is one of its indispensible and organic characteristics, and one of the most important factors in its existence. Practical activity in carrying out managerial operations being the *output of managerial activity* is, at the same time, an *input* to management science in general in the form of

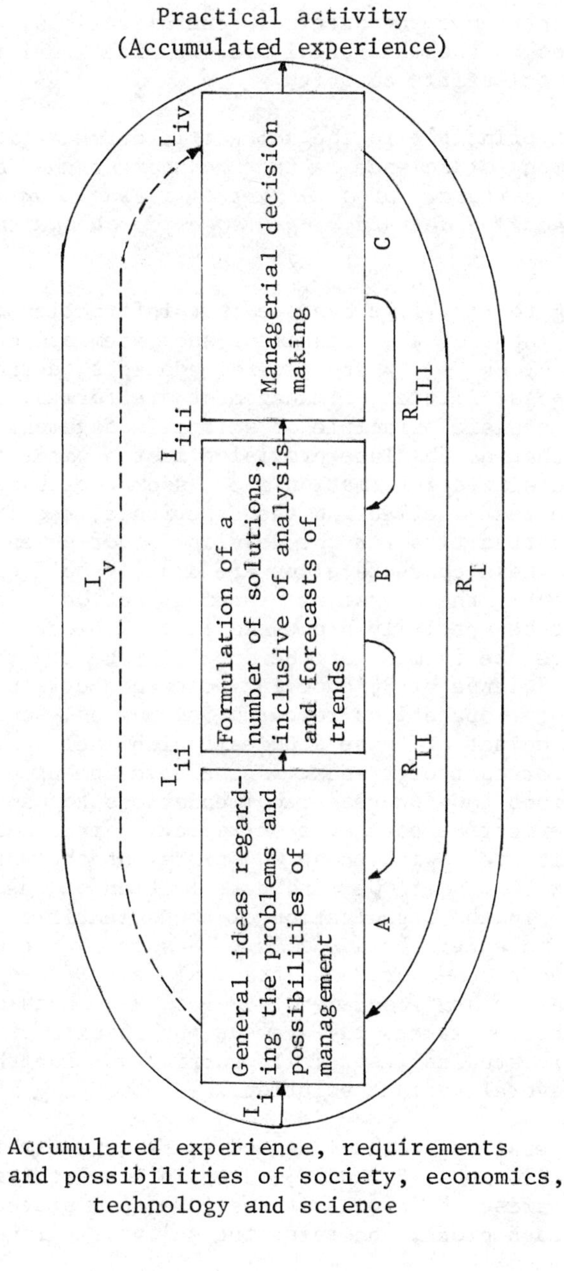

Fig. 1. Flow-chart of the organization of the management of science based on scientific principles

accumulated experience, theoretical generalizations, and proven ideas as to the managerial and organizational requirements of scientific activity.

The first principle in the scientific organization of the management of science is that *management must be regarded as an integral, organic part of a single unified system of scientific activity and a specific subsystem of the whole.*

Referring to Fig. 1 we can see that information arriving at the input of the management subsystem accumulates in work block "A", where general concepts on the problems and possibilities of management are formed. This is related to a basic principle of science management which states that an absolute provision must be made for a fund of specialized information and a source of basic and verifiable facts reflecting the experience, requirements, and opportunities for the development of science. Nevertheless, these concepts alone are usually sufficient to change directly the course of practical action. Information must be specially processed in work block "B". Here, extensive use is made of the special research techniques of the Science of Science including concepts of information theory, operations research, economics, sociology, etc. The end object of these studies, which include the analysis and forecast of trends, is the development of various solutions and concrete recommendations to management. The prerogative to accept or reject a given solution lies exclusively with those who carry out the work, shown in block "C". Here we find a reflection of the next principle in the organization of the scientific management of science: *the division of responsibilities in the management sub-system and the implementation of the so-called line and staff scheme of operation.* In this scheme the decision makers may draw on the results of many scientific studies (especially carried out for this purpose) of several variant solutions.

In the operation of the line and staff scheme which takes in work blocks A, B, and C, reversible relationships R_I, R_{II}, R_{III} arise which make it possible to improve the operation of each block, and raise the scientific level of the output of information. Here, the Science of Science (work block B, part of the work in block A, and assistance in making decisions in block C), which actually participates in the work; formulates principles based on the experience of other scientific systems. In other words the Science of Science formulates theoretical principles for increasing the efficiency of managing scientific systems. This results in a complex systems approach to the problems of managing scientific activities which is possible only through the application of the principles of the "Science of Science".

Solutions accepted by the science management subsystem change, establish and correct the whole complex of inter-relationships and ties characteristic of scientific research activity. I would like to point out that these relationships may be quite different and concrete, and the ties which embody them may be very generally reduced to the following three basic groups:

- Informational (including the influence of concrete scientific ideas)
- Organizational (including administrative-legal and moral-aesthetic relationships within a collective)
- Economic (including appropriate stimuli and responsibilities)

It should be emphasized that it is specifically these ties which unite the disparate elements and acts of scientific activity into a single unified system. The more complex and active these relationships are, the more real the scientific system and the greater the efficiency and effectiveness of its work. This is the origin of the next principle in the scientific organization in management of science — *the unity of informational, organizational, and economic ties in the process of scientific activity.*

The experience of Soviet and world science as well as the theoretical postulates of the Science of Science make it possible to state unequivocally that success comes only when all three forms of these relationships are actually operative. Underestimation or disregard of any of them leads to increasingly grave negative consequences. The Science of Science approach has been useful in several instances for improving the relationships among the U.S.S.R. Academy of Sciences and its various departments and agencies, as well as their relationships with industry. It should be emphasized that the *object of study* of the Science of Science is the organization and management of basically common problems in the scientific process, covering all stages from theoretical studies to the final development of new products and processes.

MANAGING SCIENTIFIC ACTIVITY

We can now proceed to a more detailed analysis of the general functions involved in managing scientific activity in terms of the classes of problems involved:

1. *Strategic planning* is the result of management's solving of the following basic problems:

- Determining the objectives of scientific research
- Creating effective plans for reaching these objectives
- Overall evaluation of organizational resources required to achieve the desired goals by a variety of approaches
- Establishing a system of social-economic and scientific-technical criteria for setting goals in scientific research activity.
- Selecting from a variety of possible operational plans for scientific activity and establishing basic principles of science policy.

2. *Scientific potential* is shaped by controlling and managing the process of growth in its basic component parts:

- Providing personnel capable of solving new problems
- Shaping and fixing scientific concepts and the desired level of familiarity with world science
- Satisfying material and technological requirements of the scientific research community

- Developing the right kind of organizational structure to carry out scientific research including the formation of new up-to-date organizations and the reorganization of the structure of existing organizations.

3. *Effectiveness in solving current problems* in scientific research operations requires active solution of administrative problems:

- Planning scientific research projects
- Managing the informational, organizational, and economic problems of the organization
- Managing the internal and external coordination needs of the organization
- Setting up, improving and utilizing a system of rewards and sanctions
- Managing the distribution of capabilities in an organization
- Solving problems in the organization of scientific research operations
- Managing the psychological climate of the creative organization.

These types of problems are encountered at every level of management, though their relative importance and dimensions may vary substantially. They are most typical of the management of independent scientific research institutions.

A final problem in the scientific management of research and development is the creation of a system for the monitoring of the state-of-the-art in science. This monitoring of scientific and technological change must be carried out on a continuing basis and the information developed through the monitoring system must be used in the organization and management of scientific research in any particular activity.

The preceding principles of the systems approach to the management of scientific activity is receiving ever wider acceptance in both the theory of management and in Soviet practice of science administration.

CHARACTERISTICS OF SCIENTIFIC ACTIVITY

Having discussed the management principles involved in applying the "Science of Science" to improving the efficiency and effectiveness of scientific activity we should pause to reflect upon the specific characteristics of scientific activity which must be accounted for in the application of the principles of scientific management. In many ways the essential nature of scientific research would suggest that it is unmanageable. Indeed this was the accepted "truth" for many years. Even today there is considerable resistance among scientists to the application of scientific management principles to scientific activity. However, in view of the decreasing rate of expansion of scientific resources we can no longer permit a laissez-faire attitude towards science. As the shoemakers' children have gone barefoot so, throughout history, has science gone without the benefit of the application of scientific principles to its organization and management. We have found, however, that science *can* be managed to a large extent. Management today means the timely realization of decisions and proposed solutions as well as their revision and modification in the light of changing requirements during the work process. Studies by the Siberian branch of the U.S.S.R. Academy of Sciences indicate that, for the projects they studied, a 25% time loss was incurred due to delays in making decisions an additional 10% time loss due to deficiencies arising in the supply of material and equipment. Certainly these are management problems.

On the other hand, the principles of scientific management cannot be applied blindly to the management of scientific activity. Certain aspects of science are unique and must be taken into consideration by managers. The most important characteristics of scientific activity from the point of view of managers are:

- A high degree of unpredictability in the composition and structure of possible projects
- The element of uncertainty which characterizes the basic parameters of expected scientific findings

- The difficulty of objective evaluations of planned work variations and the control of their intermediate results
- The more or less developed self-organizing and self-regulating elements inherent in research and development activities

Depending on the exact nature of the scientific activity; research or development, these elements occur to varying degrees and in a variety of forms at different levels of management. Taking these elements into account is an absolute prerequisite in the scientific management of scientific activity.

Many of the problems in managing scientific activity can be summed up as "selection" problems. Scientific activity, because of its great diversity of possible directions, is almost unique in the variety and difficulty of selection problems which it presents. The numerous Science of Science situations connected with the problem of selection may be divided into the following basic categories:

- Selecting an optimum science policy at the national level
- Evaluating a variety of approaches in achieving specific scientific objectives
- Organizing scientific work situations
- Setting up priorities *among* the various areas of scientific research
- Setting up priorities *within* each of the areas of scientific research
- evaluating progress in current research projects versus the demands of projects not yet begun. (Deciding when to drop a project)
- Systems for the collection, storage and search of scientific information
- Determining a time schedule for modifying elements within a scientific research project

All of these problems call for the application of basic scientific management principles including priority setting, resource allocation, strategy evaluation and technological forecasting to the unique problems of scientific activity as characterized above.

THE SCIENCE OF SCIENCE AS A COMPLEX SCIENTIFIC DISCIPLINE

An overview of the complex relationship between the Science of Science and the world of science is shown in Fig. 2 following. It should be emphasized that the "Science of Science" is not intended to be a "super science" encompassing all of the specific, scientific disciplines. The Science of Science is interdisciplinary in nature and although it draws from each of the major contributing scientific disciplines, it only utilizes a small part of each. Each of these disciplines provides the Science of Science with its own specific information relevant to that particular discipline and developed by its particular methodology and interpreted in terms of its own concepts. From these "building materials" used in the study of the functioning of scientific systems, plus the base of process information and general facts in the world of the concrete sciences, the Science of Science develops its own fund of facts, methods and concepts. The resulting information finds expression in recommendations and concrete proposals for the optimization of scientific systems. These results of scientific research are then fed to the appropriate decision making bodies for the formulation of appropriate science policies.

We are convinced that the success of the Science of Science at this stage of its development will be largely due to its developing ties with the already established economic disciplines. This will enrich the Science of Science with the economists' experience in solving complex problems of development and change, increase the effectiveness and efficiency of the methodology used by the Science of Science, and effectively improve the proposals and recommendations evolved. The development of a complex of economic problems in science forms one of the principle subdivisions of the Science of Science known as "the economics of research and development". On the other hand, the Science of Science itself may be instrumental (through its methods and findings) in increasing

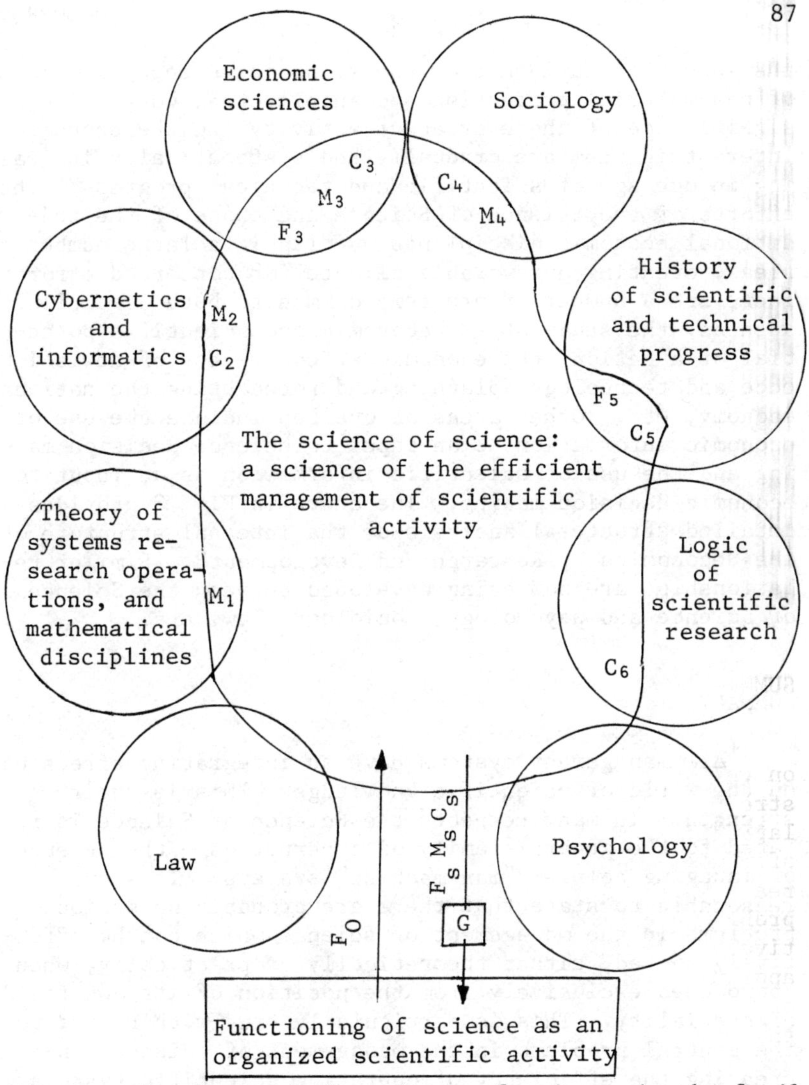

Fig. 2. The place of the science of science in the family of sciences:

- concrete scientific facts M
- concrete scientific methods
- concrete scientific concepts
- science of science facts
- science of science methods
- concepts
- decision-making element

the sphere of influence of the economic sciences in areas of non-material production and intellectual work. The significance of these areas of activity and the economic interest in them are gradually and systematically increasing in our age of scientific and technical progress. The efforts of the Science of Science and those of the traditional economic disciplines overlap in a large number of areas, creating a favorable climate for concerted effort. There are a number of problems common to both disciplines such as: the study of the economic and scientific potential of a nation, the economic effectiveness of new science and technology, planning and forecasting the national economy, etc. Other areas of overlap include the use of economic information as an input to science decision making and the use of scientific information as an input to economic decision making. The chart in Fig. 3 provides a detailed structural analysis of the internal structure of the "economics of Research and Development". Similar relationships are now being developed between the Science of Science and psychology, sociology, law, etc.

SUMMARY

All management systems have an integrating effect upon the world of scientific knowledge. This is their strength. In many respects the Science of Science is related to the general family of cybernetics — the science of managing complex "man-machine" systems. It seems reasonable to state that there are probably no serious problems in the management of science which can be effectively solved, either theoretically or practically, when approached exclusively from the position of its own field of speciality. This is particularly true with regard to the central problems in the management of science: increasing the efficiency of operating scientific systems, optimizing the increase in the scientific potential of the nation, and determining science strategy and policy. Again, it is the use of the Science of Science approach which allows us to solve these complex problems as shown in Fig. 4.

As Fig. 4. emphasizes, the Science of Science is not merely a complex of disciplines nor even a synthesis of information from various disciplines. Rather it is a science that emphasizes study of the *interactions* of the various elements which determine the development of science as a complex system. We believe that at the present stage of the development of the Science of Science, the formation of a general body of theoretical principles is of great importance. In our opinion this role should be fulfilled by a general theory on the management of research and development. The absence of such systematic concepts on science and technology as managed systems negatively affects the consolidation of the efforts of researchers on economic, organizational, social and other closely related aspects of the management of science with the overall Science of Science approach. This is one of the major reasons for the failure of many of the studies and recommendations made by the Science of Science to date.

In answer to a need for a theory and practice of Science of Science, several collective and individual works have been carried out during the last few years in the Soviet Union. Large scale efforts within the framework of the Science of Science have included: forecasting U.S.S.R. scientific policy for the next 20 years, creating an automated information service system for the management of science (the UPRAN system) and developing sophisticated projects for improving the organization and management of the larger scientific establishments and science centers. Specific recommendations have been given in all three of the above instances. These efforts have yielded a fund of theoretical concepts and experimentally verified methods which make the solution of this type of complex, large scale, important problem of raising the effectiveness of science management a reality, both at the level of the individual scientific activity and on a nationwide scale. In order to realize those potentials it is necessary to integrate the efforts of various sectors of science and to properly orient the organization of research and development by using the findings of the Science of Science approach. These recent efforts have

Fig. 3

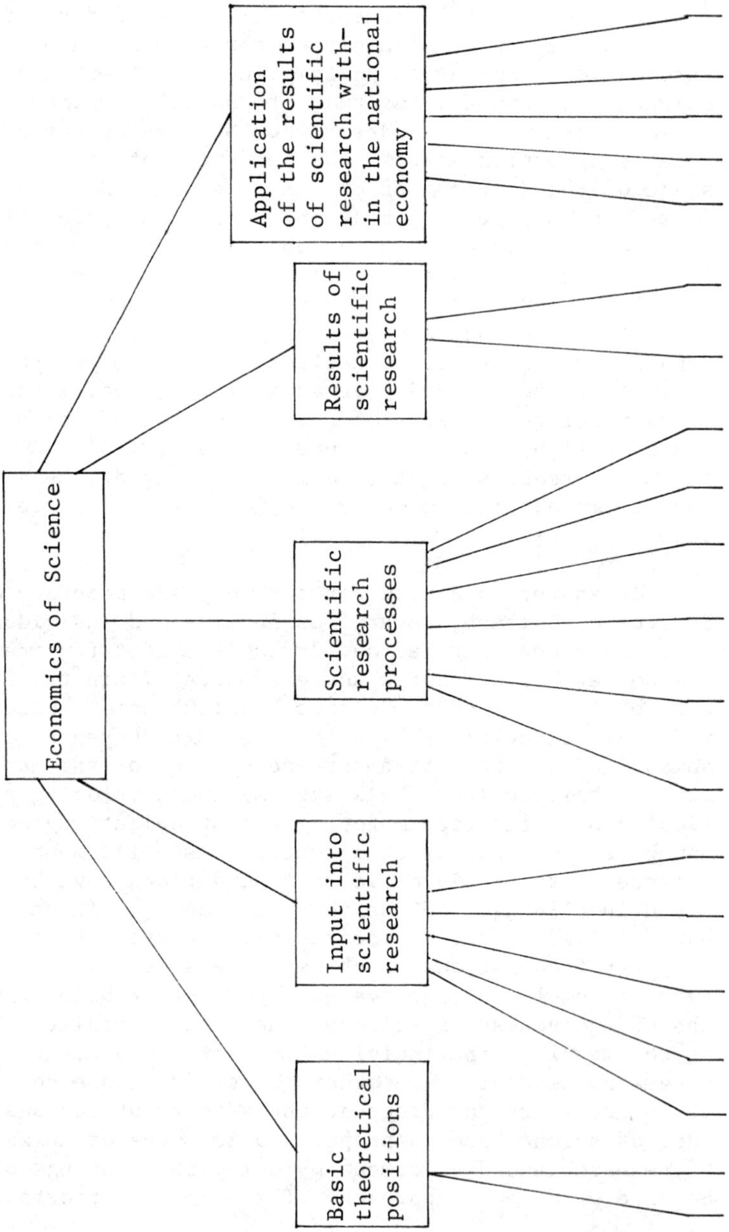
Economics of Science
Basic theoretical positions
Input into scientific research
Scientific research processes
Results of scientific research
Application of the results of scientific research within the national economy

- Mastery of latest technological development
- Serial production
- Preparation for serial production
- Effecting balanced ratios
- Problems of lag

- Economic problem in implementing the adoption of scientific research findings
- Measurement of results

- Economic analysis and basis of scientific research methods
- Stimulation of scientific work
- Economic problem in the activity of the EEP
- Rational use of the technological means of scientific work
- Developing a system of indices for monitoring current progress in scientific research
- Economic basis in selecting thematics and planning

- Capital construction
- Economic problem in maintaining the supply of information
- Economics of training personnel for science
- Economic problem in material and technical supply
- Economic problem of financing

- Scientific basis
- Political basis

Fig. 3. (cont'd)

Fig. 4. Structure of the complex relationships of the science of science

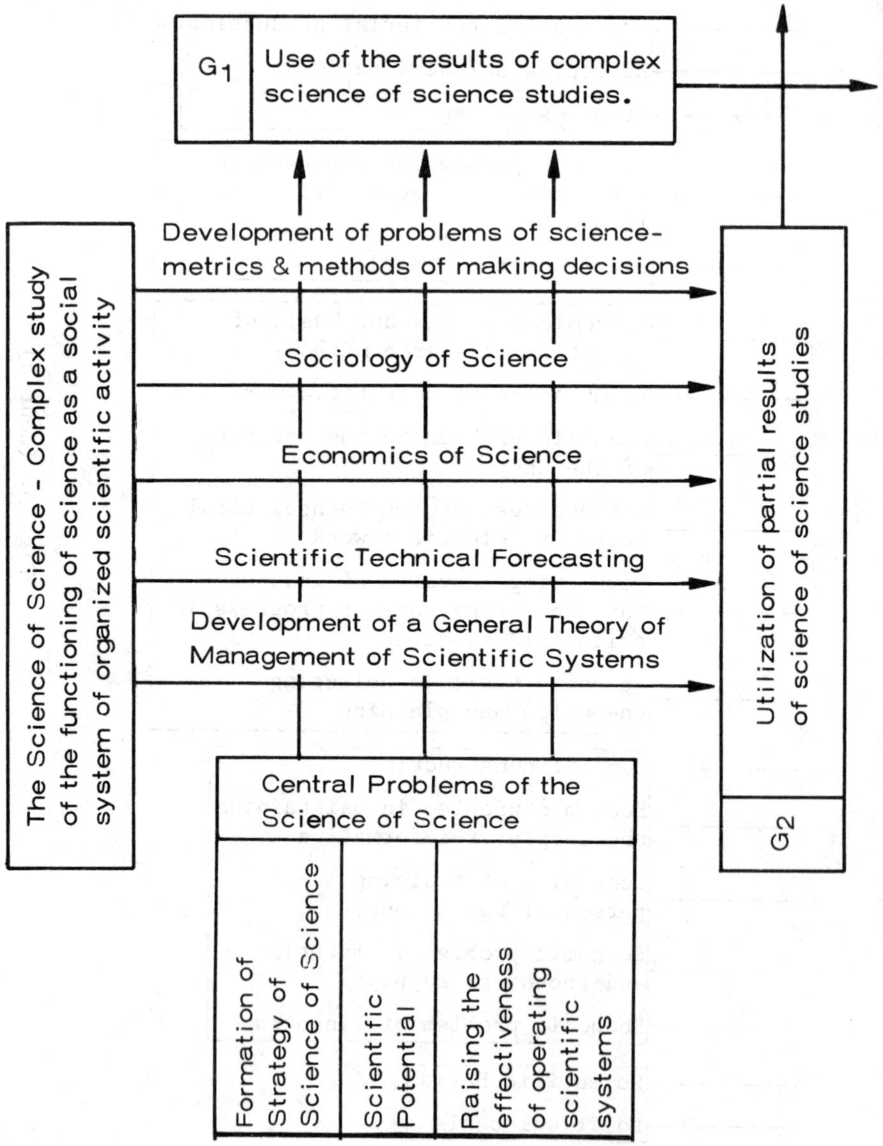

verified the importance and usefulness of the principles of the Science of Science approach which were discussed earlier in this paper.

THE FUTURE OF THE SCIENCE OF SCIENCE

Actual theoretical and experimental studies in the Science of Science during the next few years should culminate in the development of a storehouse of specific Science of Science methods for managing research and development. The most important condition for the effective realization of both existing and future approaches is the formation of an overall theoretical body of principles and information in the Science of Science, that is to say, principles in the theory of the optimum management of scientific systems. The most important aspect of this theory will deal with a transition from a complex of science models to a single systems model of science. In the coming years the Science of Science will change the research and development of complex projects offering new approaches to the most important aspects of science management (strategy, utilization, and effectiveness). Businesslike relations will arise between the Science of Science and the various administrative services, similar to those now existing between science and industry. In this, the specific task of the Science of Science will be to directly participate in the recruitment and training of personnel as science managers. This training will be achieved by presenting special courses and work leading to degrees in the Science of Science.

In the future we will see the establishment of closer and more functional relationships between both the Science of Science and the various science management services and between information services and the technical-economic analysis services of the scientific institutions. At higher levels, these relationships will lead to the establishment of "brain trusts"; that is to say, specialized institutions which will originate and develop principles for a single, unified science policy, conduct scientific-technical forecasting, create and develop automatic information service systems for the management of science. It is quite possible that the work of such institutions will be accompanied by a supplemental network of self-supporting consulting laboratories whose objective will be the solving of specific applied problems in approving and perfecting the organization and management of various scientific research activity. There is no doubt that the pressing real need for the management of science by the Science of Science will soon be officially recognized and result in appropriate action at a variety of levels including recognition as a field of specialization requiring a degree and the creation of scientific research institutes on the Science of Science accompanied by the publication of journals and periodicals.

TECHNOLOGY ASSESSMENT IN SWEDEN

MATS HÖJEBERG and DANIEL SUNDSTRÖM

Royal Swedish Academy of Engineering Sciences,
Stockholm, Sweden

INTRODUCTION

The concept of technology assessment has very recently been adopted in Sweden and basically no work has been done that has been presented under the heading "technology assessment studies". This does not mean that technology assessment in a sense similar to that used in recent publications does not exist in Sweden. In fact, quite a number of activities of this kind are going on or have been performed earlier. Some of them will be reviewed in the following pages.

Technology assessment in a narrow sense is concerned with resource allocation for research and development and the scope of the problem is connected with the subject of interest. However, a more general approach can be taken. Then the economic, environmental and social consequences of new technologies are assessed.

We will concentrate on some recent projects, studies and discussions on the impact of modern technology on Swedish society and how these problems are treated by governmental agencies and industry.

Investigations and studies on a well-qualified level are constantly performed in different sectors of the society. These studies are often initiated by the government and the results are then published as Government Official Reports, which usually include the most comprehensive information available on the treated subject. This kind of investigation is something rather unique for Sweden.

In the Government Official Reports questions concerning the evaluation and assessment of technology have been treated long since. Many of the examples given below have been studied exclusively, or among others by expert groups called in by the government.

PHYSICAL PLANNING

There is an increasing demand on natural resources in modern society. At the same time, there is a growing insight about the limits of the available resources. Even though Sweden is sparsely populated on the average, in the future there will be areas where the resources will be very limited compared to the accumulated need or demand. Most of the west coast is one such area, the Stockholm region another.

The basic requirement of physical planning is that it be concerned with resources that are non-renewable and difficult or impossible to restore. Decisions on how to use these resources must rely on estimates of the long range demand and availability. These estimates must then be combined with economic, social and other perspectives of the future development of technology and society.

A group within the Ministry of Physical Planning and Local Government has made an inventory of the demand for and the availability of physical resources. The time frame is from now to the year 2000. The work is intended to be a permanent part of national planning, and to some extent, both the time frame and the specific problems studied can be revised.

The estimates of the demand for physical resources like space, cooling water and access to transportation, as well as the risks of different kinds of pollution have been performed for industrial branches which cannot satisfy their demands in urban areas. Central and local government agencies have in parallel, made an inventory of the availability of natural resources. This inventory, which is of course by no means complete, has given quite comprehensive material, which has become an interesting source of information.

The different branches have in principle studied the maximum demand alternatives for their future development. The estimates indicate what the total demand on different kinds and quality of physical resources will be. The inventory of resources will clarify the possibilities of meeting this need. The study group will also try to classify the environmental quality of different parts of an area.

The inventory work is necessarily performed with varying levels of precision. It is, however, made on a more detailed level than the suggestions for the use of resources that will be put forward as a result of the study.

One main purpose is to identify the areas where the preferences from different sectors coincide and thus conflicts of interest will emerge. The total demand has to be balanced against the available supply. This will be done in a number of "alternative patterns for the exploitation".

The precision of the presentation will obviously differ from one part of the country to another. However, to take just one example where a definite statement must be made, we could mention the west coast of Sweden, where large areas are virgin land or used for recreation. But this part of the country also provides good conditions for industrial exploitation, especially transportation facilities with good harbour possibilities.

The first details of the considerations of the group will be published in December, 1971. The report will be publicly discussed during the spring 1972. After that a decision is planned to be taken in the autumn on how to organize work of this kind in the future.

PERFORMANCE CRITERIA IN THE BUILDING INDUSTRY

The question of replacing construction or specification criteria with functional requirements in the Swedish Building Norm has been discussed for some years. The aim is to promote innovation and development in the building industry by stating which conditions have to be fulfilled by the completed product, i.e., the building, without specifying how to produce or build it.

Investigative work carried out by the National Swedish Institute for Building Research at the request of the National Board of Urban Planning soon showed that a basic system of concepts was lacking, both in Sweden and abroad, and that the demands for a new method of drawing up building regulations could therefore hardly be met at present.

A special investigation therefore was made in order to create a coherent and usable system of concepts [2]. The report from this work will be summarized.

The systems of concepts is based on the following three terms of fundamental significance:

contingencies
user or human requirements
systems of verification

The first two concepts together constitute the functional analysis and the report limits its scope to studying "technical" functional analysis; that is, contingencies and requirements which govern the choice of materials and structural design. Bearing in mind the limitations outlined above, the following definition is proposed:

A technical functional analysis is a necessary and sufficient enumeration of requirements and contingencies pertaining to the technical design of a building.

The technical functional analysis is not, however, sufficient for an applicable building standard. The builder must also be able to prove to the scrutinizer that under given contingencies the proposed structural design, "layout", ensures the fulfillment of given requirements. This in turn gives rise to the systems of verification that will make possible the production of the evidence which it is the duty of the inspecting body to demand

A functional analytical building standard then consists of (technical) functional analysis + systems of verification.

Contingencies

Contingencies are the conditions which must in principle be taken into account, regardless of requirements, at the time of designing a building. One should beware of regarding contingencies as the counterparts of requirements, however tempting this may be.

Contingencies may be divided into four classes:

1. The natural contingencies deriving from the climate, soil conditions and so on.

2. The man-made contingencies are as a rule consequences of community decisions, e.g., vibrations due to traffic, space requirements for fire engines etc. They can therefore be altered by means of decision.

3. The use contingencies derive from the user's way of using his dwelling, and are therefore closely connected with the "instructions for use" which have been requested for dwellings in other contexts. Their quantification is often deficient.

4. The design - conditioned contingencies are totally dependent upon the technical design of a building. The noise generated when filling a bath with water is a technical issue, since the noise generated is unintentional. Had the noise been intentional, the contingency would have been classified under the heading "use" (see above).

The classification of different factors or phenomena are sometimes somewhat arbitrary. The primary question is usually whether the phenomenon should be considered as a contingency or not. The next question, usually less subtle, is what kind of contingency it is. Lightening is such a phenomenon. Some others are mentioned below.

One man-made contingency which is included in a foreign building standard though not in the Swedish equivalent, is the occurrence of power cuts. Instead, the Swedish Building Norm takes account of the risk of cuts in the supply of fuel from foreign sources. It is questionable whether burglary, shop-lifting, vandalism and sabotage come within the sphere of responsibility of building standards at all. In that case, they should be classified as man-made contingencies.

To be able to compile a list of use contingencies one must first establish what is considered normal in a dwelling, e.g., sound level, etc.

Requirements

Requirements are the user requirements (=human requirements) with regard to the dwelling. The report emphasizes the significance of keeping a list of user requirements free of references to technology, since the user is interested in the function of a building and not the engineering skill which has been used in order to ensure its function.

The user requirements refer to the completed dwelling. This is considered to be included in the term "function" but there is disagreement both in Sweden and abroad as to the implications of the word.

The Swedish Building Code mentions the "basic" requirements of safety, hygiene and comfort. These basic requirements are interpreted as user requirements (certain degree of security against collapse, fire precautions, thermal comfort etc). These requirements will constitute the requirements in the analytical performance building standard. In some cases, however, there is a practical advantage in creating what will here be called acting or subsidiary requirements. These are not functional requirements in the sense of those described above. They are instead requirements which as a rule are simply concerned with the technical aspects of materials. They justify their existence by ensuring the fulfillment of the direct requirements (under given contingencies). The report contends that acting or subsidiary requirements cannot be standardized on the basis of the building code if their fulfillment is not necessary for the fulfillment of the direct requirement.

Systems of Verification

The systems of verification constitute the third link in a building standard in that they verify that the user requirements are satisfied under given contingencies.

Contingencies and requirements must be quantified if a system of verification is to be applied. This is a condition which is not always possible to fulfill: We can then, on the basis of an earlier survey of criteria for testing methods, resort to the operational definitions. Thus, certain contingencies and requirements are quantified with the aid of the system of verification.

However, the basic idea remains the same: a functional analytical standard = contingencies + user requirements + verification system.

Performance criteria of the kind discussed above might of course be specified in many different ways, depending on which economical sacrifices one is ready to accept. The economic effects of different levels of performance criteria and the consequences for the final product have been studied for, e.g., school buildings.

A system, which includes a computer program with which the desired classroom climate can be studied as a function of many different parameters has been developed. The results give a clear indication of the economic effects of single parameters, but also show that a careful treatment of all factors makes it possible to satisfy high quality requirements within rather limited economic resources.

ENVIRONMENTAL PROBLEMS

The rapid increase of the interest in technology assessment in Sweden, as perhaps in most other countries, (especially the United States,) is combined with general concern for environmental problems. Many of these questions are the same in many countries, such as the increased proportion of carbon dioxide, particulates and sulphur oxides in the atmosphere, DDT and oil spills in the oceans and elsewhere, detergent pollution of fresh water, thermal pollution and so on.

The Baltic is a shallow sea without tides, with brackish water, and with large archipelagos in Sweden and Finland which are almost unique in the world and extensively used for recreation. Concern about the risk of large oil spills in the Baltic (and by the way also in the larger Swedish lakes) has therefore been growing in recent years along with the growing number of tankers sailing on it.

The largest tankers that can enter the Baltic are some 130,000 dwt. The question that has been discussed is whether the oil transportation in the Baltic should be made with few small supertankers of this kind or with many coastal tankers. The idea of prohibiting all tankers in the Baltic seems to be unrealistic and is only occasionally discussed.

It is argued, among others from the National Administration of Shipping and Navigation that the hazards are minimized with few big tankers. This implies fewer freights. The big ships can be supplied with more advanced navigation equipment. They can more easily be followed and led, eventually in a way similar to air traffic. Big ships do not have to let out more oil than smaller ships, if suitably constructed, in case of an accident.

On the other hand, it is argued that the hazards of oil outlets increase with the ship size. It becomes impossible to protect the waters and coasts against oil damage in case of a major accident. The size of the tankers also makes them more vulnerable and difficult to maneuver.

It is difficult to give quantitative estimates of the risk factors. It is also to be decided if the risk of a total disaster with a large tanker can be accepted at all, no matter how low the probability is.

An analysis of casualities to tankers in the Baltic during the 60's has been performed in Finland [3]. The selection of random variables and the sample size are too small and the information on human factors so diminutive that it is only possible to get some indications about tanker accidents. However, some immediate conclusions can be drawn:

1. The dominating casualty group in the Baltic area is grounding. The distribution is 56% groundings, 35% collisions and 9% fires and explosions. In harbour approaches the groundings are twice the number of collisions.

2. The highest casualty rates are found in the Danish Straits 41%, and approaches to harbour 42%.

3. The total number of casualties is slightly increasing in time relative to the total number of ships sailing in the area.

4. The probability of casualty is twice as great in loaded condition as in ballast.

5. More than half of the casualties take place in good visibility conditions.

The study does not give a comprehensive picture of the tanker casualties that could be used for planning effective countermeasures, but some reasons for what could be considered a minimum degree of countermeasures is suggested.

Passages to oil harbours should be improved and new harbours should be located at open coast. (Now, most of the Swedish part of the Baltic has a more or less developed archipelago outside the mainland coast line.) It is also evident that the number of crew members and pilots on duty has to be increased, especially in the autumn season when there is a seasonal casualty peak.

The Swedish Coast Guard Service holds the national responsibility for dealing with oil spills at sea. In 1970 it was commissioned to undertake a three year program for improving appropriate countermeasures even against major oil spills that may occur in Swedish waters or on the high sea surrounding the country.

This has led to almost a doubling of the budget for the next year. The Coast Guard Service registered 359 oil spills in 1970, compared to 321 in 1969 [4].

The first half of 1971 indicates a further 25% increase in the number of oil spills.

The first supertanker entered the Baltic in 1970 (Tiiskeri of 112,000 dwt). The maximum depth in the Danish Straits is at present 13 m, which forced Tiiskeri to unload some 20,000 tons of oil before entering. There are, however, possibilities of sailing larger ships within the Baltic and thus there is a strong interest in deepening the Straits.

This question will certainly be considered in the near future. The opponents claim that, apart from the casualty risk for large oil spills, deepening of the Danish Straits will alter the whole ecological system of the Baltic, due to inflow of salt water from the North Sea to deep water streams.

Thus, there are many different problems that must be considered when the ways and means to supply the countries around the Baltic with oil are studied. The correct and complete assessment of alternative technologies to carry this out cannot be done without a strict methodology.

URBAN SITING OF NUCLEAR POWER PLANTS

In 1968 the Stockholm Electricity Board applied for a license to build a nuclear plant only a couple of kilometers from the Stockholm city center. An expansion of the district heating system was considered necessary to improve the air pollution situation, and the nuclear plant was proposed as an attractive alternative to oil-fueled units.

This application evidently implied a drastic change from earlier licensed low-population sites. After some time it started off an official study of the economic and hygienic consequences of urban siting which proceeds along lines typical to technology assessments. The final report from this study has not yet emerged, but some preliminary information on the results are known [5].

The problem of estimating health effects from low level exposures, both from radiation and fossil fuel combustion gases, is difficult. A linear hypothesis, that the sum of all effects expressed as lost man-years is proportional to the sum of all individual doses regardless of size and dose rate, certainly overestimates the hazard and is likely to be disadvantageous to nuclear power compared to fossil fueled power.

Even with this hypothesis the comparison of power plants with regard to normal operation favours the nuclear plant. Its superiority is likely to correspond to a value closer to 20 than to 2 million dollars per year if one were allowed to put a price on health effects. However, the larger hazard potential of nuclear plant accidents is quite evident. The risk of accidents must be evaluated and balanced against the benefits from normal operation. Information on the total injury expectation from all types of accidents must be collected. One must also try to determine if there is an upper limit of accident consequences which society will accept regardless how low the probability is. Then, it must be decided if the reactors now being built (far from urban areas) create less hazards through accidents than they eliminate in normal operation, and if their maximum conceivable risk is sufficiently small.

There are some indications that the location within urban areas, as was the case in the initiating application, will not be allowed, but that siting rather close to the urban area might be accepted. This means an important liberalisation of current laws, and will eventually make it possible to use nuclear power plants for combined production of electricity and heat in the major urban areas in Sweden.

TRAFFIC ACCIDENTS

About 1,300 persons per year are killed in traffic accidents in Sweden. The direct and indirect cost for the society has been estimated to hundreds of million dollars, maybe even more than a billion dollars. In order to reduce these figures, various measures have been taken by the National Road Safety Office. Several test periods with speed limits, sharpened regulation on car conditions, more and better supervision, and so on have been tried, but evaluations of these measures are not conclusive.

The National Road Safety Office which is responsible for the work to increase road safety, will present a number of alternative action programs aimed at reducing traffic accidents. The details of the programs will be published late in 1971.

Results from test periods with speed limits and other measures and evaluation of statistics and research on traffic accidents and traffic performance will be used to estimate the costs and actions needed for different levels of reduction. The goals of the programs range from preventing a further increase in the number of accidents to accomplshing a 60% overall reduction.

The analysis of different factors that cause traffic accidents and measures that can be taken to prevent them will be the basic information used in the program formulation. Details of the methods used are not available, but this project seems to be one rather ambitious attempt to assess technological options in quantitative terms.

SPACE RESEARCH

The space program in the United States has of course had a tremendous impact on technology and industry. Without a doubt, many products that we now use have a quality and a performance capacity that can be attributed to the development work within the space program.

Obviously, some of the effects of development stimulation could be reached by direct support to the specific areas of technology rather than through the space program. One problem, however, that must be faced is the formulation of a well defined purpose for this work.

A satellite project has a clearly defined, timed and for all participants easily conceivable goal as incentive for the work. The participants are exposed to quite a number of difficult and tangible problems, which need an increased knowledge on all levels in order to be solved. At the same time, products that will correspond to social needs might be developed.

The space technology must always use the most advanced techniques, and always increase the performance capacity in the systems. This creates an innovative atmosphere. In Sweden it is sometimes argued that it is necessary to give substantial support to a national space program to keep the rate of development in society high. The opposite argument has also been put forward, namely that funding specifically directed to space research has a comparatively low impact on general technological and social development and therefore should be downgraded.

The Swedish Board for Technical Development published an investigation on space research activities in Sweden in 1970 [6]. The arguments put forward are based on qualitative estimates of the impact of such a program on different society sectors. A number of proposals on different cost levels are given with indications of what could be done in terms of research, development, industrial activities and so on.

The assessment of the alternatives in quantitative terms and concerning the total effect on society of space research might be an interesting study for technology assessment methodologies.

CONCLUDING REMARKS

The examples of technology assessment work in Sweden given in this paper are somewhat arbitrary. They are not specifically technical. No special models are used. Many references are given to qualitative factors, which may seem very loosely connected to technology. In discussing the applicability of technology assessment methodologies, however, we would like to point out a few factors that seem to emerge from our survey of the Swedish situation.

First, we do not seem to have any specific studies which use the known methods and set of definitions proposed in recent work on technology assessment. The term "technology assessment" did not have a Swedish translation until very recently and the impact of this new and ambitious way of handling problems has not yet manifested itself.

Secondly, there has long since existed an effort in both the official and private studies of this kind to analyze the different aspects of technology in relation to various aspects of social performance and development. This does not mean that the efforts have always been successful.

Thirdly, technology has been, and perhaps increasingly will be, considered as the means for the development of society. This fits into the general philosophy of technology assessment. If one looks at the examples given in this paper, they can be considered illustrations of what kind of problems are and will be treated and what will be the frame in which to formulate both the questions and the results of the analysis. In order to obtain specific results, the problems will certainly very often have to be divided into much smaller complexes, but the aim is then to put the different parts together in some kind of overall synthesis, which is comparable to the problems initially stated. And in these intermediate steps of the analysis, formal methods of the kind described for technology assessment will certainly improve the results.

REFERENCES

1. *Fysisk riksplanering Materialredovisning, June 1969.* Kommunnikationsdepartementet, Stockholm, Stencil K 1969:13, 182 pp.

2. KNOCKE, JENS, *En funktionsanalytisk byggnorm. Förslag till principer.* (A building standard expressing performance requirements — Draft). Statens institut för byggnadsforskning, Stockholm. Rapport R 21: 1970, 60 pp.

3. KOSTILAINEN, V., *Analysis of Casualities to Tankers in the Baltic, Gulf of Finland and Gulf of Bothnia in 1960-1969*, Helsinki University of Technology, Otaniemi, 1971, 7 pp.

4. *Statistik över oljeutsläpp till sjöss och i kustvatten 1970*, Kungl. Generaltullstyrelsen, Stockholm, 1971, 3 pp.

5. BERGSTRÖM, STIG, O.W., "Environmental Consequences the Normal Operation of an Urban Nuclear Power Plant," Preprint, (AE-DS-324), 1970, 17 pp.

6. *Framtida svensk rymdverksamhet. Förslag framlagda av STUs rymdteknikutredning*, Styrelsen för teknisk utvekcling, Stockholm, 1970, 71 pp.

TECHNOLOGY ASSESSMENT IN ISRAEL

SAMUEL N. BAR-ZAKAY

Manager, Center for Technological and Interdisciplinary Forecasting, Tel-Aviv University, Ramat-Aviv, Israel

INTRODUCTION

In the course of my studies on technology transfer, I distinguished between "incidental" and "organized" technology transfer [1]. It seems that the same distinction should be made with respect to technology assessment, for the following reason.

It is certain that some form of technology assessment is taking place in all the 102 scientific research organizations in Israel [2] listed in Table 1.

For example, while analyzing whether a certain agricultural research project should be funded, the directors of the Volcani Institute may look into the impact of the technology on the farmers' way of life.

Thus I define:

Organized technology assessment:
Impact analysis is a major objective.

Incidental technology assessment:
Impact analysis is a minor objective.

The term impact analysis was mentioned in the Study of Technology Assessment by the U.S. National Academy of

Engineering [3]. Congressman Emilio Q. Daddario outlined the concept of technology assessment, and of impact analysis [3]:

> "... appraising the technical, political, economic, and social influences arising from science and technology ..."

TABLE 1

SCIENTIFIC RESEARCH IN ISRAEL [2]

1971

Institutions of Higher Learning	7
Agricultural Research	10
Industrial Research	19
Regional and Environmental Studies	16
Biomedical Research	27
Nuclear Research	3
Services to Science and Technology	6
Social and Economic Studies	14
Total	102

It is interesting to note that in the recent inquiry into the organization and administration of governmental research in Israel, none of the recommendations for future work referred to organized technology assessment [4].

In this paper, organized technology assessment will be discussed. Since such activity is currently scarce in Israel, the objective here is to examine the relevance of basic assessment concepts to Israel in order to provide guidelines for future actions. Potential topics for assessment will be mentioned.

I shall start by mentioning three major problem areas and continue with more optimistic notes.

ASSESSMENT AND POLITICIANS

Organized technology assessment has two major features:

A. It is meant to be action oriented*.

B. It is future oriented.

The first item requires *a priori* that decision-makers, i.e. politicians, act upon the findings of assessment teams. Indeed a hypothetical Office of Technology Assessment was discussed by the U.S. Congress [5]. This is a feasible goal.

The U.S. National Academy of Sciences suggested that "while it must itself seek to be apolitical, any new assessment mechanism must be located close to the centers of power in the political process ... any organization less centrally situated would have no realistic hope of materially influency public policy" [6].

The second item, however, (the future orientation) seems to contradict the nature of politicians. The latter are known to be interested mostly in programs and decisions which have short-term effects (a world-wide phenomenon).

---

*The assessment will include a plan of action.

An Israeli example is given by the long delay in introducing general television into the country. Although the benefits were assessed and presented by various organizations, the government did not make any decision until 1967. (In spite of the fact that all the neighboring countries with lower income per capita introduced general television much earlier.)

This delay took place probably because the politicians did not see what short-term benefits they could derive from introducing television. In 1967, however, after the Six Day War, more than a million Arabs were suddenly within the boundaries of Israel and the government could see a clear benefit in communicating with these people. Consequently, general television started operating in Israel [7]*.

TRANSFER OF ASSESSMENTS

It would be useful if the results of assessment in one country could be transferred to another. This would:

1. decrease assessment expenditures in the recipient country,
2. enable recipient countries to use analysis capabilities of more developed countries.

Unfortunately this transfer would be difficult to achieve. For example, in a recent review of "The Environment in Israel" [8] the problem of noise pollution was mentioned. It is even stated that "noise from aeroplanes constitutes an increasingly serious problem." The U.S. National Academy of Engineering used

---

*A contributing reason was the resignation of Prime Minister David Ben-Gurion who was a consistent opponent to introducing general television in Israel.

as an experimental technology assessment the problem of subsonic aircraft noise [3].

A careful look into the strategies discussed reveals their inapplicability for Israel:

1. Israel purchases all of its large aircrafts (noise makers) from foreign manufacturers. It can not manufacture large (noise free) aircrafts.
2. Relocation of the only international airport is not feasible because of past huge investments.
3. Creation of a buffer zone would mean the relocation of a few hundred thousand citizens.
4. Soundproof residences will not be helpful since the Israeli spends a relatively large part of his time outdoors.

It is possible that this example is not typical and other assessments could be transferred from one country to another. This question calls for research.

ASSESSMENT PRIORITIES

Organized technology assessment has not yet been institutionalized anywhere in the world. Nevertheless, it can be found in many institutions.

At the Rand Corporation in the U.S., for example, research has ranged from study of the effects of nuclear war on the weather and climate [9] to communications satellites and their possible influence on the automobile [10]. The president of the U.S. National Academy of Sciences has written on biology and the future of man [11].

Once organized technology assessment is institutionalized it will be necessary to determine priorities for assessment. This will not be an easy task in any country. For example, several institutions in Israel are dealing with the social quality of the urban environment, but in the light of the country's critical shortage and ever increasing demand for water, maybe research on water quality management and pollution control should have a higher priority?

The question of priorities is just as important in a rich country like the U.S. For example, if the U.S. would have launched a national effort to exploit the sun's energy, *instead* of the program "to put a man on the moon," most of the pollution problems in the U.S. may have been solved by now.

The "solution" sometimes given to such priority problems is to introduce the term "metapolicymaking" [12] or in our case: meta-assessment."

I do not believe that anything is gained by coining such phrases. Neither do I find a solution to this problem in the literature published up to date on technology assessment [13].

The U.S. NAS added a word of caution [6]: "The panel wishes to stress the need for a cautious approach to the development of new assessment institutions ... we emphasize the importance of assuring that priorities in the assessment process be set along reasonably objective lines." This is obviously easier said than done.

ASSESSMENT POTENTIAL

In principle, technology assessment may result in great benefits. One example will be given here:

Israel's most important objective is to achieve a lasting peace with its neighboring countries. In 1946, Chaim Weizmann, who later became the first president of Israel, said:

"I feel sure that science will bring to this land peace" [14].

Unfortunately, this assessment has not yet materialized.

Recently one of the more imaginative schemes for solving the Arab refugee problem was introduced [15]. It suggested that water be brought from the Bay of Eilat-Aquaba (Red Sea) to the Dead Sea via a canal along the Arava. The flow of water from the sea level of the Gulf of Aquaba to the Dead Sea, the world's lowest lying lake, will create a natural waterfall. A hydroelectric power station would produce great quantities of very cheap electric power meeting the needs of an enormous chemical, petrochemical, and light metal industries. Water desalination plants will produce cheap desalted water for the inhabitants of an area which is currently not populated.

It is quite possible that a thorough technology assessment in this case will convince politicians both in Israel and Jordan to launch the project in spite of the tremendous political and economic difficulties.

ASSESSMENT NEEDS

In Government Decision 828 of September 12, 1971, the government has decided to appoint a committee of the Director Generals of all the government ministries to advance and coordinate government actions in matters regarding environment control.

In addition to the well publicized environmental problems which have been reviewed recently [8] and are not unique to Israel, several other problems call for in-depth assessment.

Israel is a country of mass immigration. The state is making a major effort to mold individuals who speak 70 different languages into one people. Can technology assist in achieving this goal?

A recent Harvard study [16] (which included Israel) found that "education is the most powerful factor in making men modern, but occupational experience in large scale organizations, and especially in factory work, makes a significant contribution in 'schooling' men in modern attitudes and in teaching them to act like modern men." The study did not, however, discuss whether technology per se can assist in the molding of a nation. Once such analysis is performed it may build a strong case for technology, both in Israel and in other developing countries.

A second related topic for assessment is the degree to which military technology contributes to economic growth. It is no secret that a good portion of the Israeli GNP is devoted to defense. The electronics industry in Israel has boomed in the last three years because of defense orders for military equipment. Can this non-defense payoff be measured? This question is extremely important since there is evidence that Israel is more vulnerable economically than militarily.

Among the less ambitious topics for assessment, I can cite the impact of computers on the Israeli society. It is known that the weighted average annual increase in the number of computers in Israel since 1962 was about 45% [17]. Sooner or later, problems of dehumanization, computer privacy and legal responsibility will reach a crisis situation. It would be advisable to assess this technology as soon as possible.

INSTITUTIONALIZED ASSESSMENT

In January, 1971, a new Center was established at Tel-Aviv University: Center for Technological and Interdisciplinary Forecasting [2]. The funding is currently provided by four government ministries: Defense, Commerce and Industry, Development, and Prime Minister-National Council for Research and Development.

This is the first research body in Israel to have such a broad spectrum of users.

The Center will perform, among other studies, impact analysis and technology assessment.

In its first study the Center integrated technological considerations with economic and social considerations in the field of computers [18].

Previous planning work in Israel did not perform such integration. For example, the Economic Planning Authority of the Prime Minister's Office published a detailed work entitled: "Israel Economic Development: Past Progress and Plan for the Future" [19]. Although the economics of manufacturing and mining, transportation and communication as well as construction and housing is discussed, technological development per se is not. This may not be surprising considering the fact that the Authority's team does not include any technologists.

Even the Ministry of Commerce and Industry which published a plan for the development of industry in Israel (1969-1975) [20] did not perform any impact analysis. This gap was probably not unique for Israel and is typical for many countries around the world in 1971.

The institutionalization of technology assessment should be weighed against funding assessment studies at existing institutions. A combination of the two approaches may prove to be the best way: a permanent small staff consulting the legislative body, with authority and budget to fund research elsewhere as needed. (This avenue is under consideration in the U.S.) [5]

ASSESSMENT METHODOLOGY

The above title raises a big question: can one really talk of assessment methodology? Forecasting and planning, which have been around longer than technology assessment, use certain techniques. It is not clear, at least in my mind, whether these techniques represent a separate (or new) methodology. At least five more years will have to elapse before this question can be answered. It can be stated categorically that in Israel no assessment methodology has yet been developed.

Cetron has already pointed out that "technology assessment is not official jargon. The expression 'assessment of technology' is not found listed in the table of contents or indexes on management. Nor is it identified and found in the general literature of management or in official planning, programming and policy documents of the government agencies." [21] Any effort, therefore, to bring the importance of technology assessment to the attention of the politicians should therefore be commended. Furthermore, pioneering studies such as the recent one on the "quality of life" (a Delphi approach) [22] may provide the foundations for a new methodology of technology assessment.

In order to make assessment studies useful it will be necessary to add to the analysis real political problems rather than avoid them. The United Nations publications are not really useful because of their artificial "sterility." For example, a recent U.N. paper on computers in developing countries lists ten recommendations which are as vague and detached from reality as to make them useless [23].

THE FUTURE

The National Goals Research Staff referred to technology assessment as "a movement."

Currently, this "movement" seems to be mostly staffed with pessimists. The Times, for example, wrote an editorial on: Technological gains: blessing or curse? [25] Lockheed and Rolls Royce problems have certainly shaken up many optimists [26]. The fear is that technology assessment will curtail the development technology. This fear may be unjustified.

Moreover, a point to remember is that not all that is happening in the United States will necessarily happen in other countries. It may, and yet, it may not.

In a recent article it was pointed out how 50 years of copying Britain ruined Canada's technology [27]. Copying pessimistic moods may be just as dangerous.

Countries differ in their political, social and economic structures. Technology assessment may develop to have different connotations in different countries. Negative actions, in the opinion of the technologist, may result from assessments in the United States, while positive actions may be the outcome in Israel.

The first prime minister of Israel, David Ben-Gurion, who is now retired, is currently writing his memoirs. In a paper published in 1970 he analyzed the failure of Baron Hirsch to settle Jews in Argentina versus the success of Baron Rothschild to do the same in Israel [28]. The two philanthropic barons had similar plans: purchase land and introduce Jews to agricultural technology.

Baron Hirsch could purchase at one time large areas of land in rich Argentina, while Baron Rothschild could buy only small lots over a long period of time. The land in Argentina was fertile while the land in Israel was barren. In short, in spite of the fact that all the objective conditions were favorable for the initiative in Argentina, it failed; while Baron Rothschild's initiative in the holy land, was successful, and laid the foundations for the establishment of the State of Israel. According to Ben-Gurion, Rothschild's initiative was successful (in spite of interim hardships) because of the historical and emotional significance the land of Israel held for the Jewish people. The Argentinian project lacked these two vital elements.

I note this piece of history to reinforce my opinion that technological development is governed by historic, ethnic and religious forces which are different in Spain, China, United States and Israel. The same holds true for technology assessment.

REFERENCES

1. BAR-ZAKAY, S.N., "Policymaking and Technology Transfer: The Need for National Thinking Laboratories," The Rand Corporation, P-4497, December 1970. *Policy Sciences, 2,* No. 3, September 1971.

   BAR-ZAKAY, S.N., "Technology Transfer Model," *Technological Forecasting and Social Change*, *2,* Nos. 3/4, 1971.

2. *Scientific Research in Israel - 1971*, Prime Minister's Office, National Council for Research and Development, 1971.

3. "A Study of Technology Assessment," Report of the National Academy of Engineering, Committee on Science and Astronautics, U.S. House of Representatives, July 1969.

4. "Report of the Committee to Inquire into the Organization and Administration of Governmental Research," December 1968. National Council for

Research and Development, Israel.

5. "Technology Assessment," Hearing before the Subcommittee on Science Research and Development of the Committee on Science and Astronautics, U.S. House of Representatives, Ninety-First Congress, First Session, November-December 1969, U.S. Government Printing Office, Washington, 1970. Av. Week and Sp. Tech., July 26, 1971.

6. "Technology: Processes of Assessment and Choice," Report of the National Academy of Sciences, Committee on Science and Astronautics, U.S. House of Representatives, July 1969.

7. *Israel Government Year Book*, 1967/68, Government Printer for the Central Office of Information.

8. *The Environment in Israel*, Edited by U. Marinov and J. Gale, National Council for Research and Development, Israel Academy of Sciences and Humanities, Jerusalem, March 1971.

9. BATTEN, E.S., "The Effects of Nuclear War on the Weather and Climate," Rand Memorandum, RM-4989-TAB, August 1966.

10. HALT, JOHN L., "Communication Satellites and Their Possible Influence on the Automobile," Rand Corporation, P-4164, August 1969.

11. HANDLER, PHILIP, *Biology and the Future of Man*, Oxford University Press, New York, 1970.

12. DROR, YEHEZKEL, *Public Policymaking Reexamined*, Chandler Publishing Co., San Francisco, 1968.

13. CHAMBERLAIN, ETHEL L., "Technology Assessment: A Bibliography," Rand Corporation, P-4541, January 1971.

14. CALDER, RITCHIE, *The Hand of Life, The Story of the Weizmann Institute*, Weidenfeld and Nicolson, 1959.

15. GUR, SHLOMO, *The Jordan Lift Valley - A Challenge for Development*, University Press, Tel-Aviv, 1970.

16. INKELES, ALEX, "Making Men Modern: On the Causes and Consequences of Industrial Change in Six Developing Countries," *The American Journal of Sociology*, Vol. 75, No. 2, September 1969.

17. GERTZ, AHRON, "Computer Capacity Utilization and Forecasts in Israel," *The Jerusalem Conference on Information Technology, August 1971*, Published by ILTAM Corp., for Planning and Research, Jerusalem, 1971.

18. "Computers: 1971 - 1981," Center for Technological and Interdisciplinary Forecasting, R-1, Tel-Aviv, August 1971.

19. "Israel Economic Development: Past Progress and Plan for the Future," State of Israel, Prime Minister's Office, Economic Planning Authority, Jerusalem, March 1968.

20. "Israel Industrial Development Plan 1969 - 1975," State of Israel, Ministry of Commerce and Industry, Center for Industrial Planning (Hebrew), Jerusalem, February 1971.

21. CETRON, MARVIN J., *Technological FOrecasting: A Practical Approach,* Technology Forecasting Institute, A Subsidiary of Gordon and Breach, Science Publishers, 1969.

22. DALKEY, NORMAN C., LEWIS, RALPH and SNYDER, DAVID, "Measurement and Analysis of the Quality of Life: With Exploratory Illustrations of Applications to Career and Transportation Choices," Rand Corp. Memorandum RM-6228-DOT, August 1970.

23. BARG, BENJAMIN (Office for Science and Technology, United Nations, New York), "Computer Technology and the Developing Countries: An Expanding Vista for Useful International Cooperation," *The Jerusalem Conference on Information Technology, August 1971*, Published by ILTAM Corp., for Planning and Research, Jerusalem, 1971.

24. "Toward Balanced Growth: Quantity with Quality," Report of the National Goals Research Staff, The White House, Washington D.C., July 4, 1970.

25. MORRIS, JOE ALEX, Jr., "Technological Gains: Blessing or a Curse?" *The Times*, Section B - July 5, 1970.

26. MEYERS, HAROLD B., "The Salvage of the Lockheed 1011," *Fortune*, 66, June 1971.

27. "How 50 Years of Copying Britain Ruined Canada's Technology," Technology Review, *New Scientist and Science Journal*, *49*, No. 131, February 4, 1971.

28. BEN-GURION, DAVID, "Why Did Baron Hirsch Fail and Baron Rothschild Succeed?" *Ma'ariv*, 26 June, 1970 (Hebrew).

TECHNOLOGY ASSESSMENT IN JAPAN

KEICHI OSHIMA

Professor
Department of Nuclear Engineering,
University of Tokyo, Bonkyo-Kv,
Tokyo, Japan

INTRODUCTION

During the past century, and especially in the decades after World War II, Japan has concentrated most of her efforts in reaching the economic prosperity of advanced Western countries. This goal has been pursued by industrialization of national economy and modernization of the society. As a nation with small land area and scarce natural resources, the only way to achieve this economic goal was to establish an advance technology-intensive manufacturing industry competitive in the world market.

The rapid growth and industrialization of Japanese economy after World War II are well known to the world. So-called "heavy and chemical industries", such as electricity, iron and steel, petrochemistry, electronics, ship building, automobiles, and machinery have been the main sectors of growth during this period. And, naturally, the development of technology has played the most important role in the growth.

The rapid development of technology and innovation can be understood by the following figures. The investment of research and development has increased by the rate of 6.2 times in the past 10 years, reaching 933 million yen in 1969, and the total of imported technology has amounted to 13,000 items and 774 million yen during the past 20 years.

Now, however, Japan has successfully reached the economic standard of the advances western countries, the per capita income increasing to over 1,000 U.S. dollars, and she has found herself confronting serious social problems caused by such rapid growth of economy and advancement of technoloty.

The first was pollution. Air pollution by SO_2 from oil burning power plants, CO from automobiles and water pollution by industrial waste have given rise to anti-pollution campaigns by people in many parts of the country. Though there had been in existence several regulating laws for individual pollution activities by central and local governments, the fundamental anti-pollution act, covering pollution in general, was enacted in 1968.

The act defines the basic philosophy of responsibilities by the governments and enterprises, prescribes the establishment of permissible environmental level, restrict waste disposal, etc., and promotes settlement of disputes and relief of sufferers. It also prescribes establishment of anti-pollution boards to handle such basic matters of pollution control.

However, gradually it became clear that the problems are much more profound as indicated in broader phenomena, such as increase of traffic accidents, destruction of nature, social isolation of individuals, and others. The impacts due to rapid change by new technology and industrialization on the natural environment and social life had not been properly taken into account in the past.

In nearly ten years, from 1960 to 1969, industrial production increased 3.2 times, plastics production 7.6 times and number of automobiles 4.9 times, while SO_2 pollution increased 4.7 times, plastics waste 8.8 times and traffic accidents 3.3 times.

The people have started to realize that they have to pay much more attention to the quality of life than the quantitative growth of the economy. Now, with the increase in concern over the destructive effects of present technology, it has been understood that pollution is only a part of the broader problem, and it cannot be solved basically by mere local technical treatments, but has to be approached from a much broader standpoint of assessing the technological system as a whole.

The government has been required to change the economic and technological policies in accordance with such trends in public opinion. This has been reflected in a basic change appearing in the new technology policies announced by the policy boards of the government.

TECHNOLOGY ASSESSMENT AS GOVERNMENT POLICY

The first description of the emphasis on requirements for change in technology policy and the importance of technology assessment, though the term was not used, appeared in the report on technology policy submitted by the Sub-committee of Technology Advancement of the Economic Council in 1969. The committee was established to study the guidelines of technology policy to be adopted by the government in order to pursue the social and economic development program recommended by the Economic Council to the Government. The importance of the development of technology in accord with social development is emphasized in the recommendation.

In this report, the role of science and technology in social development is pointed out and described: "technological innovation in the past has contributed to society in improving consumer goods of private sectors. However, it has become a cause of increase in social unbalance". And it is noticed that "in the future, since the scale of technology will become larger and the number of big projects for technological development will be increased, the impacts on society will become more complex and broader. Furthermore, some impacts such as food contamination, ground subsidence, radioactive waste and pesticide which do not look serious in themselves but are apt to cause serious damage to our life all together or over a long period of time, appear to be increasing. Therefore, in order to attain a well-harmonized progression of society, it is necessary to build into several levels of society a so-called 'security and welfare control system', which will control such problems within the society. A 'security and welfare control system' is to be built in at the following stages:

"1) At the stage of planning a big project, there should be a system to predict the impacts on society, and preconsider ways of dealing with them.

2) While undertaking the big project, the impacts of the project will change gradually, with changes in internal conditions, such as adoption of a new technology, and also in various external conditions. Therefore, some system which can check such changes and respond quickly to them, should be built into the project.

3) When the project is finished, there are some cases that, because of the complexity in interaction of many elements of society, some unexplained phenomena can occur and some serious impacts may be expected in the long run. Some system should be established which can predict such phenomena in the earliest stage and can take quick action to deal with them."

This system will be exactly what the technology assessment is aiming for.

In 1971, "technology assessment" has been emphasized in much more explicit terms as a basis of technology policy of the government. The First Panel of the Council for Science and Technology, which has the Prime Minister as the chairman and is the highest authority in making science and technology policy in the country, has submitted its recommendation of "The fundamentals of integrated policy of science and technology in the 1970's" to the Council. Here, technology assessment is emphasized as one of the main foundations of technology policy in the 1970's.

The beginning of the recommendation, reads as follows: "In the 1970's, there arise several new problems different in their preliminary stages and because of their importance, studies for development of methodology should be performed without delay."

Furthermore, an important change in the technology policy has come about in the Ministry of International Trade and Industry. The policy of the Ministry for industrial technology in the 1970's was announced in May, 1971, as a part of the interim report of the Industrial Structure Council, which is the most important advisory body to the Minister. Here again, technology assessment is emphasized as an important measure in the promotion of new technology and for decision of national projects of technological development.

In the policy of industrial technology, development of policy for control and management of technology is recommended as one of the main items of future government responsibility. Efforts to prevent beforehand the unbalance between technology and social environment are urged, and execution of technology assessment is recommended as an important means for such management of technology.

It was directed that "the government should place technology assessment in the total system of industrial policy in order to meet with social requirements such as better quality of life and better environment. On the other hand, private enterprises should realize that if they do not conduct technological development in harmony with society, the existence of enterprises themselves might be endangered". As for the actual program, the following actions are recommended:

"a) In general, technology assessment is to identify and assess the changes in technology and the changes of social environment accompanying them. However, at the start, subjects to be assessed will be mainly such technologies as can be well predicted and also have large impacts on social environment.

As for the method of technology assessment, a series of action, such as definition of the coverage of the technology and check points to be assessed, assessment of technology, study of counter measures, and control can be considered. However, the method should not be fixed, and the optimal means should be selected and executed in a flexible manner, depending on the technology and its impact on the society. Furthermore, regarding the development of methodology, scientific means such as systems analysis should be applied extensively to technological forecasts and evaluation, and the accuracy of the methods should be improved with the accumulation of experiences with case studies.

b) Practice of technology assessment. As for the assessment system, prediction and investigation of technology, assessment control are three necessary functions. Among these three, feedback mechanisms should function well. The organization responsible for the assessment function should be highly professional and neutral, so they can assess the impact of technology to society in an objective and integral manner.

As for the prediction and investigation of technology and control, the present legislative system and organization should be utilized to perform such functions satisfactorily.

c) Technology assessment at the present stage. As for research and development projects of the government, assessment should be conducted before they start, and if some problems are identified, the project should be abandoned or altered. After the completion of the project and before application, the technology should be assessed.

As for technologies developed in private industries, if there exists a strong suspicion that the technology may have an undesirable impact on society, the assessment should be conducted without delay.

It can be seen from such documents and trends that, in 1971, the Japanese government decided that the policies of technological development should be based on the assessment of technology as a part of the total system of the whole human activities, not only from the viewpoint of the advancement of technology itself, but much more from the effects it will cause in the interface of technology with its social, human and natural environments.

GOVERNMENT ACTIVITIES IN TECHNOLOGY ASSESSMENT

The importance of technology assessment has now come to be a consensus among the people and the policy-makers in the government, in Japan. However, it is not an easy matter to carry out "technology assessment" in actual implementation of the policy. The best organizations and procedures in the government for technology assessment are not well known or agreed upon. Even the methodology is not well established.

Under such circumstances, the first activities of the government have been directed to study procedures and organizations of technology assessment to find the proper steps and programs to be implemented in the administration. Mostly, these efforts have started by performing some case studies by *ad hoc* committees.

SCIENCE AND TECHNOLOGY AGENCY

At the Planning Bureau of the Science and Technology Agency, a three-year program of study on technology assessment was started. The points being investigated are as follows:

1) How can technology assessment be practiced in government policies on science and technology?

2) What are the proper subjects for technology assessment by the government?

3) What organization should perform technology assessments which are the responsibility of the government?

4) What kind of method is most suitable?

5) How can policies of science and technology be made consistent with other government policies regarding technology assessment.

To study these points, case studies with several themes will be conducted, because, even though methods of technology assessments are not well established, the study should be performed as early as possible.

The themes selected to be studied in 1971 were announced as computer assisted instruction, new pesticides and high buildings (skyscrapers). These are selected as typical cases with different characteristics, considering 1) field of technology, 2) character of technology (e.g. scale of system), 3) the range of impacts, 4) difference in factors to be influenced, 5) popularity of the technology.

Studies will be performed with participation of specialists from broad fields of discipline, and through consulting the examples performed in the United States, original methods and concepts will be explored. As for the process of actual study, 1) identification of problems, 2) assessment, and 3) search of counter measures from technological standpoints, will be performed. A brief description of the cases is as follows:

Computer Assisted Instruction (*CAI*)

CAI is expected to change, to a large extent, the content of education, the meaning of classes and the role of teachers. The background of CAI is primarily the change of "education from quantity to quality" and secondly, "increase in demands of education with growth of economy".

However, it is predicted that there will be a possibility that strong impact may be given to individuals or society in the phases of distortion of education, mental strains, ways of thinking and social attitudes. Therefore, the case study will be approached from broad fields as sociology, psychology, physiology, and medicine.

The number and the fields of specialists in the assessment committee are expected to be as follows:

2 CAI specialists, 1 computer specialist, 1 information specialist, 1 CAI education specialist, 2 sociologists, 2 psychologists, 2 ethicists, 2 medical men, 2 physiologists, 2 pedagogists, 2 philologists, and 2 ecologists.

The subject will be assessed in such phases as, curriculum, mental effects, contemplative power, sociality, and educational practices.

New Pesticides

With the increase in the use of pesticides in agriculture, there are reported many hazards to both men and livestock, including the residual components in agricultural products which are thought to have hazardous effects on human health. For example, poison accidents from the use of organic phosphorous compounds have happened frequently. And now the careless uses of pesticides are strongly criticized. The uses of pesticides have to be decided by their effectiveness in improving agricultural and marine production on one hand and the protection of the health of the inhabitants and creatures on the other.

Therefore, the direction of the development of new pesticides should be assessed from a broader standpoint. It will not be possible to achieve the goal merely by trying to decrease toxicity of the pesticides. Its technology must be assessed with an integral view, including improvement in use, administrative control, etc.

The pesticides will be divided into three categories by their use: insecticides, bactericides and weed killers. They will be assessed from the standpoints of ecology, medicine, animal pharmacology, bacteria pharmacology, etc.

The number of members and the field of specialists of the assessment committee are as follows:

2 synthesizers of pesticides, 1 ecologist, 2 medical men, 2 animal pharmacologists, 2 insect pharmacologists,

2 plant pharmacologists, 2 bacteria pharmacologists, 1 agricultural pharmacologist, and 2 food hygienists.

High Buildings

In recent years, with new legislation, skyscrapers have also started to be built in Japan. In 1968, Kasumigaseki Building with 42 stories and in 1970, the World Trade Center Building with 42 stories were completed and several others are now following them.

This is a natural trend resulting from technological innovation in anti-earthquake construction, and social requirements coming from concentration of population and economic activities in the cities, rises in land price and efficient use of city functions.

However, there have risen many new problems, such as supply of commodities, supply of utilities, flow of people, social environments, and residence circumstances. For optimal development of cities, such problems have to be studied in broad context. Assessment of impacts of higher buildings will be performed from the standpoint of sociology, medicine, psychology, architecture, environment engineering and so on. The assessment committee will be constructed as follows:

1 architect, 2 environment engineers, 1 interior designer, 1 residential engineer, 1 management engineer, 2 urban engineers, 1 transportation engineer, 1 system engineer, 1 psychologist, 1 lawyer, 1 economist, 1 medical man, 1 sociologist, and 1 meteorologist.

Now, separate committees for each of these three cases have started their studies by identifying and classifying the problems, and it is expected that these case studies will provide a standard example of technology assessment to be applied in other government agencies.

MINISTRY OF INTERNATIONAL TRADE AND INDUSTRY(MITI)

Following the "recommendation of policy for industrial technology in the years of the 1970's", MITI has also started a study program of technology assessment. This program is to explore basic information and conditions for applying technology assessment to development of technology in industry, which is the responsibility of the MITI. The procedure now under consideration is as follows:

1) *Selection of technological theme to be assessed:*

Subjects to be assessed will be selected by the following criteria, from among the technologies which are in the development stage and/or are expected to start development in future.

a) Legislative responsibility (whether the technology is in the field of MITI's responsibility).

b) Feasibility of technology assessment (whether the technology in its final form can be predicted well enough to be assessed).

c) Economic impacts.

d) Social impacts.

Applying these criteria by referring to the survey study carried out by MITI on "Quantitative evaluation of technological subjects for research and development", the items selected for preliminary survey are as follows:

PCM super multichannel communication system

Artificial human organs

Iron production with utilization of nuclear reactor

Recovery of minerals from the sea-bottom

Synthetic wood

Off-shore power station

2) *Assessment of economic and social benefits:*

In order to assess the positive contribution of the technology, items which will represent economic and social benefits have to be identified. The examples described in the program are as follows:

(1) Improvement of labor and social circumstances:
reduction of pollution
reduction of accidents by industrial products
prevention of natural disasters
improvement of medical care and health

(2) Improvement of life:
increase of variety and efficiency of life
extension of living sphere
improvement of commodities
enhancement of knowledge and culture

(3) Increase of productivity:
expansion of production scale
extension of economic spin-off to other sectors
reduction of cost
reduction of manpower
improvement of processing

(4) Internationalization of products:
meeting foreign needs

(5) Extension of resources:
increase of utility of resources
exploitation of resources
extension of space for industrial location

(6) Enhancement of technological standard:
meeting international standards
spin-off

Formulation of economic indicators to represent such criteria as a whole, will be attempted.

3) *Determination of undesirable side-effects to society:*

As a counterpart of the benefits, undesirable effects have to be assessed and check points for the technology from this standpoint should also be identified. The following are examples of the check points described.

(1) Points related to health and safety of human beings:
air pollution (smut, automobile exhaust, etc.)
water pollution (industrial drainage, oil waste, etc.)
noise, vibration (airplane, automobiles, machines etc.)
malodour (gasses, chemicals, etc.)
hindrance of sunshine (high buildings, etc.)
hazard of medicine and cosmetics
pesticides (residues)
food additives (accumulation effects)
luxury items (coffee, tobacco, etc.)
traffic accidents (car-overcrowding, etc.)
explosion accidents (gases, explosives, etc.)
labor accidents (mines, etc.)
leakage accidents (gas, electricity, radioactivity)

(2) Points related to preservation and maintenance of nature:
destruction of rivers (ruthless exploitation of gravel)
subsidence of ground (use of industrial water)
destruction of landscape (buildings, factories)
destruction of sea-bottom (digging at continental shelf)
eutrophication of waters
climate change

(3) Points related to social mechanism:
traffic jams (increase of cars)
disturbance in communication (high buildings)
education problems (CAI)

(4) Points related to economy:
resource problems (waste of resources)
problems of regional economy (location of factories)
problems of industrial structure (effects on other industries)

Here, again, formulation of social indicators which can represent comprehensively the combinations of such check points or criteria is expected to be attempted in the program.

The merits and demerits derived from the result of an integral assessment of the technology's effect on economy and society with criteria described in (2) and (3) will be compared.

If it is concluded that a technology will have fairly large, undesirable effects to society on its application, it is necessary to select an alternative program of development of the technology, which will not harm the welfare of human beings. And such alternatives should be studied to be most effective in the total system, including economy and society.

The program has just started, and at present a working-group consisting of members from various sections of the MITI is carrying out preparatory work for an actual study to be performed by an assessment committee consisting of specialists from various fields.

OTHER ACTIVITIES

The Science and Technology Agency and the MITI are the two government agencies which are the most advanced in attempting to introduce technology assessment in their execution of government policy in science and technology. However, there are many activities related to assessment of specific technologies under way in other government agencies, such as Economic Planning Agency, Ministry of Welfare; pollution is studied in Environment Agency, Ministry of Welfare and others.

It is expected that technology assessment will become, in the future, a general procedure in most of the government agencies which deal with the promotion and development of technology.

CONCLUSIONS

With the increase of impacts of technology on the social and natural environments and the increase of concern of people about long-term effects, technology assessment is now becoming one of the most important matters for the government in Japan. This atmosphere is also promoted by the trends in the United States and other foreign countries. Now nobody attempts to discuss the policies of technology without referring to technology assessment.

However, when it comes to the actual operation of technology assessment in Japan, many difficulties are pointed out. First of all, the exact meaning and coverage of the term "technology assessment" is rather ambiguous. Some people use it to mean an emphasis on consideration for environmental impacts in developing technology, while the others mean a quantitative assessment of technology by systems approach with consideration of the total system of technology, human society, and nature.

On the other hand, there are differences in the opinions of the extent to which technology assessment can go. Especially because the development of new innovative technology will mean basically a creation of something that is not available at the moment, and if so, it will not be possible to have exact data for its assessment. Therefore, assessment might be made only after the technology is completed; and even then, there always exists a possibility that some unknown effects can gradually come up over a long term.

Furthermore, some people are anxious that if technology assessment will be made improper ways with too much emphasis on probable harm, this will tend to discourage efforts for new breakthrough or innovations in technology and prevent progress.

These are some of the general difficulties coming out of discussions about technology assessment. When we go a step further to the detailed practices associated with its implementation, the difficulties tend to become still larger.

What will be the proper body to perform technology assessment? Is it a government agency or some independent third body which is neutral to the government and private enterprises?

What is the best methodology of technology assessment and how can we prove the validity or feasibility of the results? The difficulty arises especially because it is not only the technology that is predicted to change in the future, but also the value-system on which the technology is to be assessed.

How can the results of technology assessment be implemented in the administration of the government? This is a serious problem because technology assessment means some sort of restriction or control by the government and if this is not properly done, the basis of free economy and enterprises will be strongly distorted. This also gives rise to a question of how the impacts of technology can be monitored and the results fed back to the assessment procedure.

There may be many other points to be mentioned in this regard, however, the intention here is not to emphasize the difficulties of technology assessment, but to point out that, in Japan, the requirement of technology assessment,has risen so rapidly and intensely that the government is now urged to take some action for it, even though there exist so many difficulties and the technical basis is not well established. Since the advancement of technology and its introduction to society have been extremely rapid, the problems have become urgent and serious. If the situation is left as it is, the social strain and pressure will become destructive.

Therefore, it is urgently necessary to establish a proper organization and procedure of technology assessment and its implementation. As described above, technology assessment in Japan is in an early stage with only some case studies now starting in a preliminary way.

However, we expect that through such studies it will materialize in the not too distant future in a form which can contribute to the proper advancement of technology in harmony with human, social, cultural and natural environments.

TECHNOLOGY ASSESSMENT IN CANADA

DOUGLAS G. BAILEY

Director
Intendes Research Foundation, Toronto, Canada

While the assessment of the impacts of alternative technologies is far from new in Canada, such activities have yet to be recognized as express functions of government, and so have not warranted direct allocations of budget, material, and mandate. The evaluation of the consequences of change brought about by technologies comes under the varied guises of economic analysis, social and science policy, and public and private planning.

Nationwide, there is a growing awareness of the relationships of more immediate, popular issues to technological change. But such issues usually portray a regional character which makes generalizations about the effectiveness of technology assessment activities in this country less than meaningful in an international context. In fact, the single most visible characteristic of policy research efforts in Canada is confusion.

In the first place, political realities in Canada virtually preclude a national concensus. Not only is the Federal Government explicitly dependent upon other levels of government for the delivery of national policies and programs, but social and political issues are also tending to further differentiate contending policy viewpoints. The emergence of some public programs in Canada is more often the product of convenience than of rational decision-making.

In the second place, one cannot consider a single "market economy" in Canada by which business and industry could determine research, development, and marketing policies. Not only is the distinction between oligopolistic and competitive market conditions vast, but regional and sub-regional economic conditions also vary. Defense, industrial and social technologies considered desirable in some of the Maritime (Atlantic) provinces are socially and politically unthinkable in other provinces, owing to differences in such conditions as employment, effective cost of living, and concerns for environmental quality.

Increasingly, both the information media and the actions of private interest groups are transcending political and administrative machinery in influencing public policy with regard to such issues as economic growth, quality of public service, efficiency of transportation and so forth. Both the cancellation of some public works programs and the recent non-military use of defence facilities in Canadian cities are indicative of the growing influence of the media and interest groups upon public policies.

This over-view can do no more than review some of the machinery and attitudes of different "publics" in Canada with regard to the purposes and means for carrying out the assessment of the impacts of alternative technologies. As a conclusion, some recommendations regarding *national* as opposed to Federal criteria for the assessment of technologies may be proposed, in the context of the trends discussed here.

The administrative budget of the Federal Government of Canada reflects a decided turn away from "hardware" technologies and the "hard" sciences in allocating progressively larger amounts of Federal funds to domestic concerns and to delivery of the "softer" technologies, such as new welfare systems and cultural development incentive programs. In general federal funds are promoting a more progressive body of attitudes and actions oriented to the development of a bilingual, bi-cultural national identity. Provisions for the insurance of "more Canadian content" in the information industries (especially television and film) are somewhat more visible to the public at-large than the more subtle tax reforms introduced in the Federal Budget of 18 June, 1971. Legislation introduced in 1971 will ensure a greater proportion of French-Canadians in government than ever before, and bi-lingualism is becoming a must for all federal employees.

The current "new rationalism" of the Trudeau administration suggests that this administration has come to consider technology in its broadest sense — as any form of objective. Prior to the Trudeau administration, two organizations were established to consider technology under the guise of "science policy": The Science Council of Canada, and the Science Secretariat.

The Science Secretariat was first established in 1964 as a body within the Privy Council Office (an organization established to serve the Cabinet of the Government) to advise the Cabinet on scientific affairs in Canada. The Secretariat also serves as an advisory body to all Federal organizations, including Departments and Ministries. To-date, the Science Secretariat has maintained a small staff concerned primarily with executive affairs in government.

The structure of the Science Secretariat is patterned after that of the whole of the Privy Council Office: an operating research organization reporting to a Science Advisor. As such, its concerns change rapidly, and systematic efforts of assessment of the impacts of technological alternatives are always limited by the immediate concerns of the Cabinet. This organization must be less concerned with the rational deliberation of choices than with the delivery of concrete information concerning the technologies proper.

The Science Council on the other hand, established in 1966, is by mandate, concerned with the broader implications of technology and science policy in Canada. It too maintains a small staff of science advisors;

however, these are supplemented by consultants drawn from a variety of disciplines for undertaking major studies to develop recommendations for federal science policies. Since the start of the Trudeau administration, the Council has reported to the President of the Treasury Board in lieu of the Prime Minister.

The Science Council has been the agency most concerned with the generation of objective, rational policy recommendations. It is unfortunate that only very recently has the Council viewed the actual assessment of technologies as choices rather than imperatives. Yet, even with the recent (5th Annual Report) recommendations of the Chairman of the Council for the development of a Federal agency to undertake forecasting and impact studies, methodologies for examining and assessing the impacts of technological alternatives remain to be seen.

After the Science Secretariat and the Science Council, the next most visible agency concerned with technology assessment activities is the Defence Research Board, an organization with a long tradition (established in the 1940's) of applied research for national defence purposes. Activities of the Board have tended to emphasize the application of operational research to hardware evaluation and assessment problems, though the Board has also been concerned with the development of contingency plans for both international and domestic problems.

Recently, the Defence Research Board has taken a serious posture on environmental forecasting and the design of more rationalized organizational structures corresponding to changes in both mission analysis and the character of Canadian society. Specific criteria for the assessment of socio-economic effectiveness of alternative technologies have yet to be developed, however and are likely to be affected in defense of budget requests as opposed to a direct concern with the evaluation and assessment of hard and soft options.

A number of Federal agencies which have recently been inaugurated have direct concerns with technology assessment, even though specific criteria for such efforts have not yet been developed. These include (1) the International Research and Development Commission; (2) the new Department of the Environment; (3) Information Canada; and (4) the Transportation Development Agency; and (5) the new Ministry of State for Science and Technology. The first of these is primarily concerned with the development of new applications of agricultural and biological technologies for international development purposes. It is yet too early to comment on the effectiveness of these efforts, except to note that no explicit assessment or evaluation functions have yet been identified. The majority of research efforts undertaken are scientific in nature, and results are academic in format.

The Department of Environment was formally established in 1971, and is an aggregation of activities formerly distinct in administration and operations, such as the Fisheries Service, the Forestry Service, the Atmospheric Environment Service, and the Environmental Protection Service. While problems of coordination and articulated policy development have received highest priorities over the period of the first year of the Department's activities, it is plausible that this agency will in the future be involved in some forms of assessments. It is more likely, however, that these efforts will focus on rationalising impacts after the fact than considering potential effects prior to investment, development and the application of technologies.

Several recent issues in Canada could bring about changes in attitudes and action on the part of the Federal Government, and particularly the Departments of Environment and Energy, Mines and Resources. The potential impacts of new mines, dams, and pipeline systems are increasingly political issues which force public review prior to implementation. While some see the Department of Environment as a token response to these new issues, with no real powers other than influence over other agencies, others suggest increasing capabilities of the Agency to enforce constraints based upon detailed analysis.

Unfortunately, Information Canada has to-date taken on the characteristics of little more than a public bookstore; assessment activities concerning alternative information technologies have not emerged. Recently this agency was accused of acting as a federal propaganda device.

The Transportation Development Agency under the Ministry of Transportation represents a proposal for a significant departure from the organizational and functional mandates of most other federal agencies. While the Agency's express purposes are research applications and demonstration projects, recent proposals there suggest a broader scope and a strong potential interest in technology assessment and technology transfer activities. Unfortunately, this organization can do little more than advise its Ministry, and appears to have been placed on a "back burner" relative to other priorities and concerns of the Ministry and the Government as a whole.

The new Ministry of State for Science and Technology is a much more exciting and hopeful prospect. The Ministry's first significant action was to send the largest technical and scientific mission ever sent to confer with corresponding officials of the Government of Japan. Unlike most agencies in Ottawa, the Ministry is intent on strong involvement with private interests and industry in Canada, and has already made explicit its intent to undertake technology assessment programs. Though in early stages of development with a relatively low public profile, the Ministry promises to undertake many different kinds of activities, including technology assessments, which the Science Council does not now consider relevant.

Among the more established federal agencies, studies with a "peripheral" relationship to technology assessment activities have been undertaken by the Canadian Post Office, the Department of Communications, and the Central Mortgage and Housing Corporation.

The Canadian Post Office has completed several "environmental forecasts," which are trend extrapolations related to market demands for postal services. These efforts are some of the very few activities undertaken by the Federal Government to base organizational planning on serious inquiry concerning potential change, and to examine options with regard to directions which the agency might take. These efforts include no technology assessments to-date, since extrapolations are based on past experience in specific quantitative measures.

The Department of Communications has probably been the most active federal agency in terms of considerations of the environmental and social contingencies which may affect both the Department and Canadian society. Several task forces and conference/seminars have focused on the impacts of communications technologies with respect to geographical population migrations, the quality of urban settlements, and more generalized social concerns. Most of these, however, have been little more than discussions and the expression of public interests and opinions.

An adapted form of Program Planning and Budgeting was introduced into all agencies of the Federal Government of Canada more than three years ago. Officials at the Federal Treasury Board have noted that the greatest operational problem in implementing PPB approaches in Canada is the strong difference between structural (the traditional form) and functional (form required by PPB) organization in Federal agencies. A new development in the Central Mortgage and Housing Corporation, a "Policy Planning Function" appears to have addressed this problem directly by developing a process-oriented operating style which generates and evaluates policy options, and their corresponding alternative delivery systems. In organizational terms, it would appear that this operation comes closer than any other in Ottawa to "gearing up" for technology assessment activities. Components of the organization are charged with direct forecasting responsibilities employing relatively sophisticated techniques such as cross-impact studies, morphological analysis, and scenario development.

Though this organization's prime concern is with housing in Canada, it has also examined such policy issues as sewerage, assistance to urban areas, and the construction and finance industries. Both relatively "hard" and "sotf" (systematic judgments based upon analysis of strategic attitudes) forms of evaluation criteria have been developed for the assessment of options. As a model for organizations designed for technology assessment in the development of Federal policy options, it is likely that this organization will do much to shape future policy-planning in the federal arena.

Most other Federal agencies, including the Departments of Regional Economic Expansion, Communications, Consumer and Corporate Affairs, Labor, Justice, and Industry, all pay veritable "lip service" to the assessment of optional technological developments. Most of these activities are carried out via "Program Evaluation", which in most cases involves the rationalization of decisions usually made on an *ad hoc* basis.

This paper can be little more than a superficial examination, however, since little public information is available concerning the assessment or decision-analysis activities taken up by the Federal agencies. A summary of what appears to be the current trend in the Federal Government of Canada is as follows:

1) An increasing number of agencies are employing forms of "technology evaluation" in developing advocacy positions for particular software technologies, such as welfare and health programs. The Canadian Federal administrative budget suggests a slight lag behind the U.S. Government in the development of critical attitudes and negative reactions to scientific activities, with corresponding efforts in science policy.

2) Science policy efforts in Canada, developed by the Senate, may have decided impacts in terms of guiding the future activities of a selected few Ministries, but in general the product of these efforts has been unrealistic in terms of acceptability and implementation.

3) In part, the lag behind U.S. emphasis in this area owes itself to the Canadian tradition of expending far less per capita on hardware research and development than have the Americans. But as well, Canadian nationalism is driving a wedge against the acceptance of U.S.

policies as "the thing to watch," extending even into the sciences and professions. This means that Canada may never consider technology assessment as a serious requirement prior to making investment and development decisions.

4) "Futures studies" are becoming a popular administrative cliche in Ottawa, and though the majority of these studies to-date have employed highly questionable methodologies, they do indicate a new concern for the examination of options and the development of "strategic" plans. Results of these studies have been employed in two ways: (1) as "threat devices" to focus political attention; and (2) as explanations for current and past decision-making. Some agencies, however, are beginning to employ forecasting with more positive ends in mind, such as the exploration of policy and program options, and the analysis of socio-economic impacts of alternative program efforts. The extent to which forecasting becomes a part of decision-making processes in Ottawa will probably depend upon its incorporation into methodologies required by the Treasury Board (for budget purposes) and/or the Department of Supply and Services (for procurement purposes).

Most evaluation studies undertaken to-date have amounted to little more than reviews of the performance and efficiency of current and past programs. The development and application of effectiveness criteria for assessments is still a long way off; however, it is becoming fashionable in each of the major departments and ministries to name officials responsible for policy review and analysis of options.

5) Most Federal agencies have already developed specific policies regarding particular hardware technologies, such as air and rail transportation systems, defense technologies, telecommunication equipment and computer hardware. These tend to have been developed on the basis of more traditional cost/benefit criteria, which means that technologies which are not readily justified in terms of such criteria face serious accountability problems. The development and adoption by Federal agencies of new effectiveness criteria for assessment

activities must overcome the bias which previous methods have dictated. This will be difficult to do.

6) The introduction of effectiveness criteria which consider such dimensions as public relevance, harmony of programs, and consistency of consequences will require considerably more than the kinds of training programs and publications which accompanied the introduction of PPB. If technology assessment in its broadest sense is to be employed by the Federal Government in determining such initiatives as growth, employment, and regional development policies, it may be necessary to re-organize extant policy planning operations in agencies along more process-oriented lines.

7) It is much more likely that private interest grou groups, involved in local and national issues, will force a responsibility for the assessment of alternative technological choices upon the federal government, rather than that this activity will come about as an initiative by the government. Since Canada, and especially the Canadian Federal Government, has never really een involved in major technological innovation (usually imported from the United States), this will represent a major change in the outlook and strength of federalism.

Provincial government activities are best discussed in terms of groups of provincial governments which appear to share similar programs in which technology assessment activities are undertaken as a part or a whole of their efforts. These groups are: (1) Saskatchewan and Manitoba; (2) the Maritimes and British Columbia; (3) Quebec; and (4) Ontario and Alberta.

Neither Saskatchewan nor Manitoba have definitive provincial programs concerned with the assessment of technological alternatives. However, technological assessment in the form of cost/benefit studies have been undertaken in terms of both regional planning and agricultural research.

The populations of each of these provinces run slightly over 1 million with approximately half of the population living in urban areas. While program planning activities have been undertaken in these provinces for well over 10 years, the assessment of investment alternatives has rarely been of public concern. Activities directed to the assessment of alternative technological developments have been restricted to advocacy positions based upon such criteria as employment, agricultural efficiency and productivity, and so forth. Rational consideration of which industries and what kinds of technologies to invite into these areas must await Federal initiatives, since provincial politics demand growth of the tax base by whatever means become available.

Both the Maritimes and British Columbia have similar contexts. While the Maritime provinces have large rural populations, activities other than agriculture are principal concerns, such as fishing, boat building, forestry, and some industrial activities. In urban areas in general, shipping and transport runs high as an economic mainstay, and it is in this area that some technological assessment work has been undertaken jointly by Federal agencies, provincial governments, and some of the larger shipping concerns. Major studies have been carried out on the feasibility of large-scale container facilities in East and West Coast ports, and there were major investments in container handling equipment based upon such studies. For the most part, however, these studies have neglected consideration of larger economic and social impacts of such technologies which tend to be capital intensive and probably produce more rather than less unemployment.

Manpower development is an important dimension in these provinces in relation to any technology assessment activities. Because criteria for the assessment of benefits by federal and provincial interests have traditionally been increases in the gross regional product, and thereby the tax base of the provinces, such considerations as skilled labour rarely enter the picture. As some of the federal government officials have noted, the problem which containerization has tended to generate in the Maritimes is the costs of training and/or importing skilled labour to operate the shipping facilities. Inefficient utilization of facilities as a consequence has become a major stumbling block to the achievement of such objectives as the reduction of shipping costs and the attraction of international shipping.

Still another factor facing both British Columbia and the Atlantic provinces is that most industrial development is foreign-owned and/or directed. It would be difficult for the administrations of any of the provincial governments in these areas to assess technological alternatives without facing the risk of alienating foreign investment, especially capital investment. There is a sound "don't do it" argument for staying away from technology assessment activities in these areas; this argument is reflected by provincial government positions on industrial pollution. The standard question is, "Would you rather be dirty, or hungry?"

Quebec stands outside of this milieu because of (1) cultural differences; and (2) relative wealth and population size. Over six million people, nearly a third of Canada's total population, live in Quebec. The cultural attitude prevalent in Quebec has always been typified by stronger concerns with a balance between social and economic growth.

Putting aside, for this analysis, the political differences and difficulties which surround a bi-cultural society, Quebec's Provincial Government has undertaken activities which probably come closer to a direct examination of technological development options than any other provincial power. As a "pocket of differences" in the relatively homogeneous North American cultural setting, the population of Quebec has maintained attitudes which are an interesting mixture of French and American. The decision to maintain French as the primary language in the province is a *de facto* assessment of cultural options.

In French Canada considerable emphasis has been placed on the examination of options per se, including alternative forms of government, of industrial development, and of social programs, in order to preserve in

Quebec a character unique from the rest of Canada. Expo' 67 provided Montreal and the rest of Quebec with an excellent vehicle to consider and develop technologies of all sorts which are uniquely French Canadian in character. Perhaps one notable example is the Montreal subway system, which utilizes rubber wheels as opposed to metal wheels and tracks.

It is not possible to note any specific criteria used by one or another Provincial agency in the Government of Quebec; however, differences in provincial and Federal Incentive programs are obvious, and there can be little question that Quebec's Provincial Government is more socially-conscious than is 'conventional wisdom' in other provinces and the national setting.

Both Alberta and Ontario have taken definitive moves toward the analysis of potential impacts of alternative applications of research in the Provincial commissions and agencies. In both provinces, which are the wealthiest in Canada, there have been assessment activities in organizations to study and evaluate the future of education and education-related policies and programs. Major environmental forecasts have been developed for each of these provinces, and economic development options evaluated in social and economic terms. However, these efforts have been relatively isolated from other government activities, including planning, so that they are largely exploratory and have less than exciting impacts upon provincial planning.

In Alberta, such study programs have been embodied in new councils and commissions such as the Human Resources Council and the Commission on the Future of Education. Established with broad mandates and ambitious intent, these organizations never did manage to conduct serious reviews of provincial policies and programs with regard to innovation. In the period of the last six months, both agencies have been closed down following changes in the political administration of the Province.

Forecasts of Alberta's social and economic future have been developed, but it is difficult to ascertain as to how such studies have, do or will affect Provincial policies and programs. No explicit technology assessment studies have yet been undertaken, though with sufficient public pressure it is possible that some more serious investigations will be carried out with respect to the Alberta tar sands and their future exploitation.

The Government of Ontario has done little better, though for very different reasons, until very recently. Branches of the Provincial Government, particularly the Department of Treasury, has been given both mandate and resources for the express purposes of examining alternative social and economic development policies, which implicitly include technology assessment activities. To-date, the Department of Treasury and other agencies of the provincial government have yet to develop (1) the necessary inter-departmental coordination for making policies and strategic plans effective, and (2) methodologies for undertaking assessment, forecasting and other kinds of study operations.

A notable exception is the Ontario Economic Council which over the past year has produced some rather extraordinary analyses concerning the emerging "post-industrial" economy of the Province. The Council advises the whole of the government, but reports in particular to the Premier of the Province. Work is underway on, among other topics, industrial and economic strategy for the province, using fresh concepts and approaches to the definition of economic objectives, traditional concepts of growth and progress, and required strategy for economic change in the Provice. In the near future, the Council may well be involved in the assessment of large sets of technologies in relation to potential public investments.

In terms of education and manpower development, Ontario has also acted as a leader in the establishment of the largest educational research and development agency in North America: the Ontario Institute for Studies in Education. Until very recently, this organization has lacked (1) strong, centralized directions for the assessment of research applications, and (2) administrative structure and cooperation of other Provincial agencies and organizations for effecting policies and programs. The Institute has been characterized by a piecemeal approach to research and development, with little regard given to social and economic assessments of alternative applications.

Up to this point, it would appear that this review of provincial efforts has been highly negative. It should be noted, however, that in the more wealthy and populus provinces, efforts to assess alternative applications of results of various kinds of inquiry (not all of which are scientific) have been initiated, suggesting a growing trend to provincial, as opposed to Federal, leadership in the development of methodologies for assessment activities. In an International context, the efforts of Canadian provincial governments in evaluating alternative development directions and hard- and software technologies are highly significant, since they suggest that it may be necessary for developing nations, as opposed to "post-industrial societies", to evaluate alternative technologies on the basis of regional, as opposed to national, criteria which are "culture-specific".

In summarizing the efforts of provincial agencies in assessing technological development alternatives, the following points may be made:

1) Provincial governments of relatively deprived and/or comparatively underdeveloped provinces will continue to be reluctant to devote administrative and/or political resources to the assessment of technological development alternatives. Options will probably continue to remain scarce, and unless Federal initiatives with enforcement powers act to constrain international economic interests, these governments are likely to welcome any applications of scientific research offered them.

2) Technology assessment activities of most provincial governments will be concerned largely with capital

investment decisions tied to highly emotional, publically-visible issues of local or regional character (such as expressway systems, infrastructure, and urban renewal).

3) The Government of Quebec may continue to evaluate technological development alternatives in terms of cultural contexts; but criteria for doing so may never be explicit, nor systematic, without definite Federal leadership.

4) The Governments of Alberta and Ontario may be expected to continue to generate organizations and processes taking more serious and effective steps to forms of technology assessment, especially in view of the need for new social and educational policy directions.

5) New government administrations in provincial governments are those most likely to take international examples of technology assessment activities as models for similar activity in Canada, since the review and evaluation of alternative technological development programs, both past and current, might suggest administrative and political incompetence, in terms of decision-making.

6) The development of effectiveness criteria applicable to technology assessment activities in Canada must incorporate regional character and attitudes.

Private groups and organizations of all kinds, including labor unions, the press, film industries and special interest groups are increasingly bypassing administrative machinery in Canada, on all levels — municipal, regional, provincial and federal. It is the actions and expressions of these interests, frequently tied to localized issues, which pose the alternative means by which governments may be influenced to establish more visible and systematic technology assessment activities. These groups and private interests have already been responsible for the closure of more than a few large public investment programs, including the closing of Canada's urban renewal program and the recent Ontario decision to halt the construction of a major urban expressway.

Yet, it is interesting to note that to-date none of the major political parties in Canada have proposed the evaluation of hard- and soft-ware applications as part-and-parcel of a political platform. Such a platform might be readily developed by both provincial and federal party machinery of any of the major parties, since almost every local area in Canada has experienced at least some of the costs of technological change. Further, examination of current Federal and provincial directions in terms of employment and educational policies might provide the basis for a whole new approach to questions of a Canadian identity.

The options are certainly there, and it is possible that in the remaining months of 1971 one or another of the national parties will take up this question with strength. In the meantime, decisions which are being made without consideration of rational options will continue to come under public scrutiny — especially where the costs provide popular issue material and become vehicles for national public expression.

The following points may be noted concerning the "state of the art" of technology assessment activities in Canada:

1) The evaluation and assessment of alternative applications of the results of inquiry is relatively new as a function of political and administrative government in Canada, though cost/benefit assessments of more traditional hardware technologies have proceeded for some time.

2) Considerable confusion exists concerning the criteria to be employed in considering alternative applications and investment policies. Further, the delivery of technological applications must, by legislative and functional requirements, be the consequence of intergovernmental activities and programs.

3) As government decision-making comes under increasing scrutiny by interest groups in the public at-large, it is likely that more direct efforts, in the

form of budgetary allocations and appropriation of material, will be seen in both Federal and provincial public agencies.

4) Criteria for technology assessment activities in Canada must include both policy considerations (directions of administrative machinery) and more regional, popular considerations, such as local issues and problems, especially in a nation which actively declares a multiplicity of cultures and at least two predominant languages.

5) Some models for assessing alternative applications have been developed, but these have tended to be relatively isolated from the remaining bulk of government activities, at both the Federal and provincial levels of government.

6) More positive uses of forecasting coupled with policy and planning development and review activities may provide the required technical criteria (including social, economic, and political consideratios) together with effective delivery capability to provide the context for a re-examination of technology application policies and programs.

7) Initiatives will likely be developed at one of three levels: the Federal Government, as a product of new breakthroughs in social and science policies; (2) the provincial governments of Alberta, Ontario, or Quebec; or (3) popular interest groups, including professional associations, the press, and private interest groups.

TECHNOLOGY ASSESSMENT: SEEN FROM THE SMALLER COUNTRY

R.W. JACKSON

Science Council of Canada

The purpose of technology assessment, let us say, is to protect a society against aspects of technology that threaten its values and its well-being. From the view-point of the smaller country, the task presents special problems because of two features of the scene. First, most new technologies impinge upon the society from outside — relatively few of the innovations originate within its borders. Second, the natural laws of technological efficiency, sometimes called "technological imperatives", exhibit qualities of synergism and returns to scale, such that the smaller entities of the world tend to generate less than their "share" of innovations, and to contain within themselves less than their "share" of decision-making power. Unless deliberately countered, the forces of corporate organization (e.g., multinational corporations) and the "imperatives" of economic efficiency (e.g., the doctrine that free trade is the condition for optimum allocation of resources) reinforce these trends.

Further reinforcing the trends to technological centralization are electronic and printed communications. Canada is subject to enormous spillover of communications and advertising from her neighbour to the south. Values are influenced, wants, aspirations, and appetites are stimulated. Values that were once thought to be national characteristics, and deserving to be defended, get displaced by the communicated values, which then become the will of the people. It is a familiar lesson of history that a more vital or powerful culture tends to diffuse outward, submerging others in its path. We are here raising the question, I suppose, whether modern technological imperatives, and ubiquitous communications, (features of "scientific" or "post-industrial" civilization), add new dimensions to the time-honoured process.

Is there any defence? Must the smaller country simply trust to whatever technological assessment gets done in the originating country, in accordance with whatever social values? Or are there any countervailing powers by which citizens of the smaller country can recover some reasonable influence over their destinies? The obvious candidate that comes to mind is the power of knowledge.

Assuming that the smaller country has sufficient sovereignty to be selective, and to set terms and standards for the technologies it allows in (insofar as the citizens' own standards can prevail against the diffusion pressure), the task can only be done intelligently and consistently on a sound base of knowledge. It is not enough, however, only to try to be selective after the fact. It is essential also to have some means of influencing the decisions made elsewhere that *generate* the technologies. Again, knowledge could be the key. The authority of a sound scientific investigation is valid anywhere and should be difficult to ignore once in the public domain.

The situation of the smaller country is that of receiver or consumer of technology more than generator. Recognizing that, it would seem to follow that the smaller country can hope for more success in achieving its social goals through whatever effort it places on assessing, evaluating, choosing, and regulating technology than through such effort as it may put into the original generation game, for economic or other reasons. The smaller country might therefore spend a relatively large proportion of its R and D funds on such activities.

The smaller, or "consuming" nation has a special advantage in that, like consumers' associations, it is more often free to take an independent and objective stance than an originating nation because it need not be inhibited by self-interest or internal conflict of interest. Then, if its work is well done, its authority (Theobald's "sapiental authority") may be amplified through widespread publication and through international organizations like the OECD, to the point where even the most self-centred and independent-minded decision-makers in the largest countries may take heed.

All of which is a rather complicated, but up-to-date way of saying "the pen is mightier than the sword". However, to guard against an excess of confidence in this mighty weapon, it would be well to explore the implications, and limitations of the above thesis in terms of some particular cases, namely:

1. THE SUPERSONIC TRANSPORT

There is no evidence that the U.S. decision to discontinue development was influenced in the slightest by any Canadian representations. But it could have been if, for example, Canadian authorities had taken a strong stand on allowing overflights, or providing airport facilities. To justify such a strong stand, against the sensibilities of the Americans, British, French, and perhaps the Russians, would have required firm scientific evidence of, say, chemical effects in the stratosphere, or excessive effects of sonic booms in cold air. In fact, the Canadian Ministry of Transport has been supporting some studies, on a modest scale, limited to possible direct effects of overflights. So far, evidence of harmful effects has seemed insufficient to justify more intensive efforts, but undoubtedly the intensity of studies would escalate if the evidence seemed to warrant. What may be the main question, that of the large social investment and risk, Canada has not regarded as her concern, except to the strictly limited extent as a commercial purchase by Air Canada.

2. THE AMCHITKA ATOMIC TEST

The U.S. plan to test a five-megaton fusion bomb in a deep shaft underground on one of the Aleutian Islands excited some apprehension among environmentalists, and their apprehension was fanned by other groups who by habit campaign against all atomic weapons. There were grounds for concern, in that the explosion would be in a known earth fault zone and, however unlikely, the possibility had to be admitted that an earthquake, tidal wave, or other kind of accident such as release of radioactive products could be triggered. Effects of such an accident or catastrophe would impinge on the west coast of Canada. It was clearly a case for international consultation, perhaps before a world court. Protests were in fact registered in one way or another from and through the Canadian Government to Washington. But there were complications. Much of the technical information which would have been needed for Canada to make an independent assessment on which to base a stand, was locked up in the secret files of the AEC. Yet, even with the best of good will, it was not a case where the assessment could properly be entrusted by Canada to the U.S. experts, because the assessment involved balancing one risk against another. For the U.S. Defence Department, the test was allegedly crucial to the development of the newest anti-missile weapon system — and therefore worth fairly considerable risk to the environment and to the sensibilities of neighbours. To Canadians, most of whom hold no particular brief for U.S.-U.S.S.R. weapons competition, the risks at issue looked quite otherwise. (Canadians used to feel grateful that they could shelter under the U.S. defensive umbrella — now they are not so sure; they feel they are more likely to find themselves in the missile no-man's-land, under a rain of radioactive fragments.)

3. TELEVISION

T.V. in the U.S. is an instance of a technology where the main and original driving forces were commercial. In other countries where it has been established as a public service the problems of its use or misuse are not the same. Canada is in a hybrid situation, with its main network operated by a government corporation, in competition with private commercial networks, and with American stations received directly in most large cities. The Canadian commercial stations are highly imitative of the U.S. pattern. Assessments of the psychological effects of television have been attempted sporadically in the U.S., without much visible effect on the nature of programming. In fact, it seems that change is more likely to come about by the introduction of alternatives, e.g., a national "educational T.V." network, than by changing directly the behaviour of commercially-motivated management. Television could be a prototype of other electronic technologies growing rapidly on a free-enterprise base, e.g., computer systems. Once a pattern is initiated in the U.S., that pattern tends to grow and acquire an institutional momentum. A favorite assertion is that assessment has more chance of effect if it enters early, before institutional rigidities build up and vested interests solidify. In this case one might speculate that the opposite might be true. A definitive Canadian psychological study proving serious deleterious effects on children might have more effect on U.S. programming now, when people are becoming bored and are having second thoughts, than earlier when they were caught up in uncritical first enthusiasm. Besides, earlier on it would have been more difficult to find the evidence from which to build a convincing case. Against that, in the later stage we encounter the situation that people cannot imagine the world otherwise than it is. These observations lead to speculation on what should be the proper course for assessment of other commercially originating electronic technologies, such as cable-T.V., and computer systems.

4. THE ALASKA PIPELINE

The discovery of oil on Alaska's north shore has raised multiple problems of how to extract it and how to transport it south without deleterious environmental effects. The scheme apparently most favoured by U.S. oil interests is to build a pipeline across Alaska (crossing a major earth fault on the way) and to then ship the oil in tankers down the west coast to Seattle. Canadians object strongly to the prospect of oil spills on their beautiful west coast and particularly in the relatively populated and recreational Georgia Strait. In some respects the Canadian position is based on knowledge, and in other respects knowledge is being sought, e.g., on the feasibility of a pipeline entirely on land through Canada. The processes of assessment being followed are not entirely clear but seem to represent, in addition to some use of the power of knowledge, a search for expedient alliances among the interested constituencies. One sceptic commented that, so far, the only way that Canadian views have had any observable influence in Washington has been through convincing American environmental groups that their own self-interests were at stake (the values of "sharing space-ship earth" are still pale in their effect on decision-makers in comparison to the motives of self-interest). A more intensive Canadian effort on the knowledge front would probably be worthwhile, but the case illustrates that mere knowledge may be far from sufficient to cope with the realities of oil politics.

CONCLUSION

We conclude that technology assessment has a special importance for the smaller country, both for its own direct welfare and as a means of countervailing power over world technological developments. Canada has to some extent recognized this in her policies, but might well profit from an intensification of such work. Two types of problem confront the smaller country, one related to the assessment and social control of technologies that diffuse from elsewhere, usually under commercial pressure; examples are television, computers, psychotropic drugs (alcohol, marihuana, and others), contraceptive pills, the transistor radio, motion pictures, automobiles, etc. The other type relates to major particular decisions taken by business corporations and governments, that may affect the welfare not only of neighbours but even sometimes the entire world; examples would be the escalation of nuclear weapons, waste disposal in the sea, transportation of oil and other chemicals, major earth engineering projects that could affect climate, etc. The particular examples chosen for illustration show that, though knowledge is an essential and potentially powerful first lever, it is rarely sufficient on its own. Most real situations are complicated by other factors than the purely rational, and the smaller country must make use of various other means to increase its leverage. The situation is analogous in some respects to that of a private individual or a minority group within a larger hegemony, except that in that case the minority group has some hope of exciting the sympathy and support of fellow citizens and/or the attention of an overarching political authority. In the international situation the overarching authority is virtually absent and the bond among common citizens tends to fracture into we:they relationships at political boundaries.

TECHNOLOGY ASSESSMENT IN FRANCE

RAYMOND SAINT-PAUL

Professor,
Conservatoire National des Art et Metiers à Paris,
Paris, France — Centre d'Etude sur la Reserche et
l'Innovation. (CERI)
and

PIERRE-FREDERIC TENIERE-BUCHOT

Engineer,
Agence Financiere de Bassin Seine-Normandie, and
Director,
Centre d'Etude sur la Reserche et l'Innovation (CERI)

This article applies itself primarily to a description of what is understood in France by "Technological Assessment." Utilising a feedback concept, the system of Research and Development is analysed in the context of the firm, with the intention of illuminating the various connections which exist between technological forecasting, project selection, long term planning and the more traditional activities of the enterprise.

A practical example of technological assessment is then examined. It is drawn from the study of water pollution and the diseconomies this produces, made by one of the French Water Boards. The method used is that of an analytical structural forecast. It results in a classical operational model.

As it is concerned with products, with ideas and above all with new concepts, the phenomenon of innovation has a primary importance in French society. It is the function of R & D to promote the evolution from the fortuitous to the planned, from the exceptional to the essential and also often, the routine.

Very close to the exercise of power, such a function lends itself equally well to industrial problems as to those of public or private administration since the type of decision-making or strategic problem met with becomes more and more similar. The growing influence of the long term over the short, the pointed moral problem raised by this influence, the very large financial and technical risks run, the many socio-economic repercussions of these decisions, the accounting for external effects, are some of the main aspects which will become increasingly important in managerial decisions.

Much more than in other areas, the need for appropriate *working methods* shows itself here. But an inventory, however complete it might be, would not, of itself, be sufficient. It is a comprehensive *methodology* which allows practical action as a relation to be established between techniques that are required.

The application of remedies would seem more useful than the classification of problems. Also it must not be the aim of this paper to find a new categorisation or a new and more satisfying typology of the methods of R & D.

The concern for practical application, the desire to choose and adapt the methods to concrete cases has been the major preoccupation in elaborating what it will henceforth be convenient to call "the technological assessment of projects in research and development".

WHAT IS UNDERSTOOD IN FRANCE BY 'TECHNOLOGY ASSESSMENT'

The placing of the selection of projects or new ideas, taking account of policy choices, in their temporal reference scheme, is the essence of TA. It is, thus, to the analysis of the system "R and D" which one turns when deciding to make use of this method.

One does not choose new projects for the sole reason that one has not adequate resources. The selection is made, above all, because of long term ideas of usefulness. The items and concepts which are judged then, will be examined from the point of view of their usefulness or otherwise to this long term aim.

Everything starts with a forecast, that is to say, an initial considered idea which is used as the starting point for the feedback process shown diagrammatically in Fig. 1. This choice is arbitrary but without influence on the conclusions which can be drawn by observation and analysis from this scheme. The essential point is to choose a point of departure and to return to it after having distinguished all the loops of the scheme. This is what will now be done.

It is accepted that an initial forecast has been made. This forecast most often relates to an individual's forecast or to one produced by a small group. It is the starting idea of the research who, before putting it to his superiors in the firm, does some experiments, in order to test the initial proposal.

This "free" research, carried out very much as a marginal activity to the declared research programme, is essential since the results, promising or not, of the first experimentation will serve to establish the starting criteria to define the population of projects from which selection is to be made. Originality of idea is not in itself sufficient and its potential for success must eventually be assessed by a second experiment, a good deal more complete and painstaking. Taking the form of a prototype study, a test, a discussion, it is strongly influenced by scientific and technical knowledge (the State of the Art and Technique). It is put into train by a decision sequence described in the following loop:

- the starting idea, having been judged interesting, is chosen to be experimented upon (project selection)

- certain resources are allocated to this experiment (resource allocation)

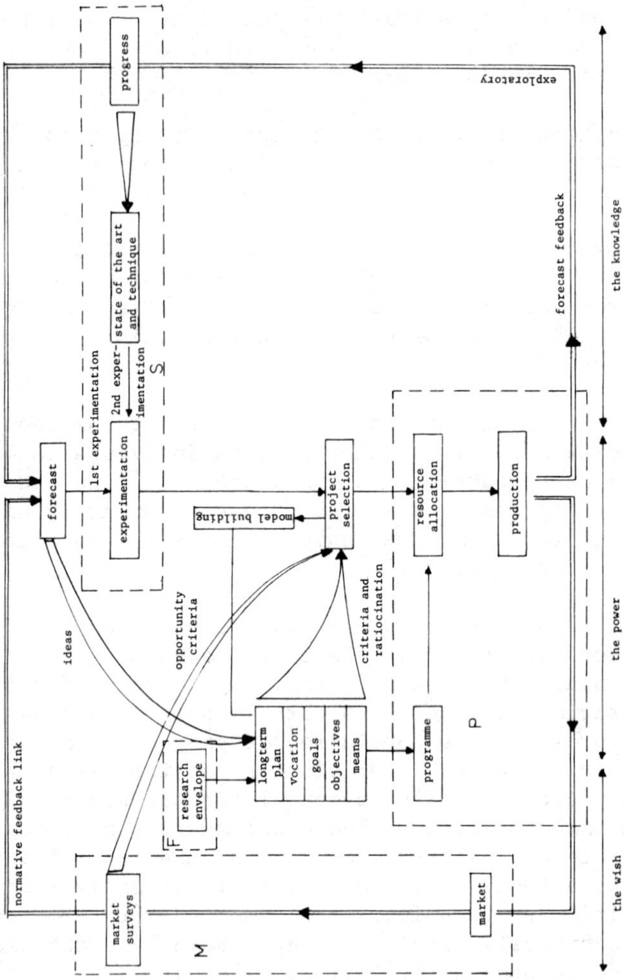
progress
state of the art and technique
2nd exper-imentation
1st experimentation
S
forecast
experimentation
exploratory
forecast feedback
normative feedback link
model building
project selection
resource allocation
production
ideas
opportunity criteria
criteria and ratiocination
longterm plan
vocation
goals
objectives
means
research envelope
F
programme
P
market surveys
market
M
the wish
the power
the knowledge

Diagram I.

- a development method is designed

- with the aid of these last results in the appropriate branches of science (progress and State of the Art) the prototype is produced

A positive result or at least the possibility of success will enable the idea to be considered as valuable or promising. Admitted to the population of projects which can be put into industrial production, the trial is submitted to the decision makers: this is the *selection of research projects*.

The central operation of the scheme, this last includes two distinct aspects:

On the one hand, the population of elements to be judged (called projects) chosen for their adequacy to the problem posed by the preceeding operations;

On the other hand, the population of judgmental viewpoints (called criteria) which derive from the long term plan (of diagram I). This latter plays a fundamental role in the technical evaluation. In effect, it is the central element of policy which will enable a direction to be given to all the elements of the plan. The long term plan must make explicit several distinct notions:

The purpose of the enterprise (taken in the largest possible terms). In the realms of the possible, the purpose must not be defined in terms of the process, narrowly specialised, but enunciated in as wide a context as possible, and without reference to the time span. Thus, an automobile maker must see himself as specialising in *transportation*. There will always be a need for transport, perhaps there will not always be a need for automobiles . . . In the same way a company concerned with information processing will see itself as a specialist in management methods and not as a user of the computer, a process which one day perhaps will be outmoded.

The definition of purpose is not a quick and easy task. It cannot be effected from nothing but only by a long and concerted effort guided, however, by other notions dealt with in the long term plan:

- the definition of goals, qualitative aims which delineate the purpose. For the automobile company above, for example, a goal might be to penetrate the travel agency market, which has a direct link with transport, but not with the automobile. Easier to define than the purpose the goal(s) has a time span: it is concerned with penetrating the travel agency market within a given period.

- With the purpose of making the aims more precise, the goals can be broken down into objectives, quantified elements relating to the goals; it is such a percentage of the market which must be captured; it is such a level of business which must be attained; it is such a production level which must be maintained.

The objectives are confronted by the means and conversely so that an obtainable compromise can be reached. The means lie strictly in the R & D domain of the "Research Envelope" (see diagram I) envisaged as supporting the system. This financial problem is considered here as an arbitrary fixed parameter.

The long term plan limits itself to the available finance. The determination of the optimal "Research-Envelope" (R-E) is thus relatively ignored here, as deriving mainly from a management decision, employing the current techniques of financial decision.

It will have been seen from the diagram that the case of the financing of the R-E is placed under the long term plan and not at the level of means. The different ideas presented are thus assembled simultaneously in the elaboration of the long term plan. The sequence, going from purpose to means, has no existence other than expository. One has, the most often, the ambition of one's means and sometimes the means of one's ambition.* The long

*This is not without some connection with the "Matthew Principle" — "to he who hath shall be given, to he who hath not even shall it be taken away." R.K. Merton,

term plan will thus have enabled the formation of a range of particular criteria, ordered by (or derived from) the general company policy. This part is important because without it no worthwhile decision can be taken, the act of selection being thus reduced to a question of the rationalisation of choices.

There is often the tendency in current usage to confuse the plan and the programme. Such confusion is regrettable since it leads very often in practice to the absence of either. While the plan demonstrates an idea and a policy translated by criteria, the programme, issuing from the long term plan and, applying most frequently over a shorter period, shows in detail the different courses to be followed and these in reference to a precise timetable.

The allocation of resources (men, money, materials) must therefore accord with this programme and with the decisions resulting from the selection of research projects. It results from the *production* process, the final and visible result of our system.

It is here where the time variable intervenes in an important way in putting into relationship the selection of projects and technological forecasting.

Two conceptions of production, in effect, can be distinguished and compared within the viewpoint of the company. The first is common, notably in France, and useful when one takes stock of the value of the research activity. It consists in the consideration of the resulting production as *progress*, the *growth* of the obtained performance being a measure of this progress.

It is to this progress that one limits oneself in searching for new methods "in order to do better the next time." The next stage, forecasting, will thus be a

---

Scientific Progress No. 136, Dec. 1969, la documentation francaise; and, Lisle, *White Paper on Innovation* (Ministry of Industrial and Scientific Development, Nov. 1971), la documentation francaise.

consequence of past successes. Related experimentation at the basic level will benefit as an incidental to the success of the main work.

This linkage between production and forecast is called "the exploratory forecast loop" (see diagram I). This loop has as its principal characteristic to draw on past experience, the most recent usually, to bring forth reasoned hypotheses about the future, systematically exploring all the possibilities of logical extension (but not all the imaginable) derived from the advances so far realised.

A second, less usual, concept that can be had of a new production is consideration this latter from the point of view of the market and the effects of putting it on to the market.

Marketing studies (motivation, inquiries, economic studies) account for and describe the reception of a new product. They indicate the wants of the users or clients and the possibilities of stimulating new needs.

Forecasting draws from these studies to orient the sequence of external operations to the end that production can be adapted more and more to the needs, natural or artificial (created by the market).

This other link between production and forecasting is called the "normative forecasting loop" (see diagram). It allows the will of the firm to be injected into the forecast arena, it is accommodated in the exterior world in the form of a "norm" (a preference). Perspectives do not then result from unstructured technico-economic trends and possibilities but come from precisely defined objective-setting.

It is clear that these two conceptualisations of the future co-exist in the majority of enterprises. They must, however, be developed simultaneously, each influencing the other, but with a clear distinction, which is possible because it is made before a didactic goal is set. This distinction corresponds to the observation of an imperative, good and simple that it is, which is not always followed.

This is that it must be remembered that a result, whatever it may be requires at the same time, knowledge (exploratory forecasting loop), purpose (normative forecasting loop) and power (long term plan which alters the development forecast to a calculated choice).

This obvious rule, however, is often disregarded.

Some companies (examples exist in electronics and computing) will have a "scientific" policy: the exploratory feedback loop will be much used. They will go from progress to progress, with exceptional results in technical ventures. The complexity of the latter will, however, be such that only specialists can perceive them. The market will lag behind, uninformed and static. To put it at its best, it can be described quickly as being ahead of its time, preceeding the needs of the market. The latecomers finish, however, by getting going and the innovating company may fail; one can only honor its unfortunate courage. The reverse failing, which is rarer, is also observed from time to time. Beguiled by the market, certain managers will not move in front of it. Exercising a veritable dictatorship over the long term plan, their policy consists in searching for the immediate maximum profitability. Everything depends on opportunity and the speed of reaction. The criteria coming from the long term plan are, in fact, the *criteria of opportunity* (see diagram I) which resolve in the advantages of the present all situations seriously hypothesized for the future; the certainty of doing well rests here on "dynamism", the element of security for those who run too fast; it will avail these captains of industry nothing if at the same time if loss of control and takeover come equally quickly.

These extreme positions have been mentioned only to demonstrate that the division into exploratory and normative isn't arbitrary. Judiciously combined, these ideas permit the realisation of the particularly pertinent forecasts, not so much on the technical side as on the commercial and policy aspects. As the prime inspiration of the whole system, the forecast which takes the name of technological forecast provides the ideas to the long term plan which can thus be progressively adapted to present or future desires.

A measure of this required adaptation can be realised by means of a model (see diagram I). Taking account of the decisions taken at the level of project selection, the model, after combining and aggregating these, re-introduces the decisions into the long term plan. Some provisional theoretical results are thus obtained and they permit the comparison with the gap which exists between the chosen course and the intended.

This "proof by nine" of the system, since it is indispensible, must not be disregarded. As the rationalising element, integrating the numerous other elements a good deal more subjective, modelling can be useful in accommodating elements which might otherwise defeat the plan.

It is clearly evident that a model will above all demonstrate any inconsistencies in the existing decisions but it cannot at the same time prove that the apparently consistent decisions are good . . . It is thus a "negative proof by nine" which must be used with caution.*

In conclusion, the technological evaluation system offers a method of going to the heart of the organisation, public or private. It is a balanced policy-forming method which takes account at the same time of the wishes, the means of effecting these and the attitudes to follow; is this policy the best one to follow? No. The *best possible* signifies nothing except in the limited context of school exercises. The policy in question is realistic and

---

*This is true for all proofs even when they are correctly established. An incorrect operation such as 15×12=45 has a "proof by nine" apparently correct: Contrarily, an operation having a negative proof is certain false. A correct "proof by nine" being got the *consistent* indications of a model are not necessarily *exact* presumptions.

realisable in as much as it has certain advantage over previous ones.

An article of faith supports it however; this is that the company policy is determined by the ensemble of R & D projects. This important point has however for its corollary that the best expression of the R & D programme is in policy terms. So one must often judge. Only twenty years ago, the phenomenon of "Research" (one still did not hear of "Development") was understood by scientists alone. This was the "knowledge" aspect shown on the diagram.

Giving no useful results, because it did not respond well to economic preoccupations, the question of "Research" was adopted by the partisans of needs, partisans of utility and efficacy, and not of knowledge. Co-existing for better rather than worse with the scientists their influence was essentially limited to certain sectors (consumer goods) but it did exist long. By an easily understandable effect it might have been feared that certain excesses were also committed: today's utility may become tomorrow's disutility, efficacy an annoying encumbrance, the gain of the few a social loss.

The thesis which is presented here supposes a certain ambivalence between industrial R & D activity and social and political activities; that is to say, the exercise of power, in the most general sense. It draws for this upon a collection of methods of managing R & D which have application in different distinct areas:

- the first is the scientific domain (zones). Forming the group, progress, the state of knowledge (state of the art and techniques), experimentation, it is a traditional user of the methods of innovation support. Without it nothing can be done, but without method, it has no efficacy. Traditionally deductive and logically exclusive, favoring recourse to experimentation and observation, the scientific domain will have difficulty in including in the methods of technological assessment those which do not strongly rely on intellectual processes. It is here that is encountered one of the difficulties in accommodating the methods in a company.

Acting more on the level of criteria than of projects, the domain (or zone) of marketing has fewer difficulties in adjusting to the methodology of technological assessment. It will indeed have even greater ability to accommodate, as was remarked above, technology assessment as a section of commercial policy. The greatest difficulty will be in maintaining a certain rationality by which the will can be distinguished from the arbitrary.

The third sector of interest is production (Zone P) which is formed of programmed elaboration, resource allocation and the actual fabrication process. It can actively participate in decision-making by means of the methods of technology assessment. The lynch pin of the system, the production, is the test of the consequences of the decisions. It matters that responsibilities are distributed at the proper level: this is why the development of budgets and to a large degree that of programming (recall that this is different from planning) belongs by right to this domain.

The results of experience, receipts, the fruits of practice, will be the elements of choice which will influence the whole system of the provisional direction of R & D and which permit it to acquire a truly technological character. It is to the concrete that such methods aspire and not to the theoretical.

The last zone appearing in the foregone scheme is Zone F (financial) and this is the "envelope of research" which we have discussed in the context of the long term plan. Appearing as a quantitative constraint the "research envelope" can demonstrate the outcomes of the methods of technology assessment used or, reciprocally, place the limits on the area of concern. Analysis of the power structure will indicate the justification for the inclusion of this latter alternative.

The presentation which has just been made of the problems of technology assessment of R & D projects had a deliberately theoretical and didactic aim. It was developed in terms of an idea which does not relate directly to any practical reality. In particular the two forecasting loops figuring in the scheme make, in fact, a unity which must be considered as such, all the operations unfolding simultaneously. This is the crux, the principal difficulty of approaching the problems of R & D and innovation. This study hopes to ease this problem a little with the reservation that it is well understood that it is not possible to attack such problems merely from one single viewpoint.

To enlarge on this last statement, it is thought useful to show how some aspects of the work of technology assessment can be put into practical terms. The example chosen has been proposed and used in a water control agency (agence financiere du bassin), para-administrative organisation recently created in France in response to the problem of water pollution.

A PRACTICAL EXAMPLE OF TECHNOLOGY ASSESSMENT: ANALYSIS OF THE "WATER" SYSTEM IN A WATER CONTROL AGENCY

The problem of water pollution is one of the questions of environmental policy which has recently occupied governments. In France, it is handled by a set of organisations of a pronounced innovative character in terms of administrative arrangements. These organisations, these water control agencies, have as their role to complement and speed the public and private actions normally taken in this area. In respect to the struggle against pollution, in particular, they are charged with receiving the charges based upon the pollution loads contributed by each water user. The funds so collected are returned, almost in their entirety, in the form of subventions and loans at low interest rates to the managements of public and private works who wish to install purification plants. The level of charges, the criteria for the concession of financial aid, do not lie within the competence of the agencies but with control organisations called water basin committees, 'comites de bassin.' These committees are a kind of little water parliament which gather together the different affected parties, ministries (environment, agriculture, equipment, finance, industry, public health), local government (councillors, mayors, MP's), industrial interests, water authorities (distribution, tourism, consumers, fishermen).

The permanent organisation for consultation and discussion has been indispensable in putting into practical use the analysis of the system which will be expounded below.

Without relevance to the natural accounts, the water basin agencies have benefited from a relatively complete financial independence. The almost total control of the financial resources in the area of water management (with deductions for the running costs) has conferred on them the character of a "mutual company" for water and allowed a particularly privileged field of experience since the area of "water" is clearly limited, as are the interests affected by it.

These characteristics make the technology assessment which has been attempted in this sector relatively modest but shows the delimited area which will permit analysis and modelling.

I. *Analysis of the system "water" in the Seine-Normandy Bassin*

The method used to analyse the system "water"* is "structural forecasting analysis". It is directly oriented towards analytic and introspective methods using matrices, network theory and operational research.

It may be broken into four steps.

1. *The first stage: the listing of variables*

Four types of variables have been distinguished:

variables characteristic of water supply
variables characteristic of water demand
external constraints, collected under "decor", acting on supply and demand
variables characteristic of the water policy pursued

*One understands by system the collection of relations which affect the given subject.

In total 92 such variables have been identified of a very assorted character.

Quantitative variables:

- state of pollution of the water
- overall distribution of water
- number of purification stations, distribution networks, etc.
- domestic, industrial, agricultural demands, etc.
- rate of economic growth
- investment cost
- price of water
- level of taxes
- aid and premiums for construction and operation of the works, etc.

Semiquantitative and qualitative variables

- fishing, leisure needs
- distribution geographically of demand
- distribution of towns and industries
- R & D effort
- state of technical knowledge

Qualitative political variables

- the policies of management and support organisations
- the policies of local organisations, professional associations, distribution companies
- pressure groups — users
- public attitudes as affected by communication media
- state aid policies
- law on effluents and use

It is perhaps essential to the method to get right this initial listing of the variables. To write down what are the effective elements of the problem is to start to see the nature of the problem and the consequent relationship of the elements is very revealing.

One is very far here from a mechanical and non-reflective operation for instance, as in maximising one variable. What must be done is to take account, at the same time, of the subject under discussion and the objectives upon which it is fixed. One sees then that a "given" is perhaps a "result" in certain cases, that a result is sometimes a constraint, as all the aspects of the problem are perceived simultaneously.

1. *The matrix of logical relations*

The list of variables being established, the second stage of the work is that of the elaboration of a matrix of the relationship between variables. This matrix sets out the items of the foregoing list in the same order in lines and columns. The influencing relations existing between the variables are indicated in the matrix following a sign convention (.). In the lines appear the functions of the variable noted in the columns (see diagram II).

A variable, such as pollution, is affected by other variables* of supply:

- flow available
- number of purification works
- network of sewars
- flows held by reservoirs

by the demand variable:

- needs of dilution for industrial wastes, domestic wastes
- heating of water from cooling plant
- geographical demand concentration

by the variable of "decor":

- degree of urbanization
- market structure for purification
- cost and performance of technologies
- cost of abstraction and functioning of works
- state of R & D in these domains
- psychological attitude of the public

by policy variables:

- para-fiscal directives
- investment incentives and premiums for operation
- legislation and rules

*The variables are most often vectors of which the components are sub-variables.

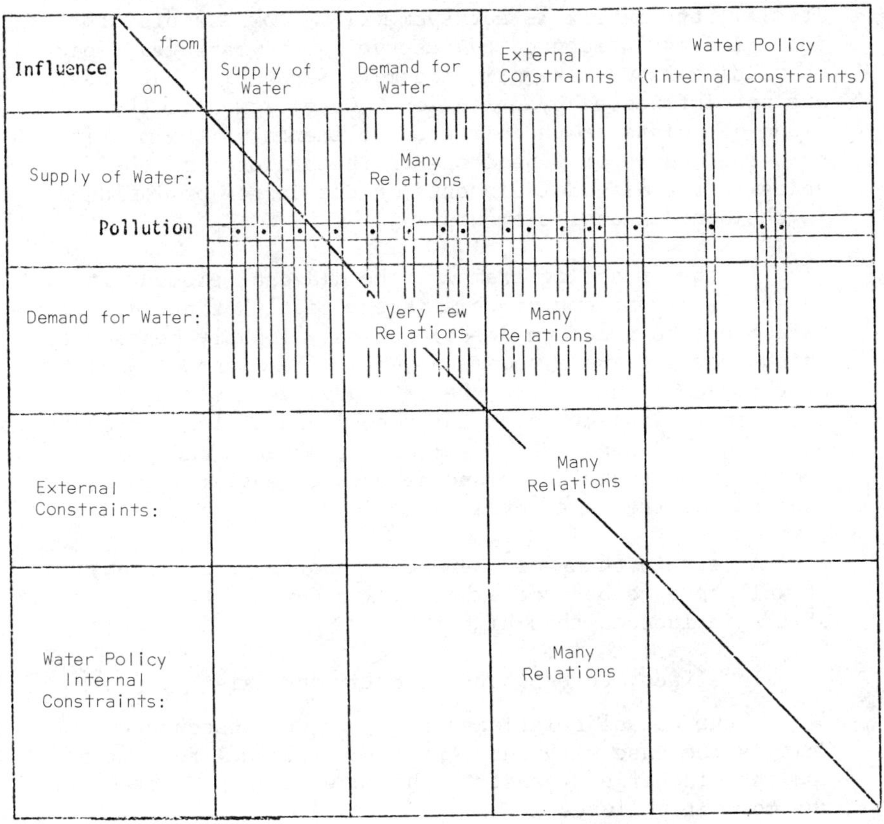

Diagram II. Matrix of logical relations

Each of these variables is itself influenced by others, which in their turn . . . , etc.

The matrix is thus filled little by little, leaving empty a certain number of elements (no influences) and the principal diagonal (a variable does not influence itself); the matrix is not symmetric. For example, the pollution is influenced by reservoirs and barrages, which on raising flows make dilution more favorable, but they are not influenced by pollution. Observation of the cloud of points in the different elements of the matrix show the degree of dependence or independence of the variables in the diagonal elements on the degree of influence in the non-diagonal elements.

In the preceding example, the diagonal element for demand is almost empty: this is the sign that the demand variable (the function needs) are only lightly dependent, which proves to be a good analysis. The diagonal element of the "decor" shows, on the contrary, very strong interrelations. Inspection of the non-diagonal elements shows that supply depends above all on the demand, which is logical, and that the demand depends essentially on the "decor", at the same level as policy.

A certain chain of phenomena thus appears already. It will have to be made exact by a study of the nature of the influences thus highlighted.

In effect, three sorts of relations exist:

- the causal relationship or direct-consequence. This is the case with industrial requirements for water and the discharge of wastes: the more industry uses water, the more it pollutes it.

- conditional relations. The capacity of the network of drains is one. If the purification plants are not linked to the discharge of pollution at source, they will not be able to abate it.

- decisional relationships. The financial incentive to purify, by means of a para-fiscal imposition taxing the polluters is an example of a possible decision. Legal regulation is another.

When it is desired to control the problem at the level of planning, a knowledge and comprehension of decision relations is capital. This is why attention is focussed here in the third stage.

3. *The matrix of fundamental relations*

The following operations are now more in the area of exploitation than that of exploration: they are concerned with the unfolding of a strategy to resolve the problem, that of pollution for example.

Some "variables of objective" must thus be designed. They represent the goals towards which it is hoped to direct the work. This "variables of objective" can belong to one of several categories which have served to classify the elements of the matrix, the given, the results and the constraints.

In such a way, it is after the study that there is decided the objectives to be realised, and not before. It is here that the structural approach has a great advantage over the classical analytical approach. It is not enclosed in hypotheses of restricted outcomes. The strategic choice which is avoided until the last possible moment when knowledge has reached a sufficient threshold.

The description of the desired results may then be totally different from the a priori ideas which may obtain at the start of the study: What matters is that the choice is made in all generality and all the variables are taken into account through the effect of the interrelationships studied.

It is thus that a seemingly unimportant objective, before the study, may be shown to be vital in assuring in every place and season the quality and quantity of water required. The analysis of the matrix of logical relations assists considerably in making precise this initial direction of research.

The real result to be obtained is,in fact,the control of water pollution. The satisfaction of the preceding demand for water is defined in this by necessary deduction: from an objective situated in the category of results (water demand) one has passed to an objective situated in one of the givens (supply of water).

All the factors influencing the item "pollution" will thus be highlighted as being one of the "driving variables" of the system. All those which do not influence the factor "pollution" can, on the contrary, be eliminated as not participating to the desired result. Here an initial selection, made apparent by the matrix of fundamental relations where nothing is included except the results — variables and the driving variables (mutually influential). To the end of orienting the work towards a concrete decision the driving variables can be separated into three groups:

- the decision variables which correspond to the will of the operator
- the variables of influence on decision, which are the expression of the real possibilities of intervention on the "decor" or scenario
- the constraint variables for which no action can be envisaged

It is by recourse to these three categories which the matrix is reduced to the fundamental relations (see diagram III).

The decisional variables which influence pollution are:

- para-fiscal incentives
- investment grants and premiums to the good operation of the works
- legislation and control

Those which are conceived as constraints are:

- industrial needs and domestic use
- the problem of "thermal" pollution
- the spatial concentration of demand

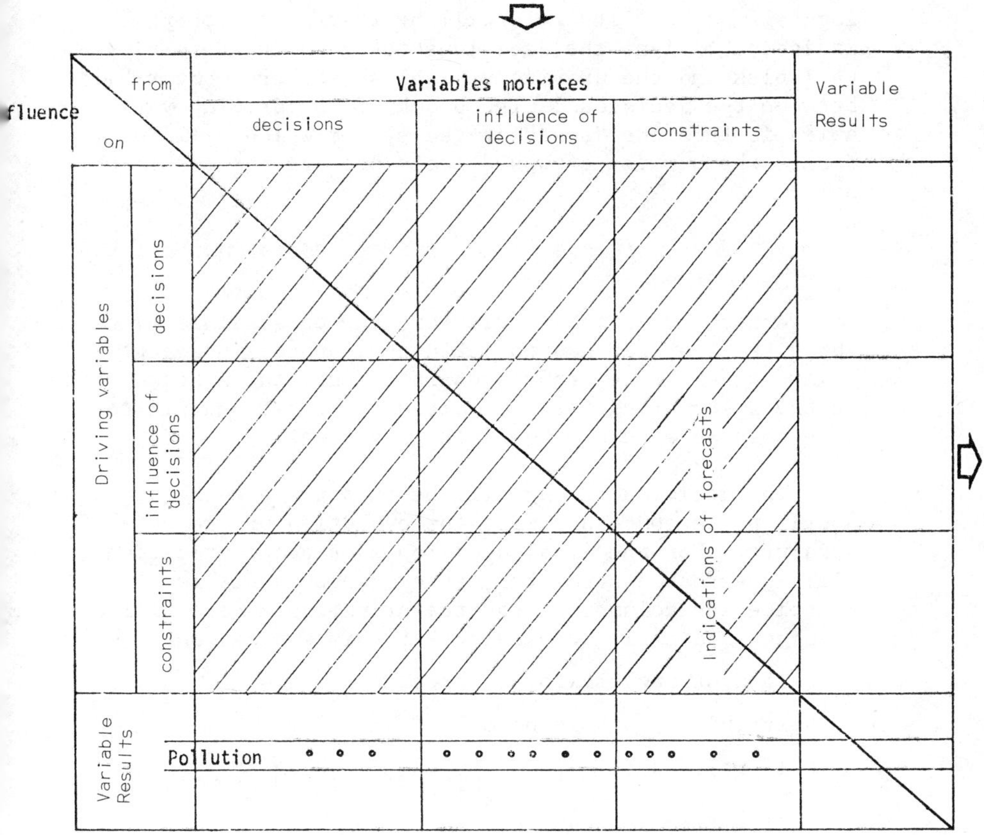

Diagram III. Matrix of fundamental relations

All the other driving variables are to be classed in the category of "decisional influence". The matrix elements showing the relationships between the driving variables (hatched in diagram III) will be filled in this new order by taking into account the relations revealed in the first logical matrix. Its role will be that of a "coherence indicator": given the decisions taken on the resulting variables and the driving variables is there compatibility between the decisions? The block of the driving variables is thus the "test" of the system which will indicate whether the proposed decisions are reasonable or not.

4. *The synthesis of actions and decisional orientations*

Among the driving variables, the constraints over which the operator has no influence, appear in a particular manner. They represent in effect, the unknown part of the system (and there is already a great benefit to have been able thus to delimit this area).

The decision making will be thus preceded by a prognostic phase concerned with the evolution of the constraints. For the final decision each prognostication will appear as a parameter, an exogenous variable. In practice the columns of constraint variables are filled by signs indicating the nature of this evaluation:

If a strong growth is foreseen

If this growth is normal

If it is weak

Signifies stagnation

indicate greater and greater decreases.

The relations (shown by the sign (.)) situated in the column of a constraint will thus be replaced by these qualitative forecasts. One may remark here that these latter limit themselves to the constraining variables only; in doing so they reduce the forecasting studies to be done and the risks of the errors which result.

In parallel, the objectives (using the same conventions) will be impressed on the resulting variables. It could be wished, for example, that the pollution would decrease rapidly.

Knowing this desire, and taking into account the indications shown in the constraint columns, decisions must be proposed for all the decisional variables and decisional influences:

Objective : Pollution

Prognostics : Needs of domestic users
Needs of industry
Cooling demand
Spatial concentration of demand

Proposed decisions:

Decisional variables : Para-fiscal incentives
Investment grants
Premiums on operation
Legislation
Education of the public

Influencing variables : Number of purification works
Number of sewage networks
Number of reservoirs
Land management
Land use density
R & D
Pricing policy of builders
State aid

The introduction of these decisions into the matrix block for driving variables must result in an overall coherence, as was indicated above. This very useful phase is in operation very complex and it results naturally in the construction of a graph associated with the matrix of fundamental relations.

II. *The computer program — "Water Pollution Policy"*

This phase consists in programming the computer with the preceding results so that it might simulate the impact of the derived forecasts and test the robustness of the arguments advanced. The name given to this phase is "POPOLE" (POlitique de la POLlution des Eaux) as is customary in these cases.

Two major stages, the construction of a network and quantification, have enabled this program to be put into use.

1. The network associated with the system

Each relationship symbolised in the foregoing matrices can perhaps be represented by a directed line joining two elements of a network, which then becomes very complex and extended. The skill of the operator consists in discovering in this net of liaisons, the feedback loops and the contrary.

By loop is understood configurations of this type:

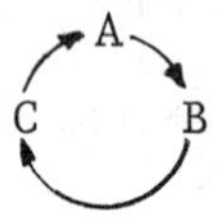

or

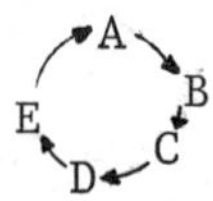

The contrary, it must be understood, is a loop reduced to its most simple expression, an opposition, as:

A

B

In each case an analysis must be carried out. In the case of an opposition, it must be accepted that this is the most important link vis-a-vis the problem being examined and it suppresses the antagonistic relation:

A

B

The loops of a high order (give elements for example) are generally less solid. It does not distort reality to open them up:

D E A B C

The loops of a lower order (3 or 4 elements) are on the contrary to be maintained. They interpret important phenomena which must be identified and understood. Two eventualities can, in effect, be discerned: the convergence of the loop towards some finite values (comparable to the supply-demand equilibrium of the classical market theory) or on the contrary the divergence towards bigger and bigger values (in military terms — escalation). These simplifications made and operation of the model having been achieved (representing an important part of the work) a network results in which only the useful relations are represented.

This network has four hierarchical levels. The lowest level is that of the inputs. The others show intermediate and final results. (See diagram IV).

2. *Quantification*

Each liaison (of causality, condition or decision) has until now been demonstrated in qualitative terms; they must then be quantified. This research on the binary relations between variables is more or less complex.

In the most favorable case (where the two variables are measurable and their evolution follows some law) one obtains a relationship of the type:

$$A = f(B)$$

In the least favorable (no known law, variables incommeasurable) a special study must be carried out* utilising in this instance a multicriteria procedure.

---

*The utility, even the necessity of such a study is

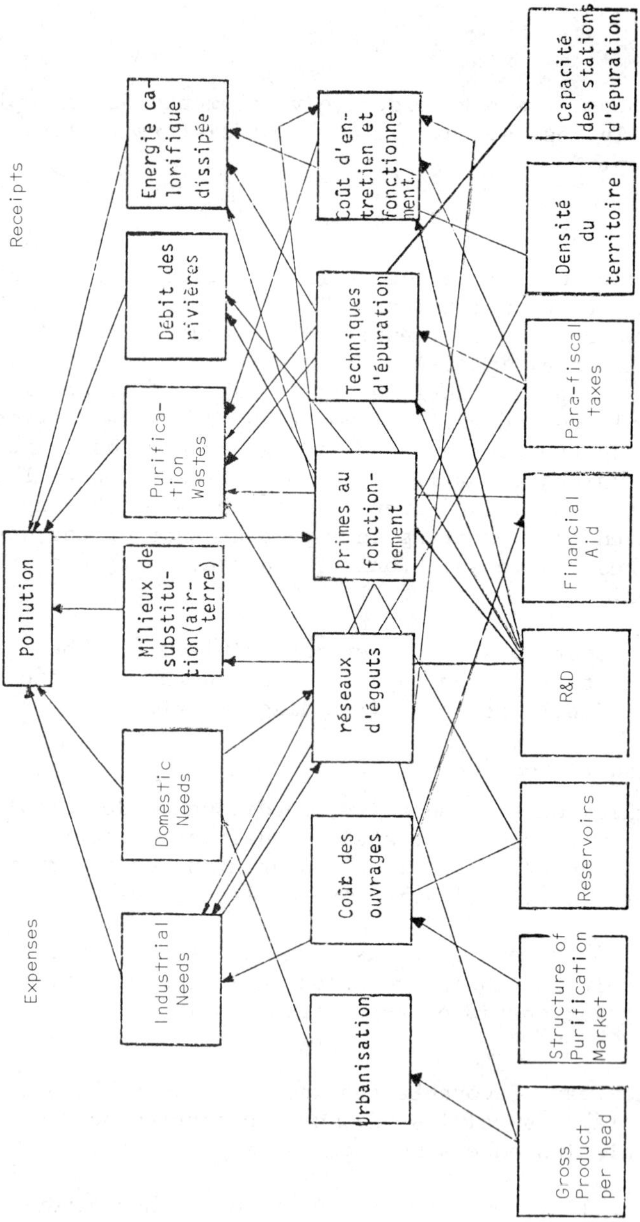
Receipts
Expenses
Pollution
Energie calorifique dissipée
Débit des rivières
Purification Wastes
Milieux de substitution(air-terre)
Domestic Needs
Industrial Needs
Coût d'entretien et fonctionnement/
Techniques d'épuration
Primes au fonctionnement
réseaux d'égouts
Coût des ouvrages
Urbanisation
Capacité des stations d'épuration
Densité du territoire
Para-fiscal taxes
Financial Aid
R&D
Reservoirs
Structure of Purification Market
Gross Product per head

Diagram IV

opposed to the preceding role, of structural analysis, as an *informer*.

The financial agency "Seine-Normandie" has deliberately opted for simplicity. Considered as experimental and preliminary the "POPOLE" model does not attempt the grand approach. The indicators which it provides are not taken as mandatory but as showing trends.

Each relation has thus been made schematic in the extreme. One can give here some of these as examples (see diagram V).

Each relation of this type (there are as many as liaisons in the network) in fact is worth a special study. The present state of knowledge of the services of the "Agence de Bassin" has, however, permitted the testing of some hypotheses. The coherence of the results obtained for 1971, as much for pollution as for the capital employed allows the belief that the method does not give false answers.

Introduced into a logical programme which they regulate the relations combine to furnish the operator with certain indications as to the probable evolution of the situation in the "Seine-Normandie" basin. This is the last phase, the exploitation, which will now be described.

III. *The exploitation of the program "POPOLE"*

The present use of this program is carried out on the IBM 360-30 (64K) computer of the Computer Laboratory of the "Conservatoire National des Arts et Metiers" in Paris.

The program is "conversational", that is, questions can be put to the machine which it answers after calculation. The inputs (see the listing 1990 given as an example below) is first set out by the computer.

At the command of the user the "physical" results and then the financial results of the "fight against pollution" account of the agency are produced by the machine.

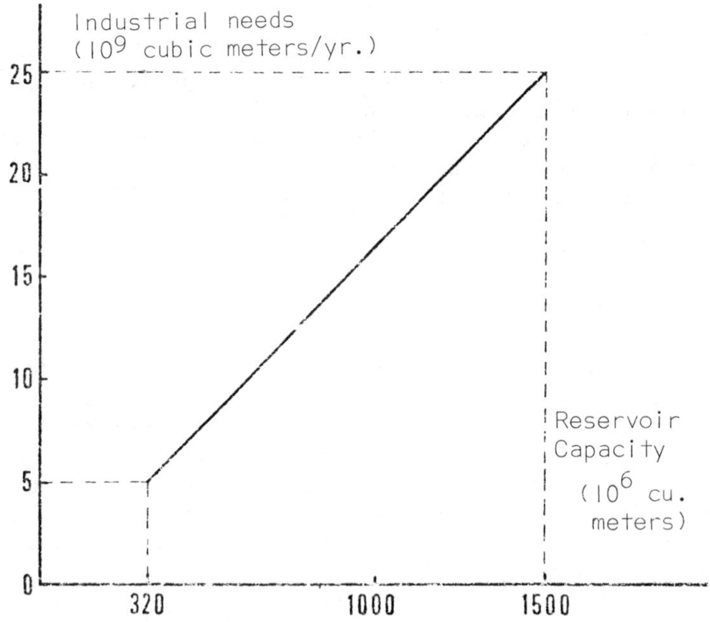

Diagram V

All sorts of difficulties, which in accumulation may affect the preceding work, then arise: can the simultaneous nature of the relations be maintained, such as appears in the model or must delays be introduced? Is it reasonable to resort to multicriteria analyses on certain variables using criteria which have not been employed in the model (one neglects, then, the possible feedback on these criteria)? Must statistical distributions be employed for the relations of which one is not sure, at the risk of overburdening the analytical programme?. . . etc. . .

Here one is, naturally enough, in the vicissitudes of operational research: it gives only as much as is given to it. Here the role of a *transmitter* of information is

---

thus a direct consequence of the structural analysis method.

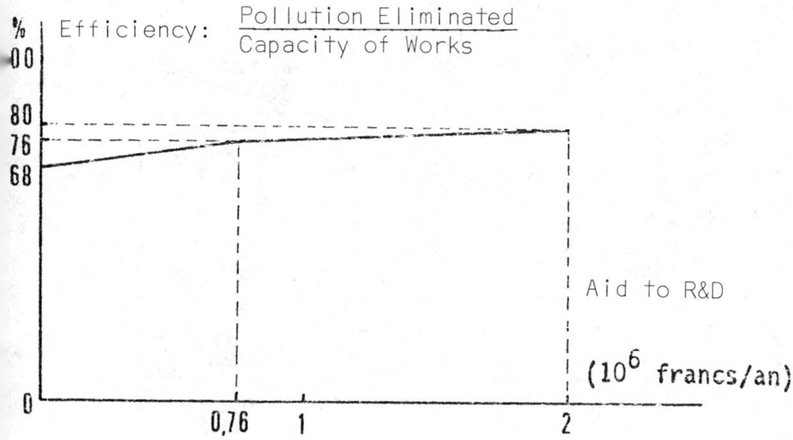

A R D = 10.000 + 75.000 x n

n : number of researchers

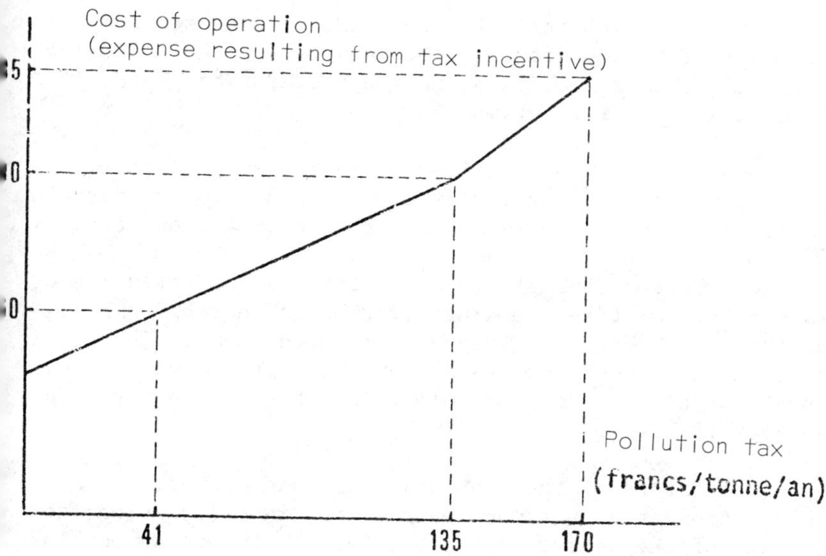

Diagram VI

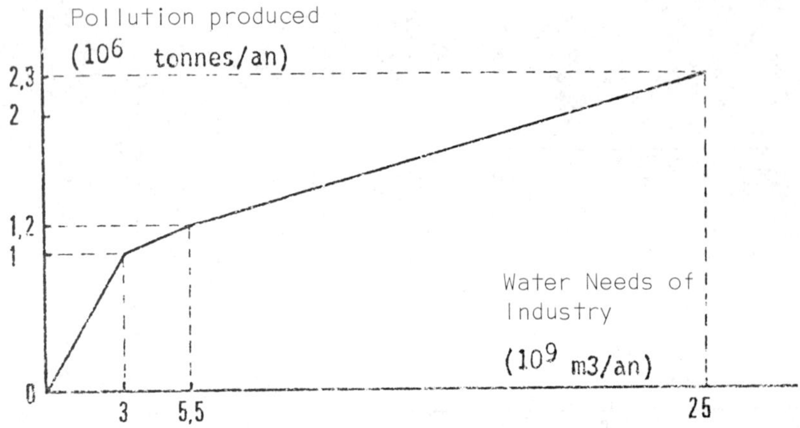

Diagram V. contd.

The new inputs are then printed out for the following year (this iteration stops at the year 2000).

If the inputs are not acceptable to the user they can be modified to his wishes. The results for the following year will thus be modified and the operator can then, if he wishes, make further modifications.

One can thus see the flexibility of such a system which permits the sensitivity of the output to the various variables to be shown (in this case the pollution level).

The method consists in demonstrating the influence of each parameter taken separately, then to search for multiplicative effects which may affect the outcome. The result will be the production of a "plausible" programmed attack on the development of the pollution and the elaboration of an overall strategy.

The nature of the results, while one cannot say that they represent an optimum, enable certain calculations to be made of the sensitivity of certain of the relationships (change in the slope of particular curves, for example).

CONCLUSION

Certain people will surely be suspicious of such methods and see in them a technocratic tendency.

The danger of such manipulation certainly exists, but it is not the objective sought. This is why it has appeared useful to present an analysis of the "water" system of the Seine-Normandy basin before the application of the programme on a computer. This has resulted in the question-answer aspect being open to public participation. The logical extension of the long term plan published by the organisation in question, it is, in use, the same type of action; above all it is a method for encouraging reflection and the testing of sensitivity.

In spite of its great lack of sophistication (it cost less than 45,000 francs) the reactions obtained from users and the pertinent authorities permitted the programme of the attack on pollution in the Seine-Normandy basin to be mounted rapidly.

BIBLIOGRAPHY

R. SAINT-PAUL Fascicules d'accompagnement du cours télévisé du Conservatoire National des Arts et Métiers: "Innovation et Développement Economique. Les méthodes d'évaluation technologique". (CERI-C 3E, 116 boulevard Péreire Paris, 17e - 1972)

P.F. TENIERE-BUCHOT L'évaluation technologique Thèse. Université Paris IX-Dauphine - 1971

F. VALIRON et P.F. TENIERE-BUCHOT L'eau en Seine-Normandie. Travaux et Recherches de Prospective N° 15 La Documentation Française, 1971

Programme "PO.POL.E.". Listing typ

ANNEE 1990

DONNEES

Produit brut du bassin	E	853.00	Milliards francs
Densité de la population	DS	200.00	Habitants /km2
Capacité barrages-réservoirs	BR	1160.00	Millions m3
Structure commerciale	POCO	2.00	Nombre de groupes
Max. aide bon fonctionnement	PBF	0,20	
Redevance pollution	RP	41.00	Francs /tonne /an
Modification budget	XDECIS	0.0	Millions francs
Remboursements prêts	REMB	8.16	Millions francs
Budget interventions pollution	AID	85.16	Millions francs
Chercheurs :			
Substitution épuration	NN 1	0	Hommes-Années
Débits	NN 2	3	Hommes-Années
Entretien des stations	NN 3	0	Hommes-Années
Refroidissement	NN 4	1	Hommes-Années
Techniques d'épuration	NN 5	0	Hommes-Années

RESULTATS

Surface urbanisée	EI	17012.34	Km2
Coût investissement épuration	CTI	2700.00	Francs /tonne /an
Besoins domestiques	EDD	2.22	Milliards m3 /an
Besoins des industries	EDA	18.53	Milliards m3 /an
Débit étiage Paris	Q	107.71	m3 /sec
Substitution épuration	TSE	0.0	Millions tonnes /an
Rendement technique épuration	TE	0.68	
Débit des égouts publics	RCT	112.97	m3 /sec
Energie à refroidir	TCI	137.08	Milliards KWH /an
Aide au bon fonctionnement	BF	28.64	Francs /tonne /an
Coût d'entretien des stations	CEF 2	149.62	Francs /tonne /an
Capacité stations épuration	SEO	2.22	Millions tonnes /an
Epuration efficace	SE	0.93	Millions tonnes /an
Proportion stations aidées	K	0.34	

Programme "PO.POL.E.". Listing type contd.

DONNEES

Pollution rejetée	P	2.52	Millions tonnes/an
Auto-épuration	COEF 1	1.00	
Echauffement des eaux	COEF 2	0.20	
Pollution résultante	P	2.52	Millions tonnes/an

RESULTATS FINANCIERS

Recettes effectives	RECEFF	98.32	Millions francs
Dépenses	DEPENS	107.75	Millions francs
Recettes - Dépenses	RMD	9.43	Millions francs

TECHNOLOGICAL FORECASTING, ASSESSMENT AND INDUSTRIAL DEVELOPMENT IN EUROPE AND THE U.S.

ANTONIO MARTELLI

Director, Società di Ricerca, Consulenza e Programmazione (CERPI S.p.A.) Milan, Italy

EUROPEAN INDUSTRY AND TECHNOLOGICAL DEVELOPMENT

As technology progresses, more up-to-date techniques are used to solve economic and social problems, while the solution of these problems becomes more complex and more difficult.

In fact, the causes and effects of actions in fields which are apparently very far apart (for example, the consequences which some industrial products have on ecology) become more interdependant. For this reason the analysis and understanding of the elements which lead to a decision have become increasingly arduous. On the other hand, the resources (especially human resources) are scarce and the consequences of frequently irreversible decisions are becoming more and more serious.

Foreseeing the consequences of decisions has, therefore, become very important, particularly where decisions regarding technological choices are involved. It is possible, given certain limits, adequate resources, and enough time, to invent and to produce practically everything. It is essential, however, to know exactly what is needed and what the consequences of the invention will be on the economy and on society in general.

In this respect the limits of classical economy are, of course, insuperable. It cannot make long range forecasts, nor can it make rational decisions except in everyday operations. This limit is a very serious one indeed in an age when technological progress renders decision making more binding for the future.

Still, in 1935 in an essay which can be considered the peak of classical economy from the point of view of style and understanding Lord Robbins issued a warning to this effect: "Even if we think that, if we know the technique, we can predict the type of economic relationship associated with it, which of course is highly disputable, how can we predict the technique?" There are still a great many people making fundamental economic decisions who still think this way.

In fact, as technology progresses the "traditional" economic (quantitative) forecasting should be combined with technological forecasting and the two should merge together. This type of forecast attempts to ascertain how, when and what the consequences will be of future changes in technologies.

There is no doubt whatsoever that in this respect Europe is far behind the United States. England, which is the European country spending the highest amount on scientific research, is way behind the United States in the practical use of the results obtained, i.e., in marketing the research and using the forecasting instruments for the orientation of the research. It can more generally be said that so far there are no practical examples of integration of technological forecasting and general economic forecasting in Europe, although efforts have been made in this direction. For example, the latest development plan of the Italian economy known as "Project 80". contains some rather timid hints at such a step.

There is a precise relationship between attempts at technological forecasting and the amount of resources invested in research. Office of Economic Cooperation and Development (OECD) has done important work on this subject and has highlighted the importance of technological forecasting, not only in establishing a suitable research policy, but also in developing industrial innovations. Table 1 gives a summary of this survey. The most striking feature is the enormous disproportion between research carried out in the United States and in other industrially advanced countries. Against 110 dollars per inhabitant spent on research and development in the U.S. in 1964, less than 25 dollars were spent in Western Europe, most of which was on individual efforts frequently incoherent and in contrast with each other. Maybe things have improved slightly since then, but not much. Both private and public organizations are only just beginning to realize that the experience which can be gained from examining technological possibilities and probabilities gives a decided advantage when it comes to applying new techniques.

The different attitude in the U.S. that technological forecasting is a decisive factor in economic development is further evidenced by the following figures: the annual amount spent by the principal 600 U.S. companies on technological forecasting is approximately 70 million dollars, whereas the leading European companies only spend about 2 million dollars.

The consequences of such an attitude on European industry are obvious. Without suitable technological and economic evaluation, it is difficult to establish the targets of research, as was the case with atomic energy. Table 2 shows the lack of balance in Europe in assigning expenses to R&D compared with the U.S.

Thus, for example, nearly 40% of the expenses assigned to R&D by manufacturing industries in the U.S. is concentrated on the aeronautic industry, whereas in European countries, excluding Britain and France, the corresponding percentage is negligible and is included under the general heading "other transport materials". In Italy, 9.1% of the total is destined for research on "textile products, rubber, foodstuffs and drinks", which only represent 2.5% of the total in the U.S.

The methods used to market the results of the researches in Europe are, to put it mildly, inadequate.

Table 1. R&D in OCSE industrial countries during the years 1963/64. (OCSE Ecarts technologique - Rapport analytique, Paris 1970.

	Gross National Expense in R&D	Per capita expense	% expense of GNP	Researchers, engineers and technicians employed in R&D	
	Millions U.S.$	U.S.$		No. fulltime operators	per 10,000 inhabitants
Unites States	21,075	110.5	3.4	696,500	35.8
Canada	425	22.5	1.1	23,850	12.6
Japan	1,060	10.9	1.4	197,225	20.3
Great Britain	2,160	39.8	2.3	159,538	29.4
Austria	23	3.2	0.3	3,220	4.5
Norway	42	11.5	0.7	3,820	10.4
Sweden	257	33.5	1.5	16,530	21.6
France	1,650	34.1	1.9	95,574	19.7
West Germany	1,436	24.6	1.4	105,010	18.0
Italy	291	5.7	0.6	30,280	6.0
Belgium	137	14.7	1.0	15,600	16.8
Netherlands	330	27.2	1.9	31,310	25.8
Total EEC countries	3,844	21.7	1.4	277,774	15.5
Total industrial countries in Western Europe	6,326	24.3	1.6	460,882	18.3

Table 2. R&D expenses in manufacturing industries

	SCIENCE-BASED INDUSTRIES				GENERAL MECHANICAL INDUSTRIES				OTHER INDUSTRIES			
	AERONAUTIC	ELECTRICAL	CHEMICAL	TOTAL	MECHANICAL CONSTRUCTIONS	BASIC METALS	OTHER TRANSPORT MATERIAL	TOTAL	TEXTILES, RUBBER, FOOD & DRINK	OTHERS	TOTAL	GRAND TOTAL
U.S.A.	38.3	24.8	13.0	76.1	8.0	2.6	8.9	19.5	2.5	1.9	4.4	100
Canada	16.9	29.1	23.6	69.6	4.2	9.8	0.9	14.9	5.4	10.1	15.5	100
Japan	(2)	30.3	27.3	57.6	5.1	9.4	11.3	25.8	8.4	8.2	16.6	100
Great Britain	29.0	24.5	14.4	67.9	8.4	8.7	7.3	21.4	6.7	4.0	10.7	100
Austria	(2)	18.6	24.0	47.6	4.0	24.9	16.3	45.2	4.0	3.1	7.1	100
Norway	(2)	22.0	21.3	43.3	6.6	23.4	5.3	35.3	6.1	15.3	21.4	100
Sweden	19.8	24.3	9.9	54.0	13.9	13.1	7.8	34.8	4.1	7.1	11.2	100
France	24.6	28.6	19.4	72.6	7.6	5.3	5.8	18.8	4.6	4.0	8.6	100
W. Germany	(1)	31.2	34.7	65.9	19.6	8.4	(1)	28.0	4.7	1.4	6.1	100
Italy	(2)	25.7	28.1	53.8	10.5	5.6	20.1	36.2	9.1	0.9	10.0	100
Belgium	1.5	20.3	43.8	65.6	5.0	18.1	0.6	23.7	3.7	7.0	10.7	100

(1) Included with "mechanical constructions"
(2) Included with "other transport material"

Source: OCDE, Ecarts technologiques. Rapport analytique, Paris 1970.

The development of new techniques has no effect at all on the economic activities until the new products or technologies are used on the market. It is up to management to market these results and, if a new discovery is not put to practical use, it is economically worthless. Europe is still making important technical and scientific discoveries, but it is far behind in marketing these discoveries.

An example of this situation is the computer industry: there are four times more computers installed in the U.S. than in Europe and many of these are not utilized on a full time basis. The problem is not in manufacturing the computers, but in convincing managers to use them.

Without this guideline, European industry does not or cannot make innovations. On the contrary, its attitude towards innovations is frequently sterile and damaging.

THE NEED FOR FORECASTING AT EUROPEAN LEVEL OF INFORMATION

European managers are faced with an enormous amount of information and for this reason it is difficult for them to orient themselves. Not only is there an enormous amount of information available, there are also the problems of judging the usefulness, the quality and significance of the information with regard to a specific problem. Paradoxically, what often matters today is not so much obtaining a few useful data, but getting rid of all the useless ones.

A selective approach to the technological information available is a must if technological forecasts are to be made which meet the needs of private and public organizations. Obviously a forecast on a multinational or European level of technological progress and its economic consequences over a period of twenty or thirty years would represent an important contribution toward the orientation of research and towards a definition of corporate goals. Such a forecast could perhaps bring about a better distribution of the available resources. But it is obvious, however, at a time like the present, that such an attempt would encounter considerable difficulties with regard to availability of suitable information on already existing technological inventions or on those foreseeable in world economy with repercussions on European economy.

The lack of suitable selected information on problems of research makes it difficult for public authorities and private enterprises in Europe to make appropriate decisions. There is no need to repeat the definitions and concepts on which technological forecasting is based but it is obvious that research efforts would gain enormous advantage if technical information was available to everyone. It would be possible to evaluate the general orientation of the research efforts and, therefore, the single companies or institutions would be in a position to avoid wasting resources in directions already pursued by others. To paraphrase Drucker's well known remark on the orientation of marketing research, it might be said that 90% of research must be concentrated on 10% of initiative which gives 90% of the results (and not 90% of the research on 90% of the initiatives which give 10% of the results).

The volume of the resources, as well as the managerial and organizational capabilities, involved in such a broad outlook of technological information makes a solution possible only on a general European level as investments are far beyond the possibilities of the single operators. However, it is not only a question of entrusting the problem of information to an international organization, (for example, the European Economic Community,) even if steps have already been taken in this direction. Another barrier which must be overcome is the mental hangups which have so far interferred with the exchange and general use of technological information. Unfortunately, the availability of technical information in Europe has so far all too often been considered in a narrow, petty manner.

We should at this point make a distinction between the strategic and tactical values of information data. From the point of view of tactics, i.e., when information refers to immediate, or short term dicisions in which the company is involved in a complicated game of relationships between competitors, reservedness or even secrecy seems absolutely indispensable. On a strictly operative level, maximum freedom is absolutely necessary and is in direct relationship with the exclusive availability of certain information data. But from the point of view of strategies, i.e., where future developments which involve the entire industrial society are concerned, secrecy is quite senseless, inasmuch as it comes between the operator and his very future, making the latter appear incomprehensible and, therefore, menacing. The present reaction to the advanced industrial society and to the large enterprises which are its protagonists probably stems from the fact that the development trends have always been kept so secret, which has had negative consequences, for example, on ecology and our natural surroundings.

Two things are probably indispensable if free access to the ever growing stock of technological information is to be assured to all alike: first, and obviously most important, is that public institutions and the entrepreneurs should make the results of their researches available for free circulation under the control of an agency which safeguards common interests. Of course, European industries, particularly the medium and small ones, would gain considerable advantage from a common policy of this type. But to give strength to this, common initiatives must be undertaken: for example, a research on the consequences which technological innovations have had on the structure of European industry and how the latter reacted to such innovations should be done. In fact, the need to face the problem of handling technical information cannot be stressed sufficiently as a basis to taking advantage of the benefits of technological forecasting, but also the need for integration of information on a European level would only be a first step towards more comprehensive initiatives.

THE PROBLEM OF PATENT KNOWLEDGE

Let us now examine one the the most significative examples of the effect of the scarcity of information, that of patent knowledge.

Even though much has been done, particularly in Europe, towards reducing the barriers which exist in the transfer of goods, capital and persons across frontiers, there is still a long way to go before the various national legislations become harmonized. In fact, unless something is done within the next few years, it can easily be expected that the entire system of protection of industrial property will collapse, because more and more time is needed to ascertain if an invention is really a novelty. It is interesting to note in this respect that the number of requests for patents throughout the whole world increased by 700,000 a year between 1954 and 1968 and it now takes from three to five years before a patent can be granted.

To return to the problem of access to information as a first step towards the realization of technological forecasting on a European level, there are approximately 16,000,000 patents presently existing in the world, many of which are mere duplicates of patents registered in other countries. Both public authorities and private enterprises spend enormous sums to find out if it is worthwhile requesting the patent for an invention. Many companies have departments whose sole task is investigating patents. In some countries it takes six months to find out if a patent has been registered. And this situation is destined to worsen as time goes on.

Similar, if not more serious, situations occur with regard to other types of information, for example studies, reports and articles on technical subjects. Speeding up the editing of scientific information presents two problems: the validity of the data gathered become shorter and shorter and the handling of the information becomes more difficult. And, of course, it is obviously becoming more and more problematic for the single operator — even in the case of large companies — to face this situation.

THE RELATIONSHIP BETWEEN FORECASTING AND INDUSTRIAL DEVELOPMENT

Let us add that in 1967 the total amount of public expense sustained by the EEC for research was equivalent to one fifth of the amount spent by public authorities in the U.S. for the same purpose even though the member countries of the Common Market (EEC) are spending more on R&D than in the past.

Obviously, the role played by the public authorities, and in particular by the EEC, becomes decisive in such a situation. The organization and management of research create particularly complicated problems and in fact very few countries have been able to overcome them. On the other hand, there are about 25 international organizations for scientific cooperation of which the European countries are members, but none seem to have had much success.

Some countries have made important individual efforts, particularly in the documentation area where there has been a tendency to organize one or more national centers working together. By this means it has been possible to face the problem with suitable methods and systems and reach good solutions, in some cases brilliant ones.

We are not faced here with the task of reviewing the information policies adopted by EEC members. Let us, however, mention the main institutions which carry out these activities. In France the scientific and technological information policy is entrusted to the Délégation Générale de Recherches Scientifiques et Techniques and to the Centre National de la Recherche Scientifique (C.N.R.S.) who have obtained important results, particularly with regard to the mechanization of services. The situation is more complicated in W. Germany where, apart from the Federal Government, the Laender (States) often have their own research and documentation centers. The documentation is, however, coordinated by the Institut fuer Dokumentationswesen which promotes research projects with financing during the starting phase and during initial phases.

In Holland, the Academy of Science has elaborated a project for the establishment of central organization for technical and scientific information. Similar activities are carried out in Belgium by the Royal Library and the National Council for Scientific Policy, whereas the situation in Italy is more backward, partly because of the continuing crisis in the National Research Council. However, one cannot help but wonder just how much more could have been achieved if these efforts had been coordinated and integrated at a European level.

What can EEC intervention do for industrial inventions? One of its essential functions could be the speeding up acquisition, spreading and use of technological ideas within companies. This is probably the most modern form of service which a public institution can offer to the industrial system. The EEC work in this field is worth remembering. In the "Prest" work group ("Policy of scientific and technical research"), a group specialized in information and scientific and technical documentation has been established. Its main aim is to define the nature and the tasks of the comparison systems between member states with regard to documentation and scientific information and it is already operating in the agricultural, metal, medical and patent sectors. Similar activities are carried out by the Center for Information and Documentation of Euratom (CIA) which has developed a computerized documentation system on the progress made by nuclear physics and this means has made it possible to collect and elaborate reference data for more than one million documents.

THE ROLE OF THE ENTREPRENEURS

However, without doubt a determining role should be played in Europe by the large private companies towards overcoming the gap between the methods of technological forecasting.

This is all the more true if we consider the enormous differences in the amount destined to research in Europe and in the U.S., which are less unfavorable to Europe when comparing the large industries and which would permit a more rapid balancing of the respective positions.

However, although it is obviously in the interests of European industry to obtain access to all the scientific and technical information available, certain preliminary conditions are lacking which make it difficult to cooperate and thereby develop technological forecasting on a par with the U.S.

Some of these conditions which are particularly significant are the scarcity of documentation specialists the impermeability of national legislations to technical and scientific changes as already mentioned before with regard to patents, and above all a general lack of preparation by European management.

Today it may be said that all three major components of invention — research, practical application and marketing of new techniques — are tasks assigned to management. The technological results are determined to a great extent by company policies. In U.S. companies the importance given to management is constantly increasing and U.S. managers are becoming more receptive to new ideas.

As there is no decisive historical proof of any European deficiency regarding theoretic scientific progress as such, the role of management becomes important in explaining the present technological gap between the United States and Europe. This probably depends more on the more extensive spreading of management capabilities than on the intrinsic qualities of U.S. managers compared to European ones. On the other hand, the only way to overcome this gap is to train European managers to use the research techniques in the field of technology and technological forecasting on a medium and long range basis.

A first step in this direction could be a contribution by European companies towards a collective effort with regard to technical and scientific documentation

Table 3. Countries where 110 important inventions have first been marketed since 1945.

	SECTORS	Austria	Belgium	France	W. Germany	Italy	Netherlands	Sweden	Switzerland	Great Britain	European Countries of OCSE	Japan	U.S.A.	USSR	Number of inventions considered
1	BASIC METALS	2	1	1	4	1		1	1	6	16		10	n.d.	
1.1	METAL INDUSTRY	1	1	1	3			1	1	4	12		4	4	12
1.2	NON FERROUS METALS				1	1				2	4		5	n.d.	9
2	ELECTRICAL COMMUNICATION INSTRUMENTS & SCIENTIFIC INSTRUMENTS			1	4			2	2	2	15	3	38	n.d.	
2.1	ELECTRICAL ENERGY							2	2	1	5		2	n.d.	7
2.2	COMPUTERS			1	1					2	4		10	n.d.	14
2.3	SCIENTIFIC INSTRUMENTS				3					2	5	1	10	n.d.	16
2.4	SEMICONDUCTORS									1	1		16	n.d.	17
2.5	ELECTRONIC CONSUMER GOODS											2		n.d.	2
3	PLASTIC				3	2			1	1	7		15	n.d.	17
4	OTHERS				3			1		5	9	1	11	n.d.	16
5	TOTAL ([1])	1	1	2	14	3		4	4	18	47	4	74	6	110

The total of innovations in all considered countries is not necessarily equal to the total in last column, as total of innovations includes those achieved contemporarily in two or more countries.
Source: OCDE, Ecarts technologiques. Rapport analytique, Paris 1970.

Table 4. R & D expenses of central governments in 1967

TARGETS	GERMANY 1.000 U.C.	%	BELGIUM 1.000 U.C.	%	FRANCE 1.000 U.C.	%	ITALY 1.000 U.C.	%	NETHERLANDS 1.000 U.C.	%	EEC 1.000 U.C.	%
1.Research Nuclear Dev.	235,619	19.5	19,997	22.6	363,354	20.3	98,990	34.6	22,584	11.1	740,544	20,7
2.Space	75,416	6.3	6,995	7.9	106,764	6.0	21,234	7.4	6,342	3.1	216,751	6.1
3.Defence	260,944	21.6	1,100	1.2	604,651	33.8	14,331	5.0	7,789	3.8	888,815	24.9
4.Natural Environment	22,123	1.8	2,389	2.7	11,869	0.7	2,050	0.7	3,993	1.9	42,424	1.2
5.Health	24,874	2.1	2,991	3.4	31,638	1.8	4,219	1.5	6,711	3.3	70,433	2,0
6.Human Environment	9,868	0.8	2,002	2.3	44,764	2.5	5,598	2.0	5,615	2.8	67,847	1.9
7.Agricolture	27,262	2.3	4,498	5.0	77,556	4.3	4,435	1.5	20,531	10.1	134,282	3.7
8.Industry	48,594	4,0	11,593	13.1	176,077	9.8	4,192	1.5	16.656	8.2	257,112	7.2
9.EDP & Automation	16,945	1.4	-	-	11,444	0.6	794	0.3	-	-	29,183	0,8
10.Human & social science	20,718	1.7	728	0,8	16,609	0.9	5,226	1.8	8,268	4.1	51,549	1.4
Sub-total	742,363	61.5	52,293	59.0	1,444,726	80.7	161,069	56.3	98,489	48.4	2,498,940	69.9
11.General promotion of knowledge excluding University teaching	81,360	6.7	9,153	10.3	138,726	7.8	36,253	12.7	9,796	4.8	275,288	7.7
12.General promotion of knowledge including University teaching	384,525	31.8	27,195	30.7	202,165	11.3	88,617	31.0	95,367	46.8	797,869	22.3
Amount not subdivided	-	-	-	-	4,173	0,2	-	-	-	-	4,173	0.1
Total	1,208,248	100.0	88,641	100.0	1,789,790	100.0	285,939	100.0	203,652	100.0	3,576,270	100.0
(amount for developing countries)	(-)	(-)	(607)	(0.7)	(24,508)	(1.4)	(-)	(-)	(1,670)	(0.8)	(26,785)	(0,7)

Source: The industrial policy of the Community, European Economic Community, Bruxelles 1970.

Table 5. R&D expenses by various companies as percentage of total sales

	Sales Millions $ (1)	Research Millions $	% of Research
Germany			
Siemens	2,187	155	7.0
Hoechst (3)	1,998	73	3.7
Bayer AG (3)	1,931	84	4.3
AEG	1,463	125	8.5
BASF (3)	1,820	66	3.6
Krupp	k,535	25	1.6
GHH/Aktienverein AG	1,410	36	2.5
BBC/Mannheim	337	27	8.0
Belgium			
Solvay	655	16	2.4
ACEC	316	8	3.7
UCB	125	6	4.8
France	(2)		
Rhône Poulenc (3)	1,384	55	3.9
CGE	1,118	50	4.5
Thompson HHB	1,025	26	2.6
CAFL (Entreprise, Nov. 68)	216	1	0.5
SFAC (SEF, May 69)	208	1	0.5
Jeumont-Schneider	137	3	2.4
Italy			
Montedison (3)	2,324	49	2.1
Finmeccanica (IRI) (4)	610	28	4.6
Carlo Erba (3)	124	5	4.0
Netherlands			
Shell (Chemicals, 1130)	8,941	109	1.1
Philips	2,685	205	7.6
AKO (3)	1,464	41	2.8

Table 5, continued

	Sales Millions $ (1)	Research Millions $	% of Research
Netherlands (cont.)			
Staatminen	300	11	3.7
VMF	276	6	2.0
Rijn/Schelds	112	2	2.0
United States			
General Electric	8,382	600	7.2
II (FD, October 69)	4,066	210	5.2
Westinghouse (FAZ, November 69)	3,296	300	9.1
Union Carbide (3)	2,686	83	3.1
Monsanto (3)	1,793	86	4.8
Dow (3)	1,653	84	5.1
Celanese	1,260	42	3.4
United Kingdom			
ICI	2,970	77	2.6
General Electric + EE	2,395	84	3.5

(1) Figures obtained from "chiffres olés des valeurs européennes" October 1969

(2) Parent company, without branches. Figures obtained from "Entreprise", November 1968 and from the balance sheets

(3) "Chemical Age", July 1969

(4) Figures given by the company

and processing of forecasts. A second more decisive step would be the training of managers for the utilization and management of R&D.

If, in fact, research and development are decisive production factors in an advanced industrial economy, it is also true that advanced technology means modifications to the structure of companies, the most important of which is the need to rationalize the structure and operations. The whole company must adapt itself to technological change. This means a change in the capabilities required to fulfill management positions. The European companies will have to adapt themselves to constant training of their managers in fields such as operation research, mathematical logic, systems engineering and also to understanding their external surroundings and collaborating with public authorities. Unless a conscientious step is taken in this direction, the technical-industrial gap between the United States and Europe will tend to stretch rather than shrink.

CONCLUSION

Use of technological forecasting is the stimulus of industrial development and, even though it is barely known in Europe and rarely applied, perhaps because it is not particularly spectacular and leave little room for imagination, they are of fundamental importance, at least as much as the other economic and marketing instruments which are normally used. In fact, it shows the real world in which companies operate and, if the effects are not always immediate, it opens the way to the future.

Technological forecasting operates on all levels *throughout* the industrial economy and implies the availability of such means that not even the largest enterprises can face such a challenge with their resources alone, anymore that the obsolete European national states can. Obviously, technological forecasting stresses the need for growing integration of resources, a need which is particularly felt at the level of European national integration, but which presumably will soon require a greater dimension. If the obligation to make suitable technological forecasts will necessitate the availability of even greater resources than those potentially available in Europe and, therefore, more extensive economic integrations and technical and scientific knowledge than those potentially existing in Europe, there will certainly not be many people who will regret it.

TECHNOLOGY ASSESSMENT IN GERMANY

J. SEETZEN

Director General, International Institute for the Management of Technology, Milan, Italy

H. SAUER

Professor, University of Karlsruhe, Karlsruhe, Germany

INTRODUCTION

The creative skill to coin new phrases such as "operations research", "systems analysis", "cost-benefit-analysis" or "technology assessment" is apparently not widely spread in Germany. Although the first three terms are becoming increasingly popular with planning and management experts in Germany, the latter is relatively unknown except to those who are familiar with the trends in American R & D politics. Apart from the problem of properly translating these expressions into German, one must realize that their origin in the military R & D field and in big science administration causes some semantic difficulties under present German conditions. Consequently one may say with respect to technology assessment, taken literally, that it actually does not exist in Germany.

On the other hand, the situation in America as reflected in the literature ([1], [2], [3]) shows that the approach termed technology assessment results from a strong impulse stemming from the political scene in relation to the increasing concern with environmental problems, and that the practical steps are in the beginning stage rather than fully established. The debate about measures for safeguarding the environment in Germany and Europe, however, is as vivid and some times even as hysterical as those in America. The reason that no one in our country has yet urged rapid movement towards setting up a special institutional authority for technology assessment is due predominantly to the fact that there is already a long historical background of protective measures to safeguard public interests vis-a-vis private interests. This means that many authorities, both local and federal, already normally play a role in assessing special developments which could be regarded as harmful to the general public. More recently, the Federal Ministry of the Interior has taken the initiative in coordinating questions of major interest.

It is important to bear in mind that due to historical development in Germany environmental questions are tackled in a broad manner. Environmental resources have long since become so obviously scarce that regulations to preserve nature, wildlife and water have grown rather naturally. It is also noteworthy that the former feudal system in Europe had a special interest in keeping forests, for example, in a condition where wood could be produced continually and hunting could take place enjoyably and often. Moreover the towns have been rather rigidly planned to make them safe places originally and nice places later, at least from the respective sovereign's point of view. Among other things, an exceptional forest administration, large national parks, a very respectable town planning capacity and a fair state of water resource conservation are results of these historical roots, although, today our highly industrialized society now has a completely different structure. Therefore questions of safeguarding the environment are neither simply related to technology nor, which would be worse, to industrial production technology. Technology is only one factor in environmental problems; growth of population and local congestion are usually important factors. Furthermore, technology has never played such a one-sided role in private industry as it has in the United States; on the contrary, in Europe very important technological systems such as railroads, airlines, water and electricity supply, telephone and broadcasting networks have been and are being operated by public enterprises, while other systems of the infrastructure such as roads and water ways are planned and maintained by public authorities.

Of course, future population growth and technology development implies serious problems which require thorough assessment. By no means do the authors feel that such an assessment is unnecessary. An earlier analysis of certain developments would have resulted in much less damage to our environment, and the present effort for its restoration which is now in a primordial stage could have been replaced earlier by marginally higher development costs. The thesis pursued in this article is that in Germany, the problems of technology assessment should continue to be handled in the future as a part of more comprehensive questions, namely social policy in general and technological policy in particular. Since we have had comparatively very good experiences with not setting up special authorities (e.g. Germany never had a special nuclear energy authority), it would not seem to be recommended in the case of technological assessment. Interministerial cooperation could actually do better by coordinating existing capacities in universities, research institutes and centers as well as in industry. This procedure is flexible and effective although perhaps not very spectacular. Moreover the special problems of today's Europe, where the coordination of certain regulations, e.g. in the technological field has to be treated as "foreign policy," can be managed only if the political executive is in a focal position.

The basic question which has to be raised with respect to technology assessment is whether there has to be — or eventually has to be established — a technology policy as a particular political field. If this were so, and we will explain strong reasons for it, then technology assessment would comprise, in many cases, the whole complex of problems denoted by expressions like benefit, cost and malefit (if the term may be allowed). In other words, depending on the political situation, technology assessment will play either the role of defending public interests against private ones or it will become a permanent and complex systems analysis effort by all who are interested in technological developments as an instrument to formulate an active policy for the future. The authors think the latter is pertinent for Germany.

TECHNOLOGY POLICY

Technology is commonly regarded as the special domain of industry and, of course, there is no denying the fact that industry is the main user of technology. The full picture of this way of thinking would mean that industry is exploiting technology to offer goods in the market which the buyer may or may not choose, and that industry after all would be only interested in its profit. This picture, however, has never been complete and tends to be less and less true. Technology plays a predominant role in all weapon systems where no real market exists; on the contrary, the state as buyer provides the specifications and chooses the supplying industry. The same role has been played by the European countries with regard to a great deal of the construction of the technical infrastructure, for example in supplying transportation and communication systems. This simply means that major parts of technology are directly influenced by political power, and that there has long been, consciously or not, a technology policy.

After the last world war, some technological developments were so expensive that private industry had practically no chance to recover the development costs in the market within a reasonable time. The prime examples are nuclear power reactors and communication satellites; aircraft and data processing machines have also needed considerable indirect support through military R & D. This means technology policy has begun to play a very important role in setting practical priorities through direct or indirect support for special developments.

The political initiative in Europe in such important fields as nuclear power and space technology has been mainly motivated by the fear of becoming technologically inferior to America in key areas. However, during the last few years the rational basis for these priorities has been increasingly questioned [4]. This has led to a process whereby the active political role in technological developments is being reconsidered.

First of all it was suggested [5], that the same pattern in technology policy be followed which was used in the development of the civil use of nuclear energy in Germany, i.e. to carry out rather large, publicly-funded civilian technological projects in cooperation with public research centers and private industry in a few key areas of the technical infrastructure of our country. The criteria for priority for such projects would be first, a broad rationalization of infrastructural sectors, and second, development costs exceeding the economic resources of industry; the benefit of such developments would lie predominantly in cost reductions through their public use rather than in sales profits. Following this pattern some projects, apart from civil nuclear energy which have already been proposed are: (1) fostering the use of computers by a more practice-oriented education and software development, as well as an intensified development of teledataprocessing, and (2) the development of a rapid intercity ground transport system. These developments are comparatively straight forward, generally accepted and well under way. The great advantage of such projects is the possibility to show the public benefit in monetary units which normally far exceeds the development costs so that a considerable positive stimulus to GNP growth can be expected.

The question about the rational basis for a technology policy is partially answered within the framework of economic considerations if it aims at fostering economic growth in fields where industry alone would not take the initiative. For example, education, research and road construction are similar patterns of public activities. The environmental debate clearly indicates however that economic growth, although vitally important, in some cases cannot be the final criterion.

CAPACITIES FOR TECHNOLOGY SYSTEMS ANALYSIS

Active technological development in the sixties has caused the creation of some new capacities in systems analysis in Germany. One should mention that because of the absence of military technological development during the post-war period systems engineering skill has not been developed as greatly as in the United States.

One should not certainly overlook the considerable effort which had been spent during the twenties and the thirties to rationalize industrial production and business administration. Furthermore, a clear-cut economic investment appraisal is standard procedure in industry, as well as the use of discounted cash flow calculations. A modern approach using somewhat more sophisticated modelling and programming techniques and computers is essentially a natural extension of these former developments. The growth of systems analysis capacity for the use of technology policy means not only developing the art but essentially the development of the right kind of groups; such teams have to work together for a considerable time and are the precondition of modern multidisciplinary studies in technology assessment. On the other hand techno-economic studies of a field involving multiple interest and with less visibility of the economic and other effects reveal some new problems.

The development of nuclear power reactors and the accompanying technological infrastructure for the nuclear fuel cycle, as well as the respective licensing and control systems, provided an excellent opportunity to apply a full systems approach to this new technology. Since Germany is no longer rich in cheap primary energy, nuclear energy will certainly play a key role in our technological development during the next decades. The technological assessment studies in nuclear energy carried out in the second half of the sixties concentrated mainly on the second generation of power reactors (high temperature reactors and fast breeders) in which Germany, together with other European countries, took a much more active part in basic development than in the first generation (light water reactors and gas graphite-reactors).

Nuclear energy from the point of view of technology assessment has a few unique characteristics which make it an almost ideal test field for this kind of analysis. First, the future supply of cheap and convenient types of energy is a must for all technological progress; nuclear energy could fill this need. Second, the deleterious elements in nuclear energy production are fundamentally understood and rigidly controllable. Because of the very harmful potential effects of radioactivity these have been studied to an extent unknown in other sectors of technology. The long-term (genetic) effects were especially carefully considered. Third, the possibility of uncontrolled pollution to air and water is practically excluded; the avoidance of thermal water pollution involves, as a practical matter, only capital investment. Fourth, without full public funding of the research and development phase of nuclear energy, this technology would not have come into existence. That is, future economic benefits both to the private suppliers of power stations and to the private producers of electricity were not high enough to justify the enormous development investment.

There are a few lessons to be learned from the systems analyses carried out already to assess the nuclear energy development in Germany. The Federal Ministry for Education and Science, as the main developing agency, initiated in 1964 general assessment studies to be undertaken by different interested groups independently and simultaneously with only information links between them. While the ministry did not carry out its own studies, the interested nuclear research centers and private firms developed studies on the possible future developments and their results with a remarkable amount of sophistication. The Interatom and Siemens [6, 7] corporations worked out especially interesting mathematical models of future powerplant systems and investigated a great variety of optional cases using economic and technical parameters such as potentials of mixed-costs reductions, fuel demand, etc. The nuclear research centers took a more strategic viewpoint, questioning the optimal choice of reactor types to be developed [8, 9]. One very interesting result of these studies was the political decision in 1969 to concentrate the developmental effort on only two advanced reactor types of complementary characteristics, i.e. high-temperature-reactors and sodium-cooled-fast-breeders.

The extraordinary dependence of results on very special technical parameters as materials or lay-out characteristics became extremely clear in these exercises. This allows one to make two general comments in technology assessment, that an umpire type of study without guidelines from the interested parties may lead to complete misjudgement, and secondly that such assessment should accompany the process of development in order that it can have feed-back on the development direction itself, rather than just being carried out when development is finished.

Of course the introduction of nuclear energy into the supply systems is only in its first phase in Germany, and one occasionally hears some arguments about its potential risks; however these risks (as well the potential benefits) especially with respect to pollution and to the general economy have been carefully assessed and will be continually re-evaluated [10].

The civil nuclear energy development has spearheaded technology assessment in Germany, but, during the second half of the sixties, other groups have pioneered other paths in this area. Again, the Federal Ministry for Education and Science contracted out an investigation on the side effects of the government space technology program. This study [11] was carried out by the Institute for Space Technology ("Institut für Raumfahrttechnik"), Technische Universität, Berlin, and led to some interesting results which have influenced subsequently political decisions. Two German industrial groups in aerospace technology, Messerschmidt-Bölkow-Blohm and Dornier, have set-up systems-engineering groups with perhaps the highest potential in Germany. The groups are increasingly involved in assessment studies of traffic systems and urban developments. At present, a set of assessment studies is being completed by these and other firms to investigate the various possibilities for future transport systems in Germany. Again, the interested groups on the basis of a governmental initiative worked independently in parallel; one might expect prototype development as a result from these studies in the next few years. Since such traffic systems must meet certain environmental considerations (noise and pollution), one sees again the necessity for a continuing and comprehensive assessment process.

There are two other institutes which have been set up during the last few years which represent rather unique approaches to a broader basis of technology assessment. The first is the Institute for Technical Development Guidelines ("Institut für technische Entwicklungslinien ITE") founded in 1969 by industry and now located in Hamburg. This institute will provide industry with a better common basis for studies concerned with both the economic factors in key technological developments, e.g. in transport systems and new materials, and the general economic influences of technology on economy. The other institute was set up in 1970 by the Max-Planck-Gesellschaft for "Studying the Preconditions of Human Existence in the Modern World", and is located in Starnberg near München. The two directors of this institute, C.F. v. Weizsäcker and J. Habermas, have set up an interdisciplinary scientific group to study the conditional scientific and sociological basis of our technology oriented present and future.

Since the general question of priorities in technology policy underlies the problem of emphasizing special developments and their assessment, the Federal Ministry for Education and Science has called for a set of studies dealing with methods of rationalizing the process of setting objectives. Along with private contract research institutes, the Study Group for Systems Research ("Studiengruppe für Systemforschung") at Heidelberg and the Berlin Center for Futurology ("Zentrum für Zukunftsforschung") supplied comprehensive studies on that question which will soon be published. Another group of industrialists and researchers have also provided proposals.

REFERENCE SYSTEM FOR TECHNOLOGY ASSESSMENT

The problem of priorities in technology policy using technology assessment as the rationalizing analytical basis, as well as the discussion on environmental questions, has shown that there is a basic difficulty in the assessment process, namely, that of the reference system. Many societal problems are commonly mapped into economic parameters because of the ease with which they can be made to fit established grounds of econometry. For instance, questions of development are usually expressed in GNP-terms in spite of the many difficulties inherent in this procedure. Thus far, science policy especially has been regarded as primarily influencing economic growth; however, more and more doubts are being expressed whether such a standard is adequate [12]. In educational policy, such doubts are even stronger because a simply economic yardstick in education contradicts its partly idealistic basis. Finally, one can suspect that even technology is not correctly regarded if it is judged solely on economical grounds. The emergence of environmental problems in technology shows clearly that economic considerations alone are insufficient [13].

The problem is that the monetary units of econometry denote imaginative values in exchange processes which are in principle inappropriate to describe economic or social states. The increasingly used expression "quality of life", especially in the environmental debate, commonly denotes an unquantified state. The authors believe that, unless some more attempt is made to quantify social states in accordance with sociological trends [14], technology assessment rests on extremely muddy ground.

In order to give some hints in which direction quantification of social states could be sought in order to get a better base of technology policy, one may formulate the following equations.

The state of a society at a given time t can be denoted by its population P (t) and the existing stock of real goods S (t). For this rough and only basic presentation we will not regard the demographic or economic microstructure of P and S however the average lifetime $\bar{L}$ in the population represents an important characteristic. Changes of the social conditions are achieved by primarily two forces, namely manpower a and energy e. The extent to which manpower and energy can lead to certain changes in a given time fraction dt is obviously dependent on the type and amount of technical instruments S^T used. On the other hand, the type of technical instruments and their efficient use is dependent on the state of technical information I^T in a society. We observe that technical instruments S^T are a subset of S : $S^T \subset S$. In case no positive force is applied to preserve the existence of P and S they decay after a certain decay function $-P_d(t)$ and $-S_d(t)$.

The potential for positive change through manpower is apparently strictly limited by the given population multiplied by the time considered

$$a_{(t)} = P(t)dt \tag{1}$$

Now evidently the manpower potential is split into fractions of use:

$$a = a^R + a^L + a^P + a^S + a^I$$

$$\text{with} \quad a^R + a^L = a_i; \quad a^P + a^S + a^I = a_s \tag{2}$$

The indices in (2) denote rest R, leisure L, care for population P, material goods S and information I; further individually available i and socially available s. This means that the total potential of the available manpower is used for rest, leisure, working to increase food production to preserve the population, safeguarding and care for children and health, further work to preserve and extend the stock of real goods, and work to preserve and extend the stock of information. The first two fractions are used for individual purposes, the other three for social ones.

The same splitting can be done with the energy used at a given time, although there is no simple limit of e(t):

$$e_{(t)} = e^R + e^L + e^P + e^S + e^I$$

$$\text{with } e_i = e^R + e^L; \quad e_s = e^P + e^S + e^I \tag{3}$$

Technology has the general function to make the use of manpower and energy more effective for whatever changes may occur. This could mean that a larger stock of technological goods either changes the relationship of the manpower fractions or leads only to an exponential growth of the technological instruments or the stock of real goods, commonly regarded as wealth or production factors or it could lead to a strong increase in available information without much increase in material wealth. Where after all manpower, reinforced by technical instruments, energy and technical information is allocated, is only a question of the motivational structure M in a society, which can be directly denoted by the fractions of a

$$M = a^R;\ a^L;\ a^P;\ a^S;\ a^I \tag{4}$$

As motivation to a certain behaviour is directly influenced by information, however, one can state that M is a function of the information in society

$$M = f(I) \tag{5}$$

The state of the stock of real goods changes with time according to the following equation

$$S_{(t)} = S_o + \int_0^t \left\{ \dot{S}_g\ (a^S,\ e^S,\ S^T,\ I^T) - \dot{S}_d \right\} dt \tag{6}$$

When $\dot{S}_g$ is the first derivative of the gain in real goods at the time t and $\dot{S}_d$ that of the decay function mentioned. These functions have to be empirically found.

$S_{(t)}$ can obviously be measured in monetary units or for more refined considerations in numbers of certain goods like houses, cars, telephones, etc., in which case a^S, e^S, S^T, I^T have to be subdivided again.

Now we are able to state a vector Q of easily quantifiable indicators for a social state denoting first order effects on the quality of life. One might criticize that the subjective consideration of the quality of life depends on many other conditions which cannot be expressed in simple terms, however the potential for a high quality of life is obviously dependent on rather simple factual conditions as lifetime, space, food, material goods, time and energy individually available. The average figures of the vector Q have to be supplemented by distribution parameters to show the average relevance to the individual of a society

$$Q = \begin{cases} \bar{L} \\ A/P \\ F/P \\ S/P \\ a_i/a_s \\ e_i/e_s \end{cases} \tag{7}$$

with A the area for living for the population, and F the available food (calories). It is also easy to follow empirically the cange of these indicators as a function of the motivational structure

$$Q(t) = f[M(t)] \tag{8}$$

and get in this way a basis of social modelling. The only problem in developing such a quantitative basis for a more comprehensive assessment of the effects of technological developments are the incomplete statistics of today.

A quantified reference system for technology assessment based on a descriptor of the quality of life shows clearly the very complex role technology plays under different sociological conditions. Technological instruments are not only the main amplifying factor in the production process, and thus technological progress contributes not only predominantly to economic growth, it also has direct first order effects on the quality of life. Of course, a social modelling of the kind sketched before should not contradict economic models dealing with the effects of technological progress [15], however one can show that economic processes are comprised.

A more comprehensive reference system for technology assessment can lead to a balanced consideration of the beneficial and "maleficial" effects of technological and other developments as urbanisation and population growth. This seems to be absolutely necessary when nowadays the deleterious second order effects become of greater concern. A one-sided assessemnt of only these second order effects might lead to an unnecessarily hostile view of technological solutions, and thereby lead to a motivational structure, which is less capable of tackling future problems.

REFERENCES

1. A Study of Technology Assessment, Committee on Science and Astronautics, U.S. House of Representatives, July 1969.

2. Technology, Process of Assessment and Choice, Committee on Science and Astronautics, U.S. House of Representatives, July 1969.

3. KIEFER, D.M., Technology Assessment, *Chemical & Engineering News, 48*, 42-56, Oct. 1970.

4. KRAUCH, H., Prioritäten für die Forschungspolitik, Carl Hanser Verlag, München 1970.

5. HÄFELE, W., SEETZEN, J., Prioritäten der Grossforschung, in "das 198. Jahrzehnt" Christian Wegner Verlag, Hamburg 1969.

6. HARDE, R., MEMMERT, G., Modellrechnungen über Aussichten und Konsequenzen der Verwendung von Kernenergie zur Elektrizitätserzeugung. *Atomwirtschaft 11*, 164, 1966.

7. MAERKL, H., Untersuchungen über langfristige optimale Reaktorstrategien unter Berücksichtigung des begrenzten Umfanges der Uranreserven, *Nukleonik 10*(4), 1967.

8. GRÜMM, H., et al., Kernbrennstoffbedarf und Kosten verschiedener Reaktortypen in Deutschland, Kernforschungszentrum Karlsruhe, KFK 366, 466, 1965, 1966.

9. WAGEMANN, K., Beitrag zu Systemuntersuchungen über die langfristigen wirtschaftlichen Einsatzmöglichkeiten verschiedener Kraftwerkstypen unter besonderer Berücksichtigung des Thoriumhochtemperaturreaktors, Kernforschungsanlage Jülich, Jül-590-RG, 1969.

10. LINDACKERS, K.H., Hrsg., Kernenergie, Risiko und Nutzen Deutsche Verlagsanstalt, Stuttgart 1970.

11. KOELLE, H.H., et al., Uber den Nutzen der Weltraumforschung aus deutscher Sicht BMBW-FB W 71 01 bis 04, 1971.

12. Report of the Secretary-General's Ad hoc Group on new Concepts of Science Policy, Science, Growth and Society, A New Perspective, OECD, Paris 1971.

13. The Goals of R & D in the 1970's, OECD, Paris 1971, (unpublished).

14. SHELDON, E.B., MOORE, W.E., Monitoring Social Change in American Society in "Indicators of Social Change" Russell Sage Foundation, New York, 1968.

15. FRITSCH, B., Forschung, Innovation und technischer Fortschritt als Gegenstand oekonomischer Analyse, Antrittsvorlesung ETH, Zürich, 1967.

A METHOD FOR PLANNING AND ASSESSING TECHNOLOGY AGAINST RELEVANT NATIONAL GOALS IN DEVELOPING COUNTRIES

MARVIN J. CETRON and LAWRENCE W. CONNOR

President, and Secretary-Treasurer
Forecasting International, Ltd., Fairfax, Va., U.S.A.

1. BACKGROUND

Proper cultivation of natural resources especially the rarest of all assets — capable people, need for environmental control, and socio-economic difficulties — makes it mandatory to seek and develop new and more effective methods for performing long-range planning and decision-making.

Further, there is a critical global need to better understand future challenges which will be brought about by current trends in technology. The effective projection of such scientific trends is essential in order that the efficient allocation of available resources can be made as early as possible. Obvious planning deficiencies in current environmental, economic, and social activities can be observed in every country in the world. In this respect and regardless of what industrial or post-industrial status some countries think they may have achieved — the fact remains that all countries are confronted with the same general set of problems. Consequentially, there is the equally urgent requirement for an integrated system of technological forecasting and planning of national goals, followed by an evaluation of the country's scientific/technological capability to meet such goals within the time frame established and, finally, an assessment of the impact that technological changes will have on the country's — as well as the world's — total environment.

Traditionally, most countries are preoccupied with the day-to-day task of conducting government business, and consider technological planning only as it applies to their specific realms of interest. Furthermore, government executives tend to react to technological developments and changes in their operating environment only when they occur. This is not enough; indications are that countries must evaluate their position and effect on the total environment. In turn, this provides the government executive the opportunity to act — not just react — in determining the country's future.

The role of technology in everyday life has come under increased scrutiny not only by government and industrial leaders, but by all sectors of society. A considerable portion of this increased awareness is due to the realization that technological exploitation can result in undesirable long-term effects. Focus at the national level in several countries is currently directed at assessing technology, but the capability required for performing a comprehensive assessment of present and emerging technologies has not been developed.

Technology assessment is an attempt to establish an early warning system to detect, control, and direct technological changes and developments so as to maximize the public good while minimizing the public risks. It is a relatively new approach to allocating scientific resources, establishing technological priorities, and seeking relevant alternatives to current technology. From this it is easy — but misleading — to conclude that solutions to problems involving technology are technological. Such an "ostrich-like" attitude ignores the human factors which may have initially created the problems, or which may inhibit their solutions.

Therefore, what remains to be accomplished is the effective integration of national goals and their related priorities, the country's scientific/technological capability to carry the national goals to a successful conclusion within a reasonable time frame and budget, the sensitivity of national priorities to specific technology assessments, and a methodology to provide the realtionship between specific technologies and areas of national interest. Such an overview of the synergistic effects of various technologies is necessary in order to provide a basis for informed decision-making at the national policy-making level.

2. QUANTITATIVE PLANNING TECHNIQUES

2.1 *Introduction*

One of the major problems besetting any organization, particularly if it involves research and development, is knowing how best to allocate available resources. There are never sufficient resources to fund every possible project envisioned by the research staff, or by management. How, then, should projects be selected, and funds allocated? The first step in this process is to try to assess what the future may hold in terms of the country's needs, competition, potential markets, international regulations and technologies [1].

There are many possible approaches to forecasting, and many techniques are available for application to specific problems [2,3,4]. However, in many cases, the technique is constrained by the quantity and quality of data available. For strictly technological forecasts, where there is sufficient historical data, a technique such as trend extrapolation provides the most accurate information in the short run (0-3 years), within specified confidence limits. Unfortunately, the other areas mentioned are not so quantitative, and must therefore rely on subjective judgment. It is important to remember that although technological forecasts predict what *could* happen, many other factors, such as forecasts of complimentary and competitive technologies and the socio-political and economic environments must be considered before a scenario, depicting what is likely to happen, can be prepared.

Once the broad area forecasts used in forming scenarios have been prepared, they can then be incorporated into a model for broad area project selection, based upon the country's five-year plan and capabilities. This model will provide a method of ordering and scaling the broad technological areas so that those of greatest potential for the developing country may be selected. However, before this can take place, the country must define and quantify its national goals (objectives) and then subdivide these goals into subtargets and rank and weigh them against each other. A model for resource allocation, using these goals and capabilities, is then developed. Specific technologies can then be assessed in an explicit and consistent manner. Also such a methodology assures the reproducibility of the national goal information for further analysis and use in subsequent years.

An important benefit to be gained from the use of such techniques is that of visibility. This is two-fold, for it provides the decision-maker with the explicit information on which he can base his decisions, and it also provides visibility to management for the needs and wants of the future — which can greatly assist in the requesting of funds for R & D.

Finally, output from the study effort should provide partial answers and greater insight to such management problems as:"Toward what ends should the developing country mobilize and allocate its resources?" "What research programs should be pursued?" "How should new programs be initiated?" "What impact should research have upon national, social, economic, and cultural goals?"

2.2 *Approach*

The matrix approach is "quantitative", but it is by no means "exact." Thus, the method used is based on subjective values [5], and therefore, instead of using numbers and measurements, the terms "indicant" and "assessment" are applied respectively. The assessment requires intuitive and creative judgment throughout the exercise. The utility of the output is therefore highly dependent upon the level of understanding of science, its interrelationship and impact on the socio-economic development in the developing country. The value of the methodology is in its facility to organize information in a formal and consistent manner, structuring thinking for reproducible results, and making visible weighted indicants to assist in the decision process. Thus, the result of the participants exercise — although it is still a preliminary exercise — is to *serve* as a starting point for using the method, and to *aid* in a more consistent formulation of science policy.

The method consists of a specific technique for addressing the attention of analysts and planners systematically to the important "things to think about" in considering, somewhat quantitatively, the effects (good or bad) specific technologies or sciences could have on the five-year targets and the developing country's environment. In short, *the quantitative management approach outlined in this exercise does not automatically produce good decisions. In fact, it only provides information upon which better decisions can be made.*

A specific method for carrying out the assessment effort has been developed and is summarized subsequently in this paper. Actually, the assessment process draws from a whole genera of fairly well defined and accepted analytical methodologies that are generally referred to as "matrix techniques." The specific matrix techniques proposed for use in this exercise, while not previously combined in this particular manner, have, nevertheless, been widely used for forecasting, project selection, and resource allocation purposes [6,7]. For example, cross-impact analysis — as a basic matrix technique — has been extensively used for developing scenarios and measuring the impact of change, whereas, cross-support analysis has been used in project selection and resource allocation techniques [8,9,10,11,12,13]. A technique similar to cross-support analysis is used by the U.S. Navy in developing their technological forecasts. Double and triple matrix techniques have been widely employed by both industry and government as weighting and scoring methods [14,15,16,17].

Finally, the method has been placed literally on the "firing line" in various countries of Southeast Asia, Africa, and Latin America. Although it has never come away from such tests totally unscathed, the resulting changes only tended to improve and refine the basic method. In its present format the method provides an extremely flexible, appropriate and powerful management *tool* for planning and assessing national policies and goals toward the ever improvement of the "quality of life."

2.3 *Cross-Support Matrix of National Development Goals*

Step 1. The initial step in this phase of the analysis is to list the major goals of the National Development Plan and rank them according to their relative importance. Such ranking might be:

Agriculture
Education
Public Works
Health
Industry
Tourism
Mining

Next, defining the total weighting to be 100 points, the major goals are then "weighted" or "assessed" based on their relative importance. This is based on the criticality of the goal and the difference between where the goal stands now and where it is expected to be in the future. This distribution of weights to major goals must be performed individually by the experts of Working Group No. 1 with minor differences being negotiated to arrive at a consensus. For example, one ranking and weighting of major goals might be:

Agriculture	25
Education	20
Public Works	15
Health	15
Industry	18
Tourism	4
Mining	3

It must be remembered that the weights assessed to the major goals were based on the subjective judgment of the participants concerning what effort is needed for each major goal during the next five to ten years. As such, they do not reflect technological costs, capital investments, the amount of budget allocated for each goal in previous years, political factors, etc.

Continuing the analysis, subtargets under each major goal are then identified, ranked, and weighted in the same manner. The total of the weights of the subtargets under each major goal is constrained to the weight of the major goal, so that the total weight of all subtargets will equal 100 points. For example, the following subtargets were identified, ranked, and weighted under the major goal, Agriculture:

Major goals and subtargets	Major goal weight	Subtarget weights
AGRICULTURE	25	
Food Crops		5
Industrial Crops		10
Animal Husbandry		5
Fisheries		2
Forests		3

Step 2. The major goals and associated subtargets together with their respective weights are transcribed on the left-hand margin of the matrix (see Annex II attached). The subtarget weights obtained from Step 1 are also to be listed under the column headed OW (Original Weight). To facilitate subsequent manipulation of the subtarget-to-subtarget matrix, it is suggested that these be called *primary subtargets*.

Step 3. The major goals and subtargets — *including Exports and Environment* — and their associated weights are also transcribed across the top of the matrix. It is suggested that these be called complementary subtargets to differentiate them from the primary subtargets in Step 2. This completes the basic structure of the goal-to-goal Cross-Support Matrix. The following is an example of the Agriculture portion of such a matrix showing major goal, primary and complementary subtargets, and their weights.

Step 4. Having developed the goal-to-goal Cross-Support Matrix, the next step in the methodology is to determine the degree to which subtargets (goals) influence

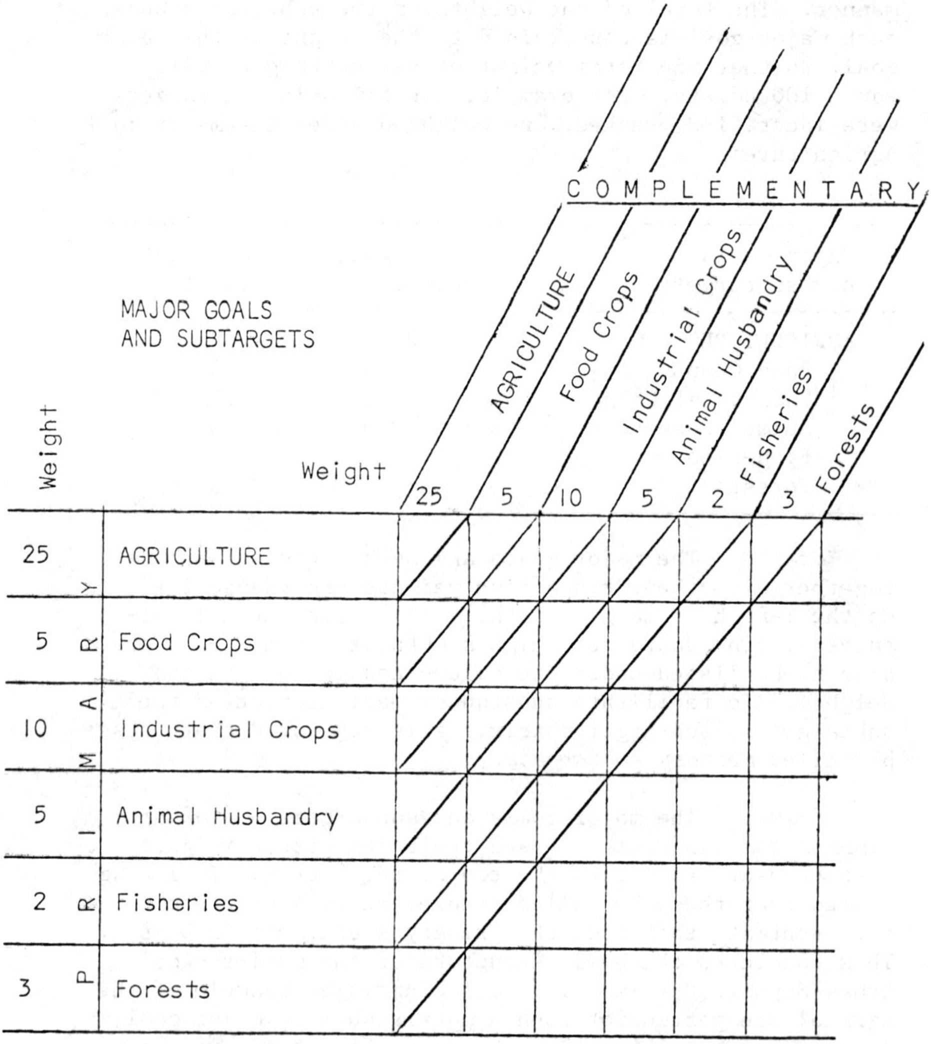
COMPLEMENTARY
MAJOR GOALS
AND SUBTARGETS
Weight
Weight
AGRICULTURE
Food Crops
Industrial Crops
Animal Husbandry
Fisheries
Forests
25
5
10
5
2
3
PRIMARY
25 AGRICULTURE
5 Food Crops
10 Industrial Crops
5 Animal Husbandry
2 Fisheries
3 Forests

each other, that is, how does the achievement of one subtarget (goal) contribute to the achievement of all other subtargets (goals). This is done by considering the effect one subtarget (goal) has on each of the others, separately, and then summing these separate contributions to obtain a total. The "contribution" that one subtarget (goal) may have on another can be either beneficial (positive) or detrimental (negative). The interrelationships between goals (subtargets) can, perhaps, best be determined from a cross-support analysis as shown in the following exhibit [18,19,20].

Exhibit: Goals Cross-Support Matrix

goals / goals	g_1	g_2	g_3	g_4	–	–	–	g_n
g_1		α_{12}	α_{13}	α_{14}				α_{1n}
g_2	α_{21}		α_{23}	α_{24}				α_{2n}
g_3	α_{31}	α_{32}						α_{3n}
g_4								
⋮								
'								
g_n	α_n							

g_1 = goal 1

g_{12} = magnitude of contribution of goal 1 to goal 2

In looking at this cross-support matrix for goals (subtargets), we have, mathematically, the value (G) for each of n goals (g), which can be given by:

$$G_i = \sum_{j=1}^{n} a_{ij} \qquad i \neq j$$

Where the a_{ij} are the assigned contribution (H, M, L or 0). This means that for each goal (g_i) there is an associated number (G_i) which indicates the degree to which that goal affects the others. This adds creditability to the method used in the ranking of goals, their order of priority, and their subsequent weighting.

The degree to which there is a contribution can be assessed by associating numeric values to an adjectival scale, as follows:

Major contribution	(H) = 8
Considerable contribution	(M) = 4
Some contribution	(L) = 2
Negligible contribution	(0) = blank

If the effects are negative, then the same scale can be used, but with negative signs (i.e., -H, -M, -L, -0).

In this cross-support matrix, and in all the other matrices of impact or support, the scale chosen is logarithmic as opposed to arithmetic [21-29]. Experience has shown that several physiological processes or functions tend to be logarithmic in nature such as the thought process, reaction of the pupil of the eye to intensity of light source, and our hearing as measured in decibles.

In each case, detailed descriptions of the numerical values are to be provided, and care taken to assure that the scales, and their arithmetic manipulation, conform to laws of scaling [30-33].

The adjectival assessment equivalents are then placed in the upper half of the bisected matrix cells (see Annex II attached). The impact, determined in the manner described and inserted into the upper half of the matrix cell, is then multiplied by the weight of the complementary subtarget in order to obtain the weighted contribution which that primary subtarget has on the complementary subtarget being considered in that particular intersecting matrix cell. The resulting indicant (contributing weight) is then inserted in the lower half of the matrix cell. For consistency and accuracy, it is essential that the procedure be iterated in precisely the same manner for obtaining each subtarget-to-subtarget contributing weight in the Cross-Support Matrix for major goals. For example: as noted in the following illustration the contributing weight that Food Crops has on Animal Husbandry is 12. This was obtained by multiplying the adjectival assessment equivalent (4) — found in the upper half of the intersecting matrix cell for Food Crops and Animal Husbandry — by the weight (5) assigned to the complementary subtarget Animal Husbandry.

Major goal and weight	(Primary) Subtarget and weight	(Complementary) Subtarget and weight	Intersecting matrix cell
Agriculture	Food Crops	Animal Husbandry	
25	5	5	4/20

Step 5. When the lower part of all applicable cells are complete, sum horizontally and place the total in the column headed Total Cross-Support (TCS).

Step 6. Based on the Total Cross-Support score appearing in the TCS column of Annex II, additional points are to be generated and listed in the Additional Points Added (APA) column using the following procedure:

(1) First — after totals have been obtained and placed in each applicable cell of the column headed Total Cross-Support (TCS) as in Step 5 above, then sum all entries down the TCS column and enter this grand total at the bottom of the column.

(2) Second — divide each of the individual totals in the TCS column by the grand total obtained in (1) above. Enter the result in the appropriate cell of the Additional Points Added (APA) column.

(3) Third — sum all entries in the APA column and enter at the bottom of the column. It all operations have been performed accurately, this sum should equal 100 points — the base standard on which this phase of the exercise was initiated.

NOTE: The above, of course, is nothing more than a "normalizing" operation wherein all of the individual contributing indicants (scores) making up the Total Cross-Support score are brought back to the base standard of unity (100 points). Essentially, this is identical to saying "Total Cross-Support score is to (:) 100 points as (::) the individual cross-support score is to (:)X". Summing all the resulting ratios (X's) will give 100 points.

Step 7. After listing the additional points in the APA column, then add them to the indicant appearing in the Original Weight (OW) column in order to obtain the New Weight (NWo) for each subtarget. For example:

Primary sub-target	Weight	Total Cross-Support Weight (TCS)	Additional points added (APA)	Original weight (OW)	New weight (NWo)
Food Crops	5	106	2	7	3.5

Step 8. By adding the additional points (APA column) and Original Weights (OW column) a New Weight (NWo) is obtained which, when summed vertically for all subtargets (or goals), provides a total of 200 — or twice the original base standard and constraining number. Therefore, it is necessary to normalize each New Weight so that the sum of the New Weights of all subtargets (or goals) equals 100 by dividing each New Weight by 2. This Normalized New Weight is listed in the column headed NNW of Annex II. However, they are not to be listed as final Normalized New Weights

(NNW) until each has been thoroughly discussed, considered, and any differences between the individual participants reconciled.

Marked differences will be observed between the Original Weights (OW) and Normalized New Weights (NNW) of many of the subtargets (and goals). This is to be expected since the Original Weights were based solely on subjective judgment of the criticality of the goal and its distance from the target, and the subtargets (and goals) were considered as mutually exclusive events, whereas, the New Normalized Weight is actually the result of the total contribution, both positive and negative, when all subtargets are considered collectively.

Again — selecting, ordering, scaling, and weighting (assessing) in the manner described above, assures the *reproducibility* of the goals and subtargets for use in cross-support analysis in this, and subsequent years.

Annex II, which incidentally is a hypothetical illustration and not an actual example displays the total output of the goal-to-goal Cross-Support Matrix. Also, this output serves as the input for Annex IV, Sciences-to-Goals Relevance Matrix which is discussed later.

2.4 *Profile of Science and Technology Relevance*

After completing the Cross-Support Matrice for Goals and obtaining Normalized New Weights for goals, the next step in the process is to integrate Goals and Science — by use of a relevance chart. For the purpose of this exercise the term relevance refers to the relative impact that each of the fields of science and technology exerts on the selected national socio-economic development goals.

Step 9. The relevance chart will be developed on a matrix base similar to that used in Annex IV.

First, list the "fields" of science and technology without weights in the left-hand margin of the matrix (refer to list in Annex IV).

Second, list the complementary subtargets described in Steps 1 and 3 along with their Normalized New Weights

(NNW) as shown in Annex IV. These "subtargets" are the same subdivisions of the socio-economic development goals selected in the national development plan and listed in the Annex II matrix.

Step 10. The above completes the basic structure of the Sciences-to-Goals Relevance Matrix. The next step in the process is to evaluate and assess the contributing impact, *as determined by the participants*, that each field of science has on each subtarget using an adjectival scale based on the following logarithmic numerical values:

8 = Progress cannot be made toward this subtarget without research, development or specialized education in this field. It is expected that this rating will be used very seldom.

4 = This field is *highly and directly* relevant to the objective. For example, soil and science is directly relevant to reafforestation. Many of the subtargets derived from the socio-economic development plan of the developing country will have one, two, but seldom more, such disciplines.

2 = There is a high probability that a breakthrough in this field would have a valuable payoff relevant to the subtarget, but more as a supporting function. For example: solid state physics is relevant to a metallurgical industry.

1 = This field is of some but remote relevance to the subtarget. Examples: organic chemistry underlies medical, agricultural and forest industries. There is no need to record "one" ratings.

0 = The value of this discipline to the subtarget is practically nil. Example: development of new materials cannot progress without associated chemistry and solid state physics. There is no need to record "zero" ratings.

The assessment factors are then transcribed in the upper half of the intersecting matrix cell. This impact (assessment factor), determined and inserted in the upper half of the bisected matrix cell, is then multiplied by the New Normalized Weight of the subtarget in order to obtain the weighted contribution of that science on the particular subtarget being considered in that cell. The resulting indicant is listed in the lower part of the applicable matrix cell. For example, one discipline of Agricultural Sciences — Agronomy — and its weighted contribution to the Agricultural subtarget Food Crops is shown as follows:

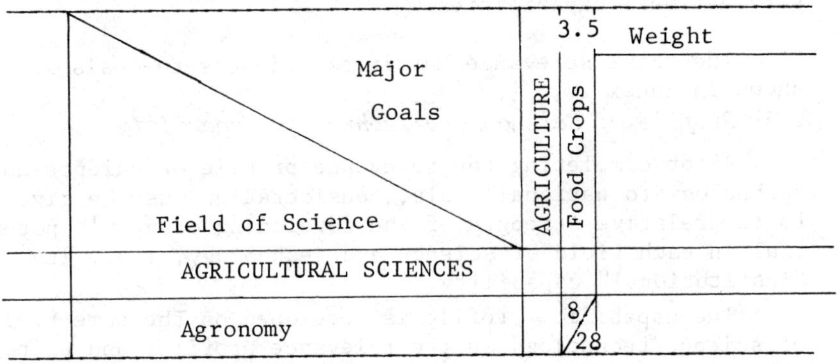

Step 11. The lower parts of the cells are then summed horizontally and the totals inserted under the column *Total Relevance Index* (TRI).

Step 12. Total Relevance Index numbers having a value of 300 or more are to be normalized to a value of 10, whereas, those having a value of 29 or less are to be normalized (set) to a 0 value according to the following conversion rationale:

TRI Number		Normalized to
300 or more	=	10
270 - 299	=	9
240 - 269	=	8
210 - 239	=	7
180 - 209	=	6
150 - 179	=	5
120 - 149	=	4
90 - 119	=	3
60 - 89	=	2
30 - 59	=	1
29 or less	=	0

The above normalized values are then transcribed in the Normalized Relevance Profile (NRP) column of the Matrix, Annex IV. This is the final relevance profile of science and technology and it should indicate the various fields of relevant science and technology needing consideration merely to reach the selected national goals in the socio-economic development plan of the developing country. It is not a measure of either existing or necessary national capabilities.

The total Relevance Matrix of Sciences-to-Goals is shown in Annex IV.

2.5 *Profile of Science and Technology Capability*

After completing the relevance profile of science and technology to national goals, consideration must be given to the relative strength of the developing country's potential in each field of science and technology, i.e., its "institutional" capability.

The capability profile is developed on the same fields of science/technology as the relevance profile, and using the same basic Annex IV matrix. Development should take into account the available physical facilities, the amount of resources spent in the area, the number of scientists actively engaged in R & D, the availability of higher education and training, the standards of excellence and other measures of both quantity and quality.

Step 13. Using the matrix of Annex IV, rate (assess) and assign a relative weight to each discipline — from 1 to 10 — according to the following adjectival/numerical scale:

10 = This field is large and well developed with a standard of quality in research, development and education comparable to any in the world.

9 = The field is strong and well funded and is capable of being recognized of world stature in a few years.

8 = A good quality specialized *programme* with a number of scientists working together in this field exists, providing research, service and education.

7 = PhD level scientists, with good contacts with the scientific community of the world, are doing research in this field. They are capable of providing strong support for programmes in related disciplines.

6 = There are a number of *researchers* in this field who publish occasional articles in scientific journals as well as providing service and education.

5 = Scientists and engineers in this field provide valuable *functions* to industry and government, in addition to teaching.

4 = *University graduates* are capable of doing advanced work in this field without further education and have received some specialized education.

3 = The educational system provides a *sound basis* for work in this field.

2 = There may be active scientists and engineers in this field, but they are not related to the research and educational institutions.

1 = Education relevant to this field is very weak.

0 = No activity relevant to this field is now planned.

2.6 *Gap Profile and Profile of Institutional Needs in Science and Technology for Each Developing Nation*

These profiles should represent a discipline-by-discipline analysis of the national needs which cannot be met by existing science and technological institutions. This profile of institutional needs cannot be obtained by a simple comparison of the "relevance" and "capability" profiles, but it should indicate important fields (education, research, public services) where scientific and technological institutions should be created or strengthened.

Step 14. Subtract ratings in the "capability" profile column from corresponding ratings — in the "relevance" profile column. This then is the "GAP" profile rating which is by and large identical with the profile of institutional needs in science and technology.

Step 15. Prepare a descriptive statement on the 3 or 4 disciplines with the highest ratings in the above relevance, capability and GAP profiles. Include information and statistics on the national potential (capability) and requirements (relevance) in the scientific and technological fields most important to the developing country.

Step 16. Identify institutions (and other means) by which the "GAP" profile may be reduced in amplitude, following the pattern of Annex 1: The *Profile of Institutional Needs in Science and Technology.*

Where possible, quantised cost estimates should be provided for strengthening existing institutions or building necessary new ones in the 1971 to 1980 decade. Also indicate possible sources of support: (a) national, (b) bilateral, and (c) international.

Step 17. Indicate disciplines of science and technology in which the nation might be interested in either (a) accepting students or scientists (high capability profile) or (b) using the resources and facilities in neighboring countries (high gap profile).

2.7 *Cross-Support Matrix for Science*

Step 18. The final step — the Science-to-Science Cross-Support Matrix (see Annex III) is generated in the same manner as the Cross-Support Matrix of Goals. The major fields of sciences (see Annex IV) are to be ranked and weighted according to their qualitative contribution to the improvement of the "quality of life" in the developing country. In this phase of the exercise it is assumed that each discipline has the same weight as it contributed to the national goals. The task here is not only to determine how science contributes to the goals, but also how each science contributes to each other and then to the goals. Inasmuch as some sciences may not contribute much directly to the goals, i.e., mathematics and engineering sciences, the amount they contribute to other sciences — and thus to goals — is an entirely different story.

In preparing a Cross-Support Matrix of Science-to-Science the internal portion of the matrix will probably be the same for all countries. However, the original weights for the Science-to-Science matrix will be the "Gap" Profile figures obtained in Step 14 of the Science-to-Goals matrix and treated in the following manner:

First, aggregate and sum the "Gap" Profile figures for all the various disciplines under each major science.

Second, obtain a simple average by dividing the total "Gap" Profile figure by the number of individual disciplines represented under each major science.

Third, normalize the average obtained above back to the base standard 100 for each major science.

Fourth, list these normalized averages as the original weights of the major sciences arrayed across the top of the Science-to-Science Cross-Support Matrix, Annex III.

Fifth, modulate the cross-support number, located in the upper halt of each cell of the matrix, by the original weight in order to determine the contribution of that science to the weighted science. Insert the contributing weight in the lower half of the matrix cell.

Step 19. Sum all rows of the matrix horizontally and insert in the column headed Total Cross-Support (TCS). Continue the same procedures outlined in the Goal-to-Goal Cross-Support Matrix in developing Additional Points Added (APA), New Weights (NWo), and Normalized New Weights (NNW) columns. See Annex III for a complete Science-to-Science Cross-Support Matrix.

3. CONCLUSION

It must be *reemphasized* that these matrix techniques do not and should not produce decisions — they will produce information that should help the decision-maker to make better decisions. Tools such as these will make visible the judgment of various individuals which, in turn, will precipitate dialogue in addition to bringing to bear the expertise of many individuals. This particular technique can be especially powerful if the decision makers themselves have participated in the exercise since they are now not only aware of the assumptions and shortcomings, but also of the strength. Finally, since they have been intimately acquainted with the analysis process they are in a better position to see that the program is effectively implemented.

ACKNOWLEDGEMENT

A few words of gratitude and appreciation are due the many dedicated individuals who contributed so unselfishly of their time and effort, and without whose help the methodology outlined in this article would not have been possible.

In particular are the authors indebted to Dr. Y. de Hemptine and Mr. Marc Chapdelaine, Director and Senior Professional Staff Member respectively, Science Policy Division, United Nations Educational, Scientific and Cultural Organization (UNESCO), Paris, France, for their excellent assistance and counsel in adapting the technique to the requirements of developing nations.

Our gratitude is extended as well to the many participants in various countries for their cooperation in helping us to modify and refine this management planning tool, and bringing it to its present stage — ready for implementation.

ANNEX I. STAGE II OF THE UNACAST WORLD PLAN OF ACTION FOR THE APPLICATION OF SCIENCE AND TECHNOLOGY TO DEVELOPMENT

Sector 1: Needs of developing countries for scientific and technological institutions [34]

Plan for annotated institutional summaries

Programme 1	National Policy-making Bodies in Science and Technology
Component 1	Central science policy-making body
Project 1	Structural build-up (or reform) of national body for overall science policy-planning (such as Ministries of Science and Technology, National Science Policy Councils, etc.)
Project 2	Structural build-up (or reform) of Centre for survey of national scientific and technological potential (STP) (including national R & D statistics)
Project 3	Other (specify)
Component 2	R & D promoting and coordinating bodies (including Academies of Sciences and Engineering)
Project 1	Structural build-up (or reform) of national body for planning and coordination of R & D in basic sciences
Project 2	Structural build-up (or reform) of national body for planning and coordination of agricultural research
Project 3	Structural build-up (or reform) of national body for planning and coordination of industrial research
Project 4	Structural build-up (or reform) of national body for planning and coordination of medical research
Project 5	Structural build-up (or reform) of national body for atomic energy research
Project 6	Structural build-up (or reform) of national body for social sciences research
Project 7	Other (specify)

Programme 2	Higher Education Institutions in Science and Technology
Component 1	University Faculties of basic sciences
Project 1	Creation (or development) of university faculties of basic sciences
Project 2	Creation (or development) of university departments in the classical disciplines of the basic sciences (such as Mathematics, Physics, Chemistry, Biology, Geology, etc.)
Project 3	Creation (or development) of interdisciplinary university departments in the basic sciences (such as Physical-Chemistry, Biophysics, Biochemistry, Geophysics, etc.)
Project 4	Other (specify)
Component 2	"Third level" polytechnic schools and schools of engineering (including university faculties of Applied Sciences or Technology)
Project 1	Creation (or development) of "third level" polytechnic schools or schools of engineering
Project 2	Creation (or development) of specialized departments in "third level" polytechnic schools or schools of engineering (such as Mechanics, Fluid-Dynamics, Electricity, Nuclear Energy etc.)
Project 3	Other (specify)
Component 3	"Third level" schools of Agriculture (including university faculties of Agronomy)
Project 1	Creation (or development) of "third level" schools of Agriculture
Project 2	Creation (or development) of specialized departments in "third level" schools of Agriculture (such as veterinary medicine, forestry, fisheries, etc.)
Project 3	Other (specify)
Component 4	"Third level" schools of Medicine (including university faculties of Medicine)

Project 1 Creation (or development) of "third level" schools of Medicine

Project 2 Creation (or development) of specialized departments in "third level" schools of Medicine (such as ophthalmology, parasitology, obstetrics, surgery, endocrinology, etc.)

Project 3 Other (specify)

Programme 3 Technician Training Institutions*

Component 1 Technological training institutions

Project 1 Creation (or development) of general "second level" technician training institutions.

Project 2 Creation (or development) of specialized "second level" technician training institutions

Project 3 Other (specify)

Component 2 Agricultural training institutions

Project 1 Creation (or development) of general "second level" agricultural technician training institutions

Project 2 Creation (or development) of specialized "second level" agricultural technician training institutions

Project 3 Other (specify)

Component 3 Medical training institutions

Project 1 Creation (or dwvelopment) of general "second level" training institutions for auxiliary medical personnel

Project 2 Creation (or development) of specialized

---

*"Technicians" here include persons who have received specialized vocational or technical training in any branch of knowledge or technology during one or two years' training beyond completed education at the "second level" or three to four years' training beyond the first cycle of secondary education.

	"second level" training institutions for auxiliary medical personnel
Project 3	Other (specify)
Programme 4	Research and Experimental Development Institutions (R & D performers)
Component 1	Fundamental Research Institutes
Project 1	Creation (or development) of University-linked (discipline-oriented) Institutes for Fundamental Research
Project 2	Creation (or development) of autonomous (mission-oriented) Fundamental Research Institutes (such as geophysical research institutes, space research centres, oceanographic research institutes, ecological and natural resources research institutes, etc.)
Project 3	Other (specify)
Component 2	Applied Research and Experimental Development Institutes
Project 1	Creation (or development) of Agricultural Research Institutes (including research institutes for veterinary medicine, forestry, fisheries, food processing and preservation, etc.)
Project 2	Creation (or development) of Industrial or Technological Research Institutes (including research institutes for transport, energy, telecommunications, etc.)
Project 3	Creation (or development) of Medical Research Institutes
Project 4	Creation (or development) of Atomic Energy Research Institutes
Project 5	Creation (or development) of multipurpose (specify) R & D Institutes
Project 6	Other (specify)

Programme 5	Scientific and Technological Public Services (STS)
Component 1	Natural resources and environment services*
Project 1	Creation (or development) of Topographic and Scientific Mapping Services
Project 2	Creation (or development) of Hydrology or Water Supply Services
Project 3	Creation (or development) of Geological Survey and Mining Services
Project 4	Creation (or development) of Energy Services (such as water, fuel, wind, solar, geothermal, atomic, tidal etc.)
Project 5	Creation (or development) of Soil Science Services
Project 6	Creation (or development) of Services for integrated land-systems surveys (including land resources analysis and conservation)
Project 7	Creation (or development) of Marine and Fisheries Services
Project 8	Creation (or development) of Meteorology Services
Project 9	Creation (or development) of Astronomical and Time Services
Project 10	Creation (or development) of Seismological and Vulcanological Services
Project 11	Other (specify)
Component 2	Information and Documentation Services
Project 1	Creation (or development) of National Scientific and Technological Information Centres
Project 2	Creation (or development) of Industrial Information Centres
Project 3	Creation (or development) of data banks and information processing services
Project 4	Creation (or development) of National Science and Technology Library
Project 5	Creation (or development) of scientific journals, bulletins and abstracts

*In French: "Services des ressources et du milieu naturels".

Project 6	Creation (or development) of Patent Offices and Clearing Houses for available patents
Project 7	Creation (or development) of Scientific Conference Centres
Project 8	Other (specify)
Component 3	Museums and Collections
Project 1	Creation (or development) of science and/or technology museums
Project 2	Creation (or development) of travelling science and/or technology exhibits
Project 3	Creation (or development) of scientific collections (anthropological, archaeological, geological, botanical, zoological, entomological, etc.)
Project 4	Other (specify)
Component 4	Standards, Norms and Instrumentation
Project 1	Creation (or development) of National Bureau of Standards
Project 2	Creation (or development) of National Metrology Services
Project 3	Creation (or development) of National Centres for the Manufacture, loan and maintenance of scientific instruments
Project 4	Other (specify)
Component 5	Extension and Innovation Services
Project 1	Creation (or development) of Agriculture Services (such as land-use development, rural engineering, agricultural extension work, forestry and livestock services, etc.)
Project 2	Creation (or development) of technological transfer and innovation services
Project 3	Creation (or development) of University polyclinics
Project 4	Other (specify)

Goals and weights (columns): AGRICULTURE (25): Food Crops 5, Industrial Crops 10, Animal Husbandry 5, Fisheries 2, Forests 3; EDUCATION (20): Primary 2, Secondary 6, Higher Education 5, Vocational 6, Adult Literacy 1; PUBLIC WORKS (15): Road Building 5, Dams 3, Utilities & Telecomms. 1, Mapping 2, Housing 4; HEALTH (15): Private Medical Care 2, Hospitals 6, Preventive 4, Eradication 3; INDUSTRY (18): Food Industry 5, Mechan./Elect. 2, Textile 3, Metallurgy 4, Cottage Industry 2, Chem., Fertzl. 2; TOURISM 4; MINING 3; ENVIRONMENT (10); EXPORTS (20)

Goal	Matrix entries	(TCS) Total Cross-Support	(APA) Additional Points Added	(OW) Original Weight	(NWo) New Weight	(NNW) Normalized New Weight
AGRICULTURE						
Food Crops	4/20, -2/-6, 2/4, 2/8, 8/48, 8/32, 2/4, 4/16, -2/-20	106	2.0	5	7.0	3.5
Industrial Crops	4/20, -2/-6, 2/6, 2/8, 2/8, 2/6, 8/40, 2/4, 4/12, 2/4, 2/8, 8/160	270	5.2	10	15.2	7.6
Animal Husbandry	2/10, 2/20, 2/4, 2/6, 2/4, 4/24, 4/16, 4/12, 4/20, 4/12, 2/4, 2/8, -2/-20, 2/40	150	2.9	5	7.9	4.0
Fisheries	2/4, 2/12, 2/8, 4/20, 2/8, -2/-20	32	0.6	2	2.6	1.3
Forests	2/10, 2/20, 2/10, 4/12, 2/2, 4/16, -2/-6, 8/16, 2/8, -2/-20	68	1.3	3	4.3	2.1
EDUCATION						
Primary	2/20, 2/10, 2/4, 2/6, 8/48, 4/24, 2/2, 2/8, 4/24, 4/16, 4/20, 4/8, 2/6, 2/8, 2/4, 2/4, 4/16	228	4.4	2	6.4	3.2
Secondary	4/20, 4/40, 4/20, 4/8, 4/12, 8/40, 2/12, 2/10, 4/12, 4/4, 4/8, 2/8, 2/12, 4/16, 4/20, 4/8, 4/12, 4/16, 4/8, 4/16, 4/12	314	6.0	6	12.0	6.0
Higher Education	8/40, 8/80, 8/40, 8/16, 8/24, 4/24, 2/12, 2/2, 8/40, 8/24, 8/8, 8/16, 8/32, 2/4, 8/48, 4/16, 4/12, 8/40, 8/16, 8/24, 8/32, 8/16, 8/16, 2/8, 2/6	596	11.4	5	16.4	8.2
Vocational	8/40, 8/80, 8/40, 8/16, 8/24, 2/2, 4/20, 4/12, 4/4, 4/8, 4/16, 8/48, 4/16, 2/6, 4/20, 4/8, 4/12, 4/16, 4/8, 4/8, 2/8, 2/6	418	8.0	6	14.0	7.0
Adult Literacy	8/40, 8/80, 4/20, 4/8, 4/12, 2/10, 2/6, 2/2, 2/4, 2/8, 2/4, 2/12, 2/8, 2/6, 2/10, 2/4, 2/6, 2/8, 2/4, 2/4, 2/8	264	5.1	1	6.1	3.1
PUBLIC WORKS						
Road Building	2/10, 4/40, 2/10, 2/4, 4/12, 2/6, 2/2, 4/20, 4/12, 2/8, 4/8, 2/4, 8/32, 4/80	248	4.8	5	9.8	4.9
Dams	4/40, 2/10, 2/4, 4/12, 2/10, 2/2, 4/20, 4/12, 4/16, 2/4, 4/8, 4/16, 8/24, 8/160	338	6.5	3	9.5	4.7
Utilities and Telecommunications	2/4, 2/12, 2/10, 2/12, 2/2, 2/6, 4/16, 8/160	222	4.5	1	5.3	2.7
Mapping	2/10, 4/40, 2/10, 2/4, 4/12, 4/20, 4/12, 2/4, 2/8, 2/6, 2/8, 8/24	158	3.0	2	5.0	2.5
Housing	4/8, 4/24, 4/20, 4/24, 4/24, 2/8, 4/20, 4/8, 4/12, 4/16, 4/8, 4/8, 8/32	212	4.1	4	8.1	4.0
HEALTH						
Private Medical Care	4/16, 2/8	24	0.5	2	2.5	1.3
Hospitals	2/10, 4/24, 4/16, 2/8	58	1.1	6	7.1	3.5
Preventive	2/10, 2/20, 2/10, 2/4, 2/6, 4/8, 4/24, 2/10, 4/24, 2/2, 2/10, 2/6, 2/2, 2/4, 2/8, 4/20, 2/4, 2/6, 2/8, 2/4, 2/4, 2/8, 8/24	226	4.3	4	8.3	4.2
Eradication	4/20, 4/40, 8/40, 4/8, 4/12, 4/8, 4/24, 2/10, 4/24, 4/4, 2/10, 2/6, 4/8, 2/4, 8/48, 8/32, 8/32	350	6.7	3	9.7	4.8
INDUSTRY						
Food Industry	2/20, 4/20, 2/12, 4/16, 4/8, 2/8, -2/-20, 2/40	104	2.0	5	7.0	3.5
Mechan./Elect.	4/40, 4/20, 4/8, 8/24, 4/12, 8/8, 4/16, 2/4, 4/24, 8/40, 4/12, 8/32, 8/16, 8/16, 4/12, -2/-20	264	5.1	2	7.1	3.6
Textile	2/4, 2/12, 2/10, 2/12, 2/8, 2/8, 4/16, -2/-20, 2/40	90	1.7	3	4.7	2.3
Metallurgy	2/6, 4/4, 8/16, -2/-20	6	0	4	4.0	2.0
Cottage Industry	2/10, 4/12, 2/2, 8/32, 2/8, 4/12, 2/40	116	2.2	2	4.2	2.1
Chems., Fertzls.	4/20, 4/40, 4/20, 2/8, 4/8, 8/48, 2/8, 4/12, 4/20, 4/12, -2/-20	196	3.8	2	5.8	2.9
TOURISM	4/80	80	1.5	4	5.5	2.8
MINING	8/32, 4/8, 2/40	80	1.5	3	4.5	2.2
		5218	100.0	100	200.0	100.0

ANNEX II. CROSS-SUPPORT MATRIX OF GOALS

ANNEX III: Science-to-Science Cross-Support Matrix

Sciences	9.2 Agricultural Sci.	3.8 Engineering Sci.	5.7 Medical Sciences	16.2 Enviromental Sci.	16.9 Biological Sci.	20.9 Chemical Sci.	7.3 Physical Sci.	13.4 Mathematical Sci.	6.6 Behav. & Soc. Sci.	(TCS) Total Cross-Support	(APA) Additional Pts. Added	(OW) Original Weight	(NWo) New Weight	(NNW) Norml. New Weight
Agricultural Sci.			2 11.4		2 33.8					45.2	2.0	9.2	11.2	5.6
Engineering Sci.	8 73.6		4 22.8	8 129.6	4 67.6	8 167.2	8 58.4	2 26.8		546.0	25.1	3.8	28.9	14.5
Medical Sciences	4 36.8				8 135.2				2 13.2	185.2	8.6	5.7	14.3	7.1
Enviromental Sci.	4 36.8	4 15.2					2 14.6		4 26.4	93.0	4.3	16.2	20.5	10.3
Biological Sci.	8 73.6		8 45.6			2 41.8			2 13.2	174.2	8.0	16.9	24.9	12.4
Chemical Sci.	8 73.6	2 7.6	2 11.4	2 32.4	4 67.6		2 14.6		2 13.2	220.4	10.1	20.9	31.0	15.5
Physical Sci.		8 30.4	2 11.4	4 64.8		4 83.6				190.2	8.7	7.3	16.0	8.0
Mathematical Sci.	4 36.8	8 30.4	8 45.6	8 129.6	8 135.2	8 167.2	8 58.4			603.2	27.7	13.4	41.1	20.6
Behav. & Soc. Sci.			2 11.4	2 32.4	2 33.8	2 41.8				119.4	5.5	6.6	12.1	6.0
										2176.8	100.0	100.0	200.0	100.0

Goals / **Disciplines**

Disciplines	AGRICULTURE: Food Crops 3.5	Industrial Crops 7.6	Animal Husbandry 4.0	Fisheries 1.3	Forests 2.1	EDUCATION: Primary 3.2	Secondary 6.0	Higher Education 8.2	Vocational 7.0	Adult Literacy 3.1
Mathematical Sciences										
Algebra						2 / 6.4	2 / 12.0	4 / 32.8	2 / 14.0	
Analysis & Funct. Analysis							2 / 12.0	8 / 65.6	4 / 28.0	
Applied Mathematics								4 / 32.8	4 / 28.0	
Computer Science			2 / 8.0	2 / 2.6	2 / 4.2			4 / 32.8	4 / 28.0	
Geometry						2 / 6.4	4 / 24.0	4 / 32.8	8 / 56.0	2 / 6.2
Logic						2 / 6.4	2 / 12.0	4 / 32.8	2 / 14.0	
Number Theory							2 / 12.0	4 / 32.8		
Numerical Analysis							2 / 12.0	4 / 32.8	8 / 56.0	
Operations Research	2 / 7.0	4 / 30.4	2 / 8.0	2 / 2.6	4 / 8.4				4 / 28.0	
Probability	2 / 7.0	2 / 15.2	2 / 8.0	2 / 2.6	2 / 4.2		2 / 12.0	4 / 32.8	2 / 14.0	
Statistics	4 / 14.0	4 / 30.4	4 / 16.0	4 / 5.2	4 / 8.4		2 / 12.0	4 / 32.8	4 / 28.0	
Topology							2 / 12.0	4 / 32.8		
Behavioral & Social Sciences										
Sensory Mechanisms										
Neural & Percep. Process						2 / 6.4	2 / 12.0	2 / 16.4	2 / 14.0	2 / 6.2
Motor Mechanisms						4 / 12.8				
Psychological Traits						4 / 12.8	4 / 24.0	4 / 32.8	4 / 28.0	4 / 12.4
Selection Method and Performance Criteria						2 / 6.4	2 / 12.0	2 / 16.4	2 / 14.0	2 / 6.2
Learning & Training						8 / 25.6	8 / 48.0	8 / 65.6	8 / 56.0	8 / 24.8
Individual Effectiveness						8 / 25.6	8 / 48.0	8 / 65.6	8 / 56.0	8 / 24.8
Group Effectiveness						8 / 25.6	8 / 48.0	8 / 65.6	8 / 56.0	8 / 24.8
Engineering Psychology								8 / 65.6	8 / 56.0	
Psychology of Individual Performance						8 / 25.6	8 / 48.0	8 / 65.6	8 / 56.0	8 / 24.8
Social Psychology and Sociology						4 / 12.8	4 / 24.0	4 / 32.8	4 / 28.0	4 / 12.4
Social Sci. Interaction						4 / 12.8	4 / 24.0	4 / 32.8	4 / 28.0	4 / 12.4
Policy Planning Studies										
Intercultural Effectiveness									2 / 14.0	2 / 6.2

Disciplines	PUBLIC WORKS: Road Building 4.9	Dams 4.7	Utilities and Telecommunications 2.7	Mapping 2.5	Housing 4.0	HEALTH: Private Medical Care 1.3	Hospitals 3.5	Preventive 4.2	Eradication 4.8
Mathematical Sciences									
Algebra									
Analysis & Funct. Analysis	4 / 19.6	4 / 18.8	4 / 10.8	2 / 5.0	4 / 16.0				
Applied Mathematics	4 / 19.6	4 / 18.8	4 / 10.8	2 / 5.0	4 / 16.0				
Computer Science	4 / 19.6	4 / 18.8	4 / 10.8	4 / 10.0	2 / 8.0		4 / 14.0		
Geometry	2 / 9.8	2 / 9.4	2 / 5.4	4 / 10.0	4 / 16.0				
Logic									
Number Theory									
Numerical Analysis	4 / 19.6	4 / 18.8	4 / 10.8	4 / 10.0	4 / 16.0				
Operations Research	4 / 19.6	4 / 18.8	4 / 10.8	4 / 10.0	4 / 18.0		4 / 14.0		2 / 9.6
Probability	2 / 9.8	4 / 18.8	4 / 10.8	4 / 10.0	4 / 16.0	2 / 2.6	4 / 14.0		
Statistics	2 / 9.8	4 / 18.8	4 / 10.8	4 / 10.0	4 / 16.0	4 / 5.2	4 / 14.0	4 / 8.4	4 / 19.2
Topology	2 / 9.8	2 / 9.4	2 / 5.4	2 / 5.0					
Behavioral & Social Sciences									
Sensory Mechanisms							2 / 7.0		
Neural & Percep. Process								2 / 8.4	
Motor Mechanisms									
Psychological Traits									
Selection Method and Performance Criteria									
Learning & Training								4 / 16.8	
Individual Effectiveness									
Group Effectiveness									
Engineering Psychology	2 / 9.8	2 / 9.4	2 / 5.4	2 / 5.0	2 / 8.0				
Psychology of Individual Performance									
Social Psychology and Sociology									
Social Sci. Interaction									
Policy Planning Studies	4 / 19.6	4 / 18.8	2 / 5.4	2 / 5.0	2 / 8.0		4 / 14.0	2 / 8.4	2 / 9.6
Intercultural Effectiveness									

Disciplines	INDUSTRY: Food Industry 3.5	Mechan./Elect. 3.6	Textile, Wool 2.3	Metallurgy 2.0	Cottage Industry 2.1	Chems, Fertz'lz 2.9	TOURISM 2.8	MINING 2.2
Mathematical Sciences								
Algebra								
Analysis & Funct. Analysis	2 / 7.0	2 / 7.2		2 / 4.0		2 / 5.8		2 / 4.4
Applied Mathematics	2 / 7.0	2 / 7.2	2 / 4.6	2 / 4.0	2 / 4.2	2 / 5.8		2 / 4.4
Computer Science	2 / 7.0						2 / 5.6	4 / 8.8
Geometry		4 / 14.4			4 / 8.4			4 / 8.8
Logic								
Number Theory								
Numerical Analysis	2 / 7.0	2 / 7.2	2 / 4.6	2 / 4.0		2 / 5.8		2 / 4.4
Operations Research	4 / 14.0	4 / 14.8	4 / 9.2	4 / 8.0	2 / 4.2	4 / 11.6	4 / 11.2	4 / 8.8
Probability	2 / 7.0	2 / 7.2	2 / 4.6	2 / 4.0		2 / 5.8	4 / 11.2	2 / 4.4
Statistics	4 / 14.0	4 / 14.8	4 / 9.2	4 / 8.0	4 / 8.4	4 / 11.6	4 / 11.2	4 / 8.8
Topology								
Behavioral & Social Sciences								
Sensory Mechanisms								
Neural & Percep. Process								
Motor Mechanisms								
Psychological Traits								
Selection Method and Performance Criteria								
Learning & Training								
Individual Effectiveness								
Group Effectiveness								
Engineering Psychology								
Psychology of Individual Performance								
Social Psychology and Sociology								
Social Sci. Interaction								
Policy Planning Studies	4 / 14.0	4 / 14.8	4 / 9.2	4 / 8.0	4 / 8.4	4 / 11.6	4 / 11.2	4 / 8.8
Intercultural Effectiveness								

Disciplines	(TRI) Total Relevance Index	(NRP) Norml. Relevance Profile	Capability Profile	"GAP" Profile	Normalized "GAP" Score	
Mathematical Sciences						
Algebra	65.2	2	2	0		
Analysis & Funct. Analysis	204.2	6	3	3		
Applied Mathematics	168.2	5	2	3		
Computer Science	178.2	5	4	1		
Geometry	207.6	6	4	2		
Logic	65.2	2	1	1		
Number Theory	44.8	1	0	1		
Numerical Analysis	182.6	6	3	3		
Operations Research	264.0	8	4	4		
Probability	222.0	7	4	3		
Statistics	345.0	10	7	3		
Topology	74.4	2	0	2		
				26	13.4	26÷12=2.16
Behavioral & Social Sciences						
Sensory Mechanisms	7.0	0	0	0		
Neural & Percep. Process	63.4	2	1	1		
Motor Mechanisms	12.8	0	0	0		
Psychological Traits	110.0	3	1	2		
Selection Method and Performance Criteria	55.0	1	0	1		
Learning & Training	233.8	7	5	2		
Individual Effectiveness	217.0	7	5	2		
Group Effectiveness	217.0	7	6	1		
Engineering Psychology	159.2	5	3	2		
Psychology of Individual Performance	217.0	7	5	2		
Social Psychology and Sociology	108.5	3	2	1		
Social Sci. Interaction	108.5	3	3	0		
Policy Planning Studies	174.8	5	4	1		
Intercultural Effectiveness	20.0	0	0	0		
				15	6.6	15÷14=1.07

ANNEX IV. CROSS-SUPPORT MATRIX OF SCIENCE-TO-GOALS

REFERENCES

1. CETRON, M.J. *et al., Technological Resource Management: Quantitative Methods*, Cambridge, Massachusetts: MIT Press ,1970.

2. CETRON, M.J. and RALPH, C.A., *Industrial Applications of Technological Forecasting*, New York: John Wiley and Sons, Inc., 1971.

3. AYRES, R.U., *Technological Forecasting and Long-Range Planning*, New York: McGraw-Hill Book Co., 1969.

4. CETRON, M.J., *Technological Forecasting: A Practical Approach*, New York: Gordon and Breach ,1969.

5. The basic idea in the Bayesian approach is that it is correct to quantify feelings about uncertainty in terms of subjectively assessed numerical probabilities.

6. CETRON, *et al., op. cit.*

7. CETRON, M. and GOLDHAR, J., *The Science of Managing Organized Technology*, New York: Gordon and Breach, 1971.

8. GORDON, T.J. and HAYWARD, H., "Initial Experiments with the Cross-Impact Matrix Method of Forecasting," *Futures, 1,* No. 2, December, 1968.

9. GORDON, T.J., "Cross-Impact Matrices," *Futures, 1,* No. 6, December, 1968.

10. DALBY, J., "Practical Refinements to the Cross Impact Matrix Technique of Technological Forecasting," CETRON and RALPH, *Industrial Applications of Technological Forecasting: Its Use in R & D Management,* New York, John Wiley & Sons, Inc., 1971.

11. RALPH, C.A., "Decision Impact Analysis Applied to the Fishing Industry," *Plessey Radar, Ltd., Report,* November, 1968.

12. RALPH, C.A., "The Beginnings of Cross-Support Analysis," CETRON and RALPH, *Industrial Applications of Technological Forecasting*, John Wiley & Sons, Inc., New York, May, 1971.

13. RALPH, C.A., "Resource Allocation Using Cross-Support Analysis," presented at the 27th JSGOMRAM, Institute of Defense Analysis, Arlington, Virginia, April 20-22, 1971.

14. CETRON, M., "Quest," *et al., op. cit.*, Chapter 2.

15. National Academy of Sciences — National Academy of Engineering National Materials Advisory Board: "Quantitative Techniques for Research Program Planning in Structural Mechanics," Report of the NMAB, NMAB-259, 1969.

16. DOBROV, G.M., "Scientific Technological FOrecasting in the USSR," *Long-Range Planning*, April, 1971.

17. RALPH, 27th JSGOMRAM, *op. cit.*

18. RALPH, C.A., "Decisions Impact Analysis Applied to the Fishing Industry," *op. cit.*

19. RALPH, C.A., "The Beginning of Cross-Support Analysis," *op. cit.*

20. RALPH, C.A., "Resource Allocation Using Cross-Support Analysis," *op. cit.*

21. BORING, E.G., *et al.* (eds.), *Foundations of Psychology*, New York, John Wiley & Sons, Inc., Chapter 11, 1948.

22. BEEBE-CENTER, J.G., "Standards for Use of the Gust Scale," *J. Psychology, 28,* 411-419, 1949.

23. BORING, E.G., *Sensation and Perception in the History of Experimental Psychology*, Appleton Century-Crofts, 1942.

24. CROCKER, F.C. and HENDERSON, L.F., "Analysis and Classification of Odors," *American Perfum.*, 325-327, 1927.

25. GUILLFORD, J.P., *Psychometric Methods*, New York, McGraw-Hill, 1936.

26. GULLIKSEN, H., "Paired Comparisons and the Logic of Measurement," *Psychol. Review*, *53*, 199-213, 1946.

27. STEVENS, S.S. and VOLKMANN, J., "The Relation of Pitch to Frequency: A Revised Scale," *Amer. J. Psychol.*, *53*, 329-353, 1940.

28. DEAN, B.V. and ROEPCKE, L.A., "Cost-Effectiveness in R & D Organizational Resource Allocation," *IEEE Transactions on Engineering Management*, EM-16, No. 4, 222-242, November, 1969.

29. RALPH, "Decision Impact Analysis," *op. cit.*

30. HALL, A.D., *A Methodology for Systems Engineering*, Princeton, N.J., Van Nostrand Co., p. 244, 1962.

31. STEVENS, S.S., *Handbook of Experimental Psychology*, New York, John Wiley & Sons, Inc., 1951.

32. MOTTLEY, C.M. and NEWTON, R.O., "The Selection of Projects for Industrial Research," *Operations Research*, *7*, November-December, 1969.

33. CHURCHMAN, C.W., ACKOFF, R.L. and ARNOFF, E.L., *Introduction to Operations Research*, New York, John Wiley & Sons, Inc., 1957.

34. See: UNACAST "Plan frame of the World Plan of Action for the Application of Science and Technology to Development" in the Sixth Report of the UN Advisory Committee on the application of science and technology to development, New York, March 1969, Annex VII and Appendix (Document E/46E/4611).

INTERDISCIPLINARY CONSIDERATIONS IN SPONSORING TECHNOLOGY ASSESSMENTS*

JOSEPH F. COATES

National Science Foundation, 1800 "G" Street, N.W., Washington, D.C. 20550 (202/632-7420), U.S.A.

"Technology Assessment" is a term applied to a class of studies intended to illuminate and hence influence public policy decision-making in the Executive and Legislative Branches of the Government. Technology assessment may be defined as the systematic study of the secondary and higher-order effects or impacts on society of a present or proposed application of technology. In its fullest form, it should anticipate, forecast, and evaluate the impacts of the technological intrusion on all sectors of society, including the economic, environmental, social, political, legal, and institutional components. Needless to say, only a few full-fledged technology assessments have been conducted, although there is a long history of partial assessments, particularly those limited to impacts on economy and more recently to impacts on the environment. The importance of emphasizing the definition's second-order impacts is threefold: 1) in the long pull, these may be the most significant impacts; 2) undesirable second or higher-order impacts often are unnecessary and may be planned or regulated out of the system; and 3) first-order impacts,

---

*Presented at the Engineering Foundation Conference on Technology Assessment, August 30 - September 3, 1971, Andover, New Hampshire.

that is, the intended technological application, usually are subject to extensive study in the planning stage. A pair of central problems is the discovery and integration of higher-order impacts, at the planning stage. For example, in building a bridge, dredging a canal, laying a cable TV system, laying an underground pipeline, introducing enzyme detergents, or electrifying a railroad, the first-order effects may be defined as those effects which are planned-for or intended as the primary goal of the effort. A test of that primacy is that effects are explicitly planned-for, accommodated and costed-out in the initial plan. Technology assessment, therefore, focusses on the question of what else may happen, be it good or bad.

By and large, technology assessment is not an experimental enterprise. It is an analytical, synthetic, systematic examination and exploration of consequences. From time-to-time, experimental investigations may have to be undertaken as part of the assessment. Also, one of the consequences of the assessment may be experimental investigations to clarify the relative importance of potential, incidental effects.

It should also be recognized that no assessment is likely to be final or completely definitive. With a complex application application of technology, it may be looked at as part of an iterative process, each iteration more clearly illuminating and defining the impacts and the related implicit or explicit policy options.

Central to the understanding of the how, the who, and the why of technology assessments is seeing it as a policy tool; that is, as an imput to a decision-making process. They are of use to government agencies in the Executive Branch in the planning and execution of projects; in reviewing enterprises over which they have regulatory responsibility; in generating new regulations; or in defining new surveillance needs to monitor environmental or social change.

Assessments should be a primary input into legislative deliberation, since legislation sets into motion, directs, or inhibits major sectors of the economy. The importance of assessments to The Congress is perhaps best illustrated by the fact that the term itself has received its most extensive advocacy under that rubric, through Emilio Daddario, former Chairman of the U.S. House of Representatives' Subcommittee on Science, Research and Development of the Committee on Science and Astronautics.

The role and impact of technology assessments on courts is somewhat more confused inasmuch as many public interest groups see the courts themselves as a technology assessment mechanism. It is quite clear, however, that the courts are in great need of more even-handed evaluations in their deliberations involving divergent advocacies over technological interventions.

A fourth client for technology assessments are those public interest groups or citizens' groups which are frequently concerned and often effectively involved in hearings and litigations over various technological interventions. They seem to be particularly effective on questions of environmental impact. In general, these groups operate in an advocacy role on an extremely limited knowledge base which, undoubtedly works in some cases to the short and long term disadvantage of society. Should they have available (or be in a position to generate) more adequate technology assessments, the public-policy decision process would be substantially enhanced.

A fifth group in need of better, more comprehensive assessments is the commercial and industrial sectors of society which often seem to get off to a myopic and, hence, bad start in planning major enterprises. The broadening of the range of considerations entering into their initial planning, e.g. in power plant siting, would eliminate much of the concern and acrimonious litigation which seems inevitably to follow plans based on factors limited to first-order effects. Technology assessment also may be significant to industry in its long-range planning by anticipating future legislative or regulatory constraints. Industry is more likely to function effectively and efficiently in an environment of competition and investment risks than in one of regulatory uncertainty.

DEFINING TECHNOLOGY

The definition of technology, which may be the subject of a technology assessment, is by no means clear and straightforward. There is near universal agreement that technology should include physical artifacts (such as power plants, telecommunication systems, airplanes) as well as those physical activities or actions which modify or alter the environment or the landscape (such as building construction, dam building, canal digging), and the siting of physical installations (such as oil refineries, power plants, highways, and housing developments). There are those, however, who would include social invention or social technology within the range of topics subject to assessments. Arguments for such parity may be made on several grounds: 1) that they are as much technology as is physical technology, since they involve the ordered, systematic application of knowledge and techniques; and 2) they are a major source of secondary impacts in our society. The following illustrate these points: 1) the introduction of pay-as-you-go income tax proposed by Beardsley Ruml in World War II; 2) the establishment of the land-grant colleges under the Morill Act; 3) the invention and application of farm subsidies; and 4) the county agent system. All had secondary effects, directly traceable to these social inventions. In the same way that a minor alteration in a physical technology may have a profound influence on side effects, a minor alteration in a social technology can have a profound secondary influence. The modification in the selection of draftees which virtually guarantees half the youth in the susceptible age bracket are removed from risk, seems to have drawn the nerve on the anti-war protest. An interesting example of a longer-term secondary effect is that of social security on the living arrangements of the elderly. Because of the loss of benefits, should the elderly on social security remarry, large numbers in that age cohort population see fit to live together without benefit of clergy or, in the jargon of the young, to "shack-up." The pasture lands of Florida seem to rival the campuses and the big cities in the frequency with which this occurs, albeit in a different age group. Perhaps a more profoundly significant secondary effect of a social technology has to do with the alleged flight of whites from the city. Some have argued that a far more profound influence on the movement of whites to the suburbs and concentration of Blacks in the core cities results from the FHA-VA policies immediately after World War II. Veterans were given great economic incentives to home ownership through mortgage legislation. However, for many those great benefits had hidden in them a virtual demand that one abandon the city, and forego renewed housing and preferentially choose new suburban housing. Twenty-five years later we can see that the whites were at least as much drawn as driven to the suburbs from the city.

As these tools of social invention become more numerous and complex in managing our society, the need to assess their consequences beforehand becomes more urgent.

WHY TECHNOLOGY ASSESSMENT NOW?

Several trends in the last two generations make it virtually mandatory that society develop new early warning techniques, and better aids to planning and decision-making. Among these trends are the following:

1. *The Complexity of Society*

Society has become so interwoven and interdependent largely as a development of physical and social technology that a false step or a failure in one component may create havoc, a major inconvenience, or some chronic nuisance. The Northeast blackout of 1965 is perhaps the most conspicuous example of a partially but inadequately cybernated system whose failure caused a major dislocation for a large part of the population. Other examples can be enumerated for virtually any of the networks of our society: waterways, canals, transportation or communication. For example, the major waterways of the Nation are so widely used, not only for traditional transportation but for sources of fresh water supply, recreation, and dumping, that the interdependency of these is often lost sight of by a particular user. Consequently, an inadvertant or a careless spill of industrial waste into a river can have an adverse if not a ruinous effect on other interdependent uses, such as recreation, fisheries, or on human health. Every complex system needs control mechanisms activated by an information system which generates either feedback or forward control, that is, mechanisms which permit the correction of growing system errors before they become catastrophic or mechanisms which anticipate future difficulties and shunt them out of the system or correct them. Technology assessment would be part of that information system of our society.

2. *Man is in Control*

Without any doubt, over much of the face of the globe nature has lost her resiliency to the attacks and impacts of man's technology. For much of recorded history, man has waged a continuing relentless war against nature in which the superior forces of nature, while challenged, were rarely assumed to be beatable. The great monuments of the ancient empires at Ankor Wat, and the Yucatan and the capital cities of past civilizations have all succumbed to the ravages of nature. Only rarely in the past has man created what on a human historical time-scale is irreparable, irreversible damage to the environment. The rare instances are noteworthy because of their rarity. For example in the time of Homer the hills of Greece were covered with trees. Similarly when the Greeks first expanded into Sicily and Southern Italy, those regions were covered with trees. The now defunct empire of Carthage flourished because it was the center of the bread basket of the Mediterranean. Man's work has made each of these areas barren.

Since the technological revolution and the great expansion, not only of population but of per capita energy utilization and material demands, man has achieved an ability to control virtually any sector of nature possibly even to overcome, if not reverse, the gross balances of nature.

The scientific knowledge which is the backbone of that control over nature also generates the ability to more reliably anticipate future consequences of human action. This ability to anticipate, engenders the exploration of the future, the anticipation of useful and adverse outcome and the call to guard against them. So, for example, while man for the first time may be able to shift atmospheric or oceanic balances, he is also for the first time able to anticipate, measure, and control these changes.

3. *The Scale of Human Enterprise*

Many human enterprises such as hydrologic control, dam building, the laying of transportation lines, the modification of weather, the giant telecommunication systems, and the exploration of space are conducted on such an enormous physical scale that they demand large investments, long planning periods, and engender substantial intractable if not irreversible consequences because of the size, volume and effects of physical facilities. Consequently it becomes important that not only the primary but incidental or secondary effects of such enterprise be integrated into their design. The cost of correcting a deficiency is often far higher than designing that deficiency out of the system.

4. *The Once-Through Economy*

Within the last two generations, as a result of the exponential growth of technological capability, the U.S. economy has moved from a parsimonious, conserving mode of production, consumption and utilization, to one in which many things are manufactured, distributed, utilized and disposed of in a continuing stream on a time scale which suggests the term "once-through economy." In contrast to the prevailing American situation, as late as the French Revolution most garments bought and sold in Europe were second hand. When finally worn out, clothing found a ready outlet in the rag paper industry industry. Not so many years ago, cigar boxes were not a waste product but a desirable commodity used by school children, housewives and home handymen. The conversion of the packaging industry to the once-through philosophy has nearly eliminated the cigar box and spawned a new commercial product - the ersatz cigar box which is sold to school children at 17 cents apiece. The glass jar and tin has often found secondary uses which, in many cases, influenced the marginal value of its contents compared to a competitor's product. For several years after World War II when the Japanese industry was getting back on its feet, one could often find *Schlitz* and *Ballentine* imprinted on the inner surfaces of cheap mechanical toys made from salvaged beer cans.

The once-through economy fostered by our technical capabilities and prosperity has created great demands on raw materials and has created a waste disposal problem. The capability to forecast the limitations of natural resources has highlighted the dangerous consequences in the next few decades if the present relentless drain on resources continues. The failure to provide adequate waste recycle promotes the drain on resources and creates the aesthetic insults of foul garbage heaps, unsightly scrap yards, and a ubiquitous patina of litter.

5. *The Shift in Societal Values*

One consequence of victory in the battle with nature has been a move to a new level of prosperity and leisure and a weakening of the old values of hard work, parsimony and "shoulder to the wheel." They have given way, as they inevitably must, to new values centering on the pleasant utilization of leisure, the demand for amenities, recreation and change, and a general concern for the quality of life. While there are still substantial areas of poverty even in American society, that cannot mask the fact that by and large most citizens have achieved a degree of prosperity which has shifted their values and their concerns for making a living to enjoying life and hence to a heightened concern for secondary effects.

6. *A Boost to Technology Assessment*

A major factor promoting technology assessment was the passage of the National Environmental Policy Act of 1969, Public Law 91-190. That law established the Council for Environmental Quality and also laid down a requirement in Section 102 for assessments of environmental impacts to be made and filed with that Council. These 102 statements, while not full-fledged technology assessments, have created a general demand throughout the Federal system for the systematic exploration of secondary impacts with a consequent demand for the development of methodologies, techniques, approaches and protocol. Much of what is occurring under the 102 requirement will benefit and influence activities of other agencies and assessments in general.

It might be noted that the report of the National Goals Research Staff, "Toward Balanced Growth: Quantity with Quality," saw fit to highlight technology assessment as a major category in its report.

SOME CONSEQUENCES OF A TECHNOLOGY ASSESSMENT

As a policy tool intended to influence decision-makers, it seems appropriate to describe some of the particular ways in which that influence may manifest itself:

1. *Project Modification*

The intention of an assessment is to explore and systematically evaluate beneficial and adverse impacts for the technological intrusion. Therefore, one might expect that unless an assessment thoroughly condemns a project and demonstrates intrinsic disutilities that far outbalance the advantages, the most reasonable result would be modification in the project plans to make it more useful, safer, effective, or aesthetically satisfying. So, for example, it would be difficult to believe that the oil will not eventually be drained out of Alaska to meet the needs of the U.S. economy. The assessments so far have led to the stoppage of that project. One may anticipate, however, that in the long term, the consequence of the assessments will be to substantially modify plans for moving the oil to make it less risky to the environment. Similarly, evaluation of the ecological consequences of thermal pollution from a power plant on a river or stream may lead to modification of a design or engineering of a power plant or may lead to its relocation, or perhaps even to substitution of an alternative source of power. But one can hardly anticipate that the supply of power either would, could, or for that matter, should be arrested as a result of such analyses. It is neither the intention nor the likely outcome of assessments.

2. *Surveillance*

Since assessments explore the unknown and the uncertain, much of the resulting analysis must be conjectural but based on plausible scientific reasoning. Consequently, one may expect that various kinds of surveillances may be suggested by a given assessment, to monitor the consequences of a particular intervention. These surveillances may be of the physical environment or a monitoring of social consequences.

3. *Research and Development*

Research and development will undoubtedly be stimulated in several different modes by many assessments. First, it may be stimulated to more reliably block out and define risks suggested by assessment; for example the effects of a given temperature change on the biota of a particular stream. Second, it may stimulate research to obviate the undesirably secondary effect of a proposed application of technology. Third, research may be stimulated on alternative methods of dealing with the primary objective of the technology. Finally, research may be stimulated on corrective measures to deal with adverse effects already generated. Consider as an example of the last, that for at least 200 years it has been known that the unchecked growth of population can lead to disaster. Likewise, it was fully predictable that the expansion of world-wide public health measures with no other changes in sociological economies would yield a great population expansion and a substantial pressure against resources. Yet it is only in the last five years that well-funded research has been undertaken in the United States and around the world on treating population growth as a public health problem. In the past it has been dealt with almost entirely as a matter of personal option and choice, as an issue in individual domestic preference. Perhaps more incisive assessments could have launched policy-oriented population research a generation earlier. However, that is not likely, as we have an interesting example of an assessment having been made in the mid 1930's of a long-term consequence of farm policy. That assessment clearly indicated that it would drive a large percentage of rural Southern farmers from the land and consequently anticipated the creation of the problems which we are witnessing the flowering of today.

4. Control Technique

Regulations, taxes, controls and prohibition may very well result from an assessment where it would seem that particular effects of technology are undesirable or even the technology itself may be undesirable. There is, for example, a fairly elaborate and growing literature on the possibly adverse effects of genetic engineering and electrophysical control of behavior. As these assessments become more defined and as the realities of the technology become clearer, it may be desirable to restrict or limit some applications of these techniques.

5. New Direction

As technology assessments unfold major potential benefits to society from a given technology or modifications in technologies, one may expect new directions to be encouraged. This may come in several ways; for example, in terms of intervention demonstrations or experiments. A capability with a given technology may suggest major benefits to society, but it may be necessary to try them out in the field in order to evaluate whether the total society should be exposed to those technologies. To illustrate this with two socio-technologies, the American society is now experimenting with income subsidies and with various forms of community control over local government functions.

Other new directions may require Federal or state government intervention, by the nature of the benefit anticipated. For example, if the introduction of two-way cable television can have a beneficial effect on the political process by putting citizens in more direct contact with public officials, it may be necessary to implement that by exercise of public authority, since it is a benefit not likely to be specifically encouraged by the private sector. There may be little profit in it. Similarly, if assessments reveal that family arrangements alternative to the nuclear family, that is, mother, father, and children, have profound benefits for society, various governmental interventions may be necessary to encourage those developed. Those government interventions may occur in terms of tax benefits or penalty, mortgage arrangements, tax incentives or penalties to employers, etc. So, for example, should it be to the long-term advantage of society to encourage multiple family living rather than single family living, these tools of government would have to be marshalled to make those developments feasible en masse.

6. New Laws May Be Required

The most extreme example of this has been the suggestion that the technologies of telecommunication and genetic engineering, may require a new Bill of Rights to bring the law into harmony with modern technologies in much the same way that the established Bill of Rights brings the law into harmony with such technologies as the mass Army, taxation, the printing press, and firearms. At another level, technology assessment may suggest and research may define differences between the structure of the law and the requirements of technology, such that new conceptual approaches to law may result. At the simplest level, an assessment may suggest a need for legislation or clarify the value of legislation under consideration.

7. *Technology Arrestment*

Any new fad or movement in government generally holds not only great promise, but also substantial fears. Any reorganization of the systems of society is bound to cause dislocation for some of its members. One of the substantial and serious fears associated with the technology assessment movement is that it may in fact become, in the words of Dr. Lee Dubridge, "...a technology arrestment mechanism." In general, many components of the industrial sector which historically seems to represent the fullest flower of the American dream have come under increasing administrative, political, propaganda, and to a lesser extent, financial pressure to modify their practices, particularly with regard to insults to the environment. So the automobile industry, the electric power industry, the mining industry, even farmers and cattle growers and processors are facing demands backed by law, to meet new standards of performance. While these undoubtedly create discomfort within each of these industries, so long as the burden is impressed and consequently borne with some even-handedness across these industries, the effects cannot be seriously disruptive to the economy. From another point of view, much of the potential legislation would only call for a redistribution of costs in many manufacturing, sales, use and disposal cycles. What economists call externalities, that is, benefits or costs incidentally generated by product, are to be more directly borne by the seller, distributor, or purchaser of a product or service. The risk of arrestment is perhaps greater in those new technologies where flamboyantly expressed fears, based more on conjecture than analysis, would seem to drive toward placing inhibiting burdens on new developments. This seems particularly true in the areas of drugs, electrophysiology, communications, and the exploitation of nuclear power.

In general, it is likely to be the case that the encouragement of technology assessment will generate a more even-handed and fair regulation of technology and will, if anything, tend to extinguish the over-reaction and the polarized positions which advocates in many current controversies seem to be driven into.

THREE VIEWS OF TECHNOLOGY ASSESSMENT

Since technology assessment is intended as the aid or guide to decision-making, three fairly distinct views of its use have grown up. The first, sees it as another tool of the advocate in supporting the case which he wishes to make. In general, those who see assessment as an advocate's tool, tend to see the proximate as well as the ultimate recourse in managing technology to be in the court or in the legal-regulatory apparatus. Since the law operates on the advocacy-adversary system, they see assessment as desirably structured to fit that mold.

A second view of technology assessment sees it as representing the neutral, even-handed application of scientific analysis to future outcomes and alternatives. In that case, the premium is on being exhaustive, comprehensive, even-handed and balanced. The premium is on the exploration of alternative outcomes and the weighing of their relative probabilities and significances. The assessment then becomes raw material for advocates of any predisposition to consider and use.

The third view of technology assessment would build on the second and suggest that the assessment should go one step further than the even-handed analysis of consequence and highlight various desirable policy options and alternatives, always making available, however, the backup analysis generated, as under view two.

The legitimate role for assessment in my view is in the second or third category. Certainly when conducted with public money, an assessment should provide an even-handed and balanced look at all alternatives, options and outcomes. If it turns out to be the case that for decision-making purposes, some policy alternatives should be highlighted, those options perhaps should be presented But the analysis and the means of getting to them should not be submerged or overlooked. Once the notion of even-handedness and fairness is removed from an assessment, it just becomes another tool in a rancorous adversary dispute. In fact, one of the major difficulties with many of the informal assessments run by public interest groups is that they in fact do not represent conclusions drawn from an assessment of consequences but rather reflect an attitude of "I've made up my mind, tell me why I'm right." Should even-handedness of analysis become a dominant trend in driving public interest group positions and activities, one may anticipate raising the level of sophistication and hence the value of public-policy decision-making to new levels of value for society.

HOW IS IT DONE?

The central feature in any assessment is that it reflects the interrelatedness and interconnectedness of the factors acted upon by a technology. Consequently, the the broad category of systems analysis is likely to be the central theme in any assessment. But there is no general method, methodology, or technique yet developed for conducting a technology assessment. It is not likely, for several reasons, that a single technique will dominate the field: 1) it is unlikely that many assessments will ever have the requirement of being totally comprehensive; and 2) the emphasis in any given ment is likely to be strongly influenced by the technology under investigation. Some technologies will find their major impacts in the environment, others in societal or economic or legal implications; consequently, the techniques applicable for pursuing these more detailed areas are likely to vary quite widely.

Among the desiderata in any assessment are first, that it be exhaustive in its scope while at the same time that it be selective in identifying areas for more intensive examination. Secondly, that it be balanced and even-handed, that it not reflect either the prejudices or preconceptions of the investigators, the sponsoring institution, or the ultimate users. This is a particularly difficult condition to meet. Third, the method should put a premium on imagination, understanding, the ability to anticipate and visualize future consequences, and the ability to step beyond preconceptions and self-interest. Obviously, these demands run up against serious human and institutional limitations.

In some cases, the problem itself in some sense strongly guides the methodology to be used. In other cases, it is less helpful. For example, in a study of the impact of snow enhancement, that is, of increased snowfall and increased runoff into rivers, the analysis might have as its core the tracing-out of two physical systems - the hydrologic system, the movement of the water; and the meteorologic system, in the downwind movement of the air masses. Conceptually, one would then spin off from these physical systems the various points of impact.

On the other hand, in working with a social invention such as modifications in building codes, or the introduction of a new physical technology such as snowmobiles or behavior-controlling drugs, the method and techniques for generating and identifying the impacts are less straightforward. But several organizations have proposed techniques of systematic examination which largely depend on the laborious exhaustion of possible points of impact. In the systematic examination of each of those possible impacts, decision theory approaches and morphological analyses have been used. The Geological Survey in Circular 645 has presented an interesting *General Procedure for Evaluating Environment Impacts*. This utilizes, among other tools, an 88×100 matrix of impacts and actions. An analogous, but more general approach to technology assessment has been developed at the Program of Policy Studies in Science and Technology at The George Washington University by L. Mayo and his associates. These methods only systematically suggest points of impact but do not indicate that an impact will occur or that it will be significant. Supplementary techniques are required for that analysis.

Other methods to bring in both expert knowledge and imagination involve the use of expert panels, the Delphi technique and cross-impact matrices. Those three methods all, of course, depend on various ways of managing and aggregating individual judgements. In some cases, modelling and simulation techniques are proven useful. Economic tools are, of course, very powerful and central to most analyses.

COST

Before turning to the question of the team requirements for technology assessments, it might be interesting to note a bit on costs. I estimate that a full-fledged technology assessment, not involving experimental work or the generation of new data, will cost between $150,000 and $250,000 or be the equivalent of four-to-six senior professional man-years of effort. Lower-cost assessments may be conducted on smaller projects with more defineable impacts of interest or in preliminary assessments designed to block-out a larger effort. Similarly, in a complex technology involving the detailed consequence of various technical alternatives, such as the laying of a pipeline or a communications system, the cost may run an order of magnitude higher. Some specific examples may help. The assessments of a proposal to extend the Kennedy Jetport into Jamaica Bay, which was conducted by the National Academy of Sciences, cost approximately $300,000. A technology assessment of the Snow Enhancement Project for the Colorado River Valley, which was conducted by the Stanford Research Institute under NSF sponsorship, cost $179,000 plus a substantial contribution from the Institute itself. A joint Department of Transportation National Aeronautics and Space Agency project, called Civil Aviation R&D Policy Study sponsored a technology assessment at The George Washington University which cost $220,000. On the other hand, the efforts to scope-out the impacts of the Alaska pipeline for the preparation of the 102 form, are estimated to have cost $9 million.

TECHNICAL ASSESSMENT TEAMS

It is almost always the case that a technology assessment will require a team effort. That team must bring to bear not only multidisciplinary talents - talents representing experience in fairly formal areas of activities such as engineering, chemistry, physics, biology, etc. - but that it also must have an interdisciplinary capability - the capability to work on those fields and on those problems for which there is not a well-elaborated academic or professional body of knowledge. Meeting the interdisciplinary requirement is obviously far more difficult than meeting the multidisciplinary requirement. There is in many assessment activities a further requirement for bringing in the insights of the practitioners in the field, as well as a need of remaining alert and aware of the range of potential policy options and technological alternatives.

In order to achieve the conditions of being comprehensive, balanced, imaginative, and broad in scope, one must put particular premium on the team leader or team leadership. It is essential that the leader not only be of broad intellectual capability but that he be capable of comfortably moving into new areas and working with divergent talents and personalities, and that he be able to make them effectively interrelate.

Any assessment tied to a decision process must be timely, balanced, and independent. Perhaps the single most difficult condition to meet in an assessment, other than that of leadership and interdisciplinary, is independence.

The team effort, as mentioned earlier should also be premised by a core of systems analysis or systems experience. In many cases, that can be quantitative, but in many other cases the system must be dealt with in a qualitative way. The team, in addition to technology inputs, whether biological, physical, or engineering, must also take adequate account of the legal, economic, political and social effects. Therefore social scientists, political scientists, lawyers and economists are central to the development of most ventures.

WHERE MAY THEY BE DONE?

The question of the organizational base for technology assessment comes at two levels: 1) the sponsor; and 2) the organization conducting the assessment. In some cases, these may be the same. First, as far as the sponsor goes, any government agency is likely to need assessments, especially in light of the 102 impact statements. With the demand for comprehensiveness and independence in assessments, the National Science Foundation is assuming a significant role as sponsor. The Congress is giving serious attention to the formation of an Office of Technology Assessment to function in a way analogous to the Congressional Reference Service or the General Accounting Office, two agencies which have well-served its interests. State and local government may also sponsor assessment on issues of more local interest. Industry, through either individual corporations or trade associations, may find it desirable to fund assessments either to block-out its own options or to block-out the anticipated public interest and regulatory constraints, it may encounter. In general, the central problem with any agency sponsoring assessment is both establishing an atmosphere of independence in the assessment itself and assuring those who are to be influenced by the assessment, that it is in fact independent. As a consequence, many agencies move toward outside groups to achieve that sought-for degree of independence. These may include universities, non-profit organizations, or even profit-making or commercial think-tanks. The further advantage of farming out the assessment is that one can often do this on a competitive basis and thereby have an opportunity to see a variety of candidates display their approaches and their talent. In general, there are major strengths and critical weaknesses in both the universities and the non-profit and profit-making think-tanks in undertaking assessment. The strengths of the universities in general are a broad base of intellectual talent, and a depth of knowledge and expertise. In general, however, universities have not proven particularly effective in broad-scale assessments because they, in general, lack the management capability to effectively marshall an integrated team to deliver a timely, comprehensive, balanced assessment.

The reward system in the university generally depends on individual performance, and publication in scholarly journals, not in policy oriented, focussed, team efforts.

It is the case, however, that at many universities the individual drive capability and leadership to field teams does exist. In other cases, there are centers associated more or less intimately with universities to provide the organization, leadership, and the environment for a focussed study. In general, it would be desirable if one could somewhat modify the reward system in the universities to recognize a value on influencing public policy equivalent to the value of publishing a scholarly work.

The non-profit and commercial think-tanks in general have a prominent advantage in conducting technology assessment in that they are usually experienced in comprehensive, systematic approaches. They are used to multi- and interdisciplinary efforts, and they generally are practiced in delivering usable outputs in a timely way. Frequently, however, they suffer from the disadvantage of not having the relatively superior intellectual capabilities of the university. While working with a more intellectually limited professional staff, they do call upon consultants for augmentation. Under proper sponsorship, they can yield an extremely valuable, useful product. A problem in working with these organizations is reinforcing independence, scope, and imagination, in spite of the fact that many of them have their initial raison d'être as the center of those very characteristics.

A third way - joint ventures involving universities and non-profit or commercial think-tanks - may, to some extent, be a best approach to utilizing the systems and and management capabilities of one and the intellectual inputs of others.

Another major kind of organization engaged in assesments is public interest groups. These are generally extremely informal, usually meagerly funded or operating on donated information and volunteer services. A major goal in pushing technology assessments should be to move the public interest group assessments to levels of more professional standards and even-handedness to make them more informed participants in the public policy process.

Government organizations, particularly under the pressure to meet the 102 form requirements, are generating both in-house capabilities in making an environmental impact assessment and also in farming out to various kinds of external organizations, and managing them.

Industry conducts some assessments, most of them, however, are not full-fledged and are closely related to market surveys. In general, industry has been very close with the results of its assessments. Consequently, relatively little is known about their quality, their methods, or their use to industry itself.

TEACHING TECHNOLOGY ASSESSMENT

Many fads or movements in government begin to proliferate activities on the campuses. Technology assessment is no exception. Dean Harvey Brooks of Harvard, School of Engineering, who led the National Academy of Engineering study of technology assessment, is now teaching a course in that subject as are Professor Goldman at The University of Maryland, Professor Ruina at MIT, and Professor Bowers at Cornell. There are also courses at the University of California, Berkeley, and York College in Toronto.

SOME OPEN ISSUES ON TECHNOLOGY ASSESSMENT

If one reflects at this stage in the development of technology assessment as a policy tool, beginning with the initial interest of the House Subcommittee in 1966, one sees a swell of activity and a growing number of useful studies, useful in the sense of either influencing public policy or on developing new approaches to assessment. As the technical end of the art expands, a number of unsettled issues come into sharper focus as critical to use in the decision-making process. Among these are:

1. Independence: The central question in developing any assessment, if it is to be comprehensive and even-handed, is the independence of a team from the preconceptions and interests, real or imagined, of its sponsor or of its participants.

2. Involving Professionals: In general, the professional societies of America have been standoffish in providing mechanisms for making their professional skills available to technology assessment, particularly in the voluntary sector. In general, professionals, whether chemists, engineers, physicists, or social scientists, tend to be tied to the interests of their university, or what is more confining, tied to the interests of their corporation. As a result, professional societies have been functionally remiss in

only giving lip service to the demands for professional responsibility in meeting society's needs.

3. Technology Assessment by the Public Interest Groups: There are literally thousands of public interest groups in America. They often operate on a non-existent or extremely lean base of information and analysis when they become involved with an assessment issue in the public policy process. They frequently seem to present their case based on private judgements, traditional prejudice, or a circumscribed outlook. The technical data adduced in their arguments is often one-sided, biased or self-serving. The voluntary sector often behaves as if it were operating in a mode of unbalanced advocacy. As I see it, a major need is to encourage voluntary organizations to become involved, if not to sponsor, professionally balanced, even-handed assessments of issues with which they are concerned, to then base their own policy positions on those assessments, and finally to impact and influence the decision process, based on that analysis. If thos three steps can be accomplished, we could go a long way in upgrading the quality of discussion of a public policy issue and hence bring a great benefit to the common weal.

4. Opportunities versus Problems. In general, assessment has grown up under the desire or pressure to deal with problems. In principal, a full-fledged assessment should expose opportunities for new developments and new benefits. It seems to me that we need to encourage more forthrightly the developments of and an awareness of future opportunities as well as an awareness of future problems and disutilities.

Closely related to this is the need for change in attitude toward potential risk. There seems to be a widespread feeling that any potential risk is a certainty and that any potential benefit is uncertainty or most dubious. This unbalanced attitude on the part of many sectors is a disservice to society.

JAMAICA BAY/KENNEDY AIRPORT: ANATOMY OF A TECHNOLOGICAL ASSESSMENT

STEVEN EBBIN

National Academy of Sciences, Washington, D.C., U.S.A.

In the fall of 1969, a draft report from a summer study group organized under the aegis of the Environmental Studies Board of the National Academies of Sciences and Engineering was being circulated among the various scientific/governmental groups in Washington. The report was in two parts, the first of which was entitled, *Institutions for the Effective Management of the Environment*. The second part, which dealt in the main with the expected environmental impact of a proposed Everglades jetport, was entitled, *Environmental Problems of South Florida*. Both these volumes were prepared under the direction of Gordon MacDonald and Marvin Goldberger, co-chairmen of the study group. They were timely and had a substantial impact on events that took place around that time. Both have been widely quoted and influential and both provided the background for the Jamaica Bay environmental study which took place during the summer of 1970. It was with specific reference to these volumes that the Port of New York Authority contracted with the Academies to undertake a study of the environmental impact of a proposed expansion of the John F. Kennedy International Airport into the waters of Jamaica Bay.

These reports also were related to other events in progress during those months, from the Fall of 1969 through the Spring of 1970.

Congress passed the National Environmental Policy Act of 1969 (PL-91-190), Gordon MacDonald was appointed to the President's Council on Environmental Quality created by the Act, and the Jamaica Bay/Kennedy Airport study and was designed to respond to Section 102 of the Act by carrying out a prototype multi-disciplinary environmental impact study, which that section makes mandatory for federally supported construction projects.

The report* was issued in February 1971 and has been widely circulated. Many people who have read it and others who have been affected by it consider it a unusually successful multi-disciplinary effort. Many have asked how it was done. The following pages attempt to tell that story, which in my own estimate is at least as interesting as the report itself.

BACKGROUND

Jamaica Bay, (JB) like most other environmental resources, particularly those which by chance lie near or within urban centers, is a much coveted site for a variety of human uses. As such, it is the object of considerable competition for its use. Though it was designated a New York City park in 1948, almost no one, least of all the City's Parks and Recreation and Cultural Affairs Department, seems to believe it. Almost nothing has been done to make it measure up to its designation and promise as a park or site for recreation. For several decades, the main use to which Jamaica Bay and its peripheral marshes has been put is as a sink for the disposal of raw sewage and solid wastes. And the prevailing view of the Bay, with some considerable justification, is that it's an "open sewer." As far back

---

**Jamaica Bay and Kennedy Airport: A Multi-disciplinary Environmental Study*, 2 Volumes, National Academy of Sciences, National Academy of Engineering, Washington, D.C., 1971.

as 1931, New York City became concerned with the pollution in the Bay as a public health hazard. Only in very recent years (early 60's), however, has an imaginative - and expensive - program to upgrade the Bay actually been undertaken and considerable progress made to meet a goal of swimming quality water by 1978. Positive concern for the Bay itself as an ecologically promising area is really much more recent. For many years man has cast covetous eyes on the Bay to support one scheme or another - a planned major seaport in the 1920's, an airport from 1938 to the present, a national park from 1964 to the present, and a sink for wastes - from time memorial. Among the many uses JB has served, and still does for the most part, are: a natural tertiary waste-treatment facility, a wildlife habitat, a sink for aircraft noise, two airport sites, housing, recreation, a source of building material (fill), transportation, commerce, fishing (commercial and sport, shellfish and finfish), solid-waste disposal and, perhaps as important as any of the above, a political football.

The single most dominant force in the recent life (35 years or so) of JB is the operation of Kennedy International Airport, still referred to by most "real" New Yorkers as Idlewild. Kennedy was created from Jamaica Bay and its 5,000 acres is constituted wholly from the sand pumped in the main from the northeast quadrant of the Bay. The average depth of the Bay has been altered from 3 to 16 feet and about 70 percent of its volume results from dredging. Kennedy airport was, at its inception, thought to reflect a highly imaginative bit of planning in the early days of commercial aviation (1938). It lay near the city in an area of then sparse population. Few protested its construction through the mid-40's and, with relatively few flights of propeller planes, an excitement with new aero-technology, a period of relative grace prevailed and lasted until the introduction of commercial jets and the rapid growth of the region's population in the later years of the 50's.

A second airport, similarly constituted on what was known as Barren Island, a marshy island lying at the southwest quadrant of the almost circular Bay, was also constructed in the early 40's. Floyd Bennett Field has served mainly as a naval air station. Only recently this 2.800-acre installation has been declared surplus by the Department of Defense and is soon to be disposed of for other, as yet undetermined, purposes; competition for its future use, like that of JB, is also heatedly disputed.

Some additional 4,000 acres along the western and northwestern rim of Jamaica Bay have been converted by a sanitary landfill, where New York garbage and other solid wastes have been disposed of for the last two to three decades. In sum, then, what was once a Bay of some 25,000 acres is now 13,000 acres; a projected half life of 35 years, at the historical rate (since 1938) of diminution of 2 percent/year.

Kennedy Airport and Jamaica Bay are included in all or part of some 44 traditional political or civil entities. These involve: Kings, Queens, and Nassau Counties; the City of New York; the boroughs of Kings (Brooklyn) and Queens; the town of Hempstead; five incorporated towns and villages; five congressional districts; five state senate districts; five city council districts; seven state assembly districts and ten school districts. Additionally, not mentioned, in the 44, are the State of New York, the State of New Jersey (half of the bi-state agreement under which the PONYA is run) and the federal government departments (including the FHA, HUD, FAA, CAB, EPA, DOI, DOT, DOD, Corps of Engineers) and the various offices within each whose missions and decision-making impact on both bay and airport.

Additionally, further complicating any study of JB/KIA are the discreet communities with special problems, needs, wants, expectations, and relationships in connection with either or both, which, while not constituting separate political jurisdictions or governmental entities, require special consideration. Some such are Broad Channel, Hamilton Beach, Howard Beach, Inwood, Woodmere, Arverne, and Canarsie. In each of these the impact of the airport, peculiarities of location, and community cohesiveness in matters relating to the airport and the Bay call for more than passing attention.

Though polluted and insulted by man and his technology, economics, and politics, Jamaica Bay is the only significant natural estuarine area left to the people of New York City. In 1954, some 2,000 acres of natural wetlands existed in the Bronx; ten years later there were 50 acres left. In Brooklyn, during the same period, half the natural coastal wetlands disappeared; a third of Queen's and Nassau County's coastal wetlands gave way to "development" and were lost to the public, to be filled for roads, parking lots, marinas, housing developments, industry and airports.

Jamaica Bay is at present moderately polluted but, despite the insults heaped upon it by man, it is still a reasonably healthy ecosystem and still serves as an efficient tertiary treatment system for the partially treated sewage effluent (220 million gallons) poured into it daily. And, though the Bay shows signs of taxation by man (i.e., large numbers of coliform bacteria, low levels of dissolved oxygen, the absence of some species normally found in most healthy estuarine systems, and the presence of some species usually found in polluted ones, a decrease over 20 years of 80-90 percent in the fish catch, and an increase from 10 to 35 days retention time of conservative substances), it still works. Some 60 species of fish and shellfish use it as a habitat and nursery, and over 300 species of birds have been recorded at the bird sanctuary which was built in 1954.

ORGANIZING THE STUDY

I have begun this paper with a short description of Jamaica Bay and its recent past because it demonstrates, at least in part, why certain basic decisions relating to the selection process of the Jamaica Bay Environmental Study Group were made. I want now to trace the history of the study and its organization and to discuss the questions originally laid before the National Academies of Sciences and Engineering.

In the Fall of 1967, just after the circulation of draft copies of the summer report on the proposed Dade County-Everglade Jetport, the Executive Secretary of the Environmental Studies Board was contacted by phone by an official of the Port of New York Authority (PONYA). This official was to play a continuing and important role in the subsequent agreement between the PONYA and the NAS/NAE. The Port official asked for an opportunity to appear before the Board to explain an "environmental dilemma" which the Board might assist the PONYA in dealing with. The appearance never took place, but a subsequent meeting in December was arranged with Gordon MacDonald, who has served as co-chairman of the Everglades Study and who was a member of the Environmental Studies Board, soon to accede to its chairmanship.

Dr. MacDonald and I discussed with two PONYA officials the various aspects of a possible study the ESB might undertake relating to the environmental impact an expansion of Kennedy Airport into Jamaica Bay would have on the Bay. The Port Officials emphasized their concern with the impact on surrounding communities of jet noise and with the mounting number and intensity of protests that it had induced.

MacDonald explained that the approach adopted in the Everglades study was probably not broad enough for the problem posed by PONYA, which would seemingly require a much more intensive and comprehensive multidisciplinary effort. The Everglades Study, he noted, was a first effort to learn whether the approach of defense-related summer studies could be applied to environmental problems - that the effort was generated by the Environmental Studies Board and did not result from a client request. The Everglades Study looked at the likely impact on the *Everglades* of a proposed jetport and also the more general problem of devising new institutional mechanisms for managing environmental affairs.

A letter to MacDonald from Austin J. Tobin, Executive Director of PONYA, followed on January 26, 1970. In it, Tobin spelled out PONYA's concerns and the questions he'd like the National Academy of Sciences to consider. Tobin expressed his judgment that

> "it would be impossible to proceed with any expansion into the Bay without first knowing what effects such an action would have on the viability of the Bay and on the people who live in its environs." He wished to know "whether such an expansion could somehow be made compatible with other plans for development of the Bay by the City of New York and the federal government, and could be designed in such a way as to upgrade the quality of the Bay's environment."

Subsequently, I prepared, with MacDonald's approval, a proposal which was submitted to the ESB for its endorsement. Informal agreement was reached conditionally, several members emphasizing that their approval was contingent on the study being broadened to include environmental problems relating to airport operation, location, and expansion that are being confronted in almost every large metropolitan area in the Agreement was reached with PONYA, including a proposed budget. On February 6, the proposal was formally submitted to the Council of the NAS and was approved, with insistence by several members of the Council that the study not limit itself to the Kennedy/ Jamaica Bay problem, but address itself to the implications of that problem in a broader national context of the environmental impact of large airports faced with increasing traffic and demand for expansion.

The NAS Council emphasized that their approval for the Academy's undertaking the study was contingent on its addressing broader, national-airport-associated, environmental problems using Kennedy and Jamaica Bay as a case study. The council was not satisfied with a telegram from Secretary of Transportation John Volpe indicating the Department of Transportation (DOT) interest in the Academy undertaking the study because of its national implications. It instructed me to secure a token participation of $10,000-20,000 from DOT in support of the study to establish the "national" interest. DOT subsequently demurred "to keep their options open" but did "emphasize" its interest in a letter to President Philip Handler of March 13, 1970. A similar expression of interest in the study, because of its national implications, came from then Secretary of the Interior Walter J. Hickel in a letter to Tobin on April 13, 1970. Both letters are published in their entirety in the final report. (Vol. I, pp. xiii, xiv.)

At the same time, the proposal was submitted to the Council of the National Academy of Engineering. A request for an explanation to that Council was not issued for several months and the NAE did not formally respond to the proposal. Almost one year later at time of publication the NAE informed the ESB that it would indeed join in co-sponsorship. Attempts had been made to encourage co-sponsorship since the ESB is a joint Board of both Academies and membership on the Board is constituted roughly equally of NAS and NAE designees.

A chairman for the study group was selected at the end of February, 1970. After a series of discussions with Dr. MacDonald and me, Dr. James Allan Fay, Professor of Mechanical Engineering at MIT agreed to serve. At that time, just after Fay had been appointed, MacDonald resigned from the ESB to become a member of the President's Council on Environmental Quality. Dr. Fay had, the previous summer, been a member of the Everglades study group and was chairman of the subcommittee in that project that looked into matters of air pollution. Fay was also much involved in and a member of the Boston Air Pollution Control Commission and had considerable research experience and on-going activity in air and water pollution problems.

Upon Fay's return from three weeks in Japan about mid-March, the organizational phase of the study was stepped up. We felt that the questions the study would have to address would require considerable pre-planning to avoid time slippage while the study group was being organized. This determination led to the creation of a Steering Committee of high-level, broadly experienced people. Fay and I also began a series of visits to the Jamaica Bay area where we met the leaders of several neighborhood associations, toured the periphery of the Bay and visited the airport. We also arranged a visit to the sewage-treatment facilities, the waste-disposal sites, the wildlife refuge, the fuel farms and, by boat, the inner areas of the Bay. We also interviewed a broad range of federal and city officials.

The Steering Committee met in a full-day session in the International Hotel at Kennedy Airport in April. PONYA officials made a detailed presentation of their problems in providing additional airport capacity in the New York Region and their history of failure to locate a suitable site for a fourth New York jetport. The committee, in executive session, discussed questions to be looked into, the fields of expertise from which participants would be sought, and the kinds of data group would need.

The next weeks (interspersed with continued visits to relevant individuals and organizations by Fay and me), were spent in contacting not only potential participants but at least three, and in some cases more, of their colleagues and others who might know about them. Three people, in addition to Fay, were invited because of their significant contributions to the previous summer study on the Everglades jetport, and because their experience and background, we felt, were relevant to the Jamaica Bay study.

Fay and I had determined early in our discussions with the ESB and the Steering Committee that the many different facets of the problem required a very broad range of talents. A rough 50-50 balance between the so-called "hard" sciences and the behavioral and social sciences was made. And a maximum number of 25 participants set, a decision arrived at in an effort to structure the group and in some part, to stay within our estimate of reasonable cost. It was stipulated in each case that full-time participation was essential.

A breakdown of the academic disciplines of the participants on the study group indicates only in part the broad range of talent represented. Included, though, were the following: public administration (public policy analysis), environmental engineering (water pollution, hydrology and hydraulics), public health (community reaction to aircraft noise) social psychology (social research-group behavior), sociology (social relations and community conflict), economics (air transportation and consumer demand and labor supply), civil engineering (public health, community water supply, waste treatment), mechanical engineering (airpollution and oil spill problems), political science (public administration of recreation and natural resources), aerospace sciences (fluid mechanics), medicine and human ecology (public health), political science (urban affairs, parties and politics), economics (urban and regional transportation), aeronautics and astronautics (gas turbine and propulsion systems), human resources development (urban economics and economic development), psychology (psychoacoustics and physiological psychology), urban and regional planning (community planning and metropolitan airport development), marine sciences (geology, oceanography), zoology (ornithology and comparative ethology), ecology (population and wildlife ecology), public administration (community relations and organizational change), psychology (psychoacoustics),, law (constitutional law, environmental law, contracts and commercial transactions), physics (science and environment), public law (intergovernmental relations and environmental law), economics (statistics and econometrics), political science (environmental law, science and public policy, legislative process).

JUNE MEETINGS

The group, at nearly full strength, was assembled for the first time in New York City June 3-8. Two members, one from England, could not be present and a third did not show up. During this six-day period the group was provided with a kaleidoscopic review of the problems of Jamaica Bay, Kennedy Airport, the neighbors, the City of New York and the environment of these areas - economic, social (recreation, housing, social unrest, etc.), physical (noise, air, water, sewage, solid waste, etc.) and political. They heard from PONYA, state, city, county, and federal officials, from special interest groups, neighborhood associations, environmental groups, politicians and private, non-profit groups who somehow related to the area. (A complete listing of individuals and organizations appears in the final report, Vol. I, pp. 27-28.) Some Steering Committee members also attended these sessions, which were structured as adversary proceedings.

Each person presenting a position before the group was vigorously questioned by members about views, facts, the past, and future projections. It was difficult to discern as a result of this that anyone in the study group agreed with any particular view or testimony, or disagreed. All those who "testified" in the closed session "hearings" were treated with equal rigor. Even in private, informal sessions at dinner or in individual hotel rooms, which frequently lasted beyond 2:00 a.m., arguments revolved around projections and included a great deal of "on the one hand ... yet, on the other hand." The complexities and challenges that could be expected throughout the study were manifested by the broad divergence of opinions about the Bay and the airport. The range from"eco-freaks" to technological/ economic determinists was represented on the study group. The range of commitments to the various environmental philosophies was as diverse and as complicated within the group as among those who testified before it.

The collective brow was knotted with exposure to a series of riddles; the parameters and countervailing arguments seemed endless. Simple answers were obviously out of sight as there were not even simple questions that could be asked.

One afternoon, at the request of the study group members, was spent in visiting Jamaica Bay. Almost all members did so on their own.

In the intervening one and three quarters months that remained before the group was to reconvene at the study center each participant was asked to:

1. Formulate questions to be examined

2. Contact persons, organizations or agencies for data, information, reports, publications, etc.

3. Visit any of the above

4. Indicate data, information, reports, studies, books, publications of any kind that should be acquired for the study library

5. Compile and suggest any consultants whose input to the study might be valuable and who might be called upon to make a presentation to the group

During the interregnum a day-long boat tour of the Bay was made by nine study group members along with the New York City Commissioner of Water Resources and three private consultants who were working with him to develop a mathematical model of the Bay and a plan for upgrading it.

The period from June to August was also useful in the generation and particularization of questions to be addressed and the sifting and integration of testimony given at the June meeting by various individuals and organizations. Several study group members arranged informal discussion meetings with one or two colleagues.

LOGISTICS

A study site had been located and arrangements made during April and May, a very difficult and taxing effort. Some mention should be made of the criteria used in the laborious search for adequate facilities. This was very important in that it played a significant role in the style, force, work pattern, and socialization of the study group and its product ... the final report. Such facilities are exceedingly difficult to locate in the New York area and the process took some weeks of searching and careful planning.

One primary requirement was to find a site suitably removed so as to isolate the study group from what was early perceived as probable pressure for information about the progress of the study from the press, the interested public, and government officials.

Another was to provide a pleasant living situation which would, for accompanying family members, at least replace summer vacations and help establish an informal atmosphere.

A third was to avoid having to turn everyone loose on his own to locate and rent living quarters in an area where such facilities are scarce and expensive.

A fourth was to provide living accommodations where the group would be together after office hours and allow for and encourage continued discussion and debate beyond the formal work day. This had a significant, planned effect of complete immersion in the study.

Every night, every weekend, often into the morning hours, loud voices in debate about one aspect of the study or another could be heard, to the considerable annoyance of families and other residents of the motel in which everyone associated with the study stayed. Moreover, the mutually supportive ambiance for wives and children relieved pressure and competition for the participants' time. As a result, every weekend was spent at work, but play was not dispensed with; participants did take early morning and evening swims and common cocktail hours. A regular tennis game convened every morning from 6 to 8 a.m. at the same time the joggers and cyclers were "doing their thing."

A fifth requirement was relative contiguity to Jamaica Bay/Kennedy Airport, as almost daily visits for small groups were part of the plan. Every member spent at least two full days, and many several days, on or at the Bay or Airport. The study center was located about 75 miles from those points. During the latter part of August a small plane supplied by PONYA ferried participants back and forth to Kennedy, where a leased station wagon transported them around the area.

A chartered shallow-draught houseboat for touring the Bay and the marsh areas and basins that feed into it was moored in the Bay and heavily used by the members.

As already mentioned, motel facilities were leased for the twenty-seven principals and supporting staff of five (four secretaries and an administrative assistant).

Seven miles away a dormitory on the campus of Southampton College, refurbished to serve as 28 offices, was leased from the college. Three rented station wagons provided transportation to a from the study center. A basement room for seminar meetings and a larger seminar room for plenary sessions in a separate but contiguous building were constantly in use. The campus, of course, also has a library which was of considerable value to the study group. Contiguity to Stoney Brook (SUNY) and Brookhaven National Laboratory was also advantageous. Courtesy, cooperation with, and isolation from the resident faculty were essential and mutually respected, given the great pressure of time for the completion of this phase of the study.

Each member of the study group was compensated at the rate of $100 for each day he worked on the study. This is not normal for Academy panel activities; however, participants in NASA-sponsored Academy summer studies have been compensated, as had members of the Everglades summer study. Decisions to pay or not rest with Academy management and are made on an *ad hoc*, pragmatic basis. Duration of the study and demands on an individual's time influence the decision. But, at least in the past, no hard rule has existed, though NAS policy in that regard is now being reviewed.

SUBCOMMITTEES

Experience at the June "hearings" provided a basis upon which to proceed with the organization of the study. Sub-groups were created to deal with some discreet, discernible problem areas basic to understanding the Jamica Bay situation. The June meeting was for most, the first face-to-face exposure to the other members of the study group. I knew, or had met, only three members prior to that time; Fay was in a similar position. Following the June meeting each individual was evaluated as to the interests manifested at that meeting. This evaluation, along with a consideration of academic discipline and a judgment about where their talents were most needed, influenced subcommittee organization and assignments. The subcommittees were organized as follows:

A. *Ground and Air Transportation Systems*

a. Aircraft noise
b. Aircraft air pollution
c. Aircraft safety
d. Aircraft traffic control
e. Ground rail link
f. Freight and passenger car links

B. *Recreation and Conservation*

a. Swimming
b. Boating and fishing
c. Shoreside parks and playgrounds
d. Gateway National Seashore Plan

e. Wildlife refuges
f. Marine ecology
g. Education

C. Land Use and Community Needs

a. Housing
b. Industry and commerce
c. Ground transportation
d. Waste disposal
e. Noise
f. Traffic congestion
g. Employment
h. Political action

D. Metropolitan Needs and the Expanding Economy

a. Central city needs
b. Fourth airport problem
c. Human need for air transport
d. Regional development and growth
e. Employment
f. Alternative uses of sites
g. Indirect costs

E. Water Management

a. Sewage
b. Flood control
c. Dredging
d. Oil pollution

The subcommittee headings are, for the most part, self-explanatory, but a word is in order about Subcommittee D. As originally planned, there was to be created, during the last week of August, a committee of authors comprised of the chairmen of the subcommittees and certain other selected individuals who would be charged with two tasks: (1) to put together a draft report, and (2) to extrapolate and generalize from the drafts of the subcommittees as the basis of a section on national issues, lessons learned, and implications of the Jamaica Bay/Kennedy Airport situation. This did not happen in this way. Rather, Subcommittee D was created as a committee of chairmen and a few others to begin with, and it addressed itself to the second task beginning in the second week of operation. In part, that reflected the propensities of the chairman and two others whose primary interests lay in the national implications sphere of concern with the active support of the other chairmen. One member of that subcommittee was assigned cognizant responsibility for dealing with metropolitan political considerations and two others were assigned to work with him. Unfortunately, he never fully came to understand the relationships of the Kennedy situation with New York City politics and bogged down. Eventually, he gave up and left with but a few hours notice, the only study member to do so.

I mention this only because it reflected on the structure of the final report, as it became necessary to reassign this aspect of the work, apportioning it among other subcommittees, where it added to the heavy workloads already being carried and required several people to do double duty. The relevant materials are not absent from the report; they are spread throughout several chapters. But this issue became an important organizational problem because with a small, carefully selected, finite number of participants and the sharp time limitations of the study, each individual had to be heavily relied on for the successful completion of the study. It is essential in a multi-disciplinary study that the apportioned work be handled by the relevant specialists. The ability of a short-lived study group to examine a broad range of problems in sufficient depth to arrive at supportable conclusions and recommendations clearly depends on each member carrying out his assignments fully and completely within deadline constraints. In short, the failure of one study member to carry his weight and fulfill his responsibility impacts severely and negatively on other members of the group who must pick up the slack resulting from his unwillingness or inability to do his job. In such an effort, there is little room for slippage. At Southampton it fell to one or two others to carry the extra load. They did so, but at considerable personal sacrifice of time and effort, which necessarily extended beyond the August 28 deadline.

A word here about subcommittee chairmen and their individual *modi operandi*. In each of the five subcommittees the chairman dealt in an individual way with the members of his group. In one subcommittee, for example, the chairman headed a grouping of three including himself, who worked together closely on their common specialty while the rest of the committee members were given their choice of assignments and worked pretty much on their own, conferring frequently with Fay and me, only occasionally meeting with other members of the subcommittee. A second subcommittee operated in light of the chairman's analysis of the problem and, based on it, assignments were handed out on typewritten sheets, with deadlines carefully spelled out. Meetings of the subgroup took place very frequently.

A third subcommittee operated with a very free-floating chairman who spent a great deal of time going about speaking to different people at the airport, in the NYC government, other area governments, and at the PONYA. Almost daily, or rather nightly, he would return with carefully spelled out assignments for the members of his subcommittee, who would then set to work on the questions he posed, gathering data and drafting papers in the particular areas designated. The chairman then took the resulting parts and put them into a whole. In a fourth subcommittee, assignments were developed by negotiation between the chairman and individual participants, the former expressing his view of need, the latter their interpretation of how to fulfill that need. Deadlines, though firmly agreed upon, were difficult to enforce and, where not adhered to by the participants, the chairman would simply take what he could get in raw data or partial drafts and work overtime to fit it into his own draft. This chairman held frequent negotiation meetings with individuals and with the whole subcommittee. As the committee of subcommittee chairmen, it fell to the chairman and a few others to do the bulk of the actual work. In essence, there were in this sub-group too many chiefs and too few Indians so the burden fell on too few.

A fifth subcommittee began operations as if each member were independent within his own bailiwick, each one doing his own thing. The chairman soon saw the dangers inherent in such a procedure and exerted his leadership. He drew up an outline paper, circulated it to the members of his group and asked them to designate the areas they wanted to deal with. He sorted the responses out, met with the group, eliminated the overlap, and with some very imaginative and hard-working assistance, pushed matters along. Each time a paper was prepared by a member of this sub-group it was put to a "scientific" peer group test, the author openly, in this limited session, apprised of the views of his colleagues. Eventually, the group emerged with strong mutual agreement and the parts put together into a whole. The ultimate drafting process, though, split this output into several different parts which were appropriately distributed throughout other sections of the final report.

These different *modi operandi* really, in very basic ways, reflect the individual chairmen and the personalities assigned to each sub-group. Some of the procedures followed were more productive than others, but there is probably no reason to attempt to achieve a uniformity of operation in this sort of study. Each group should define its own personality and methods of operation. Though a quality end product is what ultimately counted there was a considerable variance in the level of productivity of the subcommittees as subcommittees. In the end, it did work, but in at least one case, it worked because the chairman and two or three of his members were willing to assume exceedingly heavy workloads and supply, in effect, the work that should otherwise have been done by the subcommittee working as a unit.

At least weekly the entire study group met in plenary session, during which each of the subcommittee chairmen made a brief progress report. Each chairman could, at any time, ask for a plenary session to discuss his group's work or to describe the progress and nature of his committee's findings.

During the first half of August, mornings were used to hear presentations by government officials, private group representatives, political figures, or technical analysts. A few afternoons were similarly scheduled. Attendance at these sessions was left to each individual, and there was usually a host subcommittee, the members of which were particularly interested in the presentation. This procedure was mostly limited to the first two weeks so as not to interfere with drafting efforts during the latter days of the study. An occasional outside resource person was invited in that period but these were sharply limited. Group trips to the Bay and elsewhere were halted before the third week in August.

The final deadline for the submission of papers was August 28. All but two or three people made it. A plenary session was held throughout that day to consider, as a group, the findings and recommendations submitted by each subcommittee. No conclusion or recommendation was accepted for inclusion in the final report unless it received "substantial" agreement. Substantial was defined as *roughly* two thirds of those present and voting. Heated debate marked the meeting, though surprisingly the vast majority of points received overwhelming group support. Some disagreement did carry over and even mounted in the following three days.

The Steering Committee, along with the then recently appointed Chairman of the Environmental Studies Board, David Gates, who had little or no familiarity with the study or its history, came to Southampton for the three-day scheduled review session (August 29-31). Everyone was given the draft reports of the subcommittees and the conclusions and recommendations agreed upon at the 28th meeting. There was, at that time, no integrating chapter, for which the study group was criticized. Moreover, during the course of the Steering Committee meetings some study members had second thoughts about the wording of the initial and indeed principle recommendation reflecting *unalterable* opposition to *any* airport expansion into the Bay. Though nearly everyone felt that no case had been made for expansion, this group was concerned with the probable interpretation that the report was recommending against expansion *forever* under *any* possible circumstance or with any possible configuration. During the penultimate days of the study a runway configuration that would entail a very small land-taking in the Bay (2 percent) was devised by a member. Most felt, at the time, that it was really too late to give it the serious and careful consideration the others had been subjected to and that was essential for being accorded group support. As a result they voted at the 28th plenary session not to include it in the final report. The matter stewed overnight and was brought up before the Steering Committee for their intervention, *not*, repeat, not in support of recommending this configuration, but rather that the consideration be reopened and additional attention be given it, that it be subjected to the same rigorous analysis other configurations, including that representing the PONYA proposal, had received, and that it be included in the final report.

This vocal group suggested a rewriting of the recommendation to reflect a less all-encompassing opposition to Kennedy expansion. This incident provided an important insight into the operation of the study. The meeting of the entire group on August 28th was the first of its kind to be held during August. At no time previously had an attempt been made at democratic group voting on the issues. As noted above, the agreed-upon criteria for inclusion in the final report of any conclusion or recommendation, was "substantial agreement." During the vote, 5-7 members voted alike on many gut issues in opposition to the substantial majority. In doing so, they apparently established a rapport with each other that they had not enjoyed during the rest of the study period. With each voting defeat, they became more of an identifiable group expanding their area of agreement far beyond anything they had known earlier. Their disagreement with the rest of the group grew more heated throughout the day. They seemed to feel they were being steam-rollered by the "substantial" majority and it became increasingly obvious that their voting behavior was not reflecting their earlier work or expressed views. The group became sharply polarized at that point and had no time to resolve its disagreement or negotiate its disparities before the arrival of the Steering Committee.

With the exception of one person and possibly a second, there was at the closing of the Southampton review session, despite this incident, no deep schism dividing the study group. Rather, as one observer pointed out, the 28th meeting took place so quickly, with no time in the rush to get papers done to negotiate priorly the points at issue and to work out mutually agreeable language and emphasis. The 28th meeting, in a sense, "pulled the cork" and, for the first time, the group had become aware of their new understanding of the nuances of the Jamaica Bay/Kennedy Airport interrelationships and saw that they could really do the job. Moreover, and perhaps most important, they were tired and irritable. The two-room motel units had grown awfully small and their families anxious to leave for home. All this came out before the Steering Committee. The members collectively seemed to realize what was occurring. Several of them indicated they felt this was a healthy development that would ultimately redound to the benefit of a much better, less categorically "certain" but more reflective report. The final report does indeed reflect this estimate and the negotiations that followed over the next month resolved the issues at dispute.

The Steering Committee heard presentations from each of the subcommittee chairmen and a few selected colleagues during the three days. An executive session was held. There was agreement that the study was really only partially completed and that the study group had done all it could as a group in preparing individual papers, most of them very well done. A final report, of course, still remained to be written. The Steering Committee felt the need of an integrating chapter, which had been planned for, to provide coherence and wholeness to the formal report which it would not have if it consisted only of the various special studies and analyses, as they stood.

The Steering Committee recommended to the Environmental Studies Board, whose chairman was present throughout its deliberations, that a small committee of the study group be constituted to complete the report and that it be resubmitted to them for review. The subcommittee chairmen agreed to continue to work on the study until it was completed. The ESB chairman transmitted these views along with his recommendation to the NAS president. The study group was authorized to continue its efforts and additional funds were requested from the PONYA to do this. PONYA quickly agreed; PONYA officials had, from the very beginning of the study, felt that the study could not be finally concluded in the short period originally agreed to, given the complexities of the Kennedy/Jamaica Bay situation. The proposal submitted to the NAS Council in February 1970, also foresaw that eventuality.

PRINCIPAL CONCLUSIONS AND RECOMMENDATIONS

The following constitute the findings and recommendations as they appear in the final report:

> "☐ Any runway construction will damage the natural environment of the Bay and reduce its potential use for conservation, recreation, and housing. The degree of this impairment will be dependent upon the amount of Bay area taken for this airport extension. A sufficiently large land-taking, such as that proposed by the Port of New York Authority (see Figure 4, Chapter 4, Volume II) could cause major irreversible ecological damage to the Bay.

" ☐ It is possible to improve the Bay environment by technological means. Such improvements may be made independent of any airport expansion scheme, but any expansion would increase the economic costs or dilute the benefits of these improvements."

The Study Group went on to recommend:

"1. *The Port of New York Authority, in cooperation with the Civil Aeronautics Board and the Federal Aviation Administration, institute immediately a program of landing fees, consolidation of flight schedules, and other administrative devices that will eliminate existing congestion and allow for a more efficient utilization of existing system capacity. Only after complete implementation of these measures will it be possible to assess fully the need for and the timing of an airport expansion. Adoption of these measures will give priority usage of existing facilities at Kennedy Airport to commercial passenger transportation rather than to general aviation.*

Such measures will not provide indefinitely for air transport needs. Other possibilities will have to be considered in the future. We further recommend that:

2. *Since the evidence indicates that improved air-traffic-control systems would permit substantial increases in the capacity of the region's airports, such systems be developed and adopted as rapidly as possible.*

3. *If there is a proven need in the future for increased runway capacity at Kennedy Airport after the introduction of these much-needed technological and administrative improvements, then additional runways might be considered, such as those discussed in Chapter 4, Volume*

II, which would require minimal land-taking in the Bay. No new runways should be constructed or other measures taken to increase the capacity for aircraft movements unless it can be shown that the benefits exceed the costs, where benefits and costs include a full consideration of their impact upon people and the environment of the Bay area and the metropolitan region.

In the long run, planning for the air transportation needs of the New York metropolitan area will require greater federal participation than has been the case in the past. We therefore recommend that:

4. *The Secretary of Transportation, the Governors of New York and New Jersey, and appropriate governing authorities in the region should plan now for the new regional air transportation facilities that will be needed in the future. They should proceed immediately to designate and acquire a site for a new regional jetport. The development of such an airport and the timing of its construction should be consistent with a demonstrated need for additional airport capacity in the region. They should also begin an intensive study of alternative systems for intercity air and ground transport, such as vertical or short takeoff and landing aircraft and high-speed trains, which might affect future regional jetport needs.*

5. *The Secretary of Transportation, in cooperation with appropriate state and local governmental agencies, should make an evaluation of the congestion problem, both at terminals and with respect to the ground access to the three major airports in the region, and propose plans to reduce this congestion.*

Aircraft noise, and, to a lesser extent, air and water pollution related to airport operation are perceived as major environmental hazards by residents in communities surrounding commercial airports. We have examined these effects in the environs of Kennedy Airport, and have concluded that:

☐ The construction of new runways will not significantly reduce the number of residents of nearby areas exposed to intense aircraft noise. Major reduction in noise exposure can come only from use of quieter aircraft.

On both the local and federal level, measures need to be taken to reduce the impact of aircraft noise. We recommend that:

6. *Public authorities in New York City and Nassau County should establish and vigorously enforce building construction standards that protect the health and welfare of occupants against aircraft noise. These standards should apply to all new construction, especially public buildings such as schools and publicly supported housing.*

7. *In view of the present impact of noise on the community around the Airport, all relevant public agencies, especially the Port of New York Authority, should press for the development and installation of "quiet engines" on aircraft.*

8. *The Department of Transportation should:*
Require the installation of acoustically treated nacelles on all existing aircraft by 1975.

Accelerate, through increased funding and cooperation with the National Aeronautics and Space Administration and other federal agencies, the development of quiet engines, and establish a regulation requiring that all new aircraft have engines that are quieter by 10 EPNdb (effective perceived noise level) below present standards by 1975.

We have considered the many possible ways in which Jamaica Bay may be used to satisfy the needs of the people of the New York area. With respect to some of these uses, we have concluded that:

☐ The permanent conversion of any estuarine area to airport or other commercial or industrial use diminishes a national environmental asset of great potential value to future generations. Although Jamaica Bay has been greatly altered by man's activities, its ecological viability can be maintained indefinitely into the future by environmental improvements only if no additional major incursions into the Bay occur.

To achieve this potential for sustaining a livable environment for millions of people near the Bay, we recommend the following actions:

9. In the next ten years, the City of New York should develop Jamaica Bay extensively for conservation and recreational uses by its own citizens and for compatible housing. This requires completion of its existing sewage-treatment program, immediate termination of dredging and sanitary-landfill operations, and the extension of mass-transit connections to shoreside areas. The City should oppose federal programs for construction of a hurricane barrier and for creation of a national park in the forms now proposed, since these will impede the development of the Bay for the above uses.

10. The State of New York should establish a Jamaica Bay planning commission, composed of representatives of the Bay communities, the Port of New York Authority, and other relevant City, State and Nassau County agencies, which should be charged with developing a comprehensive long-range plan for compatible development of the Bay and its contiguous land areas. The planning program should provide opportunity for direct community participation in the formulation of proposals and decisions affecting the uses of the Bay and its related land area, including airport improvements and operations that impinge on the quality of life in adjacent communities.

The ability of local communities to provide for increasing air transportation needs while improving the environment in the vicinity of airports is severely circumscribed by the nature of the air transportation system. A major airport, such as Kennedy, is a component of a national and international system serving air transportation needs of the whole nation. Federal authorities should aid local communities in the resolution of their problems by appropriate actions to bring about those beneficial changes in the entire system that lie beyond local control. In addition to federal action to abate aircraft noise generation, we recommend the following:

11. The Department of Transportation, through the Secretary or its agencies, should:

Take overall responsibility for present and future air-traffic-control system development; select a new and improved system from existing technology; and request the necessary funds and authority to install such a system by 1975.

Conduct an immediate inquiry into the operation of airports that are now or soon will be congested to determine how such operations can make more efficient use of existing capacity, and whether additional facilities are required.

When a need for a new airport has been demonstrated but local authorities have failed to site and construct it, select a site compatible with national environmental and urban growth objectives after consultation with local communities, and take necessary steps to ensure its construction.

Conduct an expanded research and development program for vertical and short takeoff and landing aircraft systems, and evaluate and make recommendations as to the integration of such systems into the total intercity transport system.

Develop ground access systems that can be integrated into an airport complex.

Recommend the modification, or, if necessary, the closing and replacement, of environmentally hazardous airports.

It is our hope that public understanding and discussion of these issues and the options that exist will eventually lead to wise decisions regarding actions to be taken. It is our fear that partisan debate will oversimplify the choices to "birds versus planes" or "jobs versus pollution." In our experience, no environmental problem can be comprehended in such limited terms, nor do we expect that the environmental improvement so sorely needed by our nation can be secured through decisions so narrowly informed.

The economic and political costs of implementing these recommendations may be large, but they are bearable. These are the unavoidable costs of maintaining a livable environment for urban populations. For Jamaica Bay, the environmental bill is due now."

FINAL REVIEW

From the standpoint of the Jamaica Bay study group as a whole, the effort was, for all intents and purposes, complete at the end of August. From a historical standpoint, however, the most interesting aspects were yet to come. These are intimately related to the manner in which the National Academies of Sciences and Engineering carry out their business and to their underlying self images, which tend to impact on the outcome of studies in general and this one in particular.

A meeting of the subcommittee chairmen with Dr. Fay was arranged for the third week in September to begin a review of work necessary to complete the report. A new outline was devised and assignments made. Several other members of the Study Group also agreed to continue their participation.

In mid-November, a draft final report was sufficiently complete to distribute for review. It was distributed to the following people: 25 members of the study group, 11 members of the Steering Committee, 5 members of the Academy Report Review Committee Panel for the Jamaica Bay study, including the chairman of the Report Review Committee, 13 members of the ESB, several officers in the NAE. A total of 72 copies were sent out for review. This, as it turned out, created subsequent problems.

From the very beginning of the study it was obvious to all of us connected with it that we would be subjected to continual inquiries from the media about the study. PONYA and the Academy agreed that the former would release no information to the press except progress reports formally transmitted to PONYA by Academy officials. PONYA was true to its word throughout. A potential problem lay with the Study Group members. Each member of the group was advised by the Academy in the letters of appointment and constantly reminded throughout the duration of the study that all aspects of the project were "privileged" until its formal release by the

Academy. They too were good to their word, without exception. But, with 72 copies of the draft report out for review, control became highly problematical. In fact, we were informed that a "Los Angeles reporter" did indeed have a copy three weeks prior to publication. *Newsday*, a Long Island daily newspaper did print large portions of the draft report on Monday, February 15, two days prior to formal release. Other press operations, including Dan Greenberg's *Science and Government Report* and the *New York Daily News*, pretended to have seen the report, but it is highly unlikely that they did.

Another problem arose with the chapter on public policy, which had not gone through the same kind of intensive editorial process to which the drafting committee had subjected the others. It was in essentially draft form and was couched in highly political sounding language. The reason for this lay in the fact that internal review of it had been left to last and the pressure to circulate the entire draft report for review in time to incorporate resulting comments in the final draft prior to the scheduled (Dec. 10-11) ESB meeting, where it was hoped it would receive approval of the Board, led us to make what we thought to be an expeditious move. It was our intention to "clean up the language," by subjecting the draft to the same review procedures all the others had gone through by the drafting committee at the same time the draft was circulating for external review - a naive and perhaps somewhat stupid move on our parts. The reaction of a few outside reviewers was astounding and the heat generated tended to obfuscate even what they eventually admitted to be the quality and value of the rest of the report. Great criticism ensued and though a complete rewrite of that chapter was almost immediately circulated for re-review it was almost completely obscured by the heat of the argument already in motion.

Though the issue of the highly political nature of that chapter was resolved by an additional subsequent redrafting into more considered, therefore acceptable language, which did not in any way change the substance, the die was cast and the ESB's approval for publication was somewhat begrudgingly given. An innocent, if somewhat unfortunate error in the interests of speed and deadlines, had blown to great proportions.

While explaining this, it should be pointed out what the time urgency was all about. The agreement with the PONYA was to conclude the study on or about the end of the year (1970). By paying a premium rate to the printer it would be possible, if a final manuscript was delivered in the second week of December, to publish the report by early January. The printer, at our bidding, made all the necessary, if expensive, arrangements and we had to deliver as agreed. Pressure from the media and elsewhere on PONYA and the numerous inquiries to the Academy on the status and outcome of the study made it desirable to publish as soon as possible. As it was, however, the rewriting and re-review required by the ESB took about two weeks. The diversions of the Christmas season caused further delay. The document was set into page proof on January 26, 1971, and following some last minute adjustments in the map proofs, the final draft report was delivered to the printer on February 1, 1971.

It was not until the previous December 8 that a meeting of the subcommittee chairmen and several other members of the drafting group was held with the PONYA top staff to inform them orally of the findings of the report. PONYA received a copy of the page proofs much later, on February 10, 1971, at which time the report was in the process of final printing.

This process was a matter of considerable discussion, which had begun during the Southampton sessions during the third week in August. The original proposal stipulated that after review and revision of the report the group would "discuss the report with the ESB, the PONYA representatives, and the interested Federal agencies." Further, the proposal, which in essence became the terms of the formal agreement, stated that "the report would be transmitted as soon thereafter as possible to the PONYA, in accordance with the established procedures of the Academy for submission of such reports." During the second week of the study, questions had been raised by study members at a plenary session about the review procedures and it was stated at that time that it was the Academy's intention to transmit a rough draft to PONYA for them to read and comment on in early September prior to ESB approval. Opposition to that procedure was expressed by a large portion of the study group, and a number of participants threatened to secede if that were to be the case. It was their feeling that any such exposure of the report to PONYA prior to its completion and publication was not acceptable for an academic study and might well lead to pressure for change in the manuscript. It was the Academy's position that, because the report necessarily relied on highly technical data developed by others, it had been intended that the PONYA reading would cover only those parts of the draft report in which such information was essential to the conclusions and which could significantly affect the validity and credibility of the final report. Several members of the study group argued that the danger, though a real one, is one that they inevitably face when they publish articles in scientific journals, having relied in any part on the work of others. They were ready, they said, to assume that risk, particularly since the draft report would in normal course be subjected to rigorous review by the Steering Committee, the ESB, and the Academy Reports Review Committee.

The objections were brought to the attention of NAS President, Philip Handler by telephone to Woods Hole, Massachusetts, where summer Academy meetings were in progress. Handler concurred in principle with those petitioners and advised the PONYA by letter (8/12/70) that "By our internal regulations, no report may be transmitted to a sponsoring body until a preliminary draft has been defended before an appropriate unit of the Academy and approved by that unit ... copies of the report ... will be made available to the PONYA immediately after the Board has reviewed the report and approved it for publication."

PONYA responded (8/12/70) by stating its concern "whether within the relatively short work time the group has had available to it, a firm understanding has been developed of the complexities of airport and airspace management and the role of the Port Authority. In short, we are fearful that there is opportunity for *factual* error in this fundamental aspect of the study ... I (a PONYA official) have furnished to the committee voluminous information extending over the twenty-three years of our airport experience and have spent many hours answering questions orally. My own capacity for factual accuracy probably has been surpassed, and yet the Study Group must rely upon my statements." He referred in closing to the March 11 proposal which confirmed "your staff's oral assurance that the report in preliminary form would be discussed with the PONYA." What this letter did not reflect was that the study group had received reports, data and information from many and diverse sources and had not relied upon the factual accuracy of the statements of any individual.

A meeting of the Academy principals was held on September 17 and it was concluded, despite objections, that the draft report would indeed be discussed with PONYA immediately prior to the December 10-11 ESB meeting, to which the draft report would be submitted for approval. As stated, such a meeting was held and an *oral* presentation made. PONYA made no attempt to influence the report and, while indicating some disagreement with the Group's interpretation of some matters, seemed satisfied that careful attention to factual data underlay the Group's report.

In the interest of fairness, a word here seems in order about the PONYA behavior as a sponsor throughout the study. It was, in every respect, above reproach. PONYA made no effort to find out what directions the group was taking nor did they, in any way, at any time, interfere with or attempt to influence the outcome or findings of the study. PONYA was completely open with the study group, making data freely available. Moreover, PONYA was extremely generous with and fully understanding of the difficulties the group faced in dealing with these complex issues. And, even though under great pressure from the media, the public, the airlines, and their Board, never did any PONYA official express impatience. PONYA, was in short, an ideal sponsor.

All press inquiries to the PONYA met with the response that it, like everyone concerned, was awaiting the outcome of the study and would take no steps until the study was completed and they had had time to study and digest its findings and recommendations.

The work of the study group was unknown to PONYA throughout. Only on December 8, 1970, was PONYA afforded an opportunity in the oral presentation referred to above of knowing what the study group had found. And, further, only on February 10 did PONYA receive any written information about the study in the form of page proofs, which were at that time already with the printer for fina final publication. The report was released to the public on February 17, 1971.

An interesting aspect of this recounting relates to Academy endorsement of the report. In early November, 1970, as noted above, copies of the draft report were sent out for review, with the results alluded to above.

Essentially the Study Group was criticized for the "gratuitous" and "highly political" nature of the Public Policy section. Less than one year before (about 10 months), the Council of the Academy and the ESB had insisted that they were only interested in the Academy's undertaking the project if it related explicitly to national needs and problems. At the end, the study group was criticized for going beyond its mandate and advising the Secretary of Transportation how to run his Department and a national air transport system and doing it in a very political, therefore offensive manner. The report had explained that many of the problems at Kennedy Airport were not within the ken of PONYA's responsibility and that they could not reasonably be expected to effect change and improvement in areas outside their jurisdiction. Some of these areas are clearly in the realm of the federal government. Moreover, several of the more active plans for Jamaica Bay are federal and not local.

The Group, was accused, as well, of using the study and the report to discredit the national administration and to further the political views of some individuals. A great debate took place at the ESB meeting; the study members present threatened to walk out and release a statement to the press exposing the Academy's treatment of its report in the review procedures. It was made clear that the Academies could accept, or not, the report, its principal findings and recommendations, all clearly stated and not subject to change. A retrospective rereading of the comments of the Report Review Committee, at that time a newly created Academy committee designated to review draft reports containing public policy implications, seems to have only a coincidental relationship to the heated rhetoric that marked that meeting at which RRC members were not present as a group. Criticisms by ESB members who had read their report centered on the following, seemingly unrelated, though actually intimately connected, items:

1. That the drafting committee members continued to receive an honorarium of $100 per day for each day they worked on the study throughout the drafting period (Sept.-Oct.) after the August session. Truly, under normal Academy procedures, this is highly unusual. An exception had, though, been officially made in this case in the interest of priority treatment and efforts to publish the report as soon as possible. The decision, though not agreeable to all the principals involved in the internal Academy decision-making process, was approved by the chief officers of the Academy in whose power it lay to make the decision.

2. That one member of the group, ordinarily a staff professional with a Senate Subcommittee, with long experience, insight, and background relating to environmental matters, highly respected and known as a thorough professional of great integrity, was somehow

suspected of lacking objectivity and having a conflict of interest because of his political predilections. This individual had served without compensation and his responsibilities had little to do with any of his Senate-related activities, which are primarily in the field of intergovernmental relations. His professional responsibilities had nothing whatsoever to do with pending or past legislation relating to airports, Jamaica Bay, or any other matter, with the possible exception of intergovernmental aspects, relating to the study.

This criticism was particularly interesting in that other Congressional staff people had previously served on Academy committees as had - and still do - other governmental officials from the Executive branch. Moreover, the man in question had been duly appointed by the President of the Academy, PONYA knew of his appointment, and no one at any prior time had objected to his participation. He was an exceedingly valuable member of the study group.

3. That the study exceeded its mandate by discussing national issues having little or nothing to do with Kennedy Airport or Jamaica Bay. This has already been discussed above and needs no further explanation.

4. A presentation at a President's Science Advisory Committee (PSAC) meeting several weeks before had made the point that the state of technology for new instrument landing systems (ILS) made the adoption of a "fourth generation" system not feasible in the foreseeable future. This judgment, some reviewers who were present at that meeting felt, undermined the conclusion of the study group (Chapter I) i.e., "the technology for an improved system exists" and that "without improved air-traffic control, the aerospace will be under-utilized while airports proliferate, pre-empting land that might better serve other purposes." (Vol. II, p. 5.) Moreover, the study concluded, "since the evidence indicates that improved air-traffic-control systems would permit substantial increases in the capacity of the region's airports, such systems be developed and adopted as rapidly as possible." (Vol. I, p. 2.) Some reviewers

felt this an unrealistic assumption given their understanding of the situation as presented at the PSAC meeting mentioned above. The study group was somewhat more optimistic about the technological prospects but had not reached their conclusions dependent on this development. Moreover, they felt that the technological development would be speeded with a national decision to adopt one or another of the system choices and to push for its development.

The impasse was broken by a decision to print the report in two separate volumes, the first of which received wholehearted Academy endorsement, the second, the Academy felt they could not as adequately judge the technological explanation of how the conclusions were reached. This judgment is reflected in the language of ESB chairman David Gates' letter of transmittal, to wit:

"Volume I includes the conclusions and recommendations, and a summary report, of the Study Group, which have been reviewed and endorsed by the Environmental Studies Board. The Board recommends their publication, in view of their broad public interest and with the hope that they will prove valuable in the critical decisions to which they relate.

Volume II is the Study Group's report to the Environmental Studies Board. It has been reviewed, and the board recommends publication in view of its broad public interest and with the hope that it will prove valuable in the critical decisions to which it relates."

The matter, of course, was not resolved by resort to this subtle distinction. It did, however, satisfy the Academy, the ESB and to a much lesser extent, the study group members. The more basic issue goes somewhat deeper to the root of the Academy's mode of operation and to the way in which it is constituted as an institution. It can, for example, no longer rely, as it once could, on its very limited number and scope of membership to supply the expertise and advise in broad diverse areas of science and technology that it is increasingly called upon to give.

That implies that the Academy will, if it continues to expand its budget, staff and scope of its studies, have to invite participation of people from disciplines and sectors not traditionally represented on Academy studies. It will also be called upon to deal with questions and problems which the Academy traditionally would not have undertaken for study. For all this, the Academy's imprimatur will appear on studies and reports to which few, if any, Academy members would have contributed. Such was the case of the Jamaica Bay study.

SOME OBSERVATIONS

Many Academy-associated people have, since the study was issued, felt that this kind of project should not again be undertaken by the NAS. Technology assessments of real-life problems,because they inevitably involve economic, political and social questions of a kind the Academy would prefer not to become involved in, are not questions that scientists are particularly well-suited, by background or training, to deal with. Technology assessments can not avoid dealing with "normative," ergo difficult problems.

Once again, with reference to the Jamaica Bay study, the group addressed itself to the way it viewed its role as a multi-disciplinary technology assessment team:

"... it is evident that the problems are very complex and that their solution may eventually turn on a Solomon's choice. The members of the study group are, however, not decision makers; they serve only in an advisory role. The decision as to the disposition of the Bay is necessarily in the hands of others. Scientists alone cannot make the choices concerning alternative solutions to problems of this kind, which must ultimately depend on competing values and the weight governments assign to them in arriving at decision. Scientists can, however, isolate and consider objectively the issues involved, lay out and evaluate the effects on the physical and human ecology that may be expected to follow particular actions, and suggest and examine the various alternatives and their implications. However, we recognize that scientists are not value-free and that their concerns and predispositions will have some effect on their own conclusions and recommendations. But, whatever decision is ultimately arrived at, it will be a public one made by people whose actions must be acceptable to their constituencies." (Vol. I, pp. ix-x.)

The Jamaica Bay study, despite the institutionally-related difficulties discussed in this paper, did prove that such assessments could be done well and effectively. What the experience also illustrated is that a new institutional mechanism will be needed if technology assessments are to become a regular part of our public decision-making process.

It is probable that the subject matter of most technology assessments will tend to be controversial and involve considerations of political decision-making, economic interests, quality of life, environmental impact, diversion of resources and competing constituencies with valid claims. This seems almost unavoidable, given the scope and intensity of technological impact on human ecology. Technology assessments, of which the Jamaica Bay study may stand as a prototype, will be increasingly important to our future directions and decisions as to priorities and expenditures for Research and Development.

A FINAL WORD

The report through, was issued, as noted, on February 17, 1971. The President of the National Academy of Sciences has on several public platforms reffered to it as a landmark Academy study, an example of the kinds of things the Academy can contribute to the solution of contemporary problems relating to our environment.

The PONYA announced on February 17, 1971 that it was cancelling its efforts to win approval for any runway extension into Jamaica Bay.

TECHNOLOGY ASSESSMENT: AN INSTRUMENT FOR GOAL FORMULATION AND THE SELECTION OF PROBLEM AREAS

BODO BARTOCHA

Head, Office of International Programs, National Science Foundation, Washington, D.C., U.S.A.

1. BACKGROUND

1.1 Introduction

The increasing pace of innovative change with rapidly emerging new technologies has, in recent years, caused profound, complex, and often unforeseen results in society and the environment. The impact of these developments has become so great that it affects many fundamental aspects of national policy and our life. As a result, there have been increasing attempts to foresee and evaluate the consequences of new technological applications. The methodology of doing this is called Technology Assessment. This term was added to our socio-scientific jargon only in the late '60s, but its meaning is much broader than any encyclopaedic definition might imply. Technology Assessment as a limited art is nothing new and certainly predates Rachel Carson's "Silent Spring" (1962), a rousing book, considered by many to have been the catalyst which finally triggered our long overdue reaction to an approaching ecological catastrophe.

*The views expressed in this chapter, are those of the author, and do not necessarily represent those of the National Science Foundation.

The growing concern with the undesirable side effects of science and technology, our diminishing resources, the need for environmental restoration, and our economic difficulties have stimulated and indeed made mandatory the search for new methods that would provide a way to early detect and measure the impact - good or bad - of emerging technologies. Such a "tool" would aid in the prevention, or at least amelioration, of such far-reaching effects as the secondary consequences of the unrestrained application of certain chlorinated hydrocarbons, for instance. Which is not to say that we would not have used DDT had we been aware of its widespread ecological effects, but perhaps we would have used it more wisely.

During the last few years, since 1967 to be exact, Technology Assessment has become a household word among legislators, government administrators, the public at large, and a good many other followers who sense the importance of technology assessment without quite understanding what it is all about and simply want to become associated with the latest scientific fad. As often as the term is used, it is just as often misused.

Technology Assessment is a decision aid - no more, no less. It is not a panacea for social ills or ecological problems. As currently practiced it will hardly replace our political and bureaucratic decision making process which is unresponsive, and based on hunches,(administrators call it "intuition") and hasty generalizations. It will certainly be some time before technology assessments have reached a point where not only specialists read them, but administrators use them. To date, we do not have a surefire recipe for doing successful, complete assessments, but a large number of efforts are under way in an attempt to better understand and appraise current problems which resulted from technological progress as well as to anticipate future possible consequences of newly emerging technologies. Hopefully, these studies will result in the identification and classification of viable options for alternative futures which can be translated into national policies and constructive action.

1.2 A Definition of Technology Assessment

Technology Assessment is the process of studying the social, economic, political, and physical consequences of the application of a present or emerging technology.

It can be defined as a systematic analysis, where all impacts and implications, direct or indirect, real or potential, present and future, beneficial or detrimental, of a technology are defined, evaluated and measured, and the cause-effect relationships identified. The results of a Technology Assessment should include alternative solutions to a problem, ranked according to their social cost-benefits and recommendations for policy changes, control or mitigation options, or new initiatives.

1.3 Why Technology Assessments

Melvin Kranzberg, in "Historical Aspects of Technology Assessment," [1] has aptly described man's concern for technological change during the ages. However, there is now an increased awareness of the subtle and sometimes not so subtle changes in social values and the desire for improving the quality of life. The need for Technology Assessments is evident. The rapid increase in the rate of technological change, growing population pressures with the concomitantly faster consumption of diminishing natural resources, the resulting deterioration of the human environment, natural and social, and threatening ecological catastrophies have brought on the demand by all sectors of society for corrective action and controls to prevent further deterioration of man's habitat by technology. It follows that it would be desirable to foresee these effects, i.e., to have an "early warning" system to direct technological developments so as to maximize the public benefits while minimizing the public risks.

Out of this reasoning was born the process we now call "Technology Assessment" - a term coined in 1969 by Philip B. Yaeger, Counsel for the Committee on Science and Astronautics, U.S. House of Representatives.

1.4 Some General Remarks About Technology Assessment

It is significant for our current social climate that we almost never hear the terms economic assessment or social assessment, and very seldom environmental assessment. In the few years that the term technology assessment has been with us, it has acquired synonimity with pollution, ecological damage, and other negative aspects of technological effects. This overemphasis has led to anxieties, particularly in our industrial community, that zealots would use technology assessments to prevent the growth of R&D, hence impede invention and innovation and thus directly threaten our national security and survival.

As different as the definitions for technology assessment are, so are also the emerging schools advocating various ways to perform an assessment. They range from the methodological addicts who would like to see technology assessment reduced to a systems inspired input - black box - output model, to the fans of the intellectual-intuitive approach who claim that structured guidelines are superfluous. All that is necessary is to attract talented, creative individuals and presto! your problem is solved. It is hardly necessary to point out that a mixture of both extremes is required, and that the complexities of our problems are such that any individual, however brilliant, will feel quite forlorn when he attempts to untangle the multistructural inter-relationships and influences of our societal "Networks" such as energy grids, railroads and canals, airlines, telecommunication systems, freeways, and pipelines. If relationships were clear and easy to survey, we would not need technology assessments.

It was previously stated that Technology Assessments can be viewed as a decision aid for the policy maker.

The results of a Technology Assessment could be a series of action options arranged in the order of social cost-benefits. Unless the policy maker knows where he wants to go, i.e., is making his decision in a normative system, he will be hard put to make the right choice from the options presented to him. An effective use of technology assessments, therefore, requires the establishment of existing and future goals and objectives, thus we can view technology assessment not only as a control tool but also as the vital input to a national resources allocation system. It is in this context that the remainder of the paper will address technology assessment procedures.

2. THE STATE OF THE ART

2.1 Classification of Technology Assessment

From the points made in the background discussion, it follows that technology assessments which help us foresee future developments are most important to us. Less so are those which deal with present problems, and those which examine past happenings are of very limited interest. We can, therefore, distinguish at least among three classes:

	1.1 Anticipatory	Forecasting new trends and their possible effects: cryogenics, central data banks, genetic engineering
1. Future Oriented		
	1.2 Responsive	To already perceived problems: new dams, trans Alaska pipeline, nuclear power plants, new jet-ports

2. Present time Oriented	2.1 Reactive	To problems with unknown causes or disasters: power failures, earthquakes, oil spills, midair collisions
	2.2 Corrective	Known problems, effects are felt, causes traced, corrective action feasible: pollution, rural poverty, traffic congestion, brownout
3. Past time Oriented	Retrospective	Examination of circumstances and facts: why was a project successful, rejected, or abandoned. Nuclear plane, MOHOLE, plowshare, etc.

The subdivisions are somewhat overlapping, however, they suggest the possibility of additional stratifications.

The report of the Committee on Science and Public Policy [2] discusses classification by focal points:

1. Technology (facilitating or constraining assessments)

1.1 advance: evaluation of weather modification
1.2 transfer: satellites for surveillance of earth resources
1.3 growth in scale of application: oral contraceptives, growing size of tankers

2. Society

2.1 *system*: housing, production, education, transportation
2.2 *relationship*: employment, family, property
2.3 *category*: class structure, allocation of resources
2.4 *aggregate*: trade balance, GNP, military strength

3. Environment (physical environment or ecological objectives)

3.1 *aesthetic*: noise levels, smells
3.2 *entire ecosystems*: changes in species balance, water eutrophication
3.3 *biomedical*: lead in the atmosphere, trace residues in plants

4. The Individual (life or work experience, participation in decisions)

4.1 *effects on the child*: mass media, socialization, learning process
4.2 *adult experiences*: material wealth, mental health, companionship, solitude, risks of accidents

5. Some mix of the preceding

5.1 *synergistic effects*
5.2 *cumulative effects*

A third classification could be one by user or client [3]:

The Executive Branch

Congress

The Courts

The Public

Commercial and Industrial Sectors of Society

Interest Groups and Lobbyists

Foreign Countries

and a fourth grouping might be according to the type of recommendation made or the intended end use.

2.2 *Assessment Steps*

The diversity of problems encountered, the scarcity of full technology assessments in the past, and the lack of experience make it impossible to synthesize a generally applicable method or technique comprehensive enough to fit all circumstances.

Analysis of the very limited number of acceptable cases, acceptable in the sense of being fairly exhaustive in scope and reasonably successful in impact measurement, suggest the following approach:

1. Select and describe the subject or area.
2. Identify its elements - technical, social, legal, political, and economic.
3. Forecast, trace and assess (measure) the consequences and impacts (primary and secondary).
4. Identify alternative options (trade-offs).
5. Recommend approach or policy to solve problems.

The Report of the Committee on Public Engineering Policy of the National Academy of Engineering [4] describes seven pragmatic sequential steps:

(1) Identify and refine the subject to be assessed.

(2) Delineate the scope of the assessment and develop a data base.

(3) Identify alternative strategies to solve the selected problems with the technology under assessment.

(4) Identify all parties affected by the selected problems and the technology.

(5) Identify the impacts on the affected parties.

(6) Evaluate or measure the impacts, including the sociological.

(7) Compare the pros and cons of alternative strategies.

The same seven steps incidentally are those used by any competent systems analysis effort.

Any technology assessment approach must take into consideration a number of criteria. Those related to area selection and problem choice are perhaps the most important ones. Concurrently, the feasibility (technical or economical) of implementing a solution should be evaluated. Time and money can be saved if it becomes apparent in the selection process that whatever solution was found, it was not doable or would not result in any appreciable impact. The availability of goals of a data base, and the subject area to be assessed will influence the choice of methodologies and the sequential steps. It is, therefore, imperative that flexibility is maintained, whatever approach one develops. The whole process must be iterative. Only in the case of retrospective studies, to find out why a project was terminated or why it failed, is a one-time approach satisfactory. A continuing difficulty is the definition and description of a technology and its supplements, including social innovations and social technologies. The same holds true for the definition of social impact areas, their relationships with a major technology, and the measurement of these factors. Here checklists from PPB studies, from developing goals, or making technological forecasts are of great help.

2.3 The Assessment Tools

Technology assessment has no defined set of tools to be applied in a definite predetermined way. It borrows and adapts from all fields - the econometrician, the statistician, the sociologist, and the operations research analyst. A technology assessment should utilize the whole repertory of tools and techniques familiar to these and other related disciplines. The following is by no means a complete list and there are some redundancies, but it shows the variety of methods an "assessor" should be familiar with.

2.3.1 *Qualitative and quantitative appraisal techniques,* such as those used for the selection of research projects. Most are based on the quantitative subjective judgments of experts.

2.3.2 *Systems analysis* - better, systematic analysis, since systems analysis shows quite a number of shortcomings when applied to complex social problems. Systems analysis can be of significant help when dealing with a number of variables which all converge toward a well-defined goal, i.e., when the primary objective is the solution of a problem. Systems analysis works best when applied to development and hardware problems. Basic research problems frequently elude treatment. Cost-effectiveness and cost-benefit analyses fall into the same category. Wherever it is difficult to quantify variables or pacing parameters, these techniques become unreliable.

2.3.3 *Decision aids* such as relevance trees and evaluation matrices.

2.3.4 *Forecasts:* economic, technological, social, and, if it were possible, political forecasts (at least a scenario might be attempted). Forecasts are a necessary requirement for all anticipatory assessment and these are the ones we are most interested in. We do not want to predict the future, but endeavor to obtain a better insight for decision making. A Delphi, trend extrapolation, trend correlation, or growth analogies or a substitution method are just a few of the many necessary inputs. Forecasting becomes progressively more difficult when one proceeds from economic to technological to social forecasting, for one reason (among others) because of the lack of an adequate data basis. The absence of dynamic, causally-oriented models and the problem of testing the existing simple ones at reasonable cost under real world conditions made the anticipatory assessments, therefore, the most difficult ones. In reality no reliable political forecasts are possible. A political future can only be "divined" and that in an extremely coarse fashion. Anticipatory assessments rise and fall with the quality of their forecasts.

2.3.5 *Simulation and operations research models.* Here we are particularly interested in those able to cope with stochastic (random) phenomena. Those applicable to deterministic phenomena are of lesser interest since we are usually unable to predict the exact sequence and ranking order of future events. Mathematical and metaphoric models have been demonstrated to be useful, as well as Monte-Carlo techniques. Simulation models will most probably become our most useful forecasting and assessment tool.

2.3.6 Questionnaires - We are all only too familiar with these.

2.3.7 Public participation hearings. A process more successful and effective in the U.S. than in Europe. It is important to have the public-at-large participate in the technology assessment process. The motivating aspects of participating in the decision process and the educational value should not be underestimated.

2.3.8 Not to be overlooked is the sheer *gut-feeling* and *intuition.*

To those acquainted with the techniques I have just mentioned, the deficiencies are also obvious.

2.4 Current Assessment Deficiencies

In spite of the importance of technology assessment, very little has been done in the past to formalize and structure the procedure and to systematically analyze and identify those critical factors that form the element of a successful assessment. Various points were found to be essential but our understanding of the total process is still rather rudimentary. There seems to prevail a situation similar to that in technology transfer or coupling: Only those who have done it successfully can adequately describe it, i.e., our approach is more by intuition than by deduction from reliable empirical data. We do not have sufficient information from which a good and general approach could be synthesized.

The most critical voids in our knowledge regarding technology assessment are:

2.4.1 Criteria for choice and priorities.

2.4.2 Reference standards and pacing parameters sensitive to end objectives.

2.4.3 Indications and new value systems to express "worth" in other than just monetary terms.

2.4.4 Better simulation techniques and dynamic models.

2.4.5 Skilled interdisciplinary personnel and adequate resources.

2.4.6 Popularization and public understanding of long-range risk, expected values, and uncertainty, and

2.4.7 Imagination.

3. MANAGING TECHNOLOGY ASSESSMENT

The majority of current assessment approaches are performed or sponsored by the Federal Government, specifically the Executive Branch. State and local governments also play a certain role in technology assessments as do some private institutions, notably the "think tanks." Very little information is available about technology assessment in private industry. Most technology assessments performed outside the Federal Government are in response to governmental legislation rather than the perception of social needs or societal problems caused by a technology. It is understandable that private industry is reluctant to reveal proprietary information to a competitor. Federal Government technology assessments should be and mostly are a matter of legal and public record.

3.1 Where should Technology Assessment be Done and Who should Do It

Most Government agencies and many of the larger companies already have a staff of analysts doing cost-effectiveness or cost-benefit analyses, PPB studies, or Program Evaluations. The formation of these groups resulted from previous administrative requirements for a better planning approach and a desire to make more rational choices among the many options offered. This analytical staff could certainly form the nucleus of a Technology Assessment team and many times has.

If we remember some of the shortcomings of past assessments, we should also keep in mind that in previous efforts the social sciences were rather thinly represented. The multidisciplinary aspects and the social ramifications are very important for an assessment. We should, therefore, have economists, sociologists and lawyers on the team, not only physicists, chemists, and engineers. We will see more teams emerging where the influential role is played by social scientists.

Another problem to consider is that our in-house organization could conceivably put forward a rather biased view and also become bureaucratically stale after a few bouts with Technology Assessments. It seems, therefore, preferable to have technology assessments performed by carefully selected *ad hoc* task groups whose talents are augmented by well-qualified outside talent rather than a permanent organization. It is for this reason that I am not altogether in agreement with the proposals to have technology assessments performed within the framework of, for instance, the National Bureau of Standards or another National laboratory, i.e., to rigidly institutionalize Technology Assessments.

The paramount requirements are to establish an atmosphere of independence and to protect the assessors from undue outside influence.

However, we are not only facing institutional or organizational barriers. The primary limiting factor in technology assessments is not the lack of methodologies or managerial accommodations but the lack of personnel really qualified to undertake social and economic analyses.

3.2 *How Much Does an Assessment Cost, How Long Does It Take?*

From my experience I would estimate a partial assessment to cost between $200,000 and $300,000, the equivalent of 5-6 man-years of senior professional effort. Joe Coates in a recent paper [3] gives some examples:

Jamaica Bay Study (NAS)	$300,000	
Colorado Snow Project (SRI)	150,000	(NSF)
Civil Aviation R&D (GWU)	220,000	(NASA/DOT)
MIT Summer Study	125,000	

A study by the Mitre Corporation, Washington, D.C., "A Technology Assessment Methodology" was done for about $150,000, and a study by the International Research and Technology Corporation, Washington, D.C. being performed for the U.S. Postal Service has been contracted for $320,000. In contrast, the Alaska pipeline study is estimated to have cost $7 to $8 million.

An assessment should not take longer than one year. Otherwise the situation to be studied may have changed so much that the original assumptions will no longer hold. If the technology assessment is an anticipatory one, it should be updated periodically.

The distribution of effort which differs from case to case and depends much on the available data base and the complexity of the problem or technology to be assessed is generally divided as follows:

30% of time available for selection and description of problem and data collection

50% analysis, particularly impact evaluation

20% review and report writing

4. THE SELECTION OF A PROBLEM AREA FOR ASSESSMENT

Not even the U.S.A. can afford to tackle simultaneously all the problems she faces. Not only that, but if in over-reaction to these problems, we allocate our resources predominantly for environmental or ecological restoration instead of in a more balanced and comprehensive program to improve the quality of our life with rational choices in technology, we will surely suffer from additional imbalances and more secondary side effects.

The difficulty of priorities and that of establishing generally acceptable criteria for choice will remain with us for some time, but practical approaches to the problem are possible. The outcome may not be perfect but it is a beginning.

A small case study [5] suggesting one such approach is presented here.

A group of six senior professionals with a mixed background in terms of disciplines participated in this exercise. To begin with, a number of criteria were chosen for the general identification of problem areas. It was agreed that:

(1) The problem must be major, in that it has or will have a significant effect on a large segment of the Nation.

(2) The problem must either exist now or must be readily foreseeable.

(3) Attack of the problem should be feasible within the given technical, economic, and political possibilities and constraints.

These are simple assumptions, of course, but rather than starting with a complicated set, it is suggested that perhaps six or eight selection criteria are ample for a first try. The three given are practical and workable.

Based on these criteria, a large number of problem areas were screened, taken from various government, public, and industrial publications. About 20 were selected.

As a next step "organization criteria" were established to aid in the weighting and ranking of the list of problems.

Discussion among the task group members made it evident that the criteria were of various degrees of importance in the selection process. Consequently, the group assigned to each criteria a selective weighting factor. These factors represent averages of values in the range of 1-5 independently assigned by each group member. It is worthwhile to note that the numerical spread among the weighting factors indicated by the various participants was extraordinarily small.

Criterion	Weighting Factor
1. Successful involvement by other organizations	-3
2. Overall probability of success	5
3. Probability of short-term success	3
4. National visibility	3
5. National importance	5
6. Adequacy of present in-house capability (technical and management)	4

Each group member then rated numerically the applicability of a particular criterion to a specific problem area with ratings varying between 1 and 10.

The product is a matrix which yields a numerical suitability index. Several ratings were performed and the averages calculated [6]. A sample of the evaluation matrix is given on the next page.

The ultimate product; i.e., ranking of problem areas, is shown next: (See Problem Area Suitability Listing)

Eleven problem areas emerged as being of prime interest. The cutoff appears arbitrary but considerations with regard to resources (funding,

EVALUATION MATRIX

	C_1 (-3)	C_2 (5)	C_3 (3)	C_4 (3)	C_5 (4)	C_6 (5)	I_{P_1}
Power	3.2 / -9.6	8.7 / 43.5	6.3 / 18.9	7.5 / 22.5	6.2 / 24.8	7.5 / 37.5	137.6
Population	3.8 / -11.4	7.0 / 35.0	4.2 / 12.6	8.8 / 26.4	5.8 / 23.2	9.7 / 48.5	134.3
Human Settlements	4.4 / -13.2	7.4 / 37.0	5.2 / 15.6	8 0 / 24.0	5.1 / 20.4	7.8 / 39.0	122.8
Pollution	3.2 / -9.6	8.6 / 43.0	6.6 / 19.8	9 0 / 27.0	6.6 / 26.4	8.4 / 42.0	148.6
Crime & Delinquency	3.0 / -9.0	5.7 / 28.5	3.8 / 11.4	9 5 / 28.5	6.2 / 24.8	7.2 / 36.0	120.2
Natural Resources	5.0 / -15.0	7.7 / 38.5	6.5 / 19.5	5 3 / 15.9	5.5 / 22.0	7.0 / 35.0	115.9
Education	4.0 / -12.0	7.8 / 39.0	5.3 / 15.9	8.7 / 26.1	7.5 / 30.0	8.0 / 40.0	139.0

$$I_{p_1} = \sum \left(W_{C_1} \cdot R^1_{C_1}\right) + \left(W_{C_2} \cdot R^1_{C_2}\right) \cdots \quad \left(W_{C_n} \cdot R^1_{C_n}\right)$$

W_{C_1} = Weight of Criterion 1

R^1_C = Relative importance of Criterion 1 for Problem 1

available manpower, etc.) entered into the decision. One should also realize that the numerical listing leaves a false impression of accuracy and definiteness (and that this is a study example only and *does not* reflect programmatic priorities of the National Science Foundation).

PROBLEM AREA SUITABILITY LISTING

Rank Order	Problem Area	Index
1.	Pollution	148.6
2.	Education	139.0
3.	Power	137.6
4.	Population	134.3
5.	Human Settlements	122.8
6.	Domestic Economic Development	121.3
7.	Group Conflict & Alienation	121.1
8.	Crime and Delinquency	120.2
9.	Natural Resources	115.9
10.	Transportation	114.5
11.	*International Relations*	*114.4*
12.	Information Transfer	111.8
13.	Fire	109.3
14.	Poverty	108.9
15.	Food Supply & Distribution	104.8
16.	Artificial Stimulants	100.9
17.	Storms	95.9
18.	Health	91.2
19.	Pesticides	89.0
20.	Noise Pollution	87.1
21.	Earthquakes	75.1

5. CONCLUSION

Technology Assessments, which had a rather promising start in the late sixties, have not received the attention during the last year or two which they warrant, based on their value as decision - and policy-making aids, and in normative planning. Not only has the assessment effort within the government lacked coherence, there has also been very little activity outside the Federal government.

Many opportunities for assessments were not developed. Part of this may simply be due to the shortage of funds, but even of greater importance is the lack of a general understanding of the importance of Technology Assessments in a time of crisis when resources are scarce and good long-range planning is a must.

REFERENCES

1. KRANZBERG, MELVIN, Historical Aspects of Technology Assessments, Program of Policy Studies in Science and Technology, The George Washington University, Washington, D.C., *Occasional Paper #4*, August, 1969.

2. Technology: Processes of Assessment and Choice, *Report of the National Academy of Sciences to the Committee on Science and Astronautics*, U.S. House of Representatives, July, 1969.

3. COATES, JOSEPH F., Interdisciplinary Considerations in Sponsoring Technology Assessments, Engineering Foundation Conference, Andover, New Hampshire, August 30 - September 3, 1971.

4. A Study of Technology Assessment, *Report of the Committee on Public Engineering Policy*, National Academy of Engineering to the Committee on Science and Astronautics, U.S. House of Representatives, July,1969.

5. *Report of the ad hoc Committee on NSF Societal Problem Assessment Institutes*, July, 1970 (An internal National Science Foundation staff study).

6. This technique is explained in greater detail by CETRON, MARVIN J., in: Quantitative Utility Estimates for Science and Technology (QUEST), *IEEE Transactions on Engineering Management*, March, 1967.

THE SCOPE AND LIMITS OF TECHNOLOGY ASSESSMENT

RICHARD A. CARPENTER,*

Chief
Environmental Policy Division
Congressional Research Service
Library of Congress
Washington, D.C. 20540, U.S.A.

INTRODUCTION

This paper is presented from the viewpoint of the user of assessment results. The author represents a policy research and information service within the Legislative Branch of the U.S. Government. Similar units are developing in many organizations (governmental and private) to assist the decision maker with more comprehensive, timely, and complete analysis than would be available from conventional information resources.

The definition of Technology Assessment used in the Congressional Research Service is not markedly different from others in the literature but is repeated here in order to stress three points: TA is a continuing process, not a product; TA includes primary intended impacts along with special search for second order consequences; and effects, rather than technology, are what are being examined.

Definition:

Technology assessment is the process of taking a purposeful look at the consequences of technological change. It includes the primary cost/benefit balance of short term localized market place economics, but particularly goes beyond these to identify affected parties and <u>unanticipated</u> impacts in as broad and long range fashion

*Present address - National Academy of Sciences, 2101 Constitution Avenue, Washington, D.C.

as is possible. It is neutral and objective, seeking to enrich the information for management decisions. Both "good" and "bad" side effects are investigated since a missed opportunity for benefit may be detrimental to society just as is an unexpected hazard.

Assessments are not decisions; they have the effect of a brighter light to illuminate and reveal more obscure facts and relationships in complex matters. In the Western world, a growing number of civic action groups, official agencies, expert advisory panels, and special interest organizations are participating in policy planning and decision making. Modern communications have enhanced the ability of individual citizens to take a more direct role in discussions and choices of action. All of these user groups are served by technology assessment if they avail themselves of the opportunity.

SCOPE AND CONTEXT

Technology assessment is the current culmination of a decade or so of thoughtful development in enhancing decision making. Even earlier roots can be found in the orderly concepts of operations research and weapons systems engineering. Lord C.P. Snow proclaimed a strong motivation for improved communication between specialists and the power structure in his elucidation of the "Two Cultures" situation. The first round of science policy discussions in the early 60's addressed policies for science in order that science might better serve policies and missions of all sorts. Rachel Carson's "Silent Spring" dramatized to the public and to politicians what was already apparent before 1962 to field biologists and bee keepers — that pesticides have important effects beyond their intended targets. Admiral Rickover wrote in 1965 of the necessity for a "Humanistic Technology."

In the immediate post-Sputnik years of unquestioning public acceptance of jet aircraft, computers, nuclear deterrents, therapeutic drugs, and space exploration, a brief euphoria existed based on the premise that man could do anything — whatever he wished — with his technology;

the only question remaining was to decide what he wanted to do. Emmanuel Mesthene termed the period a "recovery of nerve" compared with the previous state of ignorance about nature and the mystery (to most persons) of science. But the confidence so recently gained was soon lost as evidence surfaced of unanticipated and unwanted consequences of applied science.

A public reaction has set in,which has still to run its course. The reaction is anti-technology, anti-science and anti-intellectual. It has resulted in:

an abrupt halt in the growth of research in many disciplines;

a decline of support for tax revenues devoted to science;

calls for a moratorium on new knowledge;

rejection of science and engineering training by students;

cancellation of major technological projects such as trans-Florida barge canal and the SST;

delay of other projects such as nuclear power plants, highways, and multipurpose dams; and

acute "embarassment" for the political leaders over their demonstrated lack of control.

These manifestations of public disenchantment pose a great challenge to rational management of society. It does not seem reasonable to face the difficulties of population growth and concentration, development of the "Third World", and preservation of a productive but finite ecosystem without drawing on all of the scientific understanding we can command. And yet if democratic political processes are not satisfied that technology is steerable, the essential tools man needs to survive may be denied for fear of unacceptable side-effects.

The O.E.C.D. report "Science, Growth and Society — A New Perspective" concludes:

"The world-wide culture of educated youth, which is deeply concerned with ecological perspectives and is increasingly anti-materialistic, egalitarian, anti-meritocratic and anti-bureaucratic, could conceivably even adopt anti-rational views..."

Optimistically, the position can be taken that the present advanced state of science and technology also supplies the means of building the steering apparatus and the feedback command and control systems. Technology assessment is the means of demonstrating to the satisfaction of society that orderly progress is possible.

Policy Analysis

Policy is the set of principles that govern action toward given ends. Technology assessment is an important new ingredient in policy analysis and planning; it should be realized that it is but a sub-unit of the larger "political assessment" which determines the action managers will take. This political assessment includes the subjective judgments of responsible authority whether in government or the private sector. A brief description of policy analysis is necessary to show how technology assessment supports the total decision making process.

Modern staff assistance to decision makers begins with the assembly and examination of factual and polemic material relevant to the problem. This input of material includes goal statements, laws and court decisions adjudicating legislation, position papers of opinion leaders and public figures, resolutions of special interest groups, resource inventories, and rules and regulations of government agencies. The policy analyst identifies objectives of society, perhaps rephrasing them to sharpen the issue. The inventions in past courses of action are compared with achievements, and means are related to goals.

The purpose is to construct guidelines for management of programs in society by government and business.

These principles will also guide the behavior of citizens in individual decisions. It is necessary to interpret what is found in the policy material in order to relate rhetorical ideals to practice.

The analyst searches for:

internal inconsistencies in policy statements

conflicts with other policies

conventional economic data on costs and benefits of certain courses of action

forecasts and probabilities of future events and situations

assessment of primary and secondary impacts on society and the natural environment

political consequences — public acceptance

problems of administrative implementation

implications for institutions and organizations including coordination problems

timetables and time horizons for implementation

priorities in the allocation of manpower, facilities and money

alternatives offered and alternatives which need to be developed

criteria and measures for program evaluation

requirements for overview by policy level.

Technology assessment results will contribute to a great portion of these considerations but the overall decision is integrated from more than just the technical ingredients.

The Transformation and Flow of Scientific Information

Most decision makers are not trained or experienced in science or engineering and some translation and transformation is necessary in order that they may use information from these fields. In the past ten years considerable effort has been made by both the management institutions and the scientific community to lubricate the interchange of questions, ideas, opinions and facts. A high degree of receptivity on the part of policy makers is a keystone in erecting the system of effective communication.

Table 1 shows the flow of scientific discoveries into management decisions. At the first stage scientists working in a variety of research laboratories and field stations generate new understanding of the natural environment. Over 350 years of development has gone into the scientific method which establishes facts and extends the boundaries of knowledge. Disciplinary specialists assemble raw data around hypotheses in order that their colleagues may replicate experiments and verify results.

At the next level these reports are combined by equally competent researchers into peer judgments of broader significance. This information analysis stage produces the traditional review articles which compare and select the most valid information and synthesize a status report on a particular field. This is an intra disciplinary validation of what is known, what is not known, and what could be known with more research in a certain time period. Information analysis feeds back into the generation stage by providing a scientific field with a constantly advancing base of accepted knowledge from which to break new ground. It is the traditional role of learned societies. Now, in addition, the analyses are tailored for further use by assessors and policy level managers.

The third stage, technology assessment, adds more context such as social and economic considerations. The applications of the scientific knowledge through engineering innovations and guidelines for human activities are discerned and evaluated. Options and their consequences are investigated. A systems viewpoint is assumed and inputs from other disciplines are integrated. The real world situation is the matrix upon which the scientific facts are hung for examination. Relevance is a key word.

Finally, the structured, interpreted, validated information is submitted to the political context. This stage must be supported by each of many user groups: the executive departments and agencies, the Congress, the Judiciary, private sector interest groups, conservation and consumer protection organizations and so on. Subjective values are included at this point and the alternatives offered by science are tailored to support political objectives.

The role of the public spirited scientist encompasses all four stages but he is a different person in each forward progression. It is very important for the lay-user of scientific information to establish the level at which he is communicating with his scientific assistance. Is it as a generator of new facts, a validator of disciplinary information, an interpreter of relevance or an advocate of a particular option?

LIMITATIONS AND CAVEATS

Technology assessment has been called a social science or political science invention — if so, its innovation and diffusion into everyday use is likely to be far more tortuous than any of the natural science applications (e.g., sliced bread, television, spray deodorant). The following difficulties and limitations are apparent with the present experience.

(A) Most important of all, TA does not make decisions; the decision maker (duly constituted or not) cannot abdicate his duty since assessments only provide and structure alternative actions according to the available factual information. The comparison of options may show that one is clearly outstanding. However, the

Table I. The flow of Science Information.

Stage	Functions and products	Institutions and performers
I. Generation	New facts and knowledge from research, monitoring, surveys, testing. Publication in the scientific literature after peer review and editing. Discussions in meetings of learned societies.	Universities, government laboratories, industrial research operations, independent laboratories and foundations
II. Analysis	Information processing: collection, organization, indexes and bibliographies, storage, retrieval. Critical review: synthesis and comparison of independent studies, evaluation and validation, confidence and error limits, digestion and repackaging. Publication as critical reviews, state-of-the-art reports, compilation, standard reference data.	Specialized information analysis centers. Ad hoc panels from professional societies. Government agencies, commercial information organizations and publications. Recognized scientific leaders. Universities with strong faculty in certain fields.

III. Assessment	Economic and social context added. Applications, alternatives and options for engineering innovation. Social indicator and environmental quality indicator concepts. Cost/benefit, risk/benefit analysis. Interdisciplinary systems considerations with long time horizon. Reports and studies may be closely coupled to particular user or may be freely available.	Regulatory agencies, industrial marketing organizations, policy advisory apparatus. Ad hoc studies, and impact assessments. Specialized assessment organizations may be developed.
IV. Decision Making	Policy analysis directly related to authority. Political judgment added. Management action and implementation in the real world. Integration of all aspects. Timetables and priorities are set. Different decision makers respond to their own constituencies and policy objectives.	All governmental departments and agencies. Special interest groups, consumers, conservationists. Business organizations. Individuals and the collective public make many decisions.

results should be presented without advocacy permitting the political process to add its context to a neutral, objective assessment.

(B) Priorities must be assigned among candidate assessments. TA is an expensive undertaking. No decision making organization, nor indeed society itself, can perform indepth analysis on all the problems which it confronts. There are degrees of assessment of course, but since the very nature of TA is to look beyond the primary impacts, a superficial study could well be worse than none at all as it might engender the feeling that no unexpected results would occur. Some of the criteria for assigning priority are:

- the number of persons or the area of the environment affected are large
- effects are imminent
- man-induced changes are important in comparison to natural variations or background circumstances
- effects are irreversible, may likely be irreversible or very difficult or costly to reverse
- great economic damage would result
- potential benefits to society are large and highly probable — great social utility
- the momentum of a new technology is high and will carry it into application
- a societal problem demands immediate solution involving alternative technologies.

(C) Social science techniques are inadequate. Many, if not all, assessments require the so-called "soft" data of attitudes, public opinion and preferences, social costs and benefits, social indicators, and the degree to which the value system is changing or could be changed. These inputs are vital to the enumeration and comparison of alternatives in technology assessment, yet they are the hardest kinds to collect.

(D) Industry reactions. To some in the private sector, TA appears to be the institutionalization of Ralph Nader. A more constructive response must be forthcoming.

(1) TA must be incorporated in the early stages of the research-innovation sequence. Ultimately, it cannot be an add-on activity after the application is ready for widespread use. The industrial sector, especially, must ask its scientists and engineers to consider broad implications as they select emerging lines of research for emphasis. For example, a paint company formulating new coatings would examine the toxicity of the dried material and the air pollution potential of evaporating solvents. The earlier that problems can be identified the less disruptive will be any necessary remedy. A positive finding, even more important to industry, brings the confidence with which production and marketing can proceed knowing that unwanted side effects are unlikely.

(2) Proprietary data (processes, production volume, distribution in specific uses) often may be necessary for assessment. If divulging these facts disrupts market competition, society may have to find a credible means of protecting the information, or accept an assessment performed by the affected firm, or compensate the owner. For example, Monsanto has only recently released production figures on PCB so that ecologists could estimate the environmental impact of these chemicals.

(3) Talent will be diverted from new products and processes and profit-increasing management technique into defensive assessment work. Future business opportunities may be judged, in part, as to whether or not they involve substantial assessments.

(E) TA implementation will prove to be costly to the consumer and if his willingness (or reluctance) to take risks or to pay the price of hazard avoidance is not accurately evaluated, he may rebel.

Three separate areas of risk/benefit must be kept in mind:

(1) Personal choice where no one else is likely to be directly affected. For example auto safety seat belts and shoulder harnasses are used only by a minority of drivers and the extension of these devices is meeting resistance.

(2) Accepted risks where others may be involved. Work related exposures and hazards are balanced by the necessity to make a living. To some extent it is possible for a prudent person to control his own risk. Other examples in this category are smoking, drinking, taking drugs and medicines, using exotic foods and seasonings.

(3) No choice in incurring the hazard (e.g., everyone is affected by radioactive fallout and the individual has no recourse). This area is the most important for society to erect suitable assessment and safeguards. The exposure is likely to be inadvertant, inescapable and indeterminable for the ordinary citizen. Also in this area is the protection of vital common properties such as the oceans and atmosphere.

(F) The procedures for any TA should be open to public scrutiny for the most part if credibility is to be established. On the other hand, it is necessary to give the assessors occasional privacy so that radical and untried ideas can be brought up for discussion. The desired ingenuity to generate new alternatives and the needed speculation about consequences would be suppressed in a totally exposed agenda. The record of considerations should eventually be divulged completely but inhibition of imaginative approaches is not desirable.

(G) The coincidence of expertise with vested interest is common when assessment participants are being chosen. And yet it is clear that we should opt for the expert rather than the totally disinterested person with little knowledge of the problem. One way out is to construct a professional environment for the assessment (e.g., an umbrella of a learned society) and then choose experts with a variety of inferred biases. They will tend to keep one another honest and will cancel out non-objective opinions. Dr. Edward E. David, President Nixon's science adviser, has said: "The best you can do is get a balance of interests rather than have no conflict of interest at all in such groups."

(H) The natural environment is the affected area most subject to assessment today. In the United States, the National Environmental Policy Act requires impact statements on all major government projects and these are providing the developmental opportunities for TA. It should be stressed that TA applies to all consequences and its development should be carried out on a broad front. The overemphasis on environmental quality assessment obscures urgent problems in social, cultural and unconventional economic impacts (e.g., computers and privacy, television and education, communications vs. transportation).

(I) Similarly there is an overemphasis on the use of TA by governmental agencies. In the United States, a governmental plan for society is anathema (although the act of planning is all right). No one welcomes more regulation or bigger government. TA must be woven into the free enterprise competitive system by augmenting the conventional economic measures so that the market place can have a longer time horizon and so that convenience and immediate gratification do not appear to be such obvious choices for the customer.

(J) Implementation channels must be ready when the TA results come in. A mission agency or industry must be able to adjust its activities to accord with the evaluation from a competent and thorough assessment. Political considerations may dictate a choice less than optimum on purely technical grounds. However, a constituency for TA should build as the process becomes more well known so that all types of management would have an increasingly hard time ignoring the findings. It is axiomatic that, in a free society, information combines with public education to obtain the political course of action which is desired.

CONCLUSION

Technology assessment is a social invention which will facilitate appropriate societal control of physical inventions. Technology assessment is a tool for the renewal of our basic decision making institutions: the democratic political process, and the free market economy. It is, not surprisingly, being developed just in time — remember that necessity is the mother of invention.

THE ENVIRONMENT: A SYSTEMS APPROACH WITH EMPHASIS ON MONITORING

WILLIAM D. ROWE

Head
Environment Systems Department
Mitre Corporation, McClean, Va. U.S.A.

The objective of this chapter is to present some considerations for a systems approach to understanding the nations's environment for administrative purposes as opposed to scientific purposes.

The expanding population of man and his technology has lead to the realization that we live in a closed-cycle world; and, if we are to survive, man must at least coexist with his environment if not live in harmony with it. The recognition of the degree of environmental deterioration, caused by man in recent years, has not only promulgated pollution control and abatement programs throughout the nation; but has caused great concern as to the future impact of man's collision with his environment. As programs are initiated to minimize this impact they must consider the total picture. One way of viewing this in terms of how things get done is an integrated "Envological" system which is illustrated in Fig. 1. This consists of six inter-related systems including the natural physical environment, economical, political, ecological, social, and information systems. The point is that things get done in our society when institutional, physical and political barriers can be broached as represented by the area of overlap. This overlap of systems provides a qualitative measure of the ability of society to achieve new programs.

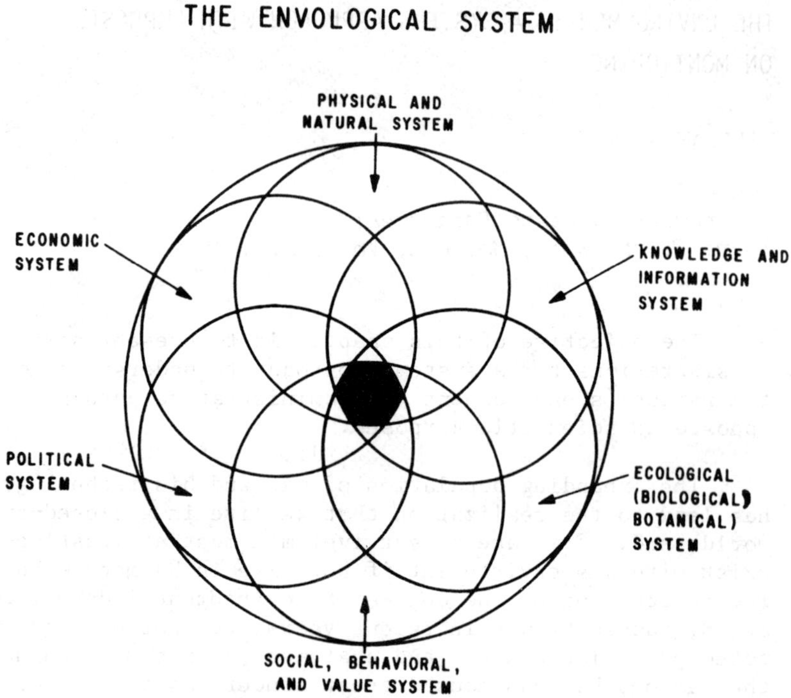

Fig. 1. Man interacting with his environment.

Physical and ecological systems are necessarily constrained and change slowly; political, social and economic systems, although they have much more flexibility, do not often change as rapidly as might be desired. This is especially true in terms of social values. The system to which least attention has been given as a motive force in reaching environmental goals is the information system. The information system includes measurement of environmental parameters at one extreme and dissemination of educational information to the public at the other end.

The information system is a powerful tool for effecting change in man's relationship with his environment. The purpose of this paper is to provide a basis for a "systems approach" to an information system that will be effective in achieving environmental goals.

"SYSTEMS APPROACH" REQUIREMENTS

The "systems approach", in general, requires certain steps to be carried out. These are:

(1) *Definition of Goals and Objectives*

(2) *Establishment of measures of goal attainment* — Express goal attainment in terms of environmental quality and relate these to real measures.

(3) *Means to measure the existing condition of the real environment* — Design a monitoring system to measure the "present" state of the environment. This will serve both Steps 3 and 6.

(4) *Establishment of a set of programs and alternatives to attain goals* — Conceptualize and develop programs for achieving environmental quality.

(5) *Means to assess the impact of programs to select an "optimal" program to meet goals* — Design an assessment capability for environmental programs.

(6) *Means to measure actual progress toward achieving goals* — Can use the measurement system of Step 3.

This paper addresses the first three aspects in terms of possible definition and measurement of goals and design of a monitoring system for the environment. A methodology for environmental assessment is developed in a separate paper [1].

GOALS AND THEIR RELATIONSHIP TO ENVIRONMENTAL QUALITY

Environmental goals can never be absolute, since they form an integral part of the national value system. They are subject to constant adjustment through the play of intellectual, social, economic, and political forces.

Goals must be related to an operational definition of environmental quality itself if they are to be useful. Both the Federal Water Quality Administration (FWQA) and the National Air Pollution Control Administration (NAPCA)* define quality in terms of standards. A standard consists of three parts: uses, criteria, and implementation plan. Implicit in the use of a standard is the existence of an agreed-upon goal. Standards, which are useful and quantitative in the real world, can be directly related to environmental goals to provide a baseline for establishing and understanding environmental quality.

A hierarchy of goals has been developed by FWQA, consisting of four levels, which take into consideration the conflict between environmental ideals and economics; and two time periods, present and future. The four levels are designated ultimate, optimum, acceptable, and current, as shown in Fig. 2.

The ultimate level is the goal of man living in complete harmony with his environment. This goal may not be even theoretically achievable, so an optimum level is set at which man can co-exist with his environment, optimum within natural constraints.

*Now both organizations are part of the U.S. Environmental Protection Agency.

LEVEL		PRESENT	FUTURE
ULTIMATE	GOAL	MAN AND NATURE IN COMPLETE HARMONY	EXPANDING POPULATION OF MAN AND NATURE IN COMPLETE HARMONY
	WATER USES	ALL KNOWN USES	ALL KNOWN AND FORESEEABLE USES
	QUALITY CRITERIA	COMPLETE SCIENTIFIC AND TECHNICAL KNOWLEDGE TO ACCOMPLISH ABOVE USES	COMPLETE SCIENTIFIC AND TECHNICAL KNOWLEDGE TO ACCOMPLISH ABOVE USES
	IMPLEMENTATION PLAN	CONTROL OF ALL MAN-MADE EFFECTS ON THE WATER ENVIRONMENT	CONTROL OF ALL MAN-MADE AND DELETERIOUS NATURAL EFFECTS ON THE WATER ENVIRONMENT
OPTIMUM	GOAL	ATTAINMENT OF USES BELOW ALLOW MAN AND NATURE TO LIVE IN MUTUAL COEXISTENCE	ATTAINMENT OF USES BELOW ALLOW MAN WITH AN EXPANDING POPULATION AND NATURE TO LIVE IN MUTUAL COEXISTENCE
	USES	PRESENT AND MAJOR POTENTIAL WATER USES	ALL POTENTIAL USES
	CRITERIA	KNOWLEDGE OF POLLUTION-USE RELATIONSHIP TO ACCOMPLISH ENHANCEMENT FOR STATED USES AND SOURCES	KNOWLEDGE OF POLLUTION-USE RELATIONSHIP TO ACCOMPLISH ENHANCEMENT FOR STATED USES AND SOURCES
	IMPLEMENTATION PLAN	CONTROL OF ALL POINT AND NONPOINT EXISTING SOURCES	CONTROL OF ALL POINT AND NONPOINT EXISTING AND POTENTIAL SOURCES
ACCEPTABLE	GOAL	MINIMIZE MAN'S DELETERIOUS EFFECT ON NATURE WITHIN PRESENT ECONOMIC CONSTRAINTS	MINIMIZE MAN'S DELETERIOUS EFFECTS ON NATURE
	USES	PRESENT AND SOME POTENTIAL BENEFICIAL USES	USES TO MEET ALL FUTURE DEMANDS
	CRITERIA	KNOWLEDGE OF POLLUTION-USE RELATIONSHIP TO ACCOMPLISH ENFORCEMENT FOR STATED USES AND SOURCES	KNOWLEDGE OF POLLUTION-USE RELATIONSHIP TO ACCOMPLISH ENFORCEMENT FOR STATED USES AND SOURCES
	IMPLEMENTATION PLAN	CONTROL OF ALL EXISTING POINT SOURCES	CONTROL OF ALL EXISTING POINT AND NONPOINT SOURCES
CURRENT	GOAL	MINIMIZE MAN'S DELETERIOUS EFFECTS ON NATURE WITHIN RESOURCES AVAILABLE	
	USES	SOME OR ALL PRESENT USES	
	CRITERIA	KNOWLEDGE OF POLLUTION-USE RELATIONSHIP TO ACCOMPLISH MAINTENANCE FOR USES AND SOURCES	NOT COMPATIBLE WITH NATIONAL GOALS.
	IMPLEMENTATION PLAN	CONTROL OF MAJOR POINT SOURCES	

Fig. 2. Present and future level of water quality standards.

The lowest level of goals, the current level, is set to establish the principle of non-degradation: that standards shall not be set less strict than the prevailing level. This level is not satisfactory to most people; a conviction exists that improvement is possible at reasonable costs. Therefore, an "acceptable" level of goals is established which will allow man to maintain and enhance his coexistance with his environment at an economically feasible level.

In many cases, environmental standards already achieve goals at the acceptable level. In these cases, the current goal and the acceptable goal are identical. Acceptable goals generally represent a reasonable balance between enhancement of the environment and the operation of society with a high standard of living. Achieving optimum goals will require the adjustment of national values to accept the additional expense of internalizing the costs of natural resources and the assimilative powers of the environment, now generally considered free, and a substantial increase in recycling and reusing materials.

RELATIONSHIP OF GOALS TO STANDARDS

Fig. 2 shows how FWQA has related water quality goals to water quality standards at these levels. The difference between present and future goals involves updating of present goals and standards to account for an expanding population at the optimum and ultimate levels. At the acceptable level upgrading from present to the future indicates that the future acceptable goal is nearly identical with present optimum goals.

Standards exist for air and water, but not for land at the present time. Air and water standards differ somewhat in their definition, but generally consist of three parts: 1. definition of uses, 2. criteria, and 3. an implementation plan leading to control of quality. Fig. 3, defines the components of standards for water quality in more detail since they differ somewhat from those for air quality.

A. USES	B. CRITERIA	C. IMPLEMENTATION PLANS
CATEGORIES SUCH AS: (1) BODY-CONTACT RECREATION (2) DRINKING WATER SUPPLY (3) INDUSTRIAL WATER SUPPLY (4) TROUT AND OTHER FISHERIES (5) SHELLFISH GROWING AND HARVESTING (6) AESTHETICS AND GENERAL RECREATION	MEASURABLE WATER QUALITY REQUIREMENTS SUCH AS: (1) CONCENTRATIONS OF DISSOLVED OXYGEN, SUSPENDED SOLIDS, CARBON CHLOROFORM EXTRACT, MERCURY, PHENOL, ETC. (2) NUMBERS OF E. COLI, FECAL COLIFORMS, VIRUSES, ETC. PER UNIT VOLUME OF WATER (3) TEMPERATURE RANGES AND RATES OF CHANGE (4) PHYSICAL DEGRADATION (5) TOXIC COMPONENTS	(1) WASTE DISCHARGE REQUIREMENTS INCLUDING SUCH ITEMS AS: TYPE OF TREATMENT, PERCENTAGE REMOVAL, AND/OR ABSOLUTE QUANTIFIABLE LIMIT OF DISCHARGE PER UNIT OF TIME. (2) SPECIFIED TIME SCHEDULE FOR ABATEMENT AS FOLLOWS: A. SUBMITTAL OF PRELIMINARY PLANS AND SPECIFICATIONS B. SUBMITTAL OF FINAL PLANS AND SPECIFICATIONS C. COMPLETION OF FINANCING D. AWARD OF CONTRACT(S) E. COMPLETION OF CONSTRUCTION F. COMMENCEMENT OF OPERATIONS G. TESTING AND CERTIFICATION H. CONTINUED COMPLIANCE AND PERIODIC TESTING TO ASSURE COMPLIANCE.

Fig. 3. The components of water quality standards.

It is desirable that standards for land and land uses be determined in much the same way as for air and water. Standards for a land program must take into account both soil and space uses of land. Designated uses of land of which there can be very many different kinds in terms of soil and space are probably not difficult to identify. For each of these uses, criteria can be established to assure that the use levels can be met to minimize effect on the environment. In a sense some levels already exist in a diffused manner for uses and criteria, i.e., zoning regulations. The implementation plan for a land standard is another thing. This is an area where almost no work has been done. There are many kinds of plans at present: redevelopment plans, master plans, recreational area plans, and so forth. None of these have attempted to develop land on the basis of plans to achieve stated uses through meeting criteria for uses designated in specific geographic areas. If this can be done, a national strategy and plan for land uses throughout the country can be estaolished for the first time.

ENVIRONMENTAL QUALITY SCALES

To obtain a general scale for environmental quality in terms of goals and related standards, a ranking of goals can be developed if negative values are added. Such a ranking is shown in Column A of Fig. 4. Column B shows how these relate to present and future objectives as well. Neither the upper limit, future ultimate, nor the lower limit, Armageddon, are absolute measures. However, they are so general that most people can agree on a definition. This assures that the end points are as close to absolute as is possible. The present state of our environmental quality in terms of these end points is itself a value judgment. Initial value judgments seem to imply that the current state of the environment is closer to the lower limit than the upper limit. An index is by definition a cardinal (interval) scale. Thus the rank (ordinal) scale of goals must be forced into an interval scale with value judgment used to describe the relative distance between rank levels. In doing this it seems that the current level ought first to be considered as a zero base level

(a) GOALS	(b) LEVELS	(c) BIDIRECTIONAL SCALE	(d) UNIDIRECTIONAL SCALE	(e) SCALE OF PRESENT GOALS
EXPANDING POPULATION OF MAN AND NATURE IN COMPLETE HARMONY	ULTIMATE-FUTURE	2.0	1.0	
MAN AND NATURE IN COMPLETE HARMONY	ULTIMATE-PRESENT	1.8	.93	
EXPANDING POPULATION OF MAN AND NATURE	OPTIMUM-FUTURE ATTAINABLE	1.5	.80	
MAN AND NATURE COEXISTING	OPTIMUM-PRESENT ATTAINABLE	1.1	0.9	
MINIMIZE MAN'S DELETERIOUS EFFECTS ON ENVIRONMENT	ACCEPTABLE - FUTURE	1.0	0.67	
		APPROXIMATE PRESENT GOAL		1.00
MINIMIZE MAN'S DELETERIOUS EFFECTS ON ENVIRONMENT WITHIN PRESENT ECONOMIC CONSTRAINTS	ACCEPTABLE-PRESENT	0.5	0.50	
MINIMIZE MAN'S DELETERIOUS EFFECTS ON ENVIRONMENT WITHIN PRESENT RESOURCE ALLOCATIONS	CURRENT	-0.0	0.33	.58
DETERIORATING ENVIRONMENT	RETREAT	-0.1	0.28	
IRREVERSIBLE CHANGES IN WORLD ENVIRONMENT	UNCONTROLLED	-0.7	0.10	
ENVIRONMENTAL CATASTROPHE AT WORLD OR REGIONAL LEVEL	ARMEGGEDON	-1.0	0.00	

Fig. 4. Environmental quality scales* expressed in terms of environmental goals. *These scales are illustrative and do not represent a final evaluation.

and a retreat below this position as a negative value. Thus the lower limit is transformed into a -1 level and correspondingly the upper limit is transferred into a +2 level to indicate that the current level is closer to the lower limit. Fig. 4, Column C, shows a first attempt at such a forced scale and is illustrative.

However, when an interval scale exists it may be transformed into any other units which may be useful. Column D shows a transformation into a uni-directional scale between 0 and +1. For display purposes a scale of 0 to 100% could be just as easily used.

The scale shown is in terms of the ultimate level goal which is not the goal that represents present desires of society.

A value judgment, which may be a consensus of the present level of goals that are acceptable to society may be made. Obviously various groups in our national society will have different goals, e.g., the conservationists and the industrialists. While a different scale for each is possible to preserve the integrity of these different value systems, it is convenient for discussion purposes to assume a single national goal at a given time. A judgment of the present goal has been estimated which places this level of goal somewhat above the acceptable level. This is shown on the scales of Column C and D as an "approximate present goal". A relative scale of present goals can then be structured with the present goal set at unity (or some other number as a limit). This provides a scale of present goals against which to measure progress as shown in Column E. The present measure of environmental quality is then shown in a more sensitive manner. This current level of quality was only 33% of the "absolute" scale, but is 58% of the "relative" scale.

The "relative" scale may change (even if quality doesn't) over time as values change, the "absolute" scale will not. This quality scale is then the primary measurement tool for defining the quality of the environment. This allows both changing national goals and priorities and progress toward environmental goals to be reflected.

The relationship between the index of quality and measured indices must be explicitly defined, but must wait for a description of the measures before an explanation can be made.

MONITORING THE ENVIRONMENT

A monitoring system to provide information on the environment can have several roles in reaching environmental goals. Data can be gathered to determine the condition of our environment as well as to provide a better understanding of environmental functions. This kind of information has a relatively long time constant. Another type of system is a reaction system aimed toward preventive or corrective action for the occurrence of environmental episodes.

The primary concern here is for a system to determine the present condition of the environment, in a manner to inform the public of present and postulated conditions, and to provide decision makers with a measure to evaluate environmental action programs and choose among alternatives. An administrative or management system of this type does not necessarily attempt to generate information for specific scientific and legal purposes. A scientific system aims at providing information in depth about our environment so that we may both understand it and as a result control it. A legal monitoring system is one that will provide data to allow law violators to be forced to comply with existing laws and regulations. Monitoring systems for all three cases are necessary, although this paper focuses on a management-oriented monitoring system. These must exist in parallel with a management system although basic measurement data generated at one point may be useful in all three systems.

Initial questions that arise are what should be measured and how shall these results be presented? In the first case, there are thousands of parameters that can and do affect the environment. It is not possible within our present resources to consider measuring all of them. This involves determining what parameters are important to be measured and what ones should be purposely excluded as opposed to being omitted due to lack of knowledge. While no specific answers can be given to this question at this time, part of the approach used in developing this system is aimed at providing at least a first approximated answer to these questions. In terms of measurements, however, the considerations is for indices and indicators for management for use by the Government, and the public and private sectors. The objective is to develop indices in the same vein as the GNP, unemployment rates, and related economic indices. An index is defined for our purposes as a non-dimensional, multivariable aggregate and an indicator is a particular measurement that is extremely sensitive to environmental change as measured in the dimensions of that measurement.

Since it is impossible to know all the parameters to be included and their effects, any set of indices and indicators developed and the parameters that affect them will change as we learn more about them. Thus, these measures must be evolutionary in that they will improve over time. This implies the need for a means to follow change over time. Thus a means to relate old to new indices must be available so that longitudinal data will survive in a traceable manner.

DIMENSIONS OF MONITORING

There are at least four separate dimensions of measurement criteria as follows:

1. Components of Environmental Quality
2. Taxonomy of measurement parameters
3. Geographic subdivision and location
4. Time.

A. *Components of Environmental Quality*

The components of quality of the environment can be divided into three general classes:

Pollution

Environmental Effects

Resources

Within each of these classes different kinds of measurements may be considered.

Pollution (P)

Amount of Pollution (A) - the amount of pollution per unit time entering the environment from all sources of pollution, e.g., millions of tons of BOD per year. This is particularly useful since measuring effluents at sources is the most straightforward way of measuring pollution.

Degree of Abatement and Control (C) - a measure of how well water, air, and land quality standards are being met.

Incidents (I) - the number of serious incidents in the environment that occur over a given period of time are a direct measure of where we stand in time in relation to the number of incidents in another instance of time. The number of inversion episodes in the air, the number of fish kills, number of species that are endangered or become extinct are all possibilities.

Resources (R)

Actual (RA) - the amount of actual damage (or gain) caused by pollution (or control) including the gain or loss of natural resources.

Potential (RP) - the potential damage that might be identified as real possibilities. Example, changes in the thermal structure of the earth due to air pollution at the 65,000 foot level by the SST is a potential threat. Another example is the possibility of a contagious epidemic if single cases are not immediately controlled.

Environmental Assimilative Capacity (S) - a measure of the amount of assimilative capacity of land, air, water, etc., is a measure of environmental reserve. Increase or decrease in this margin may be a sensitive indicator fo the environment although difficult to measure. The major natural cycles are feedback processes and the open-loop gain of these systems, if possible to measure, provide an indicator of assimilative capacity.

Effects (E) - Effects due to environment changes are also directly measurable.

Human (H) - Extrinsic effects in terms of the health of man, his economic situation, and his ability to live in the physical world. This probably includes both physical and mental health effects, such as crowding factors.

Geologic (G) - Extrinsic effects on natural environment which may or may not be caused by man.

Air - climate, etc.

Water - ocean photosynthesis, etc.

Land - erosion

- earthquake potential, etc.

Biological (B) - Extrinsic changes in the animal or plant world due to changes in the environment.

Materials (M) - Extrinsic effects on actual physical material things that are occurring, including depletion of critical natural resources.

Intrinsic (Z) - Changes in things that are important to the value systems of man in terms of social, economic, aesthetic, and recreational goals.

All of these indices contribute to describing the environmental quality index, however they are each neither mutually exclusive nor collectively exhaustive in describing quality. Just as the GNP which measures a level of economic "quality" is made up of an aggregate of other economic factors, it is seldom used alone, but in combination with other economic indicators such as cost-of-living, employment rates, and so forth. However, in this case one can consider the general quality index as aggregation of all useful indices based upon value judgments as to their relative importance.

For each parameter a measurement of each component of the index may or may not be made, depending on the validity, availability, and meaning of such measurements. For example, a measurement set for sulfur dioxide would consider the related measurement for each index. These are summarized in Fig. 5 below.

POLLUTION

Amount - Tons of SO_2 per day per geographic area
Control - Average monthly concentration of SO_2
- Number of days per year standards were exceeded
Incidents - Number of SO_2 related episodes per day per year

RESOURCES

Actual - Minimal since sulfur is not a scarce material
Potential - Minimal
Assimilative - Air stream and tail dilution factors

EFFECTS

Human - Increase in morbidity and morality both acute and chronic
Geographic - Heat of upper atmosphere and heat imbalance
Biological - Number of animals and crops affected.
Material - Corrosion, paints, etc.
Intrinsic - Irritation, odor, recreational restrictions

Fig. 5. Some examples of sulfur dioxide measurements

For sulfur dioxide amount, control and incidents all have meaning in terms of pollution. Since sulfur is a plentiful element, resources are not necessarily considered, except in terms of assimilative capacity. Possible effects may also be measured.

B. *Taxonomy of Parameters*

The classification of the large number of parameters that are or might be measured is a second dimension. Fig. 6 shows an initial classification of parameters by groups (e.g. gaseous air pollutants). There is a varying level of detail for these parameters which is best illustrated in a hierarchal structure, as shown in Fig. 7.

The macro level is defined as an index that combines all measurements into a single number or limited set of numbers that indicates the general environmental condition of the nation, a region or state, or a locality.

The meso level indices are broken down into two levels. The first level is in three parts: air, water, and land. Effects on the environment such as radiation, noise, chemicals, biological conditions, thermal changes, and erosion would be considered under each part. The second meso level breaks each of these into three parts also: intrinsic, physical, and biological. Intrinsic effects associated with water, as an example, would be aesthetics and recreational facilities. Those parameters that affect the health of animals and humans are grouped under biological; and those that affect only the physical environment are grouped under physical. Since a measurement will not always be completely covered in one category its component effects will be split to each.

The macro level is where actual measurements are made. Two levels are considered, the detailed measurement and the gross measurement level. The detailed measurement level implies direct measurement of an environmental condition such as sulfur dioxide in the atmosphere, biochemical oxygen demand (BOD) in streams, ratio, of recreation land to total land, etc. Gross level indicators take into account interaction among conditions on a higher

AIR			WATER			LAND		
PHYSICAL	BIOLOGICAL	CULTURAL	PHYSICAL	BIOLOGICAL	CULTURAL	PHYSICAL	BIOLOGICAL	CULTURAL
GASEOUS	AEROALLERGENS	AESTHETIC	HYDROLOGICAL	BIOLOGICAL POLLUTANTS	AESTHETIC	ACREAGE SURVEY	SPECIES CENSUS (DIVERSITY & NUMBERS)	AESTHETIC
INORGANIC	RESPIRATORY DISEASE RATE	CLARITY	RADIOACTIVITY	UNDESIRABLE EPISODES	CLARITY	SOIL CONDITION & EROSION	KEY SPECIES	ACRES OF URBAN BLIGHT
HYDROCARBONS		ODOR	PHYSICAL POLLUTANTS	PESTICIDE & OTHER CONTAMINANTS (AQUATIC, MARINE; ANIMAL & PLANT TISSUE)	ODOR	MINERAL RESOURCES	COMPREHENSIVE SURVEY	ACRES OF OPEN DUMPING & JUNK-YARDS
PESTICIDES		NOISE	CHEMICAL POLLUTANTS	SPECIES CENSUS (DIVERSITY & NUMBERS)	SURFACE CONTAMINANTS	WASTE MANAGEMENT & RECYCLING BUDGETS	RODENTS & OTHER PESTS	EXPLOITED, ABAN-DONED, UNRESTORED LAND (ACRES)
OTHER ORGANIC		ECONOMIC	PESTICIDES	KEY SPECIES	TASTE	ENERGY PHENOMENA	UNDESIRABLE EPISODES	AUTO - PUBLIC TRANS-PORTATION MODAL SPLIT
PARTICULATES		PROPERTY DAMAGE & DEPRECIATION	ENERGY LOAD-ING (THERMAL POLLUTION, SOLAR RADIA-TION)	SURVEY OF ALL SPECIES	INSECT INSPECTION	RADIOACTIVITY (TERRESTRIAL, FOOD)	TOXIC SUBSTANCES IN PLANTS & ANIMALS	ECONOMIC
RADIONUCLIDES		DAMAGE TO CROPS, FOREST, & LIVESTOCK			ECONOMIC		POPULATION IMMUNITY TO COMMUNICABLE DISEASE	EROSION LOSS
ATMOSPHERIC ENERGY		MAN-HOURS LOST FROM LABOR FORCE			FISHERY LOSSES		VITAL STATISTCS	TIMBER LOSS
METEOROLOGICAL		RECREATIONAL			CORROSION LOSSES		BIRTHS	NATURAL DISASTER LOSS
NOISE		RESTRICTION OF EVENTS DUE TO AIR POLLUTION			SILTAGE LOSSES		DEATHS	SHORTAGE OF ADEQUATE HOUSING
		CIVIL			PLEASURE BOAT SALES			RECREATION
		MUNICIPALITIES COVERED BY ACCEPTABLE REGULATIONS			RECREATION			URBAN PARK & RECREATION LAND PER CAPITA
		VIOLATIONS REPORTED			SHORELINE INVENTORY			INTRINSIC SPACE PER CAPITA
		VIOLATIONS PROSECUTED			WATER AVAILABLE FOR CONTACT & NON-CONTACT SPORTS			VISITORS TO NON-URBAN PARKS
					CIVIC			CIVIC
					MUNICIPALITIES WITH ADEQUATE DRINKING WATER TREATMENT			MUNICIPALITIES COVERED BY ACCEPTABLE SOLID WASTE DISPOSAL STAND-ARDS
					SEWERAGE CAPABILITIES & MUNICIPALITIES			MUNICIPALITIES WITH MASTER PLANS FOR LAND USE AND DEVELOPMENT

Fig. 6. Classification of parameters.

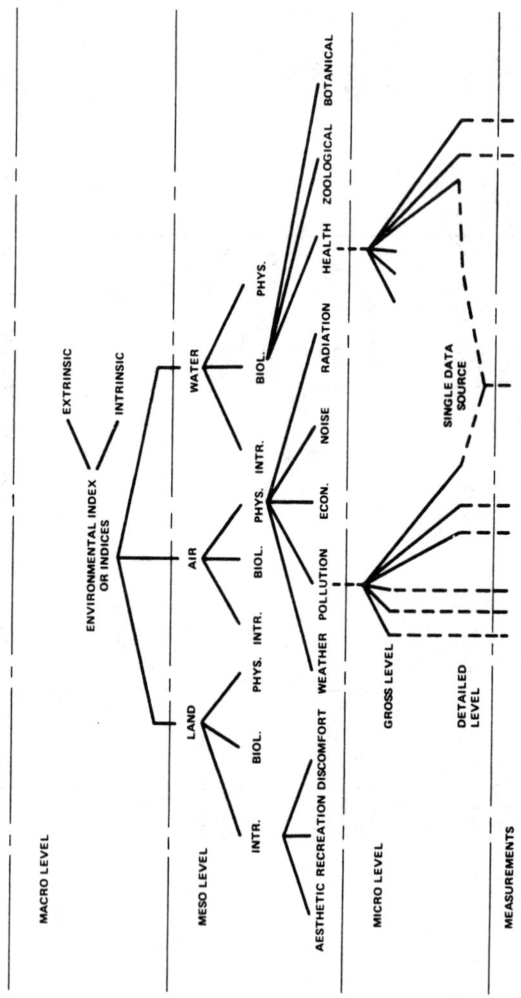

Fig. 7. Structure of the hierarchy of measurement detail.

level. Thus the number of fish kills, number of air pollution episodes, number of biological species in a stream or area, are gross measurements of more detailed conditions. An air pollution episode is not dependent on sulfur dioxide alone, but weather conditions, particulates, oxides of nitrogen, ozone, etc. Both types of measurements must be made.

C. *Geographic Subdivision and Location*

Aggregation of level of detail proceeds from the micro level where actual measurements on the environment are made through the meso level where they are aggregated into gross effects to the macro level where a single indicator is generated. A geographic hierarchy proceeds from the locality where measurements are made to a regional or state level, and then to a national level. Fig. 8 describes this two dimensional hierarchal structure, and indicates how projected indices might be used.

D. *Time*

There are several temporal aspects that must be considered. The first is the rate of change of parameters being measured in forms of long and short intervals; the second, is the manner in which effects occur in terms of acute and chronic responses; and the last is the smoothing of measurements. In general, there will be many parameters which are short term changes and a variety of changes that are long term, such as mutagenic changes in biological species. It may be desirable to have both short- and long-term indices on a separate basis, as well as an aggregate.

Harmful effects occur, especially in biological areas from two sources. The first is exposure to a disturbing factor in which the product of time of exposure to, and concentration of the factor, determine the effect. The seocnd is cumulative exposure to a disturbing factor being the deterinant of the effect. In either case there exist thresholds at which harmful effects occur. If the threshold is reached over a short period, arbitrarily set at 48 hours, the effect is called acute, otherwise it is called chronic.

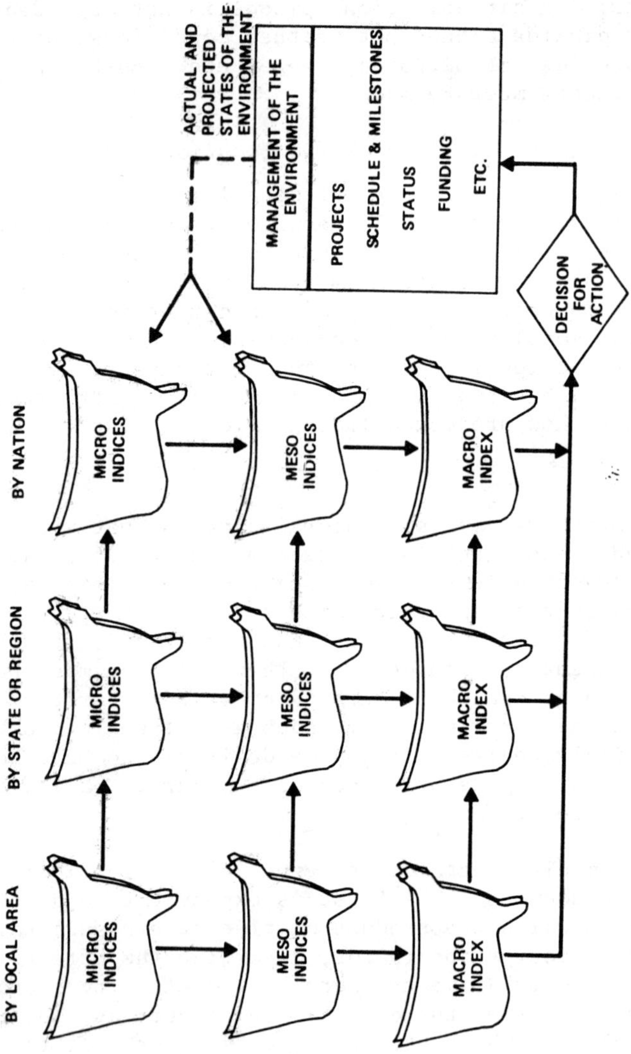

Fig. 8. A two-axis hierarchy of indices.

AGGREGATION

Aggregation from single sources upwards to an index, or indices may proceed in any dimension in any order. The problem is to find that order which is most meaningful. A given measurement, M at a source is a point in a four dimensional space.

$$M_{i,j,k,l}$$

where the subscripts indicate the four dimensions of measurement. The problem is that the order of aggregation differs with each type of measurement for each parameter as well as, differences in end use of the indices. For example, geographic aggregation to a local level as o posed to a state or national level may be different for a variety of reasons. Mutually exclusive, collectively exhaustive coverage of sources is probably not possible. The amount measure or standard for an area measure or standard for an area or region differ as to meaning and level from region to region. One region may have an entirely different problem than another, and indices must be sensitive to all major problems at the very least.

A summary of the possible levels of aggregation for all four dimensions and for measurements are shown in Fig. 9.

The three remaining dimensions are perhaps even more complex. However, this problem must be attacked and in a traceable answer. Thus a methodology, which at least descriptive, if not prescriptive is necessary to achieve meaningful indices. The methodology must provide means to relate the indices to the environmental quality scale as well as be evolutionary in nature to allow changes in the structure of indices over time. Although a methodology of this type has been developed, space restrictions do not allow discussion at this time.

As value judgments change, both in the quality index and in the indices and in the composition of the indices, the comparison with older indices on a quantitative basis

TYPES OF MEASUREMENTS:

Level 1. Measurements Aggregated to pollution, resources, and effects indices.

Level 2. Pollution, resources and effects indices to an environmental quality index

PARAMETERS:

Level 1. Micro level to lowest meso level, e.g., air, biological and physical

Level 2. Lowest meso level to highest meso level, e.g., air, water, land.

Level 3. Meso to a single micro level.

TIME:

Level 1. Aggregate short time, long time, and cyclic effects in to a single time scale.

GEOGRAPHY:

A. Aggregate from source to nation by political boundaries through locality, country, state, nation.

B. Aggregate from source to nation by natural boundaries through basis to nation.

MEASUREMENTS:

Level 1. Combine peak, average, cumulative measurements into a single measurement scale.

Fig. 9. Possible aggregation steps.

must always be made. This means the data upon which old indices were derived must continuously be retained so that new indices may be compared to historical indices to assess the implications of the change in the indices.

In general, initial aggregation over types of measurements will proceed first. From these long and short period indices for air, water, land, etc., would occur through aggregation across parameters. These would be obtained on a local geographical basis and related to environmental quality (standards) on a local basis when final aggregation across parameters takes place.

Finally, geographic aggregation to national indices for air, land, and water can be obtained and ultimately integrated, and related to a single index for the environmental quality.

AGGREGATION BY TYPE OF MEASUREMENT

The aggregation by types of measurement into a quality index is the primary relationship to be considered. Aggregation may occur at any level of the other three dimensions. For this reason, only general relationships are shown here without considering the order of aggregation.

Fig. 10 summarizes the components of measurement. A relation among environmental quality, pollution effects, and resources can be demonstrated in generic fashion.

$$\hat{Q} = a_1 \times (P \cdot E) + a_2 \times R \tag{1}$$

$\hat{Q}$ is the inverse of quality, Q, and increases as either pollution or effects increase. The relation between pollution and effects is multiplicative since cause and effect relationships are not considered, but only measured. Pollution without effect does not affect quality, effects without pollution may not be related to the environment. However, combination of pollution and effects in any combination are significant. Concern with resources is always necessary, but may be either emphasized or de-emphasized as the pollution-effect product is small or large. The weights a_1 and a_2 allow the relative importance of these two terms to be set.

The pollution index (P) may be broken into component indices as follows:

$$P = b_1(A \cdot \hat{C}) + b_2 \times I \qquad (2)$$

where the amount of pollution and the inverse of the degree of control form a product which is added to the incident index. Relative weights are also shown (b_1, b_2).

NOMENCLATURE

Q = Environmental Quality Index

F = Pollution Index

- C = Degree of Pollution Control
- A = Amount of Pollutants
- I = Number of

R = Resource Index

- RA = Actual Resource Levels
- RP = Potential Resource Levels
- S = Assimilative Capacity of Environment

E = Effects

- H = Human
- B = Biological
- M = Material
- G = Geologic
  - Air - Climate
  - Ocean - Photosynthesis cycle
  - Land - Erosion, earthquake potential
- Z = Intrinsic Parameters

Fig. 10. Environmental Quality Indices.

The resource index is a sum of resource components.

$$R = c_1 RA + c_2 RP + c_3 S \qquad (3)$$

and likewise the effects index is a sum of effects.

$$E = d_1H + d_2G + d_3B + d_4M + d_5Z. \qquad (4)$$

In this manner the component indices can be aggregated meaningfully to describe quality.

POSSIBLE STRUCTURE OF A NATIONAL ENVIRONMENTAL MONITORING SYSTEM

A system for monitoring present environmental conditions and changes over time is necessary to achieve environmental goals as indicated in the introduction. The purposes of such a system would be to support local and national government administrative bodies by providing them with:

timely alerts indicating changes in significant environmental and ecological phenomena

a basis for developing quantitative environmental impacts of publicly and privately supported alternative technological developments and programs

a uniform and systematic basis for comparing and evaluating estimated vs. actual environmental impacts of previously implemented programs

These capabilities would be achieved through:

central coordination and standardization of remote collection of data relating to air, water, and land quality

analysis and eveluation of these data to compile indices of and determine trends and interactions among environmental pollution, resources, effects and quality.

Such a concept must stress maximum utilization and exploitation of existing data collection and analysis resources. Current involvement and monitoring activities are widely distributed among organizations of Federal Departments such as Agriculture, Commerce, Defense, Interior, Health, Education and Welfare, and Transportation. The Atomic Energy Commission, Smithsonian Institution, and numerous state and local agencies are also sources of relevant data. The primary responsibility for specifying the method of data acquisition and techniques for analysis should be given to lead organizations in the new Environmental Protection Agency (EPA), who would then utilize their existing capability, augment that capability and/or task other Federal and local units to provide the necessary data and services. The Environmental Protection Agency would also contract with private naturalist groups (e.g., Audubon Society, National Wildlife Association) for acquisition of certain specialized data.

A general data flow schematic for NEMS is presented in Fig. 11.

RELATION OF CALCULATED INDICES TO ENVIRONMENTAL QUALITY

The environmental quality index must be related to goals and standandards, i.e., the computed indices for pollution, effects, and resources must be related to the goal derived in the environmental quality index. The implied structure of pollution, effects and resources is already compatible to standards, at least in a quantitative manner. For example, resources are directly related to the use component of standards. When criteria are established for defined uses in the standard, the effects of the environment are considered in the scientific base which was used to establish the criteria. Thus, effects are directly related to the established criteria. The level of pollution is also considered in the setup of uses and criteria, and the implementation plan provides a time scale measure of the deviation from criteria due to pollution along with a rate of closure toward meeting the established criteria. Thus the level of pollution especially when related to criteria measures the deviation from goals. Measured effects indicate the resultant deleterious conditions that exist when standards are not met as well as the inadequacy of the standards.

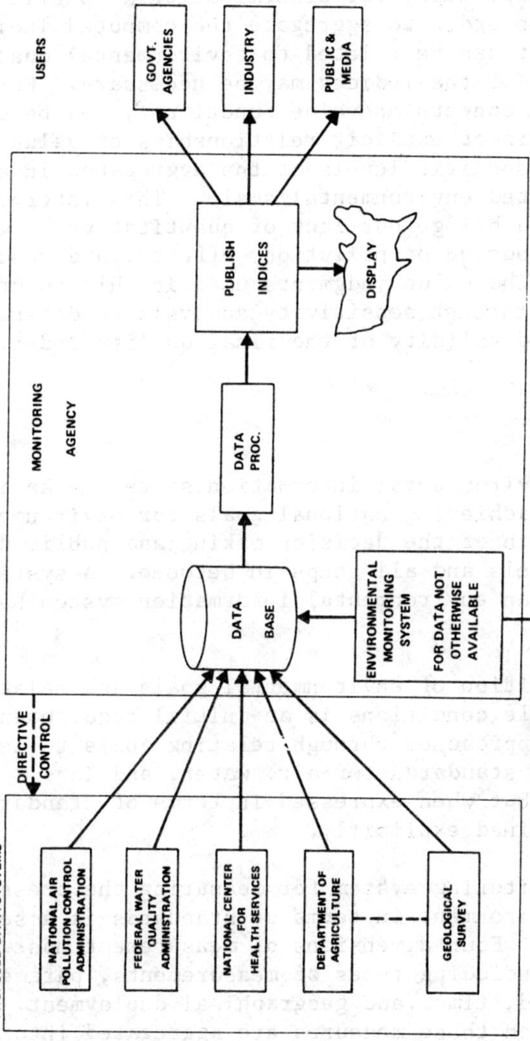

Fig. 11. Block diagram of a possible monitoring configuration.

When the pollution, effects and resource indices are aggregated into a quality index, the individual contribution of these computed indices may be obscured. Nevertheless, a single index of environmental qualiry has utility, especially for dissemination as public information. In order to aggregate the computed indices in a manner that can be related to environmental quality, some adjustment of the indices may be necessary. For this purpose, a conversion value function f, may be constructed on the basis of explicit relationships of value judgments to relate specific levels of the aggregated index to the goal-oriented environmental scale. This interpretation attempts to bridge our lack of quantitative knowledge of the contribution of pollution, effects, and resources, to quality. The value judgments used in this function may be tested through sensitivity analysis to determine the utility and validity of the final quality index.

SUMMARY

An environmental information system is an important means for achieving national goals for environmental quality both of the decision making and public dissemination levels and all steps in between. A systems approach to an environmental information system has several parts.

Definition of environmental goals and relationships to meaurable conditions is an initial requirement. This has been approached through relating goals to levels of achievable standards for air, water, and land. Goals are abstract; but when expressed in terms of standards, they may be defined explicitly.

A monitoring system for measuring the present state of the environment in terms of standards is a second requirement. Four dimensions of measurement must be considered, including types of measurements, parameters to be measured, time, and geographical deployment. The manner in which these measures are aggregated into useful indices is dependent upon the ultimate use for different indices. However, at the highest levels of aggregation these indices may be related to environmental quality.

With these indices a monitoring system can provide information on the present status of the environment, historical trends in environmental status, as well as a tool for exploratory environmental forecasting. Since interactions among environmental parameters are not covered in detail, a monitoring system must be augmented by an environmental assessment system, not discussed here.

Finally the structure of a monitoring system must take advantage of existing monitoring capability and weld this and new capabilities that may be required into a single information reporting network.

REFERENCE

1. ROWE, W.D., *An Evolutionary Normative Methodology for Environmental Assessment*, MTR-4172, October 1970.

THE FUTURE ENVIRONMENT FOR TECHNOLOGY ASSESSMENT

WAYNE I. BOUCHER

The Futures Group, Glastonbury, Connecticut, U.S.A.

The next few years may witness a radical transformation of technology assessment. Some of the important pressures for change will come from within the movement, largely as a result of several factors: a growing recognition that the basic idea of technology assessment (hereafter "TA") has been dangerously overstated; a very rapid development of methods, criteria for the choice of problems, and professional standards; a slow but increasingly significant growth in the funds being made available for TA efforts; and the almost certain emergence of centers devoted primarily or exclusively to TA. All of these factors are, of course, interrelated. Their joint effect may well mean that TA will operate in new guises and under different names, but it may also help to ensure that serious efforts are in fact undertaken to do the job that TA was invented to do.

The remaining pressures for change will come from the outside. In the end, these may prove less noteworthy and influential, but their impact will nevertheless be great. Among them will be the following: a continued outpouring of new technologies of all types (including developments in information storage and retrieval that will ease the data problem in TA); the absence of any fundamental change over the short term in how possible users of TA, especially in government, become aware of and then cope with the issues raised by technology; no change whatsoever in the decision-making process itself; a very gradual but nonetheless discernible rise in the number of potential users of TA (and related approaches) who are willing to work toward the achievement of such changes; a continued and perhaps intensified growth in *ad hoc* popular efforts to effect reforms in many of the areas that are of special interest to TA; and a concomitant shift in public sentiment and values, with the result that some concerns that might otherwise be brought to attention by TA or its supporters will, for all practical purposes, be defined out of existence - or never be noticed.

These trends, external and internal, will develop simultaneously and sometimes at violent cross purposes, with consequences that cannot be fully anticipated. Yet we can be confident that the future environment for TA will not be as happy as some of its advocates have forecasted. As suggested by the statement of the trends, the basic reasons will be conceptual and institutional, and in both of these areas the practitioners of TA can expect to find some turbulence and much frustration - accompanied, however, by the occasional satisfaction of being able to make at least a slight difference. To see why this may be so, let us look somewhat more closely at each of the trends mentioned, beginning with those intrinsic to TA.

THE INTERNAL ENVIRONMENT OF TA

Perhaps the basic problem facing TA is that it already has a formidable literature of its own, developed almost entirely by persons who have had no working experience with TA itself. Few disciplines have ever had to begin with such a handicap. Because this literature has been generated by policy researchers and academicians (who would like to become technology assessors), by potential clients (who would like TA to reflect and satisfy their special needs), by influencial friends (who would like to see TA prosper), and by journalists (who have every manner of motive), the current conception of TA seems perfectly unassailable [1]. As we lurch into post-industrialism, what could be more apparent than the task of working systematically to anticipate the full range of consequences that the choice of particular technologies - physical and social, existing or prospective - might entail? What, indeed, could be more necessary? Toffler sums up the opinions of most when he writes that although we may debate the details of how TA should be given institutional form, the need for the activity itself is "beyond dispute" [2].

An excellent idea and a patent need make a powerful combination. In the case of TA, however, the unfortunate fact is that the idea expresses a goal that cannot be attained by any known or foreseeable means. Even S.C. Gilfillan (who, along with W.F. Ogburn, must be considered one of the intellectual fathers of TA) seems to miss this point. He writes, for example, that:

> ... to predict the future political effect on a country of [a particular invention] we shall need to foresee the country's *whole* political future, taking account of all pertinent inventions as well as of the one in question and its functional substitutes, and considering all other influencial factors, both fixed ones like climate, slowing changing ones like wealth and racial capacity, and quickly changeable ones like politics. From the sum total of all these influences we may, with wisdom, information, and luck, predict how the country's general lines of development will run... [3]

Nevertheless, Gilfillan knows that there is something peculiar about attempts to list and assess all of these pieces of information, especially all of the consequences: "With the effects of the original inventive group continually attenuating in force, immediacy, probability, and verifiability, as endless other causes enter the picture, further prognostications become shortly not worth making" [4].

The obvious though nontrivial rejoinder is that the practical and psychological difficulties he notes are not decisive; the fact of the matter is that so ambitious an undertaking is impossible in principle. With all the wisdom and luck in the world, the information could still not be obtained, at least not all of it, and assuredly not all that would be required to say something comprehensive and meaningful about "the country's general lines of development." The future is too rich in possible events and relations among events for that. To believe otherwise, as apparently most writers on TA do, is to urge an absurdity.

Individuals and groups that have actually conducted assessments seem at times to know better, and a very small but now growing proportion of the literature is being devoted to their experiments and results - all of which, not incidentally, take the form of case studies. These case studies are carefully hedged in terms of time and money and procedural constraints; as in the early days of systems analysis or even futures research, they are usually structured as exercises in methodology. And out of this research is coming a recognition that while an ideal approach to TA can be described, users and analysts alike must necessarily settle for something imperfect, no matter what resources are devoted to a problem and no matter how narrowly the problem may be defined [5]. It can be expected that this perspective will be much more widely shared in the years ahead and that eventually a much more precise and reliable definition of TA will emerge.

One of the odd and important facts about TA is that every description of the ideal (or generic) approach to performing a TA makes the activity sound precisely like systems analysis. This is one of the reasons why some writers have come to the conclusion that "the principle is not new," but that "the intensity of commitment" [6] or "the emphasis being given to this function" [7] is. If this is so, as indeed it seems to be, it is likely to have significant consequences for TA in the future, when it comes to be generally accepted. It may, for example, serve to make better scholars out of its advocates and better analysts out of its practitioners, by causing them to exploit the experience of systems analysis rather than to invest their energy in re-inventions of concepts and methods that have, after all, been tested and refined in some 25 years of work. It may also lead them to benefit from major earlier efforts directed specifically toward the assessment of technology, including what is probably still the most important source of all, *Technological Trends and National Policy*, prepared in 1937 for President Roosevelt by W.F. Ogburn, S.C. Gilfillan, and others, who formed the Subcommittee on Technology to the National Resources Committee.

More important, if all of this analytic experience is thoroughly mined, we can foresee an exceedingly rapid development of new or refined techniques for coping directly with the difficulties of unique concern to TA. We may also see the appearance of some fairly objective standards of what constitutes professional excellence in this field.

On the other hand, we may find that in the course of absorbing the lessons *of* systems analysis, TA will be absorbed *by* systems analysis. As suggested a moment ago, there would seem to be no compelling reasons to speak of TA as something different, new under the sun. The aims, the methods of developing criteria and alternatives, and the use made of models in TA are virtually identical to those of systems analysis. The sole exception would appear to involve the use of cross-impact analysis, which is associated primarily with futures research [8]. Beyond that, however, what distinguishes one from the other is at best the range of problems addressed, and this is basically a distinction without a difference since systems analysts have been performing technology assessments as a regular part of their work from the very beginning [9].

In any event, as TA recognizes its limitations in principle, develops an awareness of its antecedents, begins to accept explicitly the already established capabilities of systems analysis, and then goes about its work, other developments can be anticipated. The TA enterprise will no longer be as fashionable; its publicists will no longer be making outrageous claims about its prowess, including an ability to save civilization; it will have a much better understanding of its own as to where it must develop; and possible sponsors of TA efforts will be noticeably less reluctant to support this work in practical applications. The serious and necessary business of attempting to explore alternative futures that might prompt, or be prompted by, developments in technology may then begin - realistically, pragmatically, honestly.

Success in such limited endeavors is quite likely to mean that TA activities will find appropriate institutional settings, where the results attained can affect contemporary decision-making, rather than merely clutter library shelves. This process has taken 15-20 years for systems analysis, but it has happened, at least in some major government agencies and corporations. We can reasonably expect the same to occur with TA. And it may well be that private corporations will lead the way in providing an operationally significant base for TA. Any of a number of reasons could justify such a move by business: the growth of domestic and foreign competition, the rise of the multi-national corporation, a growth in antiscientific attitudes, and the influence of citizens' groups urging a clean environment, better products, more socially attuned management.

THE EXTERNAL ENVIRONMENT OF TA

The trends and developments mentioned so far fall largely within the control of those who are part of the

TA movement. But TA will certainly be affected - indeed, buffeted - by other trends in motion today over which it has little or no control.

Certainly the most important of these is the obvious one: the continued acceleration in the growth of technology itself, both physical and social [10]. Its basic effect will be to increase the complexity of future assessments, by complicating the choice of subjects and the identification of problem elements. At the same time, it will increase the uncertainty surrounding the results of any TA. And it will also substantially increase the need for better and more timely assessments.

The social and institutional demand for better results sooner will have a particular effect on how information is gathered, disseminated, and acted upon in our society. Accordingly, we can expect to see the invention of new information storage and retrieval systems, especially new means of recording vast amounts of data on smaller and smaller amounts of tape or film. With the capabilities thus opened, augmented by a heightened awareness of need, will come the real growth of massive data banks, containing information on people, projects, institutions, and processes. Access to these data banks will transform research in this country; needless to say, TA will be among the beneficiaries and causes of this transformation, and it will have a special obligation to assess its implications.

Data banks and the sophisticated uses to which they can be put are not likely, however, to have a profound direct influence on how decision-makers who might benefit most from the results of TA will gather and cope with the information they think they need. More and more staff members may sit down before a terminal or even *read* published assessments, but it will be a good many years before the decision-maker himself will. Nor can we expect the slightest change in the decision-making process itself. Better facts, even more informed people, may be available to the decision-maker; new procedures may be invented to ease the difficulties involved in bringing relevant information, including informed judgment, to his attention. It may even be hoped that the quality of the decisions thus reached will be improved. But it cannot be expected that the process itself will change. Compromise, hunch, and hasty generalization will remain at its heart. For analysts, and perhaps for all those affected by the decisions, this may be a continuing cause of frustration and despair, though surely they will see the same process in themselves. The situation, however, should occasion a new kind of analytic support system from TA for the use of decision-makers: a system to track the sequence of events anticipated to follow particularly important decisions and to alert the decision-maker if the signs suggest that things are turning out badly. He will then at least have some opportunity to stop the process short of total failure [11].

Although the decision-making process is unlikely to change, there will be continuing pressures by possible users of TA to make it change. In the U.S. Congress, for example, there will certainly be more frequent attempts to achieve a thoroughgoing reform of the committee system - not simply to revise the seniority rules (which might be thought to solve the problem by removing the decision-maker), but also to pursue such innovations as occasionally replacing conventional hearings by processes that produce more useful information and insight for each minute invested, such as electronic conferences among interested parties or the development of scenarios that persons offering testimony would be asked to criticize explicitly in their statements. It is even conceivable that some hotheads in Congress may seek ways to bring the federal budget - the single most important document for establishing our national priorities - to the attention of the entire Congress or a select subcommittee for systematic and complete review as a statement of where the nation plans to go and how it expects to get there. At the moment, no such review is made [12]. Clearly, TA can make a contribution to such efforts; whether it will choose to do so or not, it is certain to be affected by them. And, of course, the same point can be made about pressures likely to be mounted for new information handling and decision-making approaches in other organizations, both public and private.

As all forms of technology multiply, there can be no doubt that special-interest groups will also seek a greater voice in social change. The Luddites in our midst may increase; the consumerists and environmentalists are certain to do so; and new movements will pop up as unexpectedly as the stockholders' movement did a few years ago. The net immediate effect of such developments will be a discernible rise in the influence of what President Nixon's National Goals Research Staff has called "people's technology assessment" [13]. The second-order effect of importance to TA is that TA will have to redouble its efforts to provide the kinds of information decision-makers will require to respond to the demands, the opinions, and the subversion of the *ad hoc* pressure groups. To meet this need well - indeed, to serve all of its other purposes well - may require that TA develop increasingly more sophisticated early-warning systems. Perhaps, as has been suggested by Myron Tribus, a system of "reliable" social indicators will have to be included in the kit of TA capabilities; among its uses would be to assess "priorities of assessment" [14].

Related to the possible growth in "people's technology assessment" is a final trend worth noting for its certain future impact on TA: an increasing volatility in public sentiment and values. This is usually expressed as "changes in life styles," and it is definitely that. But it is more. It suggests the possibility of more and more decisions, personal and public, that are taken without reference to goals other than perceived desires of the moment. It reflects what might be called new human imperatives: the legitimacy of exploring all forms of consciousness, all forms of expression, and all forms of institutions. In its manifestations it argues for pleasure or solutions *now*, as a means of handling uncertainty. For this reason (and because our increasing affluence is likely to give it continued momentum) it is capable of creating problems -

- e.g., drug abuse - that might best be approached by TA. It is also capable of rendering TA ineffectual, either because the changes it forces happen too quickly or because it can produce new social attitudes and desires that will be deemed acceptable, no matter how disastrous they may be *sub specie aeterni*.

H.G. Wells' forecasts in *Anticipations* (1901) regarding the personal automobile provide an early case in point. Wells was largely correct in foreseeing major consequences of that technology, but his forecasts meant nothing, since the social pressures then emerging proved irresistible. Similarly, if it had been known that the average TV set in America in 1970 would be on for some 5 hours a day and that the programming content would be so abysmally low, could TV have been stopped, redirected, better controlled? It is comforting to think so, though it is unlikely that the most perceptive technology assessor in 1927 or 1939 would even have considered the questions that now seem so obvious and important. Still less is it likely that such questions would have been asked when television was first predicted - in 1847, a mere 30 years before the first apparatus was built [15].

What are today's equivalents of TV and the personal automobile? Easy answers come to mind: some personality control drugs, computer terminals in the home, controlled fusion, video cassettes, and the like. But a technology assessor looking backward from the year 2000 would no doubt form a somewhat different list, including items we may now welcome with special enthusiasm or not really see at all: increased real leisure time, the guaranteed annual income, or an all-volunteer military, or, on the side of physical technologies, weather control, very large LNG tankers, or interactive broad-band communications systems.

Value change, whether or not it is a function of the new human imperatives, might even define some TA problems out of existence. Can atmospheric pollution be seen as a problem if everyone lives, works, and travels in a closed system provided with air purification? The example is fanciful, perhaps, but the point is not. Technology assessment, as it evolves in the years ahead, will be profoundly affected by changed values. And only very indirectly will it have any control over them.

Whatever the future environment for TA, the basic issue it will be asked to help confront is how judgment, knowledge, and a better insight into the alternatives opened through technology can further man's considered desires for security, opportunity, and fulfillment. In dealing with this issue, TA will be dealing, above all else, with the question of the power of human reason in human affairs. Bertrand Russell provides cautionary guidance to technology assessors (and all the rest of us) with these words:

> In the case of science, Reason prevailed over prejudice because it provided means of realizing existing purposes, and because the proof that it did so was overwhelming... If, in the name of Reason, you summon a man to alter his fundamental purposes - to pursue, say, the general happiness rather than his own power - you will fail, and you will deserve to fail, since Reason alone cannot determine the ends of life. And you will fail equally if you attack deep-seated prejudices while your argument is still open to question, or is so difficult that only men of science can see its force. But if you can prove, by evidence which is convincing to every sane man who takes the trouble to examine it, that you possess a means of facilitating the satisfaction of existing desires, you may hope, with a certain degree of confidence, that men will ultimately believe what you say. This, of course, involves the proviso that the existing desires which you can satisfy are those of men who have power or are capable of acquiring it [16].

NOTES AND REFERENCES

1. A neutral survey of some of the more important early contributions to this literature will be found in Walter H. Kohl, "Assessment, Transfer, and Forecasting of Technology," *IEEE Spectrum*, 70-75, January 1971. An extensive annotated bibliography on TA will be found in U.S. congress house committee on science and astronautics, *Technical Information for Congress: Report to the Subcommittee on Science, Research, and Development*, Revised Edn. (Washington: U.S. Government Printing Office, April 15, 1971).

2. TOFFLER, ALVIN, *Future Shock*, 393, New York: Random House, 1970.

3. GILFILLAN, S. COLUM, "The Prediction of Technical Change," *The Review of Economics and Statistics, 34*, No. 4, 380, November, 1952. (Italics in the original.)

4. GILFILLAN, 382.

5. An outstanding example of TA - the work done by The MITRE Corporation and summarized in M.V. Jones, *A Technology Assessment Methodology*, Report MTR-6009, The MITRE Corporation, June, 1971 - comes to roughly this conclusion, but regrettably expresses it in Gilfillan-like terms: "Modern man is *still* far from being able to understand many of the forces at work, let along to control them" (p. 3; emphasis added).

6. National Goals Research Staff, "Technology Assessment," *Toward National Growth*, 130, Washington, D.C.: U.S. Government Printing Office, 1970.

7. MAYO, L.H., *The Technology Assessment Function*, Part 1, Program of Policy Studies in Science and Technology, The George Washington University, 11, July, 1968.

8. Here, too, change is occurring rapidly. Cross-impact analysis is now being studied in the systems

analysis context (see, for example, M.C. Dalkey, *An Elementary Cross-Impact Model*, Report R-677-ARPA, The Rand Corporation, May, 1971); other work is under way to bring it into PPBS.

9. By the same token, of course, it could be argued that futures research could also be incorporated as another department of systems analysis. Were it not for the NIH factor ("not invented here") and for fancies in usage ("systems analysis" is out of favor; "technology assessment," "futures research," the "policy sciences" are in), this would have long since been generally acknowledged, as it should be. The case is made well by the various contributors to E.S. Quade and W.I. Boucher (eds.), *Systems Analysis and Policy Planning* New York: American Elsevier, 1968, esp. Chaps. 3, 10, and 18.

10. It is often said that this growth is now peaking out out, that we are entering a period of cutbacks in research expenditures, surpluses in PhD's, etc. The forecasts may be reliable, but it does not follow that the growth of science and technology relevant to TA will slow significantly. (For example, if still more intense specialization becomes the response to cutbacks, the likelihood of great numbers of more profound discoveries may well increase.) Moreover, the "technologies" that concern TA are both physical and social; there is no reason to believe that in good times or bad, man's proclivity to experiment with his social technologies will be subdued. It should be noted that TA itself is, in its modern dress, one expression of this proclivity.

11. TOFFLER proposes this idea in *Future Shock*, pp. 393-394, recommending it as a means of providing information useful *before* the decision is taken, as well as afterward. The mechanism for the purpose, as Toffler sees it, could be an independent panel of behavioral scientists constituting a "technological appraisal agency" and empowered, perhaps, to seek legislation or obtain injunctions to halt a development "until full public discussion and study is completed."

12. Cf. BENSON, ROBERT S., and WOLMAN, HAROLD (eds.), *Counterbudget*, A report of the National Urban Coalition, xii, New York: Praeger Publishers, 1971.

13. *Toward Balanced Growth*, 124.

14. TRIBUS, MYRON, Statement before the Subcommittee on Science, Research, and Development of the House Committee on Science and Astronautics, December 2, 1969 (mimeo), p. 6.

15. GILFILLAN, 368.

16. RUSSELL, BERTRAND, *Power: A New Social Analysis*, 95, New York: Barnes and Noble, 1962; first edition, 1938.

TOWARDS A HUMANIZED TECHNOLOGY - HOW RATIONAL IS TECHNOLOGY ASSESSMENT?

GERD D. WALLENSTEIN

Private Consultant
Atherton, California, U.S.A.

INTRODUCTION

Technology assessment is thought to be a very rational excercise- as rational as technology itself. But it must serve humanity, in whose life and history the rational and the irrational interact, as psychologist Carl Jung's studies have shown. It is not for lack of scientific tools alone that crystal-ball gazing and astrological forecasting have been such enduring favorites, even with men in positions of power.

Man needs irrational images when the view is long, hazy, and disturbing. Karl Marx forecast scientific socialism by conjuring up "the specter of communism haunting Europe." Today a new specter is haunting America and gradually the world: the specter of a new culture. It is a profound crisis of values, questioning the rationale of science and technology. Never before has the application of science and technology posed such a challenge to humanity, such threat to its very survival, as it does today.

To meet this challenge, leading scientists and engineers are tempted to seek ever greater perfection of their rational skills. There is a danger that this will further widen the gap between technology and humanity, perhaps provoking a new culture to arise over the ruins of technology. Yet, a catastrophe can be avoided and humanity's life be immeasurably enriched if practicioners of technology learn to rediscover the world of irrationality. Engineers are the mediators between science and humanity. Guided by humane, temperate technology assessment, they may be able to identify with the new culture. An impossible proposition? It need not be. Technical men can and will accept a reality that can be demonstrated. But, is the new culture a reality? This question needs to be answered, as we shall attempt to do here.

THE NEW CULTURE: AN EMERGING REALITY.

In a repetition of classic examples, the new culture manifests itself through many separate, largely irrational actions. Drop-outs from society, drug-taking for sensual "kicks", loud behavior in public places, irrationally destructive violence: these are some of the manifestations. A catalogue and an analysis would be out of place and redundant here. The world's mass media have been full of reports on these phenomena. Two books, in particular, have given us an inventory and an analysis: Theodore Roszak's "The Making of a Counter Culture"[1] and Charles Reich's "The Greening of America" [2]. We shall return to them for reference.

Rationally educated men bemoan all this irrationality and charge it off to youthful aberrations in our society. But is not our cool, efficient technology itself put to many irrational uses, as in the wholesale destruction of the natural environment and the wartime slaughter of civilian populations by remote-control from the air? Is the programming of computers, a rational methodology for rational machines, immune to irrational input and output data? The answers to these questions would show that irrationality is part of our, and of any, human society. Rather than treating it as an aberration, we should recognize it as a human characteristic shared by all forms of culture.

The incompatibility of the rational approach and the realities of life obstructs the discourse of otherwise reasonable men. One reviewer dismisses the substance of Reich's book because "his discussion of these issues (i.e.

poverty, pollution, etc.) and what to do about them employs none of the professional tools of any social science" [3]. Granted — but the essence of the book is precisely the birth of a new culture that disavows the reliance on professional tools along with the purely technical standards of the present society. One can disagree with the merits of such a culture; one can criticize Reich's ability to deal fairly and wisely with the old and new that he attempts to describe. But one cannot evaluate his work by ruling him out of order in terms of strict rational analysis, where the subject is outside the constraints of such an approach.

"The Greening of America" is a profoundly disturbing and somewhat unreasonable book for all those who like to believe in progress where each comfortable status-quo can lead almost imperceptibly to the next. For, "The Greening of America" does not only lament the corruption of man's ideals, but also the dehumanization that is admitted by a wide consensus of observers. This book goes much further; it claims that the State — viewed as an inhuman abstraction — is itself producing the changes in human conciousness that lead to its self-destruction. Indeed, who would not want the choice of individualism or group experiences, the freedom to "go fishing", to play music, or to do absolutely nothing, and all the many variations of the human theme that are so richly described in this book? This human longing is a reality — but it is obstructed by the more immediate reality of the actual status and organization of human affairs. To exchange the State and the machines for all-human reality involves a revolution, a revolution of values, that, according to Reich, is aided and abetted by the "establishment" itself. This would give the opponents of the revolution nothing to fight for, a situation that might make them even less amenable to give up the fight.

A much more sober assessment of the two sides of this revolutionary change in cultural values is presented in Rozak's "Making of a Counter Culture". This well-researched and documented study is much firmer in stating the new culture's objectives than in forecasting its triumph. The important point is that its objectives agree with those of "The Greening of America": "to proclaim a new heaven and a new earth so vast, so marvelous that the inordinate claims of technical expertise must of necessity withdraw in the presence of such splendor to a subordinate and marginal status in the lives of men" [4]. But how are these objectives to be achieved? The change must come by affecting men quietly from within themselves. Roszak finds an indication of the needed attitude in a quotation from the Chinese philosopher Chuang-tzu who extolled the wisdon of "knowing how to do nothing". This is a nice thought but not a very realistic one, unless men are to receive some unusual guidance and inspiration. What is the likelihood of this happening?

The guidance and inspiration may be forthcoming from such sources as Lewis Mumford, a lifelong humanistic scholar and prolific writer whose credentials have been established before our technological *mal du siecle* became a fashionable subject.

In his study "The Myth of the Machine" [5], Mumford undertakes a detailed, comprehensively researched, and erudite survey of the continued interaction of technics with human development. He judges our condition not only by its outward material manifestations, but attributes its origin to the minds and hearts of men who helped shape our culture from hundreds of years ago to the present. Thus his study appears to rise head and shoulders over most other contemporary writing on the subject. He alone seems to have dared to draw on all available sources, be they scientific, historical, philosophical, theological, psychoanalytical, sociological, or artistic. As a result, his thesis in its totality becomes almost unassailable, even if fault can be found with his interpretation of one or the other source. Mumford produces enough evidence and reasoning to assure us that our present dilemma of technocracy's dominance over humanity is no accident, that it is not a historical human imperative, and that it need not be perpetuated. He recommends:

"... a spontaneous religious conversion: one that will replace the mechanical world picture with an organic world picture, and give to the human personality, as the highest known manifestation of life, the precedence it now gives to its machines and computers" [6].

Does this mean that in the organic world picture, there will be no place for technology? No, this does not seem to be Mumford's vision; rather, he forsees a fusion of the two cultures, to quote:

"The present analysis of technics and human development rests on belief in the imperative need for reconciling and fusing together the subjective and the objective aspects of human experience, by a methodology that will ultimately embrace both" [7].

But there is a danger that technical professionals might reject Mumford's recommendation. His path to a new culture seems decidedly more difficult than Reich's which predicts an almost effortless transformation, with many appeallingly vulgar amenities making the future take shape in familiar terms. Yet for technical man who is part of the so-called technocracy, the effort suggested by Mumford is necessary. This difficult task also seems to be implied in a study by futurist Arnold Mitchell who postulates an increase in "unfolding" type humans in order to create the future society that would agree more or less with the visions of both Reich and Mumford [8].

Whether vulgarizing, esoteric, or speculative, these diverse writers agree in the coming of the new culture. There remains the question: is so much soul-searching in quest of a new culture a typical American phenomenon or are there similar signs in other countries? If Reich speaks only of the "greening of America", Mumford and others are speaking in terms of a world-embracing new culture. And there are signs in other countries; but, as French writer Jean-François Revel points out, in America the revolution has already begun [9]. It is also the only place where it can begin. By revolution he refers to the sum of all those changes, encompassing all aspects of humanity, that lead to the new culture. Revel's analysis is a most valuable contribution to our understanding, for he makes it clear that this new culture must be global, it must transcend nationalism. And

Revel's final words do invite technical man in:

"... To take the technological civilization as a means, not an end, and since we cannot be saved either by its suppression or its continuation, to countermand it without destroying it" [10].

We may now summarize. If a respected sociologist, a Yale law school professor, a leading French writer, and one of the most erudite humanists of our time all agree that *the start of a new culture is here*, and that its characteristic will be the control of technology by superior human values, then technical man had better take note. There is a further characteristic of this development: it is basically optimistic and non-political. These are two extremely important changes in the climate of the Twentieth Century.

Hitherto, technical men felt justified in essentially ignoring social unrest and attacks on the technocracy. The critical intellectual climate was so pessimistic that it seemed to leave man no chance for survival. And the only positive quest for change was associated with radical political causes. Scientists and engineers thus faced the choice of despair over the merits of their own work or revolutionary change of the society in which they worked. They had too much to lose; so the only reasonable attitude was the support for the status-quo. This meant ignoring the ferment and unrest in the society. Now, with the vision of a humanistic, peaceful new culture emerging, the keynote is positive. As more and more sectors of our society are affected by this mood, scientists and engineers are expected to do their part in this hopeful reconstruction of the human value system. Are they ready to implement these aspirations? What leadership are the scientists giving to this trend?

POSITION OF THE SCIENTISTS

Regarding science and the scientists, the controversy over their relation to human values is so deep and apparently inconclusive that it would be quite beyond the scope of this paper to enter into it. For reference, Theodore Roszak devotes one chapter of his "Making of a Counter Culture" to "The myth of objective consciousness". With quotes and documented references to scientific statements of recent years, he attempts to show that the scientific pursuit of objectivity leads to an inhuman detachment from humanity and a shirking of responsibility for the applications of science. In this context it is poignantly important and perhaps tragic that even that rare scientist who sincerely expounds "science as a humanistic discipline" [11] can serve as evidence for Roszak's point [12].

Whatever the individual scientist may rightfully claim as *his* motivation and attidues, it seems that the combination of his talents with the powers of the megamachine (Mumford's term) and the entire direction of the technocratic society result in the progressive dehumanization that has become the source of despair and the seedbed for the counter-culture.

Some leading scientists have attempted to recognize new directions in society. They propose the formation of task forces or teams to produce "social inventions", or at least improvements in man's troubled social conditions [13]. This prescription has scientists planning more scientific solutions for all human problems. There is going to be more, not less, of all that measuring, dissecting, machine-programming, that has been the critical target of Mumford and the other proponents of a new culture. These scientists — with very few exceptions — want more funds, more facilities, more public encouragement, so that so much more may be known about the secrets of man and nature that our present problems will succumb to the sheer power of knowledge. Whatever one may think of the sincerity and inspiration behind this attitude, it is clear that the scientists have nothing to share with the new culture. They are not even listening to its prophets.

POSITION OF THE ENGINEERS

While science has been in the public light, receiving much honor but also much scorn, engineering has been operating in a much lower key. The most widespread engineering attitude can be summed up in the rhetoric question: "Who, me? I just work here", an attitude reciprocated by the general public. It is an attitude commensurate with the position of hired hands.

Perhaps this attitude was bred and justified during the rapid technological growth period that coincided with a surviving business tradition of laissez-faire. During the years of our immediate past, when despair seemed to get the upper hand in public attitudes and the relationship of science and humanity became a favorite discussion subject, not much was said about engineering. Engineering was generally supposed to implement all and anything that some vague power groups, variously described as management, investors, big government, or just "they", wanted technology to accomplish. And while many questions were raised about where science would be taking us, for a long time nobody questioned where engineering would take us. This was fine with most engineers, too, because it avoided awkward questions about professional responsibility and, in any event, left unquestioned a continuous demand for more engineers.

These attitudes are changing. The hopeful aspects of the new culture and the critique of the counter-culture are companion reasons. If there is hope for a much more humanized culture, why can not engineers orient their work in that direction? Engineering has been described as the mediator between science and humanity. This means that those in engineering work on things, services, and conditions that do something to or for humans by way of applying technology. They are therefore one step closer to humanity than scientists. Should we — the general public — perhaps concern ourselves more with these mediators? In the limit, if we rejected a scientific proposal as harmful or monstrous, we would only have to stop the engineers from "reducing it to practice". Then the scientists could go on forever reading papers about it, without consequences to society.

In a climate of such questions, the profession of engineering and the individual engineers are in for some critical appraisal. If they are capable of self-appraisal and change in attitudes, theirs could be an opportunity for new leadership.

THE HUMANIZATION OF ENGINEERS

Any change or any acceptance of a new concept involves negative as well as positive efforts. To quote German humorist Wilhelm Busch:

> The Good - this we may prove direct,
> Is but the Evil we reject. [14]

The attitudes most likely to turn the engineer towards humanism are the counterpart of those that kept him away from it. We may attempt to group them as follows:

Professional Group Attitudes

Negative	*Positive*
Fancying himself a scientist (pursuit of knowledge for its own sake)	Accept role as responsible mediator
or:	
Classing himself as a hired hand (Do what "they" want)	Work through the organization (make it do what does good to society)
Overspecialization	Broadened technical education

Personal Attitudes

Professionalist in private life	Broad interests in society and the irrational creativity of arts, literature
Narrow outlook: next month's promotion, next week's project	Perspective on life and history

WORK THROUGH THE ORGANIZATION

How can the engineering professional make the transition to the role of a responsible mediator? He can do so by identifying himself more closely with the organization in which he works. This may seem a paradox to some: is not the organization per se considered one of the archvillains of dehumanization? If we take this view, we would soon return to the state of despair, for we would assume that no organization can be made compatible with humanistic values. The limitations of present organizations must be seen in the context of the limitations of people wh staff them. No organization is immutable; each engineer can contribute his influence.

The professional should work in an organization, not as a specialist of a certain engineering field, but as an integral part of that organization's effort. The individual professional contribution is equated with the company's contribution to the society's market, which in turn should have some social significance if the company's market position is at all tenable. Whatever a philosophizing social critic may think of that particular endeavor, a society with a relatively free market play determines a product's or service's social significance by the market's reaction. Therefore, engineering's social responsibility does not need to be limited to projects concerned with pollution and urban renewal. It is precisely the use of engineering in the more mundane transportation, power, communications appliances, entertainment industries — to name only a few — where increasing public inquiry and pressure raise questions of professional motivation and concern with social responsibility.

If the individual professional is the mediator between technology and its social application, then his employing organization is the mediator between the professional and the society. The professional and his organization's leaders therefore must learn to connect their respective activities, so that the society to be served becomes transparent to the professional buried in his work inside the organization. In turn, society recognizes the professional's social responsibilities by its interpretation of the products or services offered through the organization. In time, the market play will more and more reflect the standards of the prevailing culture, thereby gradually changing the market goals of the organization in step with the engineer's preparedness within the organization.

BROADENED TECHNICAL EDUCATION

Specialization is, by definition, the concentration on one part of a whole field. The greater the degree of specialization, the greater the distance from a view of the whole. At some point of specialization, the whole is no longer visible at all; in fact, it may only be possible to think in terms of many divergent items. But the system of the human senses operates by integration and synthesis. The result of overspecialization is thus an alienation from things human.

From a practical standpoint, overspecialization is apt to make the professional forseeably obsolescent, as demand for specialized skills tends to change with changing technology. For these reasons, a broader education enables the professional to understand and stay in tune with the human uses and abuses of the projects on which he works, as well as making him feel as a responsible contributor.

PERSONAL ATTITUDES

Broad interests in society: this is a deliberately vague statement. Nothing can be more truly human than spontaneity, emotional uplift or despondency in response to music, or any of the innumerable manifestations of human ingenuity, playfulness, and irrationality. It is in the description of these manifestations that Reich's book is a veritable compendium.

In an earlier article, I outlined a new perspective on life, a perspective that would be in tune with our optical perspective [15]. Such a viewpoint would be more in harmony with our humanity and would permit us to maintain our place in history without succumbing to the discontinuity that threatens us. If the professional engineer could acquire such a view, he will be able to detect the distant past and the distant future spreading before his eyes on a gentle slope. This can convince him that the utterly inhuman, steeply rising exponential curve that has been assumed to trace our human development is just a mechanical point of view, one that counts linear accumulations of facts, pieces of paper, like algae in a polluted lake. But it is not the qualtity of facts and things in abstraction that matters; it must be brought into relation with our ability to see and use them.

Continuity with perspective: that should be the engineers attitude as well as that of other humans. Every day we see anew, but we stay in sight of what we saw yesterday. And every one of us is empowered to see differently: our own selves, our families, our professional colleagues, our place in humanity.

In the words of Pablo Casals, a wise,indefatigable nonagenarian whose mastery of music may have enriched more lives than many a celebrated scientific breakthrough:

"Each second we live in is a new and unique moment of the universe, a moment that never was before and will never be again. And what do we teach our children in school? We teach them that two and two makes four, and that Paris is the capital of France. When will we also teach them what they are? We should say to each of them: Do you know what you are? You are a marvel. You are unique" [16].

To see this marvel, this unique potential in each child, man, and woman, each human must start with himself; not pretending what he should or would seem to be to others, but finding what he is. In this act and in sharing it with one close human being at a time, he will be unfolding as he helps another unfold, as Martin Buber's wisdom suggests [17]. And if the engineer can unfold, he will open "the gates of the technocratic prison" (Mumford) in which he may have become his own jailer.

CONCLUSION: HUMANIZED ASSESSMENT

The new culture, appreciated from a new personal and technical perspective, can be harmonized with scientific methods. Thus we may assess that which is technically possible in terms of a social responsibility for the wholeness of man. For this concept of wholeness we may borrow from ancient Eastern wisdom, adapting the symbol of Yin-Yang, as illustrated here (Fig. 1). Man's Rational and Irrational domains are complementary to form a whole. The two areas have long, undulating, common borders, indicating the close, though not linear, contact between them. In the extreme part of the Rational is Science, in the extreme of Irrational are the Arts. Other major human fields of activity may be classed in one of the two areas, according to the prevailing group attitudes.

RATIONAL	IRRATIONAL
Industry	Business
Engineering	Psychology
Science	Religion
Economics	Politics
Law	Crime
Medicine	Illness
Sex	Love
Information	Arts

This list makes no claim for absolute certainty, nor completeness. It and the symbolic representation are meant as an illumination of mind, a contribution to the art of understanding human purpose. Thus can the unfolding technologist seek and find his purpose. If technology assessment proceeds in this spirit, its future should be as reassuring as will be the future it helps to create.

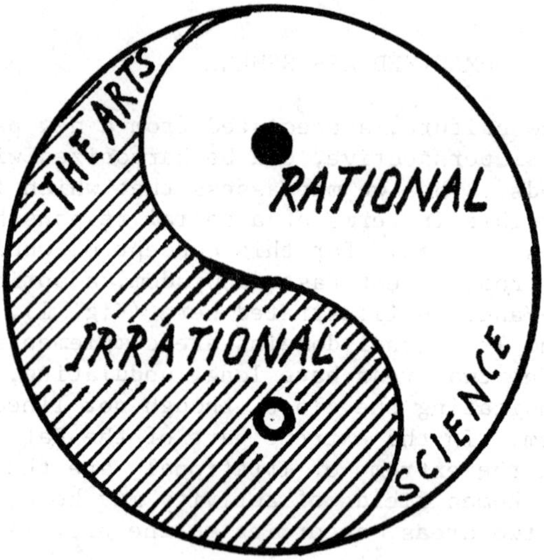

Fig. 1. Concept of the wholeness of man.

REFERENCES

1. ROSZAK, THEODORE, *The Making of a Counter Culture*, Anchor Books, New York, 1968/69. (Page references will be to the paperback edition).

2. REICH, CHARLES A., *The Greening of America*, Random House, New York, 1970.

3. *Saturday Review*, pp. 24-26, New York, Dec. 5, 1970.

4. ROSZAK, op. cit., p. 240.

5. MUMFORD, LEWIS, *The Myth of the Machine - The Pentagon of Power*, Harcourt Brace Jovanovich, Inc., New York, 1970.

6. Op. cit., p. 413.

7. Ibid., p.420.

8. MITCHELL, ARNOLD, "Alternative Future Societies," SRI Paper, SRI, Menlo Park, December, 1969.

9. REVEL, JEAN FRANÇOIS, *Ni Marx Ni Jesus*, Editions Robert Lafont, Paris, 1970. (English version published by Doubleday & Company, New York).

10. Op. cit., p. 264.

11. BRONOWSKI, J., "Science as a Humanistic Discipline", pp. 33-38, Bulletin of the Atomic Scientists, New York, Oct. 1968.

12. ROSZAK, Op. cit., pp. 252 and 300.

13. PLATT, JOHN, "What We Must Do" *Science, 166,* New York, Nov. 1969.

14. Translated by the writer from a passage in *Die Fromme Helene*, by Wilhelm Busch, first published 1872.

15. WALLENSTEIN, GERD, "The Myth of Obsolescent Kowledge," pp. 142-146, *Electronics,* June 13, 1966, New York.

16. CASALS, PABLO,(A.E. Kahn), *Joys and Sorrows*, p. 295, Simon and Schuster, New York, 1970.

17. BUBER, MARTIN, "Elements of the Interhuman," *Psychiatry*, *20,* 105-113, May 1957, Wash. D.C.

SHIFTS IN NATIONAL PRIORITIES, ECONOMIC GROWTH, AND REQUIREMENTS FOR SCIENTISTS AND ENGINEERS

LEONARD A. LECHT

Director,
Center for Priority Analysis, National Planning Association, Washington, D.C., U.S.A.

I

Changes in national priorities can have an important bearing on the utilization of scientists and engineers. The impact of rising expenditures for national defense and space goals in the early 1960's in increasing requirements for scientific manpower and the effects of the current leveling off of research and development outlays in diminishing requirements offer recent, and sometimes striking, illustrations. An analysis of the implications of shifts in national priorities for the utilization of scientists and engineers can add a significant dimension in assessing the consequences of economic, social, and technological change.

The uses of technology are closely bound up with our society's goals and its choices of priorities among them. Research and development make it possible to pursue goals previously regarded as beyond reach. The manned lunar landing is probably the classic case. Many of the nation's goals are concerned with remedying the unanticipated byproducts of technological advance, such as control of pollution. The attainment of others presupposes the development of new technology, as in coping with solid waste disposal. Moreover, the goals our society chooses significantly influences the distribution of R&D resources and the overall volume of resources made available for research and development. The experience with the high priority for space and national defense goals in the early 1960's, and with the greater priority for environmental and health goals in the later 1960's, are illustrations. An assessment of options and needs in science and technology, and of the consequences of the choices made, involves, therefore, a consideration of our goals and priorities as a force inducing change.

Many persons, in and out of the scientific community, have been concerned with the need for a reordering of national goals and priorities in the coming decade. The massive needs for physical facilities of many kinds in the large cities, the prevalence of air, water, and noise pollution, and the inadequacies of our health and educational systems have been subjects for general public discussion. As an insufficiency of suitable employment opportunities for scientists and engineers has emerged as a serious problem, professional organizations and the press have called attention in Congress to the potentials for creating new demands for scientific manpower in seeking more effective solutions to these pressing social, economic, and environmental problems.

Much of the discussion of the relationship between changes in national priorities and requirements for scientific manpower has been concentrated on the role of the Federal Government's research and development effort in the pursuit of national goals. To many it appears credible that, once we decide to provide Federal support on a sufficiently large scale for research and development, the technological thrust which has become so characteristic of our nation in space exploration or computer technology will become the order of the day in such fields as education or rebuilding the central cities. A byproduct of this effort, it is anticipated, would be a substantial increase in requirements for technical manpower.

While Federal R&D outlays have been significant in generating employment for many aeronautical and electrical engineers or physicists, the Government's R&D spending in 1971 is expected to amount to only about 1.5 percent of the Gross National Product [1]. An appraisal of the potentials of a reordering of national priorities for scientific manpower should, therefore, take into account more wide-ranging sources of demand for technically-trained professionals. These would include changes in production, in industry-financed R&D, and in government activities other than R&D induced by high-priority national programs. Assessing the importance of national programs in pursuit of societal goals in creating markets for scientific manpower also involves an overview of the demands likely to be generated in the private economy because of economic growth.

A person seeking information about future manpower requirements for scientists and engineers typically encounters a maze of estimates dealing with research needs, technological potentials, conversion possibilities, and manpower utilization. To appraise the projections, the user needs a framework relating the manpower estimates to their assumptions about national programs, R&D expenditures, and GNP growth rates.

A point of departure for an appraisal of this kind is suggested in a working paper prepared by the National Planning Association's Center for Priority Analysis and published by the Joint Economic Committee of Congress in 1968 [2]. In the paper, NPA examined the employment implications of a decline in defense purchases from industry offset by an equivalent volume of tax reductions and greater Federal spending for urban redevelopment and for increasing social welfare benefit payments. Seven-tenths of the defense cutbacks were assumed to take place in three areas, in the production of aircraft and missiles, ordnance and weapons, and communications and electronic equipment. Three-fifths of the increases in spending resulting from the offset measures were expected to represent larger consumer expenditures with the remainder largely made up of greater outlays for construction and in spending by industry for new machinery and equipment.

For each billion dollars of annual shift in production stemming from these changes, overall employment in the private economy was estimated to increase by 19,000. The reason for this anticipated increase is that the heavily defense-related sectors of the economy such as the aerospace or electronics industries are capital-intensive, durable goods industries. These industries employ relatively small amounts of labor and large amounts of capital equipment to produce their output. The construction, trade, and service group of industries would be the primary beneficiaries of the greater emphasis on social and urban goals and on private consumption. They are more labor-intensive in their operations than the aerospace or electronics industries.

This estimate of a net increase in employment resulting from the conversion from a more to a less defense-oriented economy needs tempering on several grounds. For one, it overlooks the civilian government employees who would lose their jobs in the event of a sizeable cutback in defense purchases. It also does not take into account the military personnel who would be added to the ranks of the civilian job seekers if defense spending and the size of the armed forces were to undergo a decline. In addition, the job skills of the persons released from the defense industries would often be different from the skills required in the new job openings so that employment opportunities would be substantially reduced in some occupations. Engineers, scientists, and technicians, in particular, were among the occupations characterized by an expected decline in employment in NPA's study. A billion-dollar shift in production caused by the changes considered was estimated to lead to:

4,100 fewer jobs for engineers

500 fewer jobs for natural scientists

1,400 fewer jobs for technicians.

The changes taken into account - tax reductions and greater public spending for urban facilities and social welfare benefits - make up only a few of the possible alternatives which might figure in national policy if spending for national security objectives were to decline. Concentration on other goals such as control of environmental pollution would probably create larger demands for scientists and engineers. However, it is apparent that shifts in the nation's choice of priorities are consistent with sizeable reductions as well as with increases in requirements for scientific manpower.

II

The National Planning Association's study, together with other research, suggests some of the strategic variables to be allowed for in seeking to anticipate the effects of changes in national priorities for the employment of technical manpower. One is the representation of engineers and scientists in the work force of industries affected by the redirection of objectives and expenditures.

Scientific manpower is far more heavily represented in the work force of some industries than in others. Many of the proposals for changes in priorities now receiving consideration imply a transfer of production and employment away from industries which employ a high proportion of these employees to industries with a low proportion. The aerospace, construction, and service group of industries are strong cases which illustrate the underlying problem. The representation of scientific manpower in each is summarized in Table 1.

While a breakdown of the construction and service industries into their components would lead to results which diverged somewhat from these estimates, the differences in requirements for engineers and scientists in the estimates for the three industries are striking.

Table 1 Proportion of work force made up of engineers and scientists, selected industries, 1960 and projected 1975

Occupation	Frequency per 1,000 employees in: Aerospace Industry		Construction Industry		Service Industries	
	1960	1975	1960	1975	1960	1975
Technical Engineers	107	153	21	27	6	10
Natural Scientists	10	12	0.4	0.4	4	7

Source: U.S. Department of Labor, Bureau of Labor Statistics, *Tomorrow's Manpower Needs, 4*, 1969.

Six times as many engineers are projected to be represented in the aerospace inudstry in the mid-1970's for every 1,000 employees in construction. Thirty times as high a proportion of natural scientists would be required in aerospace as in the construction industry.

Scientists and engineers are far from a homogeneous mass of technical-professional manpower whose skills are readily transferable from one type of specialty to another. Therefore, another strategic variable, one influencing the manpower conversion problems and planning in higher education, stems from the differences in specialization within engineering and the natural sciences in the industries which would be affected by a redirection of emphasis in the pursuit of national goals. The differences in occupational specialization are illustrated, for engineers, in Table 2.

Table 2 Major specializations within engineering, selected industries, 1960 and projected 1975

Occupation	Frequency per 1,000 employees in: Aerospace Industry		Construction Industry		Service Industries	
	1960	1975	1960	1975	1960	1975
All Technical Engineers	107	153	21	27	6	10
Major specializations:						
Aeronautical	55	65	-	-	-	-
Civil	3	6	19	24	2	2
Electrical	15	23	1	1	1	2
Industrial	10	22	-	-	-	1
Mechanical	18	27	1	1	1	2

Source: *Tomorrow's Manpower Needs, op. cit.*

In terms of the industries in Table 2, a lesser emphasis on national defense, or a reduced priority for the space program would primarily show up in a sizeable reduction in requirements for aeronautical engineers, and smaller reductions for electrical, industrial, and mechanical engineers. A higher priority for urban redevelopment would directly lead to an increase in demand for engineers in construction which was largely made up of a demand for civil engineers. Greater expenditures for consumer services brought about by larger social welfare benefit payments or tax reductions would generate minor increases in the service industries for civil, electrical, and mechanical engineers. The changes in direct employment for engineers would be accompanied by other changes in the firms supplying inputs to the three industries in unidentified proportions.

III

The estimates of changes in requirements for scientific manpower resulting from a redirection in priorities imply something like the recent allocation of support for research and development, and a technology and production pattern generally similar to that now in use in the industries considered. Accordingly, they minimize the extent to which programs responding to the greater national concern with social, economic, and environmental goals will involve a redistribution of R&D resources, and changes in technology and production in different industries. Similarly, it is difficult in estimates of future manpower requirements to anticipate the growth in expenditures of the Federal, state, and local agencies responsible for implementing the new programs. All of these developments can be expected to influence future requirements for scientists and engineers, and frequently to influence them in a way which outmodes projections based on extrapolations of past trends.

The Federal Government's support, especially for research and development, has been a major source of employment for technically-trained professionals. For example, in the mid-1960's, the most recent period for which comprehensive information is available, nearly 400,000 scientists and engineers, about 30 percent of the national total, derived their employment from Federal funds [3]. Federal support was responsible for the employment of over half, 55 percent, of all R&D professionals. Only about 5 percent of the employment of non-R&D scientists and engineers was attributable to the Federal Government's funding. More recently, in 1968, spending for national defense alone accounted for nearly three-fifths of all the jobs of aeronautical engineers and for close to two-fifths of the jobs of all physicists [4].

While Federal support for research and development outside the areas of national defense, the space program, and atomic energy has been increasing, about five-sixths of the Government's R&D expenditures in recent years have taken place under the auspices of the agencies concerned with the three areas. Much of the growth in requirements for scientific manpower brought about by Federal R&D outlays in the 1960's has been due to the massive development projects associated with national defence and space goals. Currently, Federal support for development exceeds that for basic and applied research by a two-to-one margin. The defense-related expenditures for development have exceeded those for research by a four-to-one margin. Federal support for development is more heavily concentrated than the spending for research. In 1970, for example, the agencies concerned with national defense, the space program, and atomic energy were estimated to account for 96 percent of the Federal outlays for development [5].

The very large development projects linked with the introduction of new spacecraft, nuclear reactors, or weapons systems have few, if any, counterparts in other fields. Public R&D expenditures in these areas, other than in health, are on a considerably lesser scale, and they are primarily for research. For instance, there has been a growth in interest in new technologies for mass producing low-cost housing and housing components. The introduction of Operation Breakthrough symbolizes the translation of this interest into development programs. However, the Department of Housing and Urban Development, the sponsor of Operation Breakthrough, is estimated to have spent a total of $22 million for R&D in 1970, and a total increasing to $35 million in 1971, including $25 million for development [6]. The projected 1971 spending for development would amount to one-fourth of one percent of the overall Federal expenditures for development. Because of this lesser role of development in the research and development programs related to the pursuit of social, economic, and environmental goals, rapid rates of increase in R&D outlays for these objectives are likely to be translated into considerably more modest increases in the utilization of scientists and engineers.

To illustrate the magnitudes involved in a rapid expansion of public R&D for programs in pursuit of societal goals, in 1970 Federal support for all research and development outside of the areas of national defense, space, and atomic energy amounted to $2.5 billion, including some $400 million for development [7]. A 100 percent increase over the 1970 level by 1975 would raise the spending to $5 billion in that year. Assuming that the average annual outlay per scientist and engineer, the "performer cost," in the programs supported by these expenditures grew to $60,000 by the mid-1970's, the additional $2.5 billion would create employment for slightly more than 40,000 scientists and engineers [8]. This would amount to the equivalent of less than one-tenth, about 7.5 percent, of all the scientific manpower employed in research and development in 1969.

IV

Growth in Federally-funded research and development contributing to social goals can be important for national well-being, and in creating opportunities for many individual professional workers. Yet it would be unreasonable to expect that a greater emphasis on publicly-supported R&D directed at eliminating urban blight, improving medical care, or reducing present pollution levels will absorb the bulk of the engineers, physicists, or mathematicians no longer required for the pursuit of national defense or space goals. Other activities, largely involving employment in the private sector, are likely to provide most of the opportunities arising from national priority programs or from economic growth for already established scientists and engineers seeking new employment or for recent graduates of colleges and universities.

As one instance of the available alternatives, research and development activities financed by industry are an important and frequently overlooked source of requirements for scientific manpower. Most of the industry-sponsored R&D is applied research or development aimed at producing new or improved products or processes. Much of this effort also contributes to social needs in improving transportation and communication facilities, in reducing pollution, or in increasing standards of living. If the experience of the recent past continues, industry-sponsored R&D will loom larger as a source of employment in the next decade than in the past one. For example, while Federal funds for R&D performed by industry increased by less than 3 percent between 1966 and 1968, company-supported expenditures rose by 23 percent in this period [9]. By 1970, nearly 60 percent of industrial R&D outlays were made up of company funds [10].

A high priority for social goals such as housing could be expected to create larger requirements for scientific manpower in production than in research and development. In the 1960's, to cite some recent history, scientists and engineers employed in construction, very largely engineers, made up 5 percent of the national total in the two occupations, but they represented less than 1 percent of all the R&D professional manpower [11]. A series of development programs similar to Operation Breakthrough could substantially increase requirements for R&D professionals in construction without affecting the importance of expansion in the production of housing as the primary source of employment growth for engineers in this field.

Programs in new priority areas will frequently involve growth in government activities which utilize engineers and scientists in many capacities. The Clean Air Act is an example. According to the 1970 report on *Manpower and Training Needs for Air Pollution Control,* state, local, and regional agencies will require 2,800 engineers and scientists by 1974 to implement the Act. An additional 1,500 would be employed by the National Air Pollution Control Administration [12]. These professionals would be largely employed in testing, establishing standards, surveillance, administration, and in providing technical assistance to other agencies.

Nearly two-thirds of all engineers and scientists are engaged in activities other than research and development. The largest source of requirements for this group stems from growth in the economy. Rapid growth enlarges the options open to scientific manpower in shifting from one industry to another, or from research and development to related activities which can utilize many of their skills. According to one standard source, the Department of Labor's study, *Tomorrow's Manpower Needs,* the economywide utilization of scientists and engineers is expected to increase from 1,050,000 in 1960 to about 1.9 million by 1975 [13]. These projections imply an annual average increase in requirements of over 50,000 a year for scientific manpower of all types in the 1960 to 1975 period. The projections illustrate the potential impact of rapid growth since they assume a pace of GNP growth sufficient to reduce the overall unemployment rate to 3 percent in the mid-1970's, about half the comparable rate in the first half of 1971. With this rapid growth, the annual increase in the utilization for scientists and engineers is projected to exceed the estimated increase in requirements for R&D professionals brought about by a doubling of research and development outlays for societal goals over a five-year period.

V

The critical-manpower problem for scientists and engineers in the early 1970's arises from the fact that the Federal Government's support for national defense and space goals involving large R&D components has been tapering off at a time when the economy has been growing at a snail's pace. The changes in national priorities which could generate new sources of employment by a greater concentration of resources on societal goals have taken place only on a modest scale. Sluggish economic growth has dampened the incentives in the private economy to embark on the large-scale capital expansion programs involving investments in the utilization of complex technology which could make use of the skills of many technically-trained professionals.

Two types of measures could serve to focus on the tendencies responsible for diminishing requirements for scientists and engineers. One is a national priorities budget indicating the consequences, in dollar and manpower terms, of alternative combinations of national priority programs aimed at devoting a greater volume of public resources to coping with an accumulated backlog of unmet social and environmental needs. The other are fiscal and monetary policies, together with the related economic controls to restrain inflation, for stimulating more vigorous economic growth.

A national priorities budget would trace through the effects of the Federal Government's outlays for programs in pursuit of national goals in increasing production and employment in the private and public economy. The budget would take into account areas where a lesser priority for particular goals could be expected to release resources for other purposes, and areas where new or pressing problems point to the desirability of a high priority for expanding programs and expenditures, including a greater research and development effort, in the next few years. The budget could show the implications of alternative courses of action such as a concentration on rebuilding the decaying physical facilities in the inner cities as compared with an emphasis on supporting the development of many more new towns and satellite cities. Where Federal Government programs lead to greater economic activity in the private economy or by state and local governments, i.e., pollution control or health programs, these effects should also be allowed for. One proposed budget along these lines, but confined to Federal expenditures, has been prepared by a citizen's group, the National Urban Coalition, and published under the title of *Counterbudget: A Blueprint for Changing National Priorities* [14]. A series of budgets reflecting different priorities could go far toward creating public awareness of the connection between the nation's priorities and manpower requirements. Probably the most significant of these budgets would be one presented by the Administration each year indicating its priorities and the expenditures and manpower needs in the public and private economy required to implement them over the next five years.

Manpower considerations enter into a national priorities budget through the use of indicators and manpower coefficients, translating the dollar figures for national priority programs into employment in many occupations, including scientists and engineers. The coefficients would show requirements for scientists and engineers, distributed by specialty, for each million or billion dollars spent for the different programs in the dollar budget. With further research, these indicators could be developed through the use of the input-output technique of analysis to estimate manpower impacts in the private economy, and by the use of "performer cost" data to anticipate requirements for scientists and engineers in government employment or in research and development. Research in the application of these techniques has been underway in the U.S. Department of Labor's Interagency Growth Project and elsewhere for several years [15]. As an initial step in making use of the manpower indicators, the Federal Government's annual budget estimates for research and development outlays should be converted into a manpower budget indicating the consequences of the Government's expenditures for the demand for scientists and engineers.

The National Planning Association's goals research can provide a basis for illustrating how manpower indicators can be used to relate expenditures for national priority programs to the requirements they generate for scientists and engineers. This research has been concerned with establishing quantitative standards reflecting current conceptions of national goals, and with projecting the dollar costs, manpower requirements, and the resources likely to be available for achieving them in the next decade. Needs in each of the goal areas and standards for their achievement have been formulated by the National Planning Association from special studies, recommendations of bodies such as the National Academy of Sciences, legislative hearings, legislation, and general national policy.

The areas of concern defined as "goals" refer to virtually all sectors of the public and private economy, and they include such areas as health and education, international aid, national defense, natural resources conservation and development, research and development, social welfare, transportation, and urban development. To bring out the manpower implications of the nation's priority choices, NPA has devised estimates of the employment in different occupations required to achieve each of the goals considered in the next decade. The estimates can be expressed as indicators based on the proportion, the frequency per 1,000, of the total manpower requirements represented by individual occupations, or in terms of the utilization of manpower in specific specialties per billion dollars of expenditures for the end products needed for the attainment of these goals. Indicators for scientific manpower, for scientists and engineers combined, are shown in Table 3 for three goal areas, natural resources, social welfare, and urban development.

The projections pertain to requirements for attaining goals, goals often anticipating substantial changes in R&D outlays or in the utilization of technology, rather than reflecting current practice. The indicators are based on an input-output analysis of the chain of production and employment in all of the industries contributing directly or indirectly to producing the output needed to pursue each goal. Accordingly, the applicability of the manpower coefficients to proposed programs in the three areas presupposes a program content similar to that entering into the goals standards.

Table 3 Illustrative indicators of scientific manpower utilization in selected goal areas in 1980

Goal Area	Indicators for 1980	
	Frequency per 1,000 employees	Utilization per $1 billion expenditure
Natural resources	80	5,600
Social welfare	15	1,175
Urban development	37	2,925

Source: Derived from LECHT, LEONARD A., *Manpower Requirements for National Objectives in the 1970's*, Manpower Administration, U.S. Department of Labor, 1968, Appendix Tables 14, 28-2, 32-2, 34-2. The original estimates refer to final demand expenditures in 1962 prices.

Allowing for these caveats, it is significant that the indicators for the natural resources goal underscores a requirement for scientists and engineers per billion dollars of expenditure roughly five times that for the social welfare goal and nearly double the comparable figure for the urban development goal. The high utilization indicator for the natural resources goal stems largely from the emphasis in that goal on an expansion of research and development programs to seek new energy sources or substitutes for raw materials in increasingly scarce supply. The indicators for urban development are substantially affected by the large-scale utilization of construction engineers in urban redevelopment programs, and by the expectation of a greater role for the mass production of building components in urban construction in the 1970's. The scientific manpower indicators for the social welfare goal are low because the expenditures for this goal largely increase the output of industries producing consumer services and consumer goods, industries frequently utilizing a relatively low proportion of scientists and engineers.

Indicators of scientific manpower utilization can be prepared showing more detailed occupational specialties or for different combinations of goals and priorities. They would make it possible to project the requirements for scientists and engineers induced by an additional billion dollars spent for pollution control as compared with an equivalent expenditure for improving medical care, for mass transit, for enlarging nuclear power facilities, or for programs leading to a manned landing on Mars in the 1980's. The indicators could also underscore the scientific manpower needs likely to be generated by alternative program choices for achieving an agreed-upon national objective. For example, a high priority for solid waste disposal could lead to new programs for developing recycling processes to convert solid waste materials into economically useful products. The solutions advanced also include proposals, as yet untested by R&D, to relieve the economic and esthetic problems associated with solid wastes by constructing special ships to incinerate wastes and dump them in the oceans. The manpower indicators derived from outlays for the two approaches to solid waste disposal could be expected to be dissimilar and, correspondingly, the number of scientists and engineers required for each, and their professional specialties, would differ.

A national priorities budget can go far toward assuring a fuller and more effective use of the nation's resources, including scientific manpower, in the years ahead. Requirements for scientists and engineers in the 1970's will be even more substantially affected by the policies adopted to achieve full employment generally in the economy. The details of these policies extend considerably beyond an analysis of the forces influencing the demand for scientists and engineers. However, their consequences are apparent. If the slow rate of GNP growth characterizing the past three years continues for the next few years, increases in public expenditures for individual, social, or environmental goals would lead to an expansion of job opportunities, and perhaps to manpower shortages, in selected occupations such as those related to medical care or pollution control. Slow growth would reduce the chances of obtaining employment in areas affected by the greater spending for these purposes by swelling the ranks of scientific job seekers with persons seeking employment because of a lack of opportunities elsewhere. In this kind of economic environment, retraining programs for scientists and engineers would largely produce newly acquired unused skills. The major condition for assuring full utilization of scientific manpower, as for others, is a dynamic economy.

REFERENCES

1. National Patterns of R&D Resources, 1953-71, National Science Foundation, NSF 70-46, 2, 1970.

2. LECHT, LEONARD A., "National Priorities, Manpower Needs, and the Impact of Diminished Defense Purchases for Vietnam," In: *Federal Programs for the Development of Human Resources, 1*, . Joint Economic Committee of Congress, 1968.

3. ARONSON, ROBERT L., "Federal Spending for Scientist and Engineer Employment", U.S. Department of Labor, Bureau of Labor Statistics, Bulletin 1663, 2, 1970

4. Annual Report of the Council Advisers, 46, February, 1971.

5. Special Analyses, Budget of the United States, 256, Fiscal Year 1971.

6. *Ibid.*

7. *Ibid.*

8. The national average cost per R&D professional in 1969, in 1969 dollars, was $47,200. The estimate, therefore, includes a substantial allowance for inflation as well as for the real earnings of scientists and engineers. For R&D performer cost data, see ARONSON, op. cit. and "National Patterns of R&D Resources", *op. cit.*, 7.

9. "Research and Development in Industry, 1968", National Science Foundation, NSF 70-29, p. 3, 1970.

10. "National Patterns of R&D Resources", *op. cit.*, p. 12.

11. LECHT, LEONARD A., *Goals, Priorities, and Dollars - the Next Decade,* 208, 1966.

12. Report of the Secretary of Health, Education, and Welfare, Manpower and Training Needs in Air Pollution Control, 1970.

13. "Tomorrow's Manpower Needs", *op. cit., 4,* Appendix E.

14. "Counterbudget: A Blueprint for Changing National Priorities, 1971-76," The National Urban Coalition, 1971. For a study translating the Urban Coalition's budget into manpower requirements, see SCOVILLE, JAMES G., and BEZKEK, ROGER H., *Manpower Implications of Reordering National Priorities,* Institute of Labor and Industrial Relations, University of Illinois, 1971.

15. See ARONSON, op. cit., and for a recent illustration of the findings of the Interagency Growth Project, see *Manpower Report of the President,* Chapter 5, 1971.

THE SOCIAL RESPONSIBILITY OF BUSINESS OR THE CHANGING ROLE OF GOALS IN CORPORATE AND R & D PLANNING

MARVIN J. CETRON

President,
Forecasting International, Ltd., Fairfax, Va., U.S.A.

"Having lost sight of our objectives, we redoubled our efforts"...said C.E.K. Mees, several years ago, and this expression still holds true today, in all walks of life, in industry, government and in personal lives. What does this mean to a company? That it doesn't know where it is? That it doesn't know where it is going? It may sound obvious, to the point of being ridiculous, but some companies really don't know what business they are in, and they don't know in what business they want to be in in the future.

SETTING CORPORATE OBJECTIVES

"Management by objectives" has become a catch phrase in the last few years, but if it is to be used meaningfully, then the objectives must be meaningful, and must be accepted by all levels of management. For management by objectives to succeed, everyone from the President down must be "on-board." The rationale for this is quite straightforward, for if everyone is working to achieve some corporate goal, then each individual is trying to help his superiors to succeed. His output, therefore, must support or compliment the work of his superiors in order to achieve the goals. However, for them to be accepted most widely, the best method probably is for the top corporate executives to develop the corporate goals by bargaining, and then to obtain the President's agreement. Once these corporate goals have been set, then each functional or division head should be allowed to develop his function or division objectives (of a more quantitative and technical nature) based upon the corporate goals. He should be permitted any approach to develop his own goals. If a division goal appears to be in conflict with corporate goals, then, by discussion, these conflicts must be overcome; if no agreement can be reached, then the corporate goals should take precedence.

Having decided upon the approach, the question then arises of how to determine the goals. There are several ways of considering the problem of goals: What in one company is a goal, is a constraint in another company, and is a plan in yet a third company. Goals can be developed in six different ways, as shown in Fig. 1.

Fig. 1. Goals of objectives can originate [1]:

1. By enunciation
2. By bargaining
3. By appeal
4. By external imposition
5. By consensus
6. By use over time

Maybe before we consider how corporate goals are classified, we should consider how personal goals are stated, since they are structured in much the same manner. If someone asked us what the most important thing about our job was, I don't believe we would say it was the salary. However, in the back of our minds, I am sure we would be thinking there was a *basic salary* below which we would not go. After this basic salary was met, we would then think in terms of *long-term professional development* in the job and other factors, such as, can I do the type of work I'd like to do? Will it be ego rewarding? Can I be my own boss? etc. Possibly after these two major goals are met, then we would think about how the job would *benefit society*, the environment, and possibly even contribute to the world as a whole. But the first thing really is a survival goal, meaning short-term survival, then long-term survival, and finally the societal aspects of the job. (This concept is covered in greater detail here in my article — The Trimatrix — An Integrating Technique for Technology Assessment.)

These same types of attitudes are present in a corporation. The first thing a firm looks for is not to optimize profit but to make a certain level of profit below which it won't go in order to survive. Then it looks for long-term growth potential for the corporation and finally, after it has survived and has its long-term goals it then starts determining how the corporation can help society as a whole. Therefore, one way to consider goals is:

1) Financial or survival goals
2) Long-term growth and continuity goals
3) Social responsibility goals

These illustrative goals are shown in Figures 2-4.

In this classification scheme, 1) represents the short-term goals aimed at short-term profit, for example, next year's P & L statement. This set is satisfied by low-risk ventures. Some managers have a very strong motive for pushing short-term profits because they believe that the P & L statement has no memory, and they won't be there when the long-term payoff arises; 2) represents the long-term goals of the company, by internal growth and development by higher-risk ventures, especially R&D; 3) the social responsibility goals are concerned with the company's image, in the eyes of its employees, the general public and the government. For example, many companies are now looking at bio-degradable or recyclable packaging for their products, while others are looking at transparent packaging, so that the consumer can see exactly what he is buying, in order to preclude such extremes as one corporation found when some of its stockholders were considering voting for safety and social goals at the expense of corporate profit, or having someone like Ralph Nader looking over their shoulders. In many companies, this third category becomes a constraint, rather than a goal. The general feeling is that it is not the company's business to take responsibility for general welfare and quality of life, although they must adhere to regulations and principles in order to retain a certain image. Too, there is the feeling that this third category of goals is, in fact, a derivative of the second category, because unless a company's image is good, its long-term potential is poor. What company can obtain financial aid from financial institutions, or from stocks, if it opposes public opinion or government regulations?

In the first two categories, we considered "risk" implicitly -- in looking at short-term profits, risk was considered to be low, whereas for long-term growth, risk was taken to be higher. However, in some companies, particularly the very large ones not heavily involved in R&D, there is another category of goals, namely, those for risk aversion. These goals invoke Savage's Law of Minimum regrets: "If everything goes wrong, how badly hurt can my company be." In many cases, this is a valid approach, but it ignores the expected value of a venture, and is thus detrimental to R&D, which is inherently risky and uncertain. A checklist of corporate goals, covering these areas, is shown in Figure 5.

Fig. 2. General corporate goals for an electronics device manufacturer

---

Survival or Financial Goals

1. Maximize return on investment, consistent with growth objectives, and operate to protect that investment. For example, the company has as its objective, 15% profit before taxes and 30% return on stockholders equity and long-term debt. In order to maximize long-term growth and profitability it must remain a leader in the (electronic component) industry in advanced technology, device development, and engineering.

2. Attain a sales volume among the top five companies in the (electronics component) industry, participating with a broad line of products in approximately 75% of all markets, and competing for a minimum of 15% in each of the markets.

Long-term Growth and Continuity Goals

1. Maintain a basic technological capability permitting the company to develop, produce, and market a specific product within one year. Development should be directed toward products which will have a significant market within two to five years. Attention should be continually focused on the balance between short-term and long-term objectives.

2. Participate in industrial, consumer and military markets. Although emphasis is on the industrial and the consumer electronic and entertainment devices market, we will accept military business, but not more than 25% in that area.

3. Develop a product line evolving from discrete components to solid-state circuits, sub-assemblies and simple equipment, including passive components. These products will include new-type solid-state devices as they begin to displace existing products.

4. Obtain more contract sales, particularly in areas where contract programs parallel company programs and product plans

5. Consider domestic and foreign markets as one integrated world market, with interrelated technical, manufacturing and marketing opportunities. The company will exploit the advantages of foreign manufacturing and marketing.

6. Growth should take place primarily through internal development, <u>not</u> through acquisition. We do not want to become a conglomerate.

7. After our minimum profit before taxes objective is met, consider products that will aid the corporation's long-term growth and continuity more than products that will have the same total long-term profit (using discounted cash flow), but that will have a much shorter life.

Fig. 3. General corporate goals for an electronics device manufacturer

Fig. 4. General corporate goals for an electronics device manufacturer

---

Social Responsibility Goals

1. Remain an ethical manufacturer of electronic devices, providing customers with full value and establishing and maintaining a reputation for fair and honest practices.

2. Provide a productive and satisfying working environment for employees, offering career opportunities for personal development and advancement.

3. Consider the societal responsibility of the corporation, especially its policies affecting the environment (air and water pollution, conservation, potential hazards, etc.).

4. Hire a minimum of 5% minority groups and 1% disabled workers.

5. Provide time off from work to allow capable individuals to get involved in civic and social problems.

6. Comply with government regulatory agencies' recommendations, especially regarding employee safety and competitive practices.

Fig. 5. Corporate goals check list

1. The business the company wants to be in
2. Method of growth intended
3. Direction of growth
4. Rate of growth
5. Allowable dependence on suppliers
6. The desired capital structure
7. Degree of stability desired
8. Company image desired
9. The company's desired size
10. Allowable degree of government control
11. Percentage of the total market to be held
12. Percentage of the market in a given geographical area
13. Degree of technical flexibility desired
14. Degree of product diversification desired
15. Balance of price-volume and profit-volume markets
16. Organizational decentralization intended
17. The minimum and desired rates of return on investment
18. Desired participation in technological opportunities
19. Desired balance between short-term and long-range objectives
20. Number of new products during next 2, 5, 10 years, etc.
21. Product and production safety
22. Desired social achievements
23. Desired employee environment

TRANSLATING CORPORATE GOALS INTO ACTION

When the choice of corporate goals has been made, then meetings should be held with all subordinates to explain the goals common to all parts of the organization so that the individual translators will be aware of the corporation as a whole.

An example of the translation of corporate financial goals into functional or division objectives is as follows:

Functional, Division or Areas Objectives	Corporate Financial Goals
Corporate Headquarters	Profit
Production or Manufacturing	Cost of product or component
Service	Cost of Service
Sales	Revenues or margins
R&D	Level of effort budget
Administrative (or other) Staff	Fixed budget based on sales

As this paper is directed primarily at technologically dynamic companies, and in particular to their R&D departments, the next step should be to show the relationships between the company goals and the R&D departments. It was stated earlier that the divisional managers should be made aware of the corporate goals. In a similar manner, the top executives should be informed of the expectations from R&D -- not the details of how it is to be performed, but scope, timing, and degree of need of the R&D programs -- for the corporate goals and the research objectives must be feasible before they can be used in the planning process.

USING R&D OBJECTIVES IN CORPORATE PLANNING

There are four levels of plans: policy, strategic, tactical, and operational.* The R&D manager should participate in all four levels. (Figure 6 shows where the different types of planning fit in the planning procedure.)

Fig. 6. Basic steps in planning

---

1. Analysis (where are we?)
2. Policy planning. Formulation of goals and objectives. (Where do we want to be, or where should we be?)
3. Strategic planning. (How are we going to proceed?)
4. Tactical planning. (What are the steps we are going to take, and when?)
5. Operational planning. (Which organizations and individuals will carry out the day-to-day work and the schedules?)
6. Appraisal of Results. (How well did the results meet the expectations, should we change or modify our goals and objectives?)

---

* *Policy planning*, or the formulation of functional objectives for the future. *Strategic planning*, or the formulation of a set of alternative ways of achieving objectives, based on normative forecasting and resulting in goals or targets falling within the range of what is considered feasible. *Tactical planning*, or the delineation of the sequence of actions necessary to implement a particular strategy. *Operational planning*, or the specifications that the particular organization or individual is responsible for in carrying out day-to-day work. When strategic planning involves significant technological developments (determined through normative technological forecasting), we then speak of technological planning (at strategic levels).

Illustrations of the relationships between strategic planning, functional planning and R&D planning are shown in Figures 7-9. Focusing our attention on Figure 9, it can be seen that there are three basic requirements for setting R&D objectives: a forecast of technological change and requirements for new technologies; the sources of the organization (personnel capabilities, facilities, equipment, budget); and the corporate goals and strategy. If the R&D objectives developed from these requirements are well written, they should act as a catalyst in stimulating research toward results, while leaving the researcher the flexibility of selecting his own technical approach in achieving the objectives. Because of the inherent uncertainty in research, many people feel that it is extremely difficult to set objectives for research, and equally difficult to develop a plan for it. However, if the research department is to make a contribution to the company as a whole, then research must be planned and the results measured.

As can be seen from the illustrations of the planning processes, planning is not a "one-shot" affair, nor is the development of goals and objectives. The environment is constantly changing, breakthroughs occur in certain technological areas, other technologies become obsolete, and so plans and goals should be updated accordingly. That does not mean that programs should be turned on and off like a faucet, but (as most companies are currently doing) goals and plans and programs should be reviewed at least annually so that changes can be gradually incorporated.

CHANGING CORPORATE GOALS

This type of iterative goal setting and planning represents quite a shift from that of a few years ago. Up to about 1964-65, corporate goals were primarily profit and Return On Investment oriented, and planning was mainly financial planning. This type of planning had several drawbacks. When the only goal that mattered was

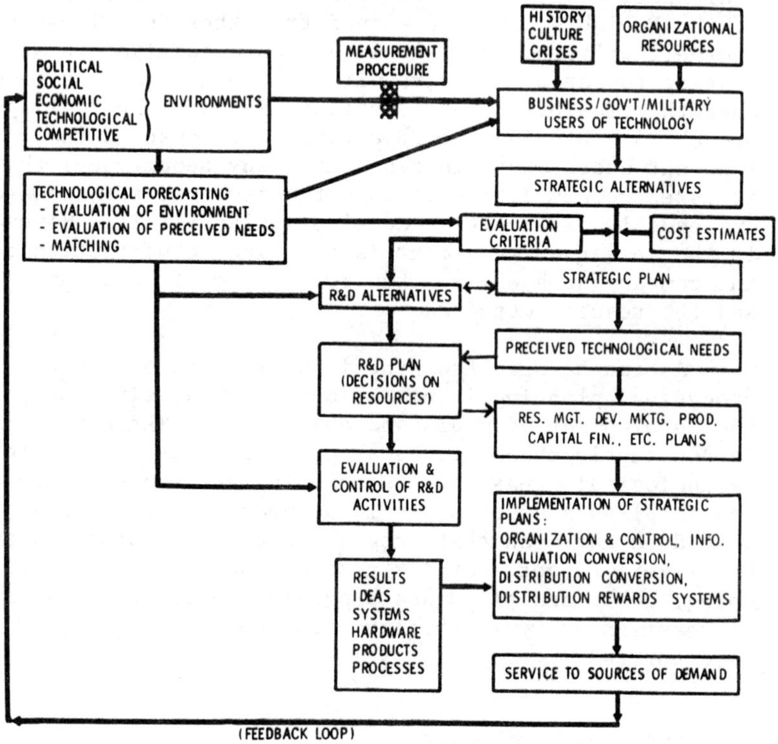

Fig. 7. Setting goals and corporate strategy

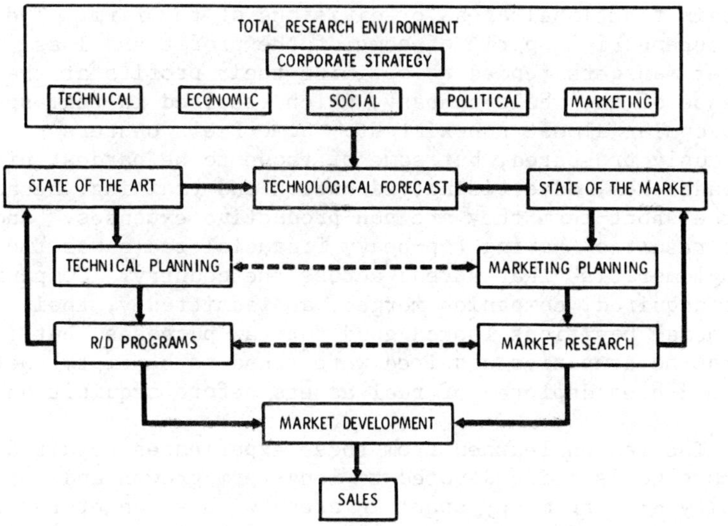
TOTAL RESEARCH ENVIRONMENT
CORPORATE STRATEGY
TECHNICAL
ECONOMIC
SOCIAL
POLITICAL
MARKETING
STATE OF THE ART
TECHNOLOGICAL FORECAST
STATE OF THE MARKET
TECHNICAL PLANNING
MARKETING PLANNING
R/D PROGRAMS
MARKET RESEARCH
MARKET DEVELOPMENT
SALES

Fig. 8.

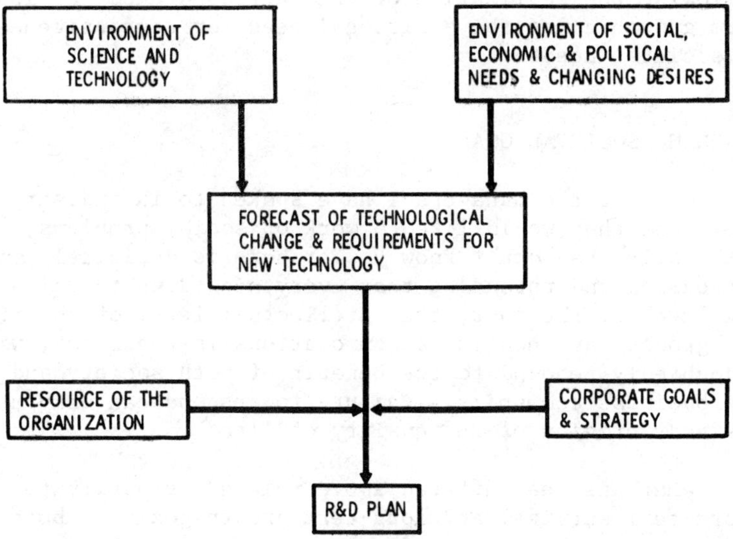
ENVIRONMENT OF SCIENCE AND TECHNOLOGY
ENVIRONMENT OF SOCIAL, ECONOMIC & POLITICAL NEEDS & CHANGING DESIRES
FORECAST OF TECHNOLOGICAL CHANGE & REQUIREMENTS FOR NEW TECHNOLOGY
RESOURCE OF THE ORGANIZATION
CORPORATE GOALS & STRATEGY
R&D PLAN

Fig. 9.

to make X billion dollars in five years, trouble arose in certain functional areas or divisions of the firm. The entrepreneurial spirit of some of the profit and loss center managers tended to optimize their profits at the expense of the whole company, which resulted in team spirit becoming almost nonexistent. Some P & L centers obviously prospered, but some of those to be hardest hit included R&D, advertising, education and training, because in the short-term they are non-productive expenses. Another result of having top-heavy financial goals was the "conglomeritis" that spread across the country. Companies were acquired, companies merged, and admittedly, their financial positions looked good for tax purposes, but often the companies acquired were found to be empty shells and had been depleted of real assets before acquisition.

The lesson learned from these experiences resulted in more goals being devoted to long-term growth and continuity -- more money spent in areas with a longer-term payoff; individual expertise being acquired, instead of companies; and services being hired instead of bought. Now, of course, more lessons have been learned, and following the realization of the increased need for long-term growth comes the additional need for responsiveness to societal needs.

PLANNING SOCIETAL GOALS

Most of the managers I have spoken to in industry feel that they would like to work on social problems, but really they don't know how. There is definitely an increasing understanding that everything done to raise the level of the poor, the intellectual level of the minority groups, and housing accommodations in the slums, will undoubtedly rebound to the benefit of both society and of the business community. Failure to respond can only mean a steady erosion of the quality of life.

What has been offered above related primarily to short-term survival and long-term growth goals -- both of which are internal to the welfare of the corporation. The remainder of this paper will address the external aspect of the problem, namely the environment.

Technology forecasting/assessment is a new emerging management tool, the purpose of which is to assist with such problems as "Toward what ends should my company mobilize and allocate its resources?" "What research programs should be pursued?" "How should new programs be initiated?" And finally, "What impact should our company's research have upon national, social, economic, and cultural goals (and vice versa)?" There are several firms that are rather deeply involved in tech. forecasting/assessment for industry, including two "not for profit" firms, The Hudson Institute and Institute for the Future. Three independents are Forecasting International, Ltd., International Research and Technology Corporation, and The Futures Group. These five organizations make available to various corporations environmental forecasts and show the corporate managers how they can best meet societal responsibilities and maintain their profit posture at the same time.

As can be seen by just one look at the titles of new books being published, almost everyone is getting on the "social responsibility of business" kick. Every day items show up in the news reflecting industry's abdication of social responsibility: the companies are polluting the environment, profit is their most important product, and so on; and now some academic institutions are voting their large blocks of stock in industrial organizations *against* what was considered to be management's best interest.

One of these new books, *Business and Social Progress*,* is really a collection of views of two generations of executives, the traditional role of American business in our society and the challenges now facing it. The book was the outgrowth of discussions held by the Committee for Economic Development (CED) in 1967, and is best summed up by Emilio Collado, the chairman of the CED Research and Policy Committee: "American business will have to do a good deal more to help solve the nation's social problems. The younger business executives clearly feel -- and our CED commentators concur -- that the corporation as such should identify those social problems on which its particular resources and skills can be most effectively brought to bear and make these virtually as much a part of its business objectives as traditional commercial activities."

The contributing authors are trying to change the old adage that "what's good for business is good for society" to "what's good for society is good for business." To use Walton's words, "There is increasing understanding that anything done to raise the economic level of the poor, the intellectual level of minority groups, and housing accommodations in the slums will rebound to the good both of the society and of the business community. Failure to respond can only mean a steady corrosion of the quality of American life. To the assumption that technology is good because it induces growth and productivity is added the further assumption that growth itself is good. According to the American gospel, growth reflects the dynamism of an achieving society, and achievement always means social progress. To be bigger is to be better. Today, a different message is being urged upon us. Increasingly questioned is the easy identification of growth with progress. Countries (and cities within countries) view with apprehension the influx of people to their borders; colleges set limits on enrollment and expansion; businesses grow bigger, but lawmakers frown. When, for example, the electric power industry responded to public demands for more energy by announcing a 6 to 7% increase annually in production, vociferous complaints were quickly heard regarding the dangers of thermal pollution and other problems."

Corporations were, up until a few years ago, rated only by their profit performance, and the "best" companies were those which had the highest earnings and paid the highest dividends. Today, there is an additional standard of corporate performance: to be best, a firm must be a good citizen and try to protect and enhance the environment, hire minorities and the disabled, train the unemployed, and insist that their executives get involved in civic and social problems.

Accordingly, the pressure is as great for companies to perform in the social sector as it once was only in the profits segment. In fact, some brokers recommend the stock of companies that have demonstrated good citizenship as well as good earnings. The federal government is making the hiring of minorities a condition of renewing contracts. Of course, no pressure is necessary for many executives; they have grasped every opportunity to make their companies not only profit machines but vehicles for social progress. Many among them even are seeking a method of "social accounting" to match their annual financial accounting.

For example awards have been given to Western Electric Company for aiding minority business, Levi Strauss & Company for hiring and upgrading minority employees, Coca-Cola Company for establishing a salary program, housing and day care centers for agricultural workers, and Campbell Soup Company for efforts to rehabilitate Camden, N.J., its home town. Awards for improving the physical environment have gone to American Metal Climax, Inc., for its system of handling tailings from a new mine, to Rouse Company for Columbia, the new town, to Weyerhaeuser Company for recycling and controlling the waste from its pulp and paper plants, and to Teller Environmental Systems, Inc., for new concepts in environmental control. Individual awards were won by William H. Wendel, president of Carborundum Company for heading urban renewal in Niagara Falls, and by Paul W. Driscoll for implementing a successful hiring program for Navajo employees of the Fairchild Camera & Instrument Corporation's

Shiprock (N.M.) plant, where he is general manager, and for creating a plan for an urban center for the Navajos.

A NOTE OF CAUTION

When I finished reading *Business and Social Progress* [2], and the latest newspaper, I must confess I felt that maybe social pressure would make the pendulum swing too quickly too far. Maybe a lesson from the past will serve to illustrate my concern.

Fifty years ago, United Fruit was one of the 100 top companies in the United States. United Fruit became so involved in running a fleet of ships, setting up company towns for the workers, building schools and churches, carving our railroad spurs in the wilderness, that it began to think of itself as a country rather than as a company. It suffered from a sort of self-perpetuating megalomania. The firm's management began to believe that it had a mission in life to provide bananas to the hungry world. It lost sight of the fact that it was in business to make money. Year after year, the high banana trees of the United Fruit Company were blown down by tropic wind storms or shriveled by tropical diseases. Year after year the company picked up the pieces, politely cursing these acts of God, when it might have been looking about for a banana that grew closer to the ground and was less susceptible to tropical diseases. After dropping from their perch in the top 100, they started a major research and development effort. Their scientists have now come up with such a banana. But United Fruit faces an uphill fight to regain a place among the nation's top 100 companies.

WRAPPING IT ALL UP

American business will have to do a great deal more to help solve the nation's social problems. This feeling has been voiced, not only by the young, but also by the older managers, in the study I have just completed. The corporation should identify those social problems on which its particular resources and skills can be most effectively brought to bear, and make these a *part* of its goals while maintaining its business objectives and traditional commercial activities. There is a general feeling too, that to a much greater extent, social problems could somehow be converted into profit opportunities in the best sense of the term. This may be easier in some areas than in others.

In conclusion, the business community must generate a new set of goals that makes sense both to business and to society of which it is a part -- let us hope it can achieve this delicate balance.

REFERENCES

1. Adapted from an unpublished paper: "Developing Applied Research Goals," delivered by Bodo Bartocha, as part of a seminar of Technological Forecasting, American Management Association, New York, November 1970.

2. *Business and Social Progress: Views of Two Generations of Executives*, edited by Clarence C. Walton, Praeger, 1970.

HOW MEN CAN SHAPE THEIR FUTURE*

JOHN PLATT

Associate Director, Mental Health Research Institute, Ann Arbor, Michigan, U.S.A.

A "wagon-train model" of pioneers moving into unknown country helps to illuminate the collective cybernetics of social goal-seeking and social change. The future is divided into an inertia period (of 2-10 years?), T_I, followed by a choice-and-control period (of 10-20 years?), T_C, followed by the uncertain future, T_U. In the historical process of leadership and collective decision for the T_C period, maps of alternatives with costs and benefits ("contour maps") are seen as important power sources and conflict reducers.

The Protestant theologian, Harvey Cox, has a recent book entitled, *On Not Leavint It To The Snake*. It is a plea for mankind to take responsibility for the world and its future. The whimsical title is derived from the situation in the Garden of Eden. Adam and Eve had been given dominion over all the animals in the Garden but, as Cox says, when they came to the moment of truth, they let the snake decide.

Today this is no longer a parable of the past but of the present. Mankind now has the technological power to make the world a Garden of Eden, a Garden of abundance. And with our poisons and our protection, we now have dominion over the numbers of all the plants and animals in this Garden, including ourselves. In this

---

*Reprinted from *Futures*, March 1971, pp. 32-47.

analogy, some might identify the snake with our ancient habits and attitudes and our failure to control population or human conflict. But the real snake in the Garden today is our misunderstanding of human history and human responsibility.

As Cox goes on to say, there are three major attitudes towards history. The first is the apocalyptic attitude, that the human race is doomed. The second is the chiliastic or Messianic attitude, that it is destined to a glorious future. In many of us there still lurk these or some other deterministic variations, such as the view that history is random or that it is cyclical or that it is determined by economic or technological imperatives or innate human aggression, so that in any case there is nothing we can do about it.

Against all these, however, Cox would set the third attitude, which he calls the prophetic attitude towards history, and which he identifies with the attitudes of the prophets of the Old and New Testaments. They held that history is conditional, that is, that it is an "if ...then..." process, with the future shaped by what we choose and do. *If* you steal or kill or bear false witness, *then* you will roast in hell. *If* you love God with all your heart, *then* you begin to create around you the Kingdom of Heaven.

GUIDING THE WAGON TRAIN OF HISTORY

This comes close to the existentialist view which sees the future as continually being shaped by human responsibility and action. Indeed, the if-then view is the view we take every day as we plan supper or a vacation trip or a career. Nations plan social security and economic development. Why should it seem so hopeless to plan or shape a world? The reason is that in the past, with warring nations and uncertain catastrophes, we had to leave it up to God or chance to determine the large-scale future. Random surprises, or statistical stabilisation, or the feedback stabilisation of the Invisible Hand of the market-place, or self-stabilising attitudes or organisational structures, were likely to thwart any deliberate plan. But today we are beginning to have the world organisation and the technological power and understanding to actually achieve some successes in large-scale planning.

As we do so, we begin to need a coherent picture of the process of social design and change. Social processes should no longer be compared to a railroad train running on fixed tracks to an inevitable destination, but rather to a wagon train of settlers moving across the country towards the frontier, in an on-going collective search for a better place to live.

It is worth exploring this last simile in some detail. For example, it is certainly true that for any well-organised living organism, or decision-system, or social group, the process of change is in many ways like the driving of a single vehicle. At a given instant of time, a vehicle — or a society — is at a particular place with a particular direction of motion determined by its previous history. (Initial position, x, and velocity, $\dot{x}$, as the physicists would say.) For some short time ahead, which we might call the correlation-time or the inertia-time, T_I, little can be done to change its course. We hit the dog on the road, or our society gets into the war, regardless of the most heroic efforts to swerve away. This interval goes over gradually into a second time period, the choice-period, T_C, within which our present steering will be increasingly effective in avoiding danger or achieving an intended path or goal. (In physics, our deliberately chosen steering force of acceleration, $\ddot{x}$, now as we turn the wheel, will finally produce a calculated change of time-course or path.) This interval finally goes over into a third period, T_U, the time-period of uncertainty, which is too far off and too dependent on intervening hazards and our intervening responses, for present planning or steering to be relevant except in terms of very general directions. (The "I-C-U" model.)

THE PERIODS OF PREDICTION AND CHOICE

The three periods are not sharply separated, and of course their lengths may be different for different social phenomena. Nevertheless, the relation of prediction, choice, and planning, is very different for the three intervals. In the first period or inertia-period, T_I, social science is deterministic like the natural sciences, because non-interfering predictions or extrapolations are possible, to the extent that the causal chains in society are understood. The predictions of course are not necessarily linear, and they may have oscillations or exponential explosions or decays. And they may be probabilistic and uncertain because of extra-societal intrusions (like assassinations or external attacks), thus resembling probabilistic predictions in physics.

But the whole nature of prediction changes for the second period ahead, T_C, which is governed not by physics but by cybernetics. Thus, in driving a vehicle, the momentum-physics of the next half-second from now is steadily replaced, in anticipating future seconds, by the cybernetics of goal-directed steering. The "weak interactions" of the natural sciences, where the observer or his prediction does not change the phenomenon, become replaced by the "strong interactions" of organism-environment feedbacks, where prediction changes to a range of alternatives with cybernetic choice. It ceases to be prediction and becomes "prophecy" in the sense used by Harvey Cox. It becomes "if-then" prophecy, which does not tell us what will happen, but rather what is the range of directions we can go in, and where we will go if we turn the wheel of initiating-action by a particular amount one way or the other. "Predictions" cease to be predictions and become something more like "advice" or "warnings" so that the alert society, or the alert driver, can avoid the foreseen dangers or choose the desirable directions that have just become apparent. For this period, the problem is not to predict the future but to change it.

It is sometimes supposed that these choices by a driver or a group of drivers as they scan the alternatives, are themselves predictable by psychology or social science. This, however, is false, since science is a public affair, and if the social scientist is a responsible and participating human being, he will surely communicate his forecasts of great opportunities or great dangers, and he will thus become an interacting part of the decision system. All social science "predictions" of such decisions therefore begin to share the logical fallacy of "self-predictions", which are not predictions at all but cybernetic intentions or warnings, with uncertain outcomes depending on the resultant choice and action. (Thus, in economics, the notorious unpredictability of the stock market may be a simple result of the strong short-time interaction between "predictions" — which are warnings — and decisions, with group effects and leader effects preventing even a statistical stabilisation.) The final choice and action after such "predictions", are certainly not predictable from any simplistic or "rational" assumptions. We, or our leaders or protesters, will often insinuate amplified fluctuations in our value-systems, for the most personal and subtle of reasons — so that instead of choosing the easiest or the shortest path towards a goal, we often choose a different one for aesthetic reasons or for exploration or to test ourselves or just for the sake of variety.

When we look at the T_I and the T_C periods in this way, we see therefore that we must distinguish something like five futurist categories: predictions, uncertain predictions, prophecies (or ranges of alternatives), intentions, and interacting decision-making. These distinctions give us a new view of our controversies about the "descriptive" or "normative" character of the social sciences, and about their ultimate ability to make accurate predictions as in the physical sciences. These disputes are the result of a failure to separate prediction and choice in the different time-periods of reference, coupled with a widespread failure to appreciate the cybernetic character of well-organised social dynamics. The dynamics of society and social change, as we see here, is rooted not primarily in physical sciences, but rather in the organism-environment cybernetic interactions characteristic of the biological sciences.

The third future period, the ultimate period of uncertainty, T_U, is like the period after the next bend in the road. It is a period for which only general goals — or the preparation for general goals — can be chosen now, and the details of the path have to be worked out as we come closer and can see better. This is a period for which, because of our inability to see details, our present choices often can be guided only by some very general heuristics or values or moral rules that have worked in the past, or by some kind of *Sitzgefühl* as in chess. When the uncertainties are great, as in times of war or revolution, the T_U period comes very close to us, a day or an hour away, and the moral rules shift towards the necessities of immediate survival, so that a man is a fool if he starts building a house between the trenches. On the other hand, in times of peace, when T_U is far off, we can begin to make long-range savings or long-range plans for economic development or for feeding the world.

Many of our most violent disagreements in social attitudes, such as the conflict between personal gain and social morality or between short-range and long-range planning, are not disagreements over ultimately desirable goals, or ultimate possible gains, but simply different estimates of the current level of uncertainty and how far T_U is away. Some of our liberal-conservative disputes would be less bitter if both parties could see this and could spend their energies on making better estimates of the uncertainties and the real possibilities of effectiveness. But with our multiple accelerating changes today, clearly it will be a long time before T_U can be pushed much farther away than 30 years or so, or before our social understanding of the secondary consequences of our actions will be great enough that we can see the alternatives ahead for more than a generation.

THE TIME-CONSTANTS OF CHANGE

The length of the inertia period, T_I, is commonly over-estimated in the social sciences. Both conservatives and radicals often hold that things will not change, except by violent revolution. The determinist school seems to be more scientific in holding that there is a dynamic of history, with compelling long-range collective forces, so that with a better understanding of social dynamics and with larger computers, the course of society should be in principle predictable indefinitely into the future, like a railroad track. This view would hold that the choices in the choice period, T_C, are only apparent or surface choices which do not change the main thrust; so that the T_I period of ongoing momentum actually extends through the T_C and T_U periods for a generation or more, or perhaps up to 1,000 years in the Toynbeean cultural dynamics.

But it can be shown that, at least in self-conscious societies with centralised planning, for many phenomena, T_I is in the range of no more than 2 to 10 years. Beyond this, choice becomes increasingly effective, as in the time scales of full economic mobilisation for war (1-2 years), of the change of opinion in the USA to legalise abortions (perhaps about 2 years), of the rise and fall of the Vietnam war (5 years), of the creation of the international monetary system of Special Drawing Rights (5 years), or of the race to reach the moon (9 years). It will hardly be denied that these are large-scale social acts of conscious choice. The last three are typified by individual acts of leadership which would have been almost impossible to predict very far in advance, by the shrewdest students of social dynamics, or even by the leaders themselves!

A recent study with Karl Deutsch and Dieter Senghaas has shown that half of 62 great achievements or inventions in social science in this century had widespread acceptance or social effects within about 12 years, and that this time appears to be getting shorter since World War II. In each social system, there is an underlying style to these changes, of course — the social "vehicle", capitalist or communist, is still the same social vehicle. But these styles have changed, too, in the last 20 years — and partly by design, as with the growth of agro-industry, automation, service industries, credit, medical care, universal higher education and government-supported industries, which transform the whole mechanism of social incentives, methods of operation, and class structures.

Of course, there are some social phenomena where change is more sluggish and difficult, like attempts to change the direction of a clumsy ox-cart. World population growth is in this category, for it can hardly be levelled off in less than 10 to 25 years, even if a perfect contraceptive were available tomorrow. This means that world population will almost surely increase in the next 25 years by some 3 billion people, or roughly a factor of 2. And who can deny that we must bring up all these new children — wanted or unwanted! — so that they can reach as nearly as possible their full potential as creative citizens of the world? To do this and to make up for our present deficiencies, we must have an increase in world food in this period by roughly a factor of 3, in world housing by a factor of 4, in world health care by a factor of 4, and in world education by a factor of 5 or 10. We have not yet mobilised our abundance and know-how to meet these massive needs, and in fact it is only in the last year or two that we have begun to realise the full scale of these advancing problems. Nevertheless, analysis and action is beginning to be taken on many fronts, and it seems possible that we can approach many of these goals of world well-being in 20 years if we begin effective action now.

Other social phenomena resistant to change would include "lock-in" situations, such as Federal bureaus, or the military-industrial complexes of both the USA and Russia, which are themselves tied in to the conflict of lock-ins of the arms race. Nevertheless, some of these problems yield to cost-benefit analysis, as in the proposed restructuring of the US Post Office system. Others can be pushed towards re-adjustment by the growth of world interdependence, as well as by abstract analysis of the barriers to change in such systems and the damage they do to the larger system of which they are a part. The lock-ins slow up change — like wagons with fixed wheels — but they can themselves be changed, usually suddenly, when the pressure for a new direction gets great enough.

It would appear that the T_C period, when our present social choices create fairly predictable changes, might run to about the year 1990 or 2000. The period beyond that almost surely is in our T_U interval, because of our great uncertainties as to the consequences of our present rapid ferment and change. By that time the world might have been transformed to any of several extreme forms, such as a post-nuclear desert; or a world dictatorship by nuclear blackmail; or a *1984* type of permanent war, sustaining competing dictatorships; or just possibly a group of open pluralist democracies in a world of growing abundance for all. The probability of the latter is terribly small. It will only be by enormous efforts of social invention and action in the next few months and years that the other dangers can be averted and that it can be achieved at all. Not until it is clear whether we are moving in that direction will it be clear whether we can predict or even prophesy the shape of the year 2000.

THE DESIGN OF CHANGE

We can see these problems better if we enlarge our simile, going beyond the problems of the motion and guidance of a single vehicle, to the problems of overseeing the progress and the collective guidance of a whole wagon train exploring new country. As we shall see, we need to explore the analogies to the problem of looking over the country ahead, sending out scouts, and mapping the alternative routes; and the analogies of making sure the vehicles are designed well and are working properly and are responsive to the steering mechanism, and have rules of the road so they do not get in each others' way.

We will need especially to explore the role of technical experts (scouts, repairmen, advisors) and managers (captains, quarter-masters, drivers) and the process of collective decision-making and agreement by all the parties.

For this purpose, the wagon train analogy is better than the strongly authoritarian analogy of a ship, where the ignorant passengers have no choice but to trust their safety to the authority of the experienced captain and his crew. The role of managerial elites is diminishing today; and with increased education there is an increased demand by all individuals and groups, as with the families in a wagon train, for direct participation in all decisions affecting their destiny.

How a large society of men could set up a structure that would enable them to solve their decisions collectively without some royal or authoritarian hierarchy was the problem first solved explicitly by the Federalists in 1789. They felt that for the first time they were putting into practice theoretical and experimentally tested principles of social design. They said, "It seems to have been reserved to the people of this generation to decide the important question, whether societies of men are really capable or not, of establishing good government from reflection and choice, or whether they are forever destined to depend, for their political constitutions, on accident and force".

The Marxists later criticised this 18th century attempt at social design on the grounds that it favoured a privileged economic group, and they tried to make a new social design that would fulfil instead the goals of the class struggle and permit the workers or the majority of mankind to participate in the fruits of production. On the other hand, such technical features of social design as feedback correction ("checks and balances") to correct bureaucratic abuses and inadequacies, or allowance for diversity of tastes and styles of life, or the provision of incentives for improved systems organisation, or the adjustment of social structure to take account of change, seem to have been given inadequate attention in Marxist formulations and practice.

Today it would appear that we could make social designs and world structures greatly improved over any of these previous models. We have a new range of technological powers and abundance and a new understanding of technological change. We have a new knowledge of feedback stabilisation, cybernetic goal-seeking, and hierarchical decision-systems. And we have a new knowledge of the biological and psychological bases of individual and group behaviour. With our new abundance, it is possible to make new designs — systems designs for the benefit of the whole society — that do not need to favour any one group or class in society over any other one. It is possible to make designs that will satisfy both urban and farm, suburb and central city, workers and intellectuals, managers and consumers, blacks and whites, males and females — designs that can permit and encourage pluralism and diversity of tastes and life styles, with abundance for every group. Some societies today are closer to this kind of world than others are, but even the best of them have grave misdirection of energies, unresponsive administrations, oppressed groups, and great blights and crises; and the world in its overall structure is still very far from such a goal.

Can we make a successful transition from where we are, to this kind of improved future design for everyone? Can we do it without any break in the social continuity of food and goods and services that we need to sustain us all? It is almost certainly possible, but we can achieve it only if we understand and work out in detail what kind of new designs we want, both for our societies and for the world, and if we understand the processes of change and institutional restructuring that can take us from here to there. This kind of improved social design for the future of mankind must be begun now if we are not to be destroyed by the crises of the next few years. It will require the very best scientific and analytical, administrative and political, legal and philosophical talents we have. It can go, and it must go, in its competence and detail of design and in its general satisfaction of human needs, far beyond any social design we have ever had in the past.

MAP-MAKING FOR THE FUTURE

When we look more generally at the if-then formulation of history, we see that what we are talking about is a *map* of all the possible directions of change. "If we do *a*, then we reach *A* in 30 years," specifies one path on the map. "If we do *b*, then we reach *B*", specifies another. All the alternative possibilities together could be laid out on a real map, in our T_C range of times (or distances) although it might have to be a multi-dimensional map to fit the multi-dimensional character of real social decisions and actions.

As soon as we recognise the map analogy, we see how over-simplified is any single if-then statement. A complete map is not one-choice-one-goal, like firing a gun at a distant target. Rather it is a set of roads or paths of least energy, which may not be straight, because they have to avoid mountain barriers and precipices. And these roads have no single goal, but march on and on across the map, with every point a decision-point for the drivers in terms of direction and speed through the next section, and with occasional intersections where radical new directions can be chosen if desired. Sometimes we may have to go up in order to go down, or go around in order to go forward. Such subtleties are probably beyond most of our present social prophecies, which are still in the linear-forecasting stage, but they will come with increasing knowledge.

Of course, our map will be very uncertain and full of *Terrae Incognitae* in regions of social space we have never explored, like those before us at the present time. We can only see the future, so to speak, from a high point near the trail, when what we really need is an exact aerial survey of the various paths ahead. We must develop these improved surveys, and quickly; but since we are travelling on and cannot wait, at any given instant we can only choose what seems the best route at that time while keeping careful watch to see that we do not pass some point of no return or get trapped in a narrow valley from which there is no way out. Ancient Egypt, with the burden of its royal-tomb economics, was stuck in such a social valley for 2,000 years. Many scenarios today could lead to a world dictatorship or a new Dark Ages with much the same result for us.

Of course, a good map of our future would not be a "flat" map showing completed roads, but rather a "contour map" of the terrain, across which we make our own road as we go, like the wagon train. The heights and depths on the contour map are fairly exact metaphors for the social costs and consequences of any given path of travel. The regions with high social costs have to be avoided in planning the future, because society generally tends to "go downhill" or in the direction of least resistance except when it is carried into a high-cost region, so to speak, by its own momentum.

Society may be able to get over a small barrier, such as organisational problems or vested interests, by such mechanisms as capital investment or national re-allocation, provided there is promise of a downhill "gain" on the other side. A hilly path, if short, has frequently been preferred to a "more natural" longer path in conscious human decision-making, because of our determination to make visible progress in the foreseeable future in satisfying human needs and aspirations.

Probably the step-by-step mapping of the costs and consequences of different paths of social change will become one of the main contributions of "prophecy" or "futurology" to large-scale human affairs. All the forecasting techniques of industry, with product assessment and market research, will need to be applied on a far larger scale. It is not "advice" we want, so much as an impartial mapping of costs and consequences. This is one of the most valuable things that parents, teachers, psychiatrists, economic counsellors, or other intellectual elites have to offer to the young or to other sectors of society. We do not want to be told or coerced or manipulated into doing some particular thing — even if it is supposed to be for our own good — so much as we want to know as accurately as possible what the consequences are of the different alternatives, so we can make the decision for ourselves. This nourishes our sense of dignity and responsibility and helps the growth of knowledge and freedom that we need to become complete men and independent citizens. Perhaps we can add important additional facts to the map that direct persuasion would never have uncovered! So the teacher who offers, not advice, but a map of all the consequences, has already started to wean the student and to make him a chooser and teacher himself. An accurate map is basically nonpolitical, or can be made so, so that it spreads by desire to those who want it, affecting all the decisions of society much faster than the best self-interested attempts at persuasion.

MAPS AS CONFLICT REDUCERS

Improvements in our maps of costs and consequences also can often help in reducing disputes and differences. Someone has said that men do not differ in their goals so much as in their choice of competing heuristics or paths towards reaching the goals. A better calculus of costs can help to straighten out misinformation or false cliches, and make one strategy or the other much more convincing. Trustworthy maps reassured the families in the wagon trains and kept them from splitting up. Probably the same will be true for many social groups today facing disputed social decisions.

Better maps may not only bridge divergent pictures of reality but may even do something towards bridging divergent self-interests. If one route can be shown to be clearly more promising than another in terms of total social costs in reaching a generally agreed-upon goal, then that total social advantage can be partly used to give compensating personal advantages to groups whose interests are damaged by taking that route. Thus we compensate landowners displaced by a highway, or workers displaced by automation. It is only when the total advantage is uncertain that the disputes rage on. Much wider use of this principle of pre-assessment and compensation would help many of our needed social changes to go faster and with less disruption.

Conflicts of values or general life-styles for whole communities are harder to deal with, because a culture can scarcely be both communal and capitalist, both tribal and nuclear, or both existentialist and Puritan at the same time. Such conflicts can only be resolved in a society of space and abundance, which does not need to co-opt all of its members into any particular kind of consumerism or housing or work-pattern. In such an economy, we can see that Diversity will be one of the most valuable directions on the map of change. (In a multi-dimensional map of social "phase-space", Diversity would be represented by all the regions "away from the axes", where there are components from many different axes of variation; while Uniformity would be represented by the much narrower regions along one axis or another, with no components in other directions permitted.) Such a society would begin to be a society of tolerance and experiment, in which different personality types and different cultural traditions have a chance to be protected in trying out their preferred ways of life on a limited scale, although they might not be imitated by the rest of society as their costs and values become more apparent.

Why should we say that men are born equal or unequal? The truth is rather that all men are born incommensurable. They live their lives on different axes. The encouragement of experiments in diversity would be a great aid in reducing our attempts to make ourselves one-dimensional and in reducing our present divisive conflicts over goals and our coercion of minorities by majorities.

It is worth noting how far Western society today has gone in the creation and tolerance of this kind of diversity. In the USA today, in spite of all its continuing problems of mutual adjustment of different groups, there is probably more diversity accepted in such things as architecture, jobs, dress, books, films, records, music, poetry, and life-styles than in any society in history. It would be easier for many nations to accept the idea of One World if they saw it as both having and encouraging this kind of variety and self-expression for every cultural tradition.

THE ROLE OF TECHNICAL ELITES

Probably it will always be technical elites who have to perform the "look-out" operations and make the maps. But they are also needed to keep up with the over-all preparations for the trip, checking the state of the vehicle, making needed repairs and improvements, and managing the carrying-out of the group's decisions after they are made. The "vehicles" in our analogy are the social institutions involved in change, their organisational leadership and responsiveness and stability. The "repairs and improvements" are the small "seed operations", such as needed research and development, which can greatly change the organisational stability and responsiveness to change.

What needs particular attention is the steering mechanism — our collective decision-making mechanism for collective social problems. Technical behavioural science is badly needed, to show us how to debate and reach decisions more quickly and satisfactorily in groups and parliaments, how to make use of needed technical advice, and how to tolerate diversity and provide mechanisms and incentives for conflict-resolution between dissenting parties. We may need new methods of information-handling and information reduction, new types of social contract in co-operating groups, and new parliamentary decision-rules.

Specialists are also needed for carrying out technical decisions, and for mediating the social processes and the feedback channels of the system to be sure they are working in the directions of change intended by the group. This involves continuous monitoring and information-sharing, with expediters and ombudsmen to keep the wheels greased on the social vehicle and to look out for malfunction or "overheating". This role goes over continuously into the larger goal-assessment and direction-assessment and progress-assessment and expediting for the whole on-going process of change.

Will the specialists get out of hand and become dictators? Not in a high-education high-information society where common sense and personal concern and widespread know-how can monitor the monitors. Not if we design our rules of collective decision-making so as to assure a rotation of technical elites and equal technical help to the "loyal opposition", and to permit counter-analyses and counter-plans when any group feels it needs them to prevent preemption of humane alternatives by technocratic arrogance or stupidity.

COLLECTIVE DECISION-MAKING

After the technical elites have made their maps, and provided their information and inventions and assessments and advice, it is still the participants in our planetary wagon train who must make the final choices of their future direction or directions.

Several features stand out as requirements for satisfactory group decision-making in the world ahead. The first is that all social decisions from now on must be participatory. Every individual or sub-group must have as large a share as is practically attainable in the decisions that affect his destiny. Any society that violates this rule, as by subjecting individuals to discrimination or injustice that its affluence and education could avoid, or by subjecting them to administrative decisions in which they or their representatives have had no responsive part, is in grave danger of riot and breakdown. This may or may not lead to the counter-response of a police state, but whether it does or not, the cost in economic and social abundance from such a malfunctioning system is enormous. Equality costs very little compared to these costs or compared to the enormous payoffs for everybody from a well-functioning society whose human beings feel that their needs and complaints have adequate and equal satisfaction and prompt redress.

The second requirement for satisfactory group decision-making is general recognition of and education about the non-zero-sum nature of all our social decisions. We are now involved with each other in a total system where we will all win together in effective system-operation, or all lose together in malfunction and breakdown. The total result depends on whether we offer each other tolerance or hatred, reward or punishment, trust or distrust, independence or coercion. The collective advantages in a satisfactorily operating social and political system, with its present blocks and malfunctions removed, could give all the world an abundance almost beyond imagining. It would be enough to make the present poor as well as the present rich far richer in goods and personal satisfactions than anyone is now, in our present conflicting and maldirected social groups.

THE SENSE OF HISTORICAL PROCESS

A third requirement for collective decision is a general understanding of the historical process of change. Everyone involved in debating the collective future needs to realise that, as we have shown here, change cannot be instantaneous and is rarely smooth. There will be advances and retreats as existing feedbacks come into play to resist encroachment on their domains. Locked-in systems will resist change until they are restructured. There will be successive stages of education of those involved in change. Sometimes old or growing abuses can be wiped out at the stroke of a pen, but at other times they are ingrained deep in personal habit patterns or in slow patterns of committee work or legal rules or large-scale organisation or economics that have to be attacked indirectly.

A sense of historical process helps us maintain our patience as well as our pressure. A lost battle or an important crisis, like the California oil spills, may dramatise the human problem; and if we are prepared to make use of it, it can be an important step in re-educating the public for the next step. Crises, if not completely catastrophic, can be incentives to faster action, as well as useful signals of the imminence of more dangerous crises and of the level of needs and demands that we still have to satisfy and the distance that we still have to go.

Change can also be accomplished more easily and successfully if all those involved can be kept aware of the goals and the distance to them, and of whether the mechanisms of change are on schedule as the wagon train of history swings around to new directions. In particular, the sense of process helps us keep in mind that no new direction is ever final. As with any vehicle, the direction must be continuously corrected and corrected again to adjust for the bumps in the road or the errors in the last correction, or to swerve towards an easier and swifter route even though it is not the straightest.

The reason is that the directing of change is a feedback system, a process of successive approximations. First solutions, like early motor cars or early detergents, are frequently crude, with unforeseen side effects, and they require successive later inventions to improve on them or to correct the first-order side effects, the second-order side effects, and so on. The new technology devised for a new age will rarely seem as easy or familiar or rewarding at first as the age-old practices of farm and kitchen — which were themselves once new inventions. So social therapy or change, like individual therapy, must not be thought of as a one-shot affair but as a continuous process of "working through", with the consequences from one step leading to a new evaluation or awareness, which then makes possible the moving on to the next step, in a process that never ends.

Nevertheless, the sense of process must not deprive us of the sense of urgency. We must never forget that men who have been handicapped in reaching the full potentiality that the world should first have offered them, will want to see visible progress and real hope for getting their problems solved at maximum speed, so they will not be driven to despair and destruction. While we recognise the long run historical process, we must still find an emergency mechanism to feed the starving family today, not a year from now.

THE NEED FOR TENSION

It would also help us if we all recognised that the process of change necessarily involves tension in the society. The tension occurs because there is a personal and a social cost in changing our institutions, just as there is a personal and social cost in letting them go unchanged. The only way to achieve the minimum cost and the least disruptive speed of change is to let the debate go on between those who are more hurt by the speed of change and those who are more hurt by its slowness.

More fundamentally, the creation of change requires prophets and leaders who see the social ills and the new goals and the practical steps towards them. They are the ones who produce the creative tensions that pull the wagon of history. But these prophets must not go unchallenged. Prophets may be wrong, and their insights damaging to large groups or ways of life. Real conservatives, who see the danger in the new ways, need to make their counterclaims for holding on the old, or for modifying or slowing down the new proposals so that older people or institutions with their own values will have more time to adapt or to protect themselves.

As long as this tension of ideas and values does not reach the breaking point, it is a necessary part of a healthy society and healthy progress, and is something to be applauded, even in one's opponents. The prophet's case, if it is a good one, is not damaged but helped by the pressure on him to make it convincing, or to prove its necessity to those who feel the need to reconcile it with older values. And if he cannot prove it to the satisfaction of others, everyone is the gainer for having settled it by discussion and analysis instead of by expensive trial. These debates will often lead to useful pilot studies to clear up the uncertainties, like scouts that go out ahead to survey the terrain.

It is not good if the social motor can pull the social vehicle in new directions with only trivial pressures; it must exert a serious stress on it if the existing ways that are to be changed have any ongoing value at all. (Some kind of stress or cognitive dissonance is the pre-condition for learning, as the psychologists say.) This means we need a society that responds, but not too easily; because we also need one that is careful to conserve whatever values and effectiveness were built in earlier. So the vehicle needs to be sturdy, and must not have such flimsy steering, so to speak, that it is blown off course by every wind or wanders all over the road with no continuity or stability in the direction that has been chosen.

This need for tension as well as for resistance to tension must not be viewed in our usual glib moralising way in terms of "good" and "evil". The solutions we must look for are total systems-solutions that offer "goods" to both the conservative and the prophet, both to this man's need for continuity and that man's need for urgent change. The difference between a successful system and a less successful system or combination of competing sub-systems is not an opposition between positive and negative like "good" and "evil", but rather like the difference between two positives, like "good" and "less good", or richer and poorer, or higher and lower temperatures.

From the "systems" point of view, every man or organism, by the fact of being a viable organism, a living system, is always pursuing some kind of good for his own system, even if it is only a personal or short-term good, or a false good perhaps representing some past reinforcement or reward that is now mistakenly pursued. No man is wicked to himself; what the small man is pursuing is his own small version of the good, even if he regrets it afterwards; and it is rare that our old religious notions of good-and-evil confrontation or punishment can help him to enlarge that version. It would be more effective if we could always see these smaller or short-range goods of the individual or the group or nation as being, not contradicted, but greatly enlarged for every single one as soon as we begin to pursue the larger long-range goods of our human world system as a whole. The good of Man does not conflict with the goods of men, but can enlarge them and can provide them more easily and abundantly than any men can provide by themselves in smaller isolated or competing groups.

So, we should perceive the "wicked" reactionary, or the "wicked" anarchist, as rather pursuing his own version of the good, or acting out his own doubt and hopelessness that larger goods can ever be achieved. Those of us who want more satisfying social systems for the world would have more persuasiveness and more effectiveness in producing change if we could remember this. Instead of damning these errors as evil, we should learn to recognise that our worst opponents are only holding smaller versions — or perhaps only different versions — of idealism and the good. By accepting the creative part of their vision, we could begin to show them, and ourselves, how that vision can actually be encompassed and enlarged by mutual support and mutual change.

THE PROFIT OF UTOPIAS

These considerations on the total value of better overall human design lead to two useful theorems on the motive force for social change. These theorems show how our collective decision and action can actually be powered in the needed directions. The first theorem is that Utopias will be enormously profitable for everybody, in economic as well as human terms. What we mean by a "future good" is some real hope for ourselves or our community or our world — some hope tangible enough and probable enough to be worth working for, which means worth investing in, in time or effort or money. Why else do men work at all, except to reach such anticipated goods? We see that the gap between what is and what might be is the only real source of power in the world. Megawatts of technical power and organisation will be turned off if we do not think they are bringing us closer to the good that "might be".

But if improvements are worth working for, it means that we expect them to bring a human or economic return that will more than "pay for" the effort. Why then do we not pursue larger Utopias more often? The reason is that we do not see how to get there, or we have a low estimate of success in an untried enterprise in an uncertain world, so that we fear we will lose our effort and be taken for fools. Better maps of the costs and consequences can reduce this uncertainty and make the investment of effort and money in new directions more attractive, as several nations have shown by their investments in housing, medical care, education and many other constructive areas. Does the control of pollution cost too much? No, it actually saves money, as can be shown immediately by a better social accounting, that measures, say, the domestic savings in reducing dirt and lung disease, or the total public and private savings from re-cycling and the better systems-handling of pollutants.

A good society is a society with its blocks and malfunctions and inequities removed, so that every transaction and operation becomes smoother and more profitable. Men will enjoy and support such a society more and will be better educated and supported by it than they are at present. Can any believer in the effectiveness of well-integrated organisms doubt that such a society would be immensely richer and more successful than ours?

In the economic calculation, the future profits that a better system could give us will need to be "discounted", of course, like any futures that take investments away from present pleasures. But this discounting, or interest to be paid, simply means that the business of planning and working for the future can be made part of the ordinary economic process, and that alternative goods and strategies and uncertainties can be balanced against each other by familiar economic mechanisms. Banks invest in the development of dams and housing. From these, it is not such a far step to realise that we can all invest in the development of the world, and so bring it into being.

The fact that Utopias are profitable means that most of the barriers to reaching them can be surmounted by analysis and design, because we can find investment money for doing the analysis and work to get over the barriers, as long as there is a clear gain on the other side. It is like a social siphon for going over hills; or like a social enzyme, to use the biochemical analogy, to get around or tunnel through barriers as long as the overall direction is generally "downhill". The problem of barriers to needed change, whether they are barriers of organisational complexity, or vested interests, or group conflicts, is therefore generally solvable if some "social entrepreneur" — some individual or group — will analyse the ultimate human and economic profits and the local difficulties, and will then apply that investment potential to pay-offs that will help get around the bottlenecks and the human resistances that stand in the way.

POWERING CHANGE

This leads to the second theorem, which is that the profitability of Utopias can be transformed into human and economic profit for every individual or group involved in the change. In fact, strangely enough, the poorer the group is, the greater is the benefit or pay-off from a restructuring of society to give them more. The reason why downtrodden and underprivileged groups are willing to make such an enormous personal investment in protest and violence is not only because there is nothing to lose but because there is everything to gain. Any society that does not realise this fact of Utopian power, and lets inequities grow without redress, is in the gravest peril because of the violent transforming potential of the difference between "what is" and "what might be". The ghettoes will be heard, and the "third world" will be heard, by hijacking and assassination if not by armies, because they are the dry tinder for "the fire next time" if their demands for humanity and hope and abundance are not met. The world is now too dangerous for anything less than Utopia.

But the positive side of this theorem is that analysis in advance can find investments and pay-offs sufficient to move the most obstructionist individuals or institutions. The ghetto child, if well-nourished and well-educated, begins to be a fully contributing citizen who will pay back in taxes alone far more than the cost of his food and education. There is profit enough for the Rajahs and the steel owners to be paid off, for the Post Office unions to be offered real raises, and for the oil-spill conservatives and the abortion opponents to be shown the political error of their ways. Sometimes these pay-offs seem like rewards for obstructionism, but this is often the cheapest and simplest way to remove the blocks to needed change. It is not entirely inequitable to pay a small midwife fee to the old system to let the new world be born.

These mechanisms of real profit in creating Utopia mean that we have an additional way of healing disputes and obstructions in our collective decision process. These theorems might even permit helpful "meta-agreements" on the ground rules of the debate — like the meta-agreements between opponents in Parliament, who may agree to always speak in turn even if they violently disagree on everything else. The meta-agreements we could establish today might include, first, a general recognition by every party of the *non-zero-sum character of our world*, a recognition that we will all win or all lose together depending on whether we build up mutual satisfaction or mutual hostility. This means a recognition of the profitability of Utopia, and of the possibility of an adequate adjustment for those who might be hurt. A second meta-agreement might be an agreement to tolerate and support *diversity*, including even obnoxious opinions and life styles, for the sake of pluralism and to encourage the exploration of new regions of the social map, even those that may not be satisfying in the end. A third meta-agreement might include a kind of *moral imperative* in all our attitudes, whether Establishment or anarchist, or rich or poor, that even if we differ on details, each of us will agree to keep searching and working for larger and more long-range social goods for everyone on the planet.

These thoughts suggest that it might be possible to arrive at a generally accepted and coherent picture of the process of social design and change. The process of shaping history is not a predetermined physical mechanism but an on-going cybernetic process, powered by practical Utopian constructs, with large-scale collective decision and action which have already been shown to be able to reach particular goals in times that are surprisingly short. For creating a more stable and more rewarding society, the human and economic profits are great and the investment for getting over the barriers is economically feasible. The mechanisms and pay-offs for reduction of conflict and for the increase of large-scale and more successful systems organisation are numerous. In short, the future lies open to our collective design and action.

The problem is only to begin.

U.S. EXECUTIVE AGENCIES AND TECHNOLOGY ASSESSMENT

VARY T. COATES*

George Washington University,
Washington D.C. U.S.A.

Modern political theory has revolved around issues that arise, in large part, from the interface of modern technology with public policy. Can democratic systems control and direct the development of technology toward achievement of societal goals without an immense sacrifice of individual liberties? Can we escape environmental disaster only at the cost of a highly controlled society? Must the complex technical decisions that now determine the quality of men's lives be made within an invisible corporate structure of by an elite corps of technologists not responsible to a democratic process, while the mass of men are unable to participate in, or to understand, the issues and forces that shape their future?

Federal agencies and public administrators seldom consciously concern themselves with such theoretical questions, yet in a complex post-industrial society technology and public policy constantly interact. All Federal agencies effect in some degree the public interest, that is, the health, safety, and welfare of the citizenry. All

---

*The author is Research Scientist (Political Science) for Program of Policy Studies in Science and Technology, the George Washington University, Washington, D.C. Much of the information in this chapter is derived from research supported by National Science Foundation Grants No. GQ-4 and GI-30422.

agencies make use of or deal with modern technology, and some are directly charged with either fostering or regulating technology. Some agencies indeed have both responsibilities.

Technology assessment is by definition, a systematic consideration, analysis and evaluation of the real and potential impacts, whether beneficial or detrimental, of a technology on social, political, cultural, economic, and environmental systems and processes. Since it is primarily intended as an input into decision-making, technology assessments should consider all alternative technological routes to achievement of social objectives and alternative control options for minimizing the social costs of the chosen route.

If a workable methodology of technology assessment can be developed and made to penetrate and permeate day-to-day decision-making in Federal agencies. it may provide one important tool by which democratic systems can hope to evade the overwhelming threats identified by contemporary political theory. It is clear that present methods of evaluation used by Federal agencies do not constitute adequate wide-impact technology assessment. Yet there are some hopeful beginnings.

Federal agencies foster technology in many ways: by active R&D programs, subsidies, financial incentives, communication networks, promotional activities, or simply by non-interference with the market. They do so because certain "benefits" are expected to result, either for citizens as a whole, or more often, for some segment of society. Similarly, regulation of technology comes about because undesirable consequences are recognized, although generally only after the effects have become widespread and severe, often irreversible.

Rarely, however, have Federal agencies been able either to perform or to use an adequate total impact assessment. Where the responsibility for fostering and regulating a technology is divided between several agencies, the imperatives of organizational maintenance and protection will prevent either agency from sponsoring, or even recognizing, a balanced, neutral evaluation of social benefits and costs. The result will often be espousement by the Federal government of mutually inconsistent policies — for example, subsidizing tobacco farmers while (feebly) attempting to discourage cigarette consumption. Sometimes, in fact, the same agency is given by Congress two mutually incompatible but equally sancrosanct directives — for example, development of an efficient agricultural system that will lower the cost of food to consumers on the one hand, and the maintenance and protection of that great American institution, the family farm on the other.

Those agencies which have responsibility for both regulating and fostering technology, have scarcely done better at balanced evaluation, as witness the dismal record of the Department of Interior in protecting the environment while licensing the extraction of mineral resources from, or their transmission across, public lands. To mention another example, the FAA has been far more concerned with fostering the development of new air systems such as SST, or V/STOL, than with forecasting either their environmental consequences or indeed their social utility.

Paradoxically the most dismal record of all may well belong to those agencies who have virtually full responsibility over a unique technology (i.e., NASA and AEC). Again, the demands of organizational protection and survival hopelessly bias assessments toward active promotion and away from undue regard for potential social costs, or for competing societal priorities.

There are of course other strong constraints on the development of a governmental technology assessment process. The deeply ingrained American suspicion of "central planning" tends to make agencies cautious of interpreting their mandate more broadly than necessary (in spite of the popular stereotype of power hungry bureaucracy). Indeed, proliferation of positions is more avidly sought than increase in responsibilities, especially where diligent performance of these responsibilities is apt to draw critical fire, and no agency willingly sticks its neck out for a constituency so nebulous as "the public." The agencies are strongly affected by the attitudes of their most powerful constituency, which is likely to be industry, with its usual defensive over-reaction to impertinent questioning of corporate actions, policies, or decisions.

Beyond this reluctance, however, there is a real and justifiable fear that popular enthusiasm for "the environment" or "the community" (with its neo-Luddnite overtones) may stifle invention and innovation, destroy public support for badly needed basic and applied science, and — carried beyond reasonable limits — prejudice necessary attempts to prepare for the inevitable demands of the future.

Public administrators also, in most cases, regard themselves as professionals, and their professional integrity, though perhaps too narrowly conceived, seems to impel them to view the future as a continuation of the recent past, in terms of ever-increasing demands — for power, for water, for highways, for mineral resources, and for the technological base of an ever-rising standard of living.

Perhaps in the long run the greatest obstacle to the development of a technology assessment process will be the demand on agency resources of money, time, and personnel. In most cases a demand for technology assessment will constitute either a real and large addition to the legislatively assigned charter of an agency, or at least a significant expansion of the existing planning and evaluation of a wide range of social, institutional, cultural, economic, and environmental factors — information that most administrators, and for that matter most technical specialists, are not equipped by training or experience to either provide or use. As will be discussed below, a meaningful technology assessment is time consuming and demanding in terms of research and analysis resources. As the public, under the influence of community action groups and a growing communications network among environmentalists and conservationists, becomes more sophisticated about issues, its expectations and demands for full disclosure of information will be harder to satisfy.

In spite of these inherent limitations several recent events and trends have provided stimulation for development of governmental technology assessment. Since World War II we have had steadily growing expectations of the ability of science and technology to provide: "better things for better living," coupled with growing demands by minority groups and the chronically disadvantaged for a more equal distribution of these "better things." This demand has been extended to (but often conflicts with) a demand for a better environment. During the 1960's, successively enacted statutory requirements, particularly in the area of highway and airport buildings and urban redevelopment, drastically widened the scope of comprehensive and community participation in decision-making, and led to a rapid growth in the number and effectiveness of citizen action groups.

Often acting through the courts, such citizen action groups were successful in a considerable number of cases in defeating or delaying construction projects and in changing or influencing the planning of others. Such groups, crystallized around local issues (an airport, a highway project, an urban redevelopment area), may become dormant once that issue is concluded either by success or defeat. But in many cases such a group is reactivated by the emergence of a similar conflict elsewhere, or the more active members become units in a growing network of informal communication on what are loosely termed "environmental issues." In other words, such citizens have become "politicized" on issues related to technology and public policy.

Such citizen politicization was evident in the controversies surrounding, for example, the formerly proposed Everglades jetport and the subsidy of prototype SST's. Eventually it contributed to the passage of the National Environmental Policy Act of 1969 (P.L. 91-190), which was the next major impetus to the development of technology assessment methodology.

This act, declaring it the Nation's policy to "fulfill the responsibilities of each generation as trustee of the environment for succeeding generations," and to "assure for all Americans safe, healthful, productive, and esthetically and culturally pleasing surroundings," gave certain specific charges to the executive agencies. They were directed (in Title I, Section 102) to (1) utilize, in planning and decision-making, a systematic interdisciplinary approach, to insure the integrated use of the natural and social sciences and the environmental design arts; (2) to develop methods of giving appropriate consideration to "presently unquantified environmental amenities and values;" (3) and to provide an environmental impact statement for every major government action "significantly affecting the human environment." This statement must set out the impacts (particularly those likely to be adverse) of the proposed action, any irreversible commitments of resources involved, the relationship of the action to the long-term productivity and enhancement of the environment, and finally, alternatives to the proposed action. Further, the agency is directed to develop and make available to the Federal government, states, local governments, and citizens "advice and information useful in restoring, maintaining, and enhancing the quality of the environment."

Hundreds of these environmental impact statements, generally spoken of as "102 Statements" have been filed by the Agencies. They may be written by the agency or by the local sponsor of a project or program, and drafts are circulated among other interested agencies for review and comment. After a further period for public comment during which public hearings may be held, a final version is submitted to the Council on Environmental Quality. Despite elaborate guidelines promulgated by the CEQ, by agencies and by their bureaus, there is still considerable confusion as to which programs and projects require statements, and considerable diversity between agencies as to the depth of analysis, the extent of critical review, and the amount of effort which goes into these statements. Most of course, are only of limited and local significance; for example, proposals to extend a runway at a county airport (using Airport/Airways Development Grants), for a housing project (using HUD funds), or for a tracked air cushion vehicle link from Washington to Dulles Airport (a proposed twenty-five mile demonstration project of the

Urban Mass Transit Administration). In such cases, since most of the information comes from the project developer or designer, only a few professional man/days may be required on the 102 Statement and in many cases no public hearing is requested.

Other 102 Statements deal with projects or proposals which are of wider impact and which arouse a great deal more concern, for example, the Trans-Alaskan Pipeline or the leasing of oil tracts off the shore of western Louisiana. In the case of the pipeline, a Federal court enjoined the Department of Interior from granting a lease until the requirements of the National Environmental Policy Act are met. It is estimated that the preparation of the Pipeline Statement has already cost more than $7 million. This includes the costs of several series of field studies of the tundra ecology, the negotiation of two successive sets of environmental stipulations to the proposed lease, and the deliberations of an Interior Department Task Force, an Interagency Committee, a Technical Advisory Board, and Public Hearings in both Washington and Alaska. There is no way to include in the estimate the cost of staff time for the many other agencies which reviewed the draft statement, which in the end was unsatisfactory to almost all concerned. A "Final" Statement is now being written (Summer of 1971), but whether the court will decide that this constitutes in itself "compliance" with the Act, no one can say.

As against this very rough estimate of $7 - 8 million, we have no estimate at all of the social costs which might have resulted from proceeding with the pipeline as first planned. The Trans-Alaskan Pipeline is the biggest single construction project ever undertaken by private industry. While the Department of Interior's original stipulations were designed to reduce foreseeable environmental costs to "acceptable levels," many of the hart questions had not been asked or even conceived. Once a large capital investment has been made, by either public or private sectors, for political reasons it becomes extremely difficult to terminate or restrict a technological development when adverse secondary consequences begin to appear. Economic power is direct and focused; the power of "the people" even in democrateic systems is diffuse and hard to arouse until detrimental impacts reach a stage of crisis or disaster.

We are not yet able to measure or (in the language of Section 102) to "give appropriate consideration to presently unquantified environmental values or amenities." Nor, although the Environmental Policy Act is broadly framed to express the intent to protect the quality of life as well as the physical environment, have we yet begun to consider the real value of conserving varigated cultures and life styles, protecting privacy and individual choice, fostering esthetic excellence, and other non-physical components of "quality of life." Until we can in some way measure these values it will remain difficult to make rational choices about national needs and priorities.

The 102 Statements have greatly added to the work load of Federal agencies, especially in the case of Agriculture, Interior, and DOT, which have prepared about 500 Statements. The change requires is one not of quantity only. Public administrators are being pressed to address questions which they had not been asked previously and to adopt an interdisciplinary approach which is foreign to their experience and training.

However inadequate and fumbling the implementation of the Act at present, some lessons may have been learned through this attempt to foresee the consequences of social actions. The Statements provide an entering wedge for technology assessment, in making explicit some of the needed data, in providing a mechanism for multiple reviews of planned actions, and in raising the expectations of public participation in decision-making. When agencies become able to integrate environmental evaluations, economic evaluations, and social utility evaluations, technology assessments worthy of the name will be produced. At present however, an agency — under pressure — may produce the required information, but most often economic factors (as defined in classical economics, i.e., excluding "externalities") dominate the actual decision-making. The development of Program, Planning and Budget Systems (PPBS) in the Sixties, it should be noted here, had already provided an earlier step in teaching administrators to set objectives and goals for programs. How these goals shall be defined, to include "social benefits," is a task for technology assessment to clarify.

In addition to the effects of the movements toward comprehensive planning and environmental protection, impetus toward development of technology assessment has come from increased interest in forecasting and growing use of modeling and simulation. To the extent that an assessment is "anticipatory" it is an attempt to forecast *potential* impacts of a new or developing technology, or the effects of a much more intensive use of an existing technology. Conversely, technological forecasting and, indeed, all kinds of prediction, must take into account the probable physical and social environment of the future, which will reflect the consequences of today's technology. Thus there is a symbiotic relationship between the development of technology assessment and technological forecasting. Many Federal agencies, particularly those which are concerned with transportation or with water resources and power generation are increasing their attention to long-range planning and to projecting future needs, demand, and production of goods or services. In general, however, agencies limit their formal projections to ten years or less for purposes of planning research programs, agency programming, etc. In a more informal way, many attempt some look beyond this to a period of 25 to 30 years, perhaps to that favorite target, "the Year 2000," for better anticipation of problems and contigencies. As yet, however, agencies have made little significant contribution to the much needed methodology of framing alternative scenarios and options for society's futures. A few agencies are beginning to experiment with such techniques as Delphi, but in general the upper management levels of the bureaucracy have yet to be convinced of the value of such exercises, and "planning" continues to focus largely on allocating appropriations for the next fiscal year or two.

A handful of agencies are making extensive use of modeling, simulation or gaming in their programming and evaluation activities. There is much wider use of computerized data collection and retrieval, which can provide an information base for technology assessors.

Technology assessment may be focused on a technology or a technology application, or on changes in the level of use of a technology, and may attempt to evaluate the secondary impacts on social systems. An example is the study under the Surgeon General of Television and Social Behavior. Another kind of technology assessment which might be called problem-oriented assessment - focuses on a problem area (perhaps pollution or regional underdevelopment) and looks at technologies as a cause of the problem and/or as a method of alleviating it.

Until recently the closest approach to technology assessment in agencies (with the exception of a few comprehensive studies to be mentioned later) had been as a component of narrower studies intended to be utilized:

(a) to create or expand the internal information base on which departmental responsibilities, programs, or policies are formulated — contract or grant programs, rule-making, program administration, capital investment;

(b) to support or influence Executive policy formulation — long-range planning, need/supply studies;

(c) to influence Legislative decision-making — project feasibility studies, programming, and program evaluation;

(d) to supply information or services to Congressmen, industry, professional groups, client or user agencies, state or local agencies, or other agency constituents.

Such "assessments" are most often narrowly defined in terms of technical performance, economics, or utilization of resources; and they are most often concerned with primary or planned consequences of an action rather than with secondary social impacts. In part, this inattention to secondary consequences of public actions results from the pervasive American philosophy that governmental power must be narrowly construed and jealously watched, in part it may be laid to the lack of development of the social sciences which are only beginning to identify and understand the process of social change. For example, the consequences of highway location and building in terms of disruption of ethnic neighborhoods and the powerlessness and alienation of minority groups, were long ignored by engineers, not out of callousness entirely, but also from lack of understanding and knowledge which social scientists might have identified and supplied, but did not. The same failure to project and evaluate potential secondary and tertiary social consequences was all too evident in our lack of preparedness for the urban migration of workers displaced by rapid mechanization of agriculture, and by the critical pollution of water and air in large scale industrial development. Relatively few agencies utilize social scientists as such in their planning, programming, and evaluation process. In a large number of agencies recently sampled in this regard, fewer than one-third had sociologists, psychologists, or anthropologists involved in evaluation.

The National Science Foundation is now undertaking support of some technology assessments through its "Research Applied to National Needs" program. Because this program has the broad charter of funding basic research designed to make more germane to society's problems, it can perhaps take a somewhat broader and more effective view of assessment needs than the traditional line or mission agencies. RANN'S Office of Exploratory Research and Assessments therefore undertakes to support research activities in the areas of (a) "problem assessment," including analysis of social problems, definition of the issue areas involved, and identification of the public policy nexus; (b) "exploratory research" including laboratory, field, and paper feasibility studies and evaluations of innovative technological applications; and (c) "technology assessment," defined as the study of secondary and higher order impacts of a technological intrusion, construing technology to include social inventions and institutional innovations. NSF is also unique in supporting some compilations of technology assessment and some assessments of assessments, including the research underlying the writing of this chapter.

To the extent that technology assessment is performed by or for Federal agencies, several organizational mechanisms are used. Many assessments are carried out within an agency, by engineers, scientists, economists, systems analysts, or "generalists." Such efforts may be the work of one staff member or of a team, and often represent a continuing interest of the agency. They may of course be biased by the institutional needs of the agency or of its constituents. On the other hand, staff members may be shielded somewhat by their anonymity from direct pressure by industry or other interest groups, and the results may provide a continuing store of information, expertise, and new insights on which the agency may draw. (This is less true if the assessment is done by an outside organization under contract or grant, even though this arrangement has the advantage of making it possible to assemble a more interdisciplinary group than is possible in most offices or bureaus.) All such in-house assessments can be fairly easily suppressed for political and institutional reasons and at best may have relatively little visibility, political clout, or public credibility.

Interagency committees, at cabinet or subcabinet level, are sometimes used for assessments, particularly those which are focused on a problem area such as aircraft noise or regional underdevelopment, or on a very large project such as the Ramparts Project or the Trans-Alaskan Pipeline. Such interagency assessments have high level visibility and may provide a mechanism for continuing monitoring of a problem and the potential approaches to alleviating it. However, agency viewpoints and interests are seldom superceded or overridden, and therefore fresh approaches seldom emerge; nor is such an assessment likely to contribute much in the way of new information or new understanding.

Agencies sometimes recruit "blue-ribbon" panels of distinguished citizens or of specialists from universities and industry, to provide an assessment. The Commerce Technical Advisory Board uses this mechanism, for example, for its policy-oriented studies such as the report of the Panel on Electrically Powered Vehicles, called "The Automobile and Air Pollution." The National Academies of Science and Technology also provide such panels, for example, for the Everglades Jetport Study and the Jamaica Bay-Kennedy Airport study. The Surgeon-General's Scientific Advisory Committee on Television and Social Behavior, a recent example of a "blue-ribbon assessment" consisted of ten specialists from the faculties of colleges and universities, and two directors of research for broadcasting networks. Congress itself frequently appoints "blue-ribbon" assessment groups in the form of National Commissions chartered to conduct problem-oriented assessments, e.g. the National Commission on Product Safety. Some "blue-ribbon panels" are provided with a staff which may in fact turn out a separate report of its own, as was the case with the Surgeon-General's television and social behavior effort. This assessment, in fact, was accompanied by a one million dollar program of basic research, and resulted in a report by the Scientific Advisory Committee, a report by the Committee's staff (located within National Institute of Mental Health), and five volumes of reports summarizing the basic research which supported the assessment.

Blue-ribbon panels are able to mobilize for public service high-level talent which could otherwise not be concentrated by one organization, and to do so at minimum cost to the public since members ordinarily serve without compensation. They have the advantage of representation of multiple points of view and interests. However some such panels must be questioned on the grounds of bias since there is heavy representation of industries founded on the technologies being assessed. The National Academy of Engineering is particularly susceptible to such charges of institutional bias. By the same token, blue-ribbon panels may be used deliberately as a way of "educating" or even "co-opting" influential members of society for the implementation of public policy. This method, however, has the drawback of not generating new expertise within Federal agencies, nor can it constitute a mechanism for continuing assessments. Finally, conclusions from such assessments may be ignored with relative ease if they run counter to agency policy or interests.

Ideally, assessments should be anticipatory and continuing. That is, new technologies or significant changes in the utilization of technologies should be assessed before large capital investments are made either by the private or the public sector, with full consideration of alternative courses of development or of alternative ways of controlling and minimizing detrimental impacts or maximizing secondary beneficial impacts. Developments in R & D and in utilization of technology should also be monitored to detect unsuspected side effects in the implementation and operational phases before these become critical; pollution from long-lasting pesticides, and health hazards from asbestos are pertinent examples.

The use of technology assessment methodology, that is, of identification, analysis, and evaluation of secondary and higher order implications and impacts, whether environmental, economic, cultural, social, or political, should be incorporated into feasibility studies for all projects and action programs. Such anticipatory assessments are one major category of technology assessments. Another kind may be called "reactive assessments" — those that are reactive to perceived social problems and trends, with attention to future projections of what may at the present be minor problems, but have the threat of becoming serious or critical (given present trends in population growth, congestion, or spread of the technology). Finally, some assessments are initiated only after a disaster has occurred (e.g., earthquakes from unwise construction practices, "dead" lakes, power failures, or the discovery of unsuspected epidemics of lead-poisoning among center city children).

As part of a study supported by the National Science Foundation, 73 case histories of technology assessment and related future studies were collected and analyzed during 1970 and 1971. Fifty-eight of these were characterized as having a large component of technology assessment, although only 12 could generously be called full impact technology assessments. Another 13 were problem-oriented assessments, focusing on problem areas such as center city congestion, air pollution, environmental odors, water pollution, noise, waste management, and power failures. One was a methodological study of technology assessment techniques. Thirty were very narrow assessments that is, concerned with only selected impacts such as environmental effects. Of the 15 "futures" studies related to technology assessment, most were projections of future needs, demand, supply, and/or prices of resources, goods, or services; none made any significant effort to predict possible social environments or to frame alternative scenarios of future conditions of life.

It may be useful to make some broad generalizations about these studies since they were mostly those put forward by the agencies as examples of their assessment activities. Eight of the 71 were interagency studies. Of the rest, roughly half were in-house efforts and half were done for the agencies under contract or research grant. Some agencies tend to have assessments done by outside groups often university teams or consortium of universities and research organizations on the grounds that such groups have greater creditability than the agency itself, with its organizational bias could offer, and that it is possible to put together a more multidisciplinary team in this way. On the other hand, about as many agencies say that they have had bad experience with contractors and grantees, cannot rely on their product, or in some cases that internal assessments have greater internal creditability, that they have the advantage of being able to take into the assessment departmental objectives and values, and thus are more likely to lead to institutional change and modification of programs.

Of the 58 T.A. studies, 30 were categorized as "anticipatory," or dealing with new or evolving technologies and anticipated impacts. Twenty-three were initiated in reaction to long standing trends and perceived problems; 4 were in response to specific disasters (power failure, oil spill, midair collisions).

Wide-ranging full impact studies are best carried out by interdisciplinary teams brought together in a kind of total immersion or intense interaction, preferably on site. This method was used by most of the studies here categorized as full-impact assessments. Another method which appeared was an in-house analysis of results compiled from broad and long standing programs of basic and applied research and previous partial assessments. Overall, the 73 studies analyzed contained large components of the following techniques:

Analysis of all existing data	34
New research or field investigation	29
Public participation or survey	14
Conference of non-govt. experts	12
Systems analysis	7
Modeling, simulation	3
Panels of distinguished citizens	6
Technology forecasting	2

Technology assessment indeed involves a "mix" of basic and applied research. Some "field research," or the gathering of additional data, may be necessary. Often a compilation and re-analysis of all existing data is sufficient, or indeed all that is possible. Assessment methodology is largely a matter of asking questions which have not been asked before, conceiving the possibility of consequences which have hitherto been only dimly glimpsed, estimating the probabilities of cause-effect relationships, and attempting to strike a desirable balance which will optimize on broadly defined social benefits and costs. This implies that such analysis should be as systematic and objective, or "neutral," as possible; yet the question then arises as to how an exhaustive list of benefits and costs can be assured without full representation of interests. But representation of interests inevitably introduces a "political" factor into assessments. The relationship of value premises to "scientific" inquiry is one which has not been resolved either on the theoretical level nor on the operational level, certainly not by Federal agencies.

Of the 14 studies from the sample described which made use of "public participation," nine included public hearings, four made extensive use of interviewing, and one involved a survey of concerned citizens. Public hearings are at present not a fully satisfactory form of public participation in that they often stimulate and harden opposition, but seldom contribute to extending the base of relevant information for either the public or the decision-makers, seldom contribute to the building of a concensus, and rarely evenly represent the diverse points of view of "the public." Much research is needed on how techniques of popular participation can be improved and made more effective in a broad sense.

Somewhat more detail might be useful on the dozen studies which have been called here "full impact." This term "full impact" is used only in contrast to others which considered a far narrower range of impacts; it does not imply that the whole range of potential consequences were considered. Eliminating one of the studies which is still in the planning stage, the others can be described as (1) taking into account, with varying degrees of rigor, some real and potential impacts on social, ecological, and economic systems and in some cases ethical and political implications; (2) involving teams of investigators drawn from a mix of disciplines and interacting with each other throughout the study; and (3) an attempt to evaluate a project, or a technology, in order to assist in or influence public policies and decisions. Five dealt with proposed high technology projects which would require significant resource allocation and promise significant benefits and costs to a local area, with significant implications for the rest of the nation. Two of these were studies of the proposed jetport in the Florida Everglades, a third was the proposed extension of Kennedy Airport into Jamaica Bay, a fourth was the interagency study of the proposed Ramparts Project in Alaska (water resource and power generation), and finally, there is the continuing evaluation of the Alaska Pipeline leading to, but not limited to, the 102(c) Statement. The other five studies were broader in the sense of the technology evaluated, though not necessarily in the range of impacts considered. The first is the report of an ad hoc Task Force to the Director of the National Heart and Lung Institute on the medical, ethical, psychological, and economic implications of cardiac replacement. Second is an assessment of potential effects of an operational weather modification in the Colorado River Valley. There is the six-year long study called the Northeast Corridor Transportation Study, which included many technology assessment components. Another study is an evaluation of forest management technology, especially clear-cut timbering, by the Forest Service. Finally, there is a study which provided an input to the joint DOT/NASA Civil Aviation R & D Policy Study, and dealt with the potential impacts of civil aviation in 1985 and 1995. With the exception of the forest management studies, all of these were anticipatory that is, they dealt with projects which were proposed, or with technologies which are expected to have far greater application in the future than at present. Other studies could be mentioned which are quite broad, such as the television and social behavior study, the earth satellite study by NAS for NASA, an assessment of the impacts of mechanizing the tobacco agriculture, or a study of the future of the world's fishery resources, to name a few. These have been categorized, however, as narrow assessments dealing with a pre-selected type of impact.

The most difficult question to grapply with in governmental technology assessment is that of level of effort, or cost. No agency questioned as yet does program or project accounting to the extent that they can fix a "cost" to an in-house study. Inter-agency and "Distinguished Citizen" panel assessments are equally as hard to cost. Converging lines of evidence however suggest that an adequate technology assessment should be budgeted at a minimum of $100,000 for six to eight months work, or preferably about $200,000 - $300,000 for a year-long study with an interdisciplinary team making use of consultants to further expand their capability.

Perhaps the most important category of technology assessments is that of assessments which have not been made. For example, if the full consequences of the mechanization of agriculture had been set before us in the 1930's, had the benefits of an efficient, rational system of agriculture in terms of abundant food and the freeing of manpower and resources for invention and industrial development been set out against the social costs of displaced agricultural workers, migration to the cities, and the human miseries of the urban ghettos, would we have been able to maximize and more equitably distribute those benefits, while avoiding or controlling those costs, through better planning and preparation? In fact, such an assessment was made, or at least begun, in the early 1940's, by the Bureau of Agricultural Research, before the extent of the changes being induced was fully visible. This study dealt with the devastating secondary impacts of improved agricultural technology and Federal agricultural policies on Mississippi rural blacks, and reportedly it correctly foretold many of the tertiary and higher order consequences for the nation as well. Before this assessment was completed, it was "leaked" to a chairman of a Senate appropriations subcommittee who was from Mississippi, and the resulting upheaval in the Department of Agriculture has dampened the enthusiasm for technology assessment in that, and perhaps other agencies, to the present time.

Public administrators in Federal agencies, when asked about areas where additional technology assessments are needed, often are floundering, unimaginative, and short-sighted, perhaps reflecting the need for more technology forecasting to help them see beyond the next organizational crisis. Among the technologies, or technological applications which are mentioned as needing assessment are the following:

- Uranium enrichment and gas centrifuge
- Satellite broadcasting
- New innovations in weather prediction
- Fabrication of structured proteins (synthetic substitutes for meat and milk
- Integration of pork farming, similar to that which occurred in chicken farming
- Non-lethal police weapons
- Genetics engineering

Cures for cancer
Extensive organ replacement; cardiac pacemakers; respiratory devices
New contraceptives
Modification of disease organisms to prevent infectious disorders
New high-speed transportation systems; intermodal systems
Dual powered locomotives
Vacuum-flush and other residential sewage collection systems
Inter-basin water transfers
New methods of extracting mineral resources
Underground power transmission lines
Automated and/or underground irrigation systems
Industrialized housing
STOL or VTOL aviation systems
Multifamily housing
Scattered site public housing

Frequently, however, when public administrators were asked about needed assessments, they insisted on responding instead in one of two ways: by listing problems to which they are desperately seeking technological "quick fixes" (most often, underemployment and regional underdevelopment, urban congestion, and pollution control, particularly solid waste disposal); or areas where guidance or information is needed to improve present assessment activities. By far the most often mentioned need was for a workable method of quantifying and comparing environmental and social impacts. The second frequently mentioned need was for a method of assessing the "Value" of research and development; that is, of calculating the pay-off in terms of program objectives of resources allocated to research. It may well be, therefore, that the next significant stimulus to governmental technology assessment, or the next critical constraint, will depend on the resolution of these two problems.

The further acceptance and development of technology assessment in Executive Agencies will also be influenced in part by actions of Congress in requiring the kind of information technology assessment can provide, both directly — possibly by establishing an Office of Technology Assessment to serve Congress — and indirectly through demands for information to justify agency budgets. More and more, Congress is forced to concern itself with the level of risk which a technological society must tolerate as a by-product of a high material standard of living. Recently, jogged by rising public concern, Congress has frequently taken the initiative in requesting groups such as the National Academies of Science and Engineering to advise the Federal agencies on the consequences of government programs (for example, on automobile emissions, radiation standards, and the long-range effects of the use of herbicides in Vietnam). Thus, the major elements concerned with governmental technology assessment, the public, the research community, Congress, and the Federal Executive Agencies, are beginning to interact and move toward the goal of integrating technology assessment into the democratic decision-making process.

THE REQUIREMENTS OF TECHNOLOGY ASSESSMENT

MILLER B. SPANGLER

Director,
Center for Techno-Economic Studies, National Planning Association, Washington, D.C., U.S.A.

SOCIAL REFORM AND THE CRISIS OF SCIENCE AND TECHNOLOGY

There is universal agreement that these are troubled times. An unpopular war, the so-called "mess in the cities," mounting insults against the environment, and a variety of other issues or grievances have inspired a widespread mood of social reform including a youthful urge for a counterculture or a new life style. This call for reform is directed toward our established institutions which are blamed for real or fancied failures in the planning and management of their responsibilities. Science and technology, which have made invaluable contributions to an unparalleled economic growth in the United States, have shared in these attacks. According to Lawrence Lessing [1]:

> The assaults on science and technology have developed a variety of themes and come from many directions. Technology is pictured as a powerful juggernaut, with an internal momentum of its own, relentlessly pressing on with irresponsible developments, out of control. On the other hand, this juggernaut is also seen to be so weak that it is suffering all sorts of breakdowns — in power, transportation, and telephone services in our cities, for example. In one breath, science is said to be progressively more abstract and irrelevant to society; in the next, it is urged that this "irrelevant" activity is a threat that must be curbed.

Environmental issues, of key importance to the planning of water resources utilization in urban areas dealt with in this study, rank high among the attacks on the misuse of science and technology. Largely as a result of these attacks or misgivings, the dramatic rate of growth in public support for research and development achieved in the decade of the 1960's has sharply diminished and even suffered cutbacks in the level of funding in certain areas. The crisis of confidence in science and technology is illogical in the sense that scientists and engineers who engage in R & D are certainly not to be blamed for the social misuse of the tools they produce.

A less emotional and more thoughtful response to these attacks is the movement now under way, called *technology assessment*. This approach seeks to sharpen the analyses and forecasts of the desirable and potentially undesirable consequences of the application of science and technology *prior* to the use of these tools and, indeed, to evaluate and identify socially superior alternatives in the design of R & D projects or programs. This chapter will review the why's and wherefore's of the technology assessment movement, its methodological aspects and difficulties of implementation, and its relevance to decisions in the R & D planning mode affecting water and related land use developments.

Too Much or Too Little Research and Development?

A major shortcoming is that we have expected too much of science and technology to solve any and all problems. A magic-wand view of the accomplishments of science and technology in meeting social needs has never been a very healthy attitude. If a measure of disillusionment in science and technology has purged this view, a much more realistic and constructive understanding of the role of science and technology in our society — its difficulties as well as its potentials — can hopefully be established. A valuable insight into one such difficulty is provided by Daniel Moynihan, who sought to explain why it is our Nation could land men on the moon and yet not succeed in cleaning up the mess in our cities [2]:

> At least a partial answer is that two quite different problems are involved. Getting to the moon involves men overcoming a struggle with certain gravitational forces. Those not involved in the struggle are neutral in it. The condition of the cities — or the countryside, or the suburbs, or whatever — involves situations in which men have, or think they have, conflicting interests, and act in ways to thwart one another just as often as they act in concert to achieve some common end.

A conclusion to be drawn from this observation is that if society were more homogeneous or unified in its interests — or, more realistically, if an acceptable compromise in goals could be decisively established — then it would be vastly easier to plan the appropriate contributions of science and technology in solving society's problems or fulfilling its aspirations. Since such a utopian state of affairs is not likely soon to emerge, the dimensions of research and development (R & D) planning must go considerably beyond its traditional concentration on the physical and engineering sciences to include a balanced emphasis of the social and management sciences [3]. This is particularly essential in developing the information needed to define social problems and to establish goals in a broad systems analytic framework. This framework must include not only the conceptualization of the hardware characteristics to be developed by the R & D, but also the technology's likely social, economic, political and ecological impacts, and the explicit institutional arrangements that will facilitate and regulate the implementation of the technology in serving these goals.

Rather than research and development itself being irrelevant in solving many of society's most difficult problems, it would appear that the main shortcoming has instead been an insufficiency of research, particularly in the area of the social and management sciences. Nor does an increase in the latter imply a reduced requirement for R & D in the physical or engineering sciences. Quite the contrary. However, as will be discussed below, the very process by which funding for research resulting in advances in technology is justified will often require advance information or evaluations provided by the social and management sciences. While coordination of efforts between different branches of physical, engineering, management, and social sciences are not easily consummated, there is little doubt that such a prescription must be fulfilled if many of our more difficult social problems — and particularly those of water resource developments — are to be effectively dealt with by the tools of modern science and technology.

The Technology Assessment Movement

Technology assessment has been defined in the Report of the National Goals Research Staff as "nothing more than a systematic planning or forecasting process that delineates options and costs, encompassing economic, environmental, and social consideration (both external and internal) and with special focus on technology-related 'bad', as well as 'good', effects" [4]. The phrase "nothing more" is a curious one, for as will be amply seen in the remainder of this and subsequent chapters of this study, technological assessment, including its planning and forecasting components is, if anything, exceedingly difficult. The complexity of cause-and-effect relationships that are often poorly understood, the commonplace encounter of the technological assessor with monumental information gaps, and the inherent difficulties in treating uncertainty in the preparation of technological and related forecasts reflect a challenge requiring more than a little energy and courage.

A study in 1969 by the National Academy of Sciences indicates the broad dimensions of concern influencing the technology assessment movement [5]:

> In recent years concern has mounted over society's seeming inability to channel technological developments in directions that sufficiently respect the broad range of human needs. Whether rightly or wrongly, the belief is now widely held that the continuation of certain technological trends would pose

> grave dangers for the future of man and indeed that the ill-considered exploitation of technology has already contributed to some of the most urgent of our contemporary problems; the specter of thermonuclear destruction; the tensions of congested cities; the hazards of a polluted and despoiled biosphere; the expanding arsenal of techniques for the surveillance and manipulation of private thought and behavior; the alienation of those who feel excluded from power in an increasingly technical civilization.

Largely under the spark plug of Emilio Q. Daddario, former Congressman, and Chairman of the House Subcommittee on Science, Research and Development, there have been since 1966 numerous staff studies, seminars, and Congressional hearings on the subject of technological assessment that have provided a valuable body of literature on this subject [6]. However, as pointed by the Report of the National Academy of Engineering, concern for the broader implications of technology on society is scarcely new [7]. In 1937 the House Subcommittee on Technology of the National Resources Committee issued a report on *Technological Trends and National Policy, Including the Social Implications of New Inventions*. Indeed, one can go back at least as far as the writings of Karl Marx, who, in 1867 in *Das Kapital*, raised concern over adverse social effects of mechanization in the new technologies of that era [8]:

> In so far as machinery dispenses with muscular power, it becomes a means of employing labourers of slight muscular strength, and those whose bodily development is incomplete, but whose limbs are all the more supple. The labour of women and children was, therefore, the first thing sought for by capitalists who used machinery. That mighty substitute for labour and labourers was forthwith changed into a means for increasing the number of wage labourers by enrolling under the direct sway of capital every member of the workman's family, without distinction of age or sex. Compulsory work for the capitalist usurped the place, not only of the children's play, but also of free labour at home within moderate limits for the support of the family.

Although Marx must be credited with being an astute diagnostician of certain undesirable side effects of technological change, his inability to forecast correctly the institutional adjustments of democratic capitalistic society in solving these ills did not adequately serve the forecasting and planning requirements of technology assessment as defined above. It would appear that herein lies an important lesson for the recent movement in technology assessment: unless it is done well, society may be worse rather than better off for the exercise.

The National Academy of Engineering laid considerable stress in its report on the need to base technology assessment on a series of assumptions and a broad range of associated forecasts including technical, economic, social, and political futures [9]. However, this report also cautions that forecasts covering more than five years are likely to be unreliable because of unforeseen events and scientific discoveries — although longer-term forecasts are nevertheless valuable for planning and for "setting the stage" to evaluate the impact of unforeseen events when they do occur [10]. The National Academy of Science also notes the difficulty of foreseeing and quantifying the secondary consequences of new technology, which discourages their consideration in decision-making processes [11]. Yet, it warns that unless our society selects those technological alternatives that minimize unwanted conwequences, it is "entirely possible that, frightened by the untoward side-effects of technological change and frustrated by their inability to 'humanize' its direction, people with much power and little wisdom will lash out against scientific and technological activity in general, attempting to destroy what they find themselves unable to control" [12].

The problem of forecasting bad effects of introducing new technologies is seen, in that these effects sometimes wait for many years to become severe, as noted by Dr. Lee A. Dubridge [13]:

> ...it is relevant to note that only successful technologies have bad effects. Others are not permitted

> to develop and are forgotten. As the successful technologies develop, by the time the bad effects fully show, the industry may be too big and the effects too severe.
>
> Henry Ford, for example, did not cause a pollution problem with his Model T, nor did the Wright brothers create aircraft traffic problems. Yet the seeds of future problems were present in both of these pioneering efforts, but who could have foreseen them, let us say, in 1906? How to identify and then flag, at the appropriate time, the relevant set of a multitude of potential problems is the real challenge of a technology assessment mechanism.

The technology assessment movement has frequently drawn attention to water pollution problems. Water pollution at different times and places has impaired aesthetic values, recreation, commercial and sport fishing, breeding grounds for aquatic birds and other wildlife, and tourism — not to mention the increased costs of treatment of intake waters for industrial and household uses. This class of technology assessment has been referred to by the National Academy of Engineering as "problem-initiated" in contradistinction to the "technology-initiated" class [14]. The latter involves the need for assessment of both new and existing technologies — space satellites, lasers, genetic modification, and digital computers — focusing on the potentially good and bad effects on all citizens. This contrasts with problem-initiated technology assessment, which seeks to discover the best technological approach in solving a particular problem of society that has already become more or less urgent, as in the case of many forms of water pollution.

However, a study of the role of R & D in meeting water resource requirements of urban areas will also involve the "technology-initiated" type of assessment. Examples of technology-initiated alternatives include desalination of brackish or sea water, or increasing precipitation available to a drainage basin through new techniques of weather modification. Yet, "technology-initiated" cases of innovation sometimes involve possibilities of socially undesirable by-product effects, even though the primary interest in them is to create positive social benefits. Technology assessment, then, is needed in the broadest possible sense to eliminate or minimize the undesirable consequences. For instance, obtaining fresh water supplies through desalination poses the problem of how to dispose of the concentrated bitterns without creating adverse ecological effects. Likewise, weather modifications to increase snowfall on mountains — and hence the water supply of a drainage basin — may increase the risk of avalanches which need to be considered in the implementation of the new technology.

Although much of the concern in the technological assessment movement is over the unintended negative byproducts of innovation rather than over its sought-after positive benefits, both aspects are of considerable importance. That technological developments have sometimes been accompanied by undesirable consequences is a problem primarily of the institutions that utilize and control — or fail to control — the new technologies, rather than any inherently negative moral values in science or in the technologies themselves [15]. Particularly in regard to water resource developments, a fairly large number of industrial and governmental institutions are involved in decisions that determine whether the application of science and technology is beneficial to the broad interests of society. These institutions exercise their responsibilities for water resource developments with different goals or objectives in mind, and these goals are often incompletely considered in the R & D planning phase.

DEFINING THE RELEVANT SYSTEMS

The complexity of interrelationships that must be analyzed to improve coordination in R & D planning for water resource development invites chaos unless appropriate systematic procedures are pursued. To deal with such analytical matters various responsibilities have been established in private and public institutions under such functional titles as systems analysis, corporate planning, R & D planning, market analysis, and Program-

Planning-Budgeting System (PPBS). The systems approach, which cuts across these various functions, is defined by Dr. Simon Ramo as [16]:

> ...an intellectual discipline for mobilizing science and technology to attack complex, large-scale problems in an objective, logical, complete, and thoroughly professional way. It starts by definition of goals and it ends with a description or a design of a harmonious optimum ensemble of the required men and machines, with such a corollary network of flow of information and materials as will cause this system to operate to solve the problem or fill the need.

A major aspect of the systems approach is that of defining the system's external boundaries as well as identifying its internal variables and their interrelationships. The problem is clearly stated by Dr. William F. Pounds [17]:

> Everything in the world is, in principle, related to everything else in the world. But, because of our limited rationality and ability to cope with complexity, none of us — scientists, engineers, managers, or anyone else — can cope with the whole world at once. We, therefore, use the concept of systems to help us out of our dilemma. We talk, for example, about fuel systems, or economic systems, or production control systems, or solar systems, or information systems, or social systems, and so on. We do so, not because we think these things do not interact with one another, but because, for the time being, we want to ignore their interactions and concentrate instead upon what is going on *within* some particular system. We want to limit ourselves to analyzing and possibility of influencing the interrelationships within this one particular piece of the world.

This study deals with the system variables that need to be analyzed for decision-making purposes by three kinds of planners: (1) the R & D planner, (2) the water resources planner, and (3) the urban planner. Thus, we are dealing with three separate, if overlapping, systems. By identifying the system variables of each — particularly their goals and supporting strategy alternatives — we shall be in a position to explore how these systems relate to each other, or *ought* to be related or coordinated with each other, to serve the broad needs of society. Only in this manner can the major concerns of the technology assessment movement regarding water resources and related environmental aspects be assured of a systematic treatment in an operationally significant framework.

There is, of course, a marked difference between the private and public approaches to systems analysis. According to former Budget Director Charles L. Schultze, who examined the role of PPBS and program analysis in the Federal decision making process [18]:

> Unlike the private sector, the public sector does not have a clear definition of the problem. Definition of the problem depends on the particular interest of each group concerned with it. The political process deliberately clouds public program objectives in order to achieve the broadest possible support for each program proposal. In Schultze's words, "A good political decision is one on which it is possible to achieve consensus." By contrast, PPBS focuses on the "goodness" of a program in terms of well-defined end objectives; hence, political decision making is process oriented while PPBS is performance oriented.

Since R & D planning of relevance to water resources planning is performed in both the private and public sectors, the methodological approach will be different for these two kinds of planners. Nevertheless, as will be seen in the subsequent section, there are some commonalities in the systems analysis approach used in both private and public R & D. For one thing, although computers and computer simulation have found useful application in water resources and other kinds of planning, the successful application of a systems approach requires thoughtful, insightful human judgments as well as technical and analytical skills [19]. Indeed, subjective boundaries and subjective structures are the basis for all systems analysis, according to Dr. Pounds [20]. Systems are carefully selected fragments of the real world whose analysis is of value only to the extent that it leads to the payoff of successful action. This is equally true for systems analysis in the public and private sector, despite the contrasts noted by Schultze above.

EVALUATING R & D ALTERNATIVES

Criteria for the Private Funding of R & D

As noted above, the challenge of technology assessment not only holds implications for establishing appropriate institutional controls to limit socially harmful effects in the introduction of new technology. It also requires that the system of appraising research alternatives should seek to forecast the good and bad effects of the potential advances in science and technology (including assumptions as to institutional arrangements). If this is done in the design of a well-balanced R & D program, then institutions will find it easier to warn and protect society against potentially harmful effects, while enlarging the overall benefits from society's use of the technological advances thus produced. To gain a clear understanding of what is involved in improving the social responsibility of R & D planning, it is desirable to explore the criteria and their interrelationships by which R & D is justified in both the private and public sectors.

The primary motivating force of private enterprise is to make those investments and to conduct operations in such a way as to increase profits. There is, of course, often a willingness to sacrifice short-term profits to enhance the attainment of growth objectives under the assumption that the strengthened competitive position inherent in high growth rates will be reflected in higher profits in the long run. Except for short-reach improvements in technology that have near-term payoffs, most R & D investments reflect a willingness of the private corporation to sacrifice some of the near-term profits to secure higher growth and competitive status whose payoff may not be realized for five or ten years after the initial R & D investments. Most of this kind of R & D entails some substantial measure of risk.

Accordingly, the appraisal of the merits of undertaking this kind of R & D involves three basic criteria: 1) the cost (C) of the R & D, 2) the net returns (R) if the research is successful, and 3) the probability (P) of success. These decision criteria have been combined in a simple equation called the appraisal ratio [21]:

$$\text{Appraisal Ratio (A)} = P\frac{R}{C} \tag{1}$$

In general, the higher the appraisal ratio for a research project, the higher the priority that is given to the project [22]. Despite the simplicity of the criteria incorporated in the appraisal ratio, the difficulty of estimating its components with satisfactory reliability is so serious that considerable controversy is inherent in R & D planning decisions where a high probability of success is not assured or the cost is of a substantial order. The estimation of net returns to the corporation from the R & D, if successful, often involves a team approach including corporate planners and controllers, and manufacturing, marketing, engineering, and research personnel. Grant Dove, Vice President of Corporate Development of Texas Instruments, stresses the need for sufficiently challenging objectives [23]:

> One of the criteria of a good Business Objective is that it be challenging enough, even shocking enough, to force a radical re-thinking of strategies and tactics. Whenever we have enough well-defined strategies to give us a high confidence level in exceeding the goals stated in the objective, then that Business Objective is probably not ambitious enough and the probability of truly innovative strategic thinking is likely to be low.

Technology assessment, which must consider broader social as well as corporate interests, enters into the evaluation of the appraisal ratio in several ways. If a technology-initiated assessment is involved, the forecasted corporate returns from the successful R & D may be reduced if society moves to restrict the areas of permissable application of the technology in order to curtail certain socially harmful effects. For example, the use of a newly developed pesticide that does not readily degrade may be permitted in some areas, but not others where significant ecological injury might result. In the case of problem-initiated technology assessment, the main focus of the R & D effort will be in finding a more cost-effective way of reducing or eliminating certain socially adverse conditions, whether of natural or man-made origins. Here the primary technology resulting from the R & D will be solely directed to the problem in question, though it may itself have some socially undesirable side effects. For example, the development of a new herbicide to control rampant algal growth in certain lakes or watercourses may have injurious effects on certain marine organisms it is desirable to protect. It is evident that problem-initiated R & D can be greatly stimulated by the passage of laws or other political actions that would promote the purchase of the innovative output of the R & D.

Under these circumstances the market research essential to R & D planning may prove to be one of the most difficult and critical elements to the decision. The evaluation of net returns in the appraisal ratio of R & D requires analysis of the legal and political environment, present and future, that would favor or disfavor the form of innovation being planned. This, in turn, will influence the design objectives of the R & D itself. Likewise, the possibility that society may subsidize the conversion of older plants, or be lenient in enforcing new antipollution standards for certain of these, may also affect the outlook for returns from the proposed problem-initiated R & D.

Criteria for the Public Funding of R & D

The overwhelming majority of publicly sponsored R & D is funded by the Federal Government, which in FY1970 spent $15.2 billion on R & D [24]. By contrast, state and local governments spend very little for R & D and much of this ($225 million in 1968) is channeled through university and college programs [25]. Although data is not available, a small amount of R & D by state and local governments is conducted in the form of field experiments or field survey work associated with operating agencies dealing with natural resources such as fisheries, forests, soil or beach erosion control, minerals, and fresh water. Caught in the squeeze of rising demands for services and redevelopment projects and a shrinking or inadequate tax base, many urban areas can afford to spend very little for R & D. Instead, reliance is placed on using off-the-shelf technology whose improved effectiveness over the years results from R & D conducted by other public or private agencies.

Because of this limited input of non-federal spending for R & D, much of the discussion in this section will deal with the methodological aspects of federal justification of R & D projects or programs. Moreover, this approach is felt to have parallel methodological implications for funding R & D at the state and local levels. Despite the significant difference between the process-oriented political sector planning and the performance-oriented private sector planning (see Schultze, *supra*), the criteria for public sector funding of R & D bears a close resemblance to those cited above for the private sector. The national appraisal ratio for federally funded R & D takes the following form [26]:

$$\text{National Appraisal Ratio} = A_N = P_N \frac{R_N}{C_F} \qquad (2)$$

where

R_N = the net returns to the nation from the developments being promoted by the R & D being funded

C_F = the cost in federal expenditures for the R & D

P_N = the estimated probability of the net returns to the nation, R_N, being realized

Even though the elements of equation (2) are closely parallel to those of equation (1) applying to the private sector, there are important, if somewhat subtle, differences. One difference is that the net returns to the corporation from its R & D programs are of common economic units measured in dollars of profits, whereas the net returns to the nation of federally sponsored R & D are often a mixture of economic, social, and political values. Some of these values cannot be measured, or even expressed, in common units that are consistent with the notion of the addition of positive values and subtraction of negative values to obtain an aggregate "net" return to the nation from the R & D. Thus, the public decision-maker must grapple with a mixed logic of apples and oranges — a rather untidy and controversial process, to say the least.

Moreover, many governmental programs affect different groups of people in different ways [27]. That is to say, the cost of the R & D (the denominator of equation 2) and its effectiveness in serving national interests (the numerator) do not affect various groups or persons in the same ways. This is particularly important in water resources research planning. The cost of the publicly sponsored R & D is generally borne by receipts from taxpayers, only some fraction of whom may enjoy the benefits of the R & D as it affects water resource problems. Issues in water resources development vary considerably from one region of the country to another and hence involve different needs for R & D. Thus, if each group (i.e., the benefactor group and the larger, non-rewarded taxpayer group) presses their self-interest to the fullest, very little R & D may be sponsored as a result of the ensuing controversy, even though it may be possible to indicate a favorable ratio from the standpoint of an impersonal appraisal of "net national interests" relative to costs and the probability of a successful outcome of the R & D.

This difficulty can be resolved if various segments of society show restraint in not pushing their self-interests to the fullest, displaying willingness to aid disadvantaged or injured parties improve their status. In still other cases, congressional log-rolling or administrative pork-barrel practices can promote programs deemed to be in the broad national interest — as well as some that are not — as a mode of overcoming opposition to desirable legislation. The potential legitimacy of this procedure is that at some time or other each segment of society may find itself in a minority position. Moreover, if majority groups feel that a particular measure will have only a *slightly* adverse effect on their interests and can be persuaded to yield in favor of some lesser group whose interests are *strongly* at stake, then overall national interests may be served. This principle is essential to achieving social justice and may serve as a basis for decision in certain environmental issues.

To date, the researcher has, for the most part, deferred the justification of proposed R & D programs to administrators and the political process. However, the specialized insights of scientists and research engineers are certainly needed if the criteria of national interests inherent in the appraisal ratio are to be satisfactorily evaluated. Former Congressman Emilio Daddario has made a plea for scientists to contribute in devising new intersections with governmental and public factors that affect the future of science. Specifically, Daddario recommended the following features of a new organizational structure for justifying public investments in research [28]:

(1) It must provide strategies for eliminating the causes of complex crisis problems we are *now* facing.
(2) It must formulate these strategies in such a manner that they will not interfere with each other's implementation or create new problems.
(3) It must project or foresee undesirable consequences of a new technology and provide methods for eliminating these consequences before they reach crisis proportions, or simultaneously with implementation of the technology.
(4) It must not interfere with the current system of checks and balances in the Government, nor with the independence of scientists and engineers in the academic world.
(5) It must intersect the established power structure in such a way that it wields enough *practical* power to ensure its effective functioning. Merely influencing opinion in a uniform manner is probably too idealistic a methodology, as well as being an almost impossible task, and, further, would not supply the other needed features of the proposed organization.

(6) It must serve to evaluate, correct, and coordinate programs that are now in progress.
(7) It is probably essential that it be interdisciplinary (governmental and nongovernmental) and that it makes extensive use of concepts such as systems analysis and systems theory.
(8) Finally, it probably must use and promote the most recent advances of social science, but also it should incorporate what can be learned, on an empirical basis, from the natural sciences concerning the evolution and optimization of existing complex physical systems. Studies such as the latter might provide guidelines for identifying areas of the social sciences that need greater clarification, as well as guidelines for social systems engineering.

Despite the soundness of this advice, its implementation will undoubtedly proceed in a halting fashion. A major difficulty lies in the probability coefficient of the appraisal ratio. As noted by Donald Schon, R & D (except of the short-reach improvement variety) has experienced a high rate of failure — estimated at between 7 to 1 and 20 to 1, depending largely on definitions [29]. While one can establish reasonably accurate failure rates by hindsight, the estimation of the chances of success or failure of an R & D project prior to its start is a most difficult, subjective process. In the first instance, human nature seems to require a certain amount of optimism or self-delusion in order to undertake a difficult task. Without this capacity for optimism the progress of mankind would undoubtedly have been slowed, given the numerous sources of resistance to change in society.

While evaluation of the probability of success by a peer group of scientists provides some assistance in this problem, even this is likely to be strongly influenced by the reputation among the peer group of the scientists proposing the research, as well as by the competition for R & D funds within an overall budget that is regarded as fixed. The result is that the chief political considerations are those worked out primarily between the scientists themselves rather than the external political considerations of society at large.

This does not mean that the scientist is immune to the broader interests of society. Rather, his generally limited exposure to the social or political sciences places him at a disadvantage in appraising the net national returns of his R & D and in devising organizational or institutional strategies (i.e., the supporting systems) for making the innovation more socially acceptable by limiting undesirable by-product effects. Also, the scientist, being cognizant of the historically high failure rates of R & D and not wishing to undercut the chances of acceptance of his proposed R & D by a low — and possibly more realistic — estimate of the probability of its success, may tend to compensate by being over-modest in making advance claims of benefit to society if the project were successful. Fortunately, this modesty seems to work well for the long-term interests of scientific progress in general, since the failures of the R & D do not seem to be so disappointing if expectations of social gain were not raised so high and, conversely, the achievements of science will seem more dramatic and exhilarating if they exceed advance expectations. Either way, the general reputation of science is likely to be stronger, and the national interests in the long run may be better served, than if a more accurate appraisal were made of both the chances of success of the R & D and the net benefits to society if it were to succeed.

From the standpoint of technology-initiated R & D, the increased uncertainty of social acceptance (or delays in acceptance) due to the need to foresee and safeguard against undesirable by-products of the innovation would seem to have an appreciably adverse effect on the appraisal ratio of R & D. In general, the probability coefficient of the appraisal ratio would seem to require some fractional reduction because of the difficulty of foreseeing all the potentially adverse ecological or social effects that may result from the innovation, including its misuse or mismanagement. To provide this sort of assurance — though complete assurance is never possible — will require additional R & D, swelling the costs of the denominator of the appraisal ratio. Moreover, the numerator of the appraisal ratio would tend to suffer some reduction, particularly if serious delays result in the application of the new technology. This is because of the need to discount future benefits to present day values according to an imputed interest charge, variously assumed in federal accounting to range from 5 to 10 percent.

Overzealous demands that all technological innovations entail *absolutely no risk* of social harm may stultify technological progress. This must certainly be reckoned with in both publicly-and privately-sponsored R & D. The Technology Assessment Panel of the National Academy of Sciences spoke out against a danger of this kind [30]:

> The panel does not, for example, favor any proposal that would routinely impose upon each manufacturer or producer the burden of proving to some public agency's satisfaction that every new technological application he contemplates will meet a prescribed standard of social or environmental "harmlessness." Any such scheme would ignore a major unfortunate prospect — new and socially desirable technologies lying fallow because of an excessive presumption in favor of the *status quo* — and would necessarily be based on the untenable notion that there exists some objective measure by which a sufficiently wise organization can assess the net social utility or toxicity of a given proposal.

What is needed, of course, is a comprehensive technology assessment procedure that will permit potentially large benefits to society to outweigh risks of lesser social costs or harm, particularly if the harm is not an irreversible or uncurtailable one. Philip Handler, in a Technology Assessment Seminar conducted in 1967 by the House Committee on Science and Astronautics, made the following point [31]:

> My concern is ... with unwillingness to approve new technologies because, despite their advantages to society, some relatively small disadvantage may be apparent. And as evidence, let me indicate that were aspirin or beer unknown and invented tomorrow for the first time, I doubt that either could find approval in the Food and Drug Administration. Please

understand that this is not intended as a criticism of this important and well-managed agency but it is a creature of its origins, legislative authority, and tradition.

The Cooperative R & D Roles of the Public and Private Sectors

Generally, the government does not desire in its R & D programs to compete with private enterprise sponsorship of R & D. Only when it appears unlikely that private enterprise will perform R & D required by the public interest, or at an adequate scale and timeliness to support this interest, is it deemed desirable for the government to sponsor R & D. Even then, there are numerous alternative measures by which the government may seek to encourage private sponsorship of R & D. Over fifty such measures were analyzed in a recent study [32]. A significant group of such stimulative measures have to do with non-competitive forms of R & D sponsored by the government. For example, government-supported R & D in one aspect of broad systems technology may encourage private enterprise to perform R & D in the remaining aspects of the systems technology. This is also the case in government sponsorship of basic research or of the provision of certain kinds of descriptive data on natural or man-made phenomena. In either case, the additions to knowledge will tend to reduce the total R & D costs of private enterprise and also permit a superior evaluation of the probability of success of the R & D.

In view of this cooperative role between the public and private sectors in providing advances in science and technology, it is appropriate that the national appraisal ratio of R & D as set forth above be modified as follows [33]:

$$\text{National Appraisal Ratio } A_N = P_N \frac{R_N}{C_{F_1} + C_{F_2}} \qquad (3)$$

where:

R_N = the net returns to the nation from the R & D sponsored jointly by the private and public sector

C_{F_1} = the federal funds expended for the R & D

C_{F_2} = the costs to the government in the form of financial or other measures to encourage private R & D investments

P_N = the estimated probability of the net returns to the nation, R_N, being realized

Trade-offs must be considered between federal costs (C_{F1}) expended for R & D and federal costs for other stimulative measures (C_{F2}). In the case of water pollution abatement technology, for example, a fairly stimulative effect on privately sponsored R & D might be accomplished by:

(1) The provision of comprehensive planning information for each water basin that would help private enterprise assess how many regions or localities would provide a market for the technologies being considered for R & D, and

(2) The early establishment of water quality standards, or a set of administrative rules for pollution control or abatement, that would provide a stable business environment for the development and promotion of pollution control technologies.

With respect to this issue of trade-offs, balanced considerations in establishing a set of rules for pollution control or abatement have been described in the 1971 *Economic Report of the President* [34]:

> While it might be tempting to say that no one should be allowed to do any polluting, such a ban would require the cessation of virtually all economic activity. Since society places a value both on material goods and on clean air and water, arrangements must be devised that permit the value we place on each to

> determine our choices. Additional industrial development, increased use of pesticides on farms, and a growing volume of municipal sewage mean dirtier water downstream and fewer opportunities for recreation. On the other hand, stricter rules for pollution control generally mean either higher taxes or higher prices for goods. What we seek, therefore, is a set of rules for use of the environment which balances the advantages of each activity against its costs in other activities foregone. We want to eliminate pollution only when the physical and aesthetic discomfort it creates and its damage to people and things are more costly than the value of the good things — the abundance of industrial or farm products and efficient transportation — whose production has caused the pollution.

The Council of Economic Advisers also notes that the rules and procedures should not themselves entail a higher cost of administration and enforcement than the cost of having no rules [35]. Another difficulty in establishing an optimal set of rules, noted by the Council on Environmental Quality, is the deficiency in measurement tools, which "prevents an assessment at this time of the status and trends of the major environmental classes of the Nation" [36]. In President Nixon's Message to Congress on environmental quality, he made it clear that the basic causes of our environmental troubles are deeply imbedded and cannot promptly be overcome [37]:

> It should be obvious that we cannot correct such deep-rooted causes overnight. Nor can we simply legislate them away. We need new knowledge, new perceptions, new attitudes — and these must extend to all levels of government and throughout the private sector as well: to industry; to the professions; to each individual citizen in his job and in his home. We must seek nothing less than a basic reform in the way our society looks at problems and makes decisions.

Thus, considering the complexity and difficulty of the task, it seems clear that progress in establishing a stable set of rules and procedures for water pollution control or abatement will be difficult to attain. Cooperative efforts between the public and private sectors in establishing an appropriate regime for monitoring and regulating pollution and other aspects of water and related land uses will greatly facilitate the selection and planning of R & D alternatives.

DECISION VARIABLES OF THE WATER RESOURCES R & D PLANNER

Three Classes of Variables and Their Relationships

In principle, the decision variables of the water resources R & D planner are the same as those for all R & D planners. However, each kind of R & D planning will differ in subject content and, particularly, the nature of the issues evoked in consideration of these variables. As seen in the appraisal ratios of R & D set forth above, both the private and public R & D planner must appraise three classes of decision variables: (1) the cost of the R & D and its supporting systems; (2) the net returns to the corporation or society, as the case may be; and (3) the probability that the net returns will be realized. Because of the need to deal with a mixture of quantitative and qualitative appraisals of these three classes of decision variables, the principal utility of the above equations is as a shorthand notation for affirming that the desirability of the R & D alternative being evaluated is directly proportional to the estimated net returns from the R & D if successful and the probability of a successful outcome. Also, the desirability of the alternative is inversely proportional to its cost.

Yet it must be noted that these decision variables cannot be assessed independently of each other. As suggested in Fig. 1, the probability of the net returns will vary with the scale of efforts of the R & D program [38]. The probability of the R & D being successful begins to increase from C_0 where funding is insufficient, up to the point (P_1, C_1) where increased costs of the R & D outweigh further gains in the net returns (i.e., gross returns

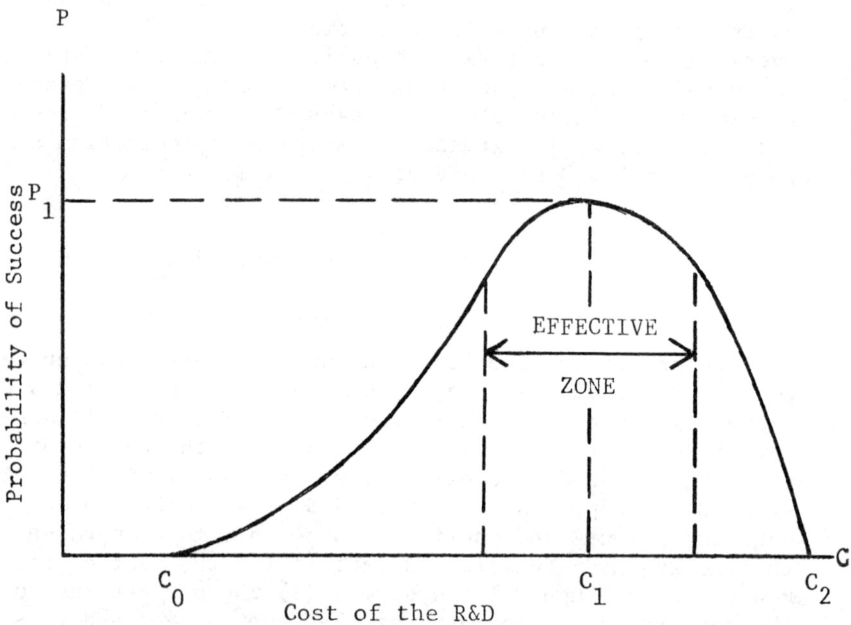

Fig. 1. A hypothetical illustration of the probability of the success of R & D as a function of level of effort or costs.

minus costs including R & D costs). Beyond this point the probability of success will decline until it again becomes nil at C_2. Establishing a curve as shown in Fig. 1 involves a process of subjective probability estimation [39]. Since this process is difficult and costly, it is practical to evaluate only a few alternative levels of R & D effort or cost.

If the relationship between probability and the cost of the R & D were considered wholly independently of the net returns, then the effective zone of consideration for decision-making would fall in the range shown in Fig. 1. The limits of this zone would be set at slopes of unity where the percentage increase (or decrease) in probability is equal to the percentage increase (or decrease) of R & D

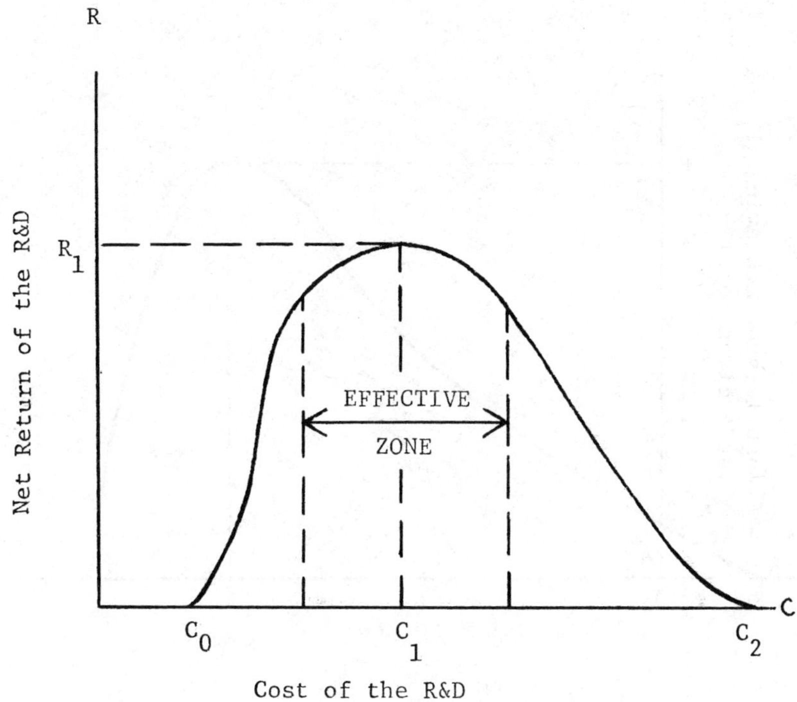

Fig. 2. A hypothetical illustration of the functional relation between net return versus the cost of the R & D.

costs. However, as suggested in Fig. 2, the net return of the R & D will normally vary with the level of effort or cost expended for the R & D. In the left part of this curve, as R & D costs are expanded it is likely that technologies of superior design and performance can be created that will increase their market acceptance or range of uses, thus increasing net return (R) accruing to the R & D investments. After the level (C_1) has been reached, net returns will begin to decline since costs, which must be subtracted from gross returns to yield net returns, are increasing faster than increases induced in gross returns.

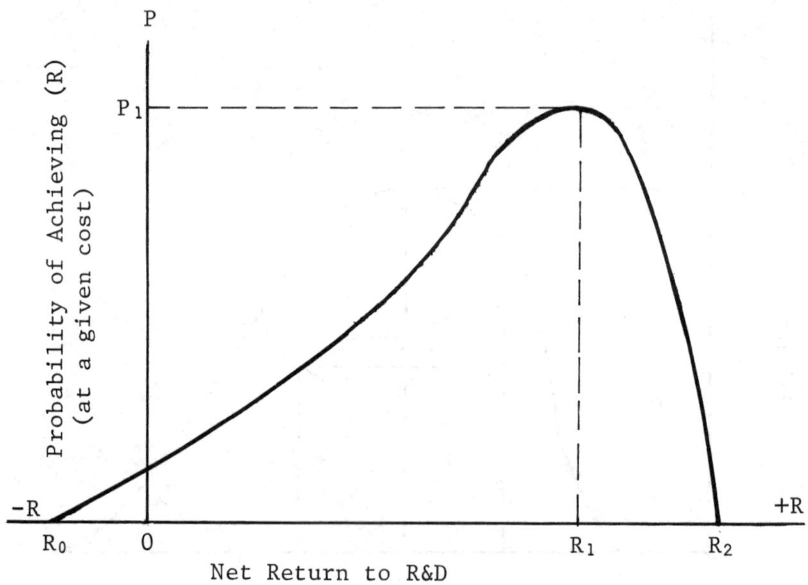

Fig. 3. A hypothetical illustration of the functional relation between probability of success versus the net returns to R & D.

The relationship between net returns and cost of the R & D shown in Fig. 2 will cause the shifting of the limits of the effective zone of decision-making slightly to the right of those shown in Fig. 1.

These two diagrams (Figs. 1 and 2) might appropriately be combined into a three-dimensional surface with a cross-section of the third dimension (i.e., at a given cost) depicting probability (P) as a function of net returns (R). This cross-section is shown in Fig. 3. It is noted that the maxima of these curves should not be expected to occur at identical values of P, R, and C, but will probably fall within the effective zones shown in Figs. 1 and 2. Nevertheless, in seeking an optimal selection among research alternatives it would be preferable to seek combinations of P, R, and C that would yield the higher appraisal ratio, thus involving trade-offs between the two-dimensional maxima. Moreover, due to the fact that R & D programs must compete with other programs in the budgetary allocation of resources, the level of effort committed to the R & D — at least initially — is frequently less than the optimal level as suggested by evaluations of the appraisal ratio of R & D for various alternative program designs. Particularly where the R & D involves innovation of "high technology," there is a marked tendency to underestimate the R & D costs and overestimate the probability of realizing the corresponding level of net return. Cost overruns may even exceed 100 percent, making many private business firms shy away from such ventures except under "cost-plus" contracts with the Federal Government. The large errors that may accompany estimates of probability, net returns, and costs in the appraisal ratio suggests that it would be preferable to express these as ranges or distributions of estimates in the evaluation process.

The Objectival Variables of Federally Sponsored R & D

As previously noted, the net return of R & D sponsored by the public sector involves a mixture of benefits (or objectives) that cannot be evaluated in common (economic) units. Moreover, there is a general tendency in water resource development programs — and no less so in R & D planning decisions — to consider less than the full range of public interests involved. The Special Task Force of the Water Resources Council has addressed itself to the problem of establishing uniform standards for the evaluation of multiobjectives for planning water and land resources. The Task Force delimited four broad classes of objectives [40]: (1) national economic development, (2) environmental quality, (3) social well-being, and (4) regional development. Each of these classes in turn reflect more detailed objectives. For instance, the national economic development objective includes the productive output of the economy, such as increases in crop yields, expanding recreational use, and peaking capacity for power systems. The environmental objective is responsive to the varied spiritual, psychological, recreational, and material needs that are served by protection or enhancement of environmental quality. The social well-being objective is enhanced by promoting: the equitable distribution of real income, employment, and population; national security; health and safety; conservation of resources; educational, cultural, and recreational opportunities; and economic stability. The regional development objective reflects numerous elements of the above national objectives but at the regional level: increases in the region's income and employment; and improvement of the economic base, environment, and social well-being.

Despite the comprehensiveness of these objectives, it is interesting to note that these do not include — or at least there is no mention of their including — such national interests as: promoting the objectives of improved international relations; maintaining a favorable trade balance in international commerce; and such important aspects of the quality of living inherent in concepts of freedom, social justice, and the preservation of domestic tranquility.

The national objectival variables omitted from the Special Task Force Report are not as remote from water resource developments and related R & D alternatives as they might at first blush appear. For example, it has been suggested that our high level of material wealth has led us to place a high value on clean air and water relative to many developing countries [41]. Imposing higher costs of production in the United States to attain clean air and water standards will increase our imports from countries with less stringent standards, which "should benefit all nations" [42]. However, the growth in imports promoted by the costs of water pollution abatement may not be balanced by an equal growth in exports which may be impaired by higher domestic costs of pollution control. This would contribute to a worsening of the U.S. balance of payments position [43], and moreover, the loss of production from certain U.S. plants may create regional unemployment. In addition, there is a potential for increasing U.S. exports of hardware associated with pollution monitoring and control, provided we attain a position of technological leadership in this field.

In discussing the generally lagging growth of productivity in our economy, Gordon Bloom comments on the inflationary impact of environmentalism [44]:

> Pollution controls, for example, will add millions of dollars to industry's cost of production and will pour additional purchasing power into the income stream without increasing productivity, as conventionally measured. Indeed, carrying on production without fouling the air and polluting the water may require an actual reduction in manhour output in some industries.

It is evident that R & D directed toward reducing the costs of water pollution monitoring and abatement can minimize this loss in productivity and reduce the inflationary impact on the prices of U.S. goods. This in turn would abet our balance-of-payments position by curbing the growth of imports while sustaining the growth of related exports including specialized equipment used in water pollution control.

In regard to promoting the quality of international relations, there is a wide variety of issues and cooperative actions between nations related to water resource development to which R & D could make significant contributions. When nations share common water supplies, as does the United States with Canada and Mexico, then problems such as riparian rights to water use and measures to safeguard water quality, enlarge or improve the reliability of supply, and regulate water flows and uses in a manner to achieve a fuller and more equitable distribution of benefits assume great importance.

For example, the United States and Mexico are engaged in discussions of possible joint follow-on studies to a United States-Mexico study report which concluded that it would be technically feasible to install nuclear power-desalting plants as a source of large amounts of fresh water and power for utilization by Southwestern United States and Northwestern Mexico [45]. Although Mexico has rights to 150,000 acre-feet of water from the Lower Colorado River, the high percentage of dissolved salts in this water makes it unsuitable for some of Mexico's fresh water requirements. There is a possibility for a trade-off of some of these water rights to be used by Southwestern United States in exchange for fresh water from the nuclear energy-desalination plant under consideration, which presumably would receive major financing from the United States if the project were to materialize.

In the management of fresh water resources the United States has enjoyed a long period of successful cooperation with Canada. Nevertheless, many potential benefits could be gained from even fuller cooperation. These include possibilities for advances in pollution control measures; an improved management regime for serving better the changing bi-national priorities in the various uses of common water resources; and promotion of greater socio-economic benefits for achieving inter-basin transfers of water, both in the Pacific Northwest and in the Great Lakes Region [46]. With regard to inter-basin transfer possibilities, the benefits would be greater to the United States than to Canada. In order to attain Canadian cooperation it may be necessary to pay their government for the import of such water or to provide offsetting benefits in trade agreements or other forms of bi-national cooperation outside the field of water resources. These inter-basin transfer possibilities represent, on the one hand, an alternative to certain kinds of R & D to enlarge the U.S. water supply from domestic sources and, on the other hand, raise the need for further research to promote the effectiveness of such transfers.

Regarding the developing countries, there is an opportunity to enhance the quality of our relations with them through sharing our advanced water resources technology and know-how. Many of these countries have problems of inadequate water supplies. Thus, the attainment or strengthening of our position of leadership in water resources science and technology through further R & D efforts will enlarge our opportunities to provide technical assistance of this kind to developing countries.

Another national interest neglected in the Special Task Force Report is the potential contribution of water and related land developments to enhancing social justice and domestic tranquility. As a result of the growing feeling that the quality of living has been sacrificed for material gain, a number of strains have been introduced in our society. One of these is the "anti-establishment" sentiment of many young people, which contributes to the so-called generation gap and which derives in part from the obvious insults that have been registered against our environment in our haste to achieve technological progress. It is foolish to think that our society will or should abandon goals of economic and technological progress. Moreover, it is evident that R & D can play a substantial role in reducing the compromise necessary in our economic goals to serve legitimate interests in improving environmental quality, thus reducing a significant source of strain or disunity in our society.

Our domestic tranquility and concepts of social justice are also threatened by the inherent conflict of economic interests of different groups in solving our problems of water resource developments. In the case of water pollution, those who benefit and those who "pay" for pollution abatement measures are often not the same parties, or at least different groups pay and benefit to unequal degrees. For example, those who bear the greatest share of the tax burden may feel they are not gaining benefits commensurate with their payments from enhanced environmental quality and recreational opportunities resulting from pollution abatement measures. But as a matter of social justice, it can also be said that higher income groups can better afford than lower income groups to select a residential environment of high quality and to travel farther and wider in seeking recreational outlets in a less polluted environment.

Another decisive factor relating to pollution is that those plants whose profitability is adversely affected by high costs of pollution abatement measures will require certain parties to bear this cost, while still other parties may be the primary beneficiaries of these measures [47]. One should hasten to add that in the plants adversely affected in this regard the division is not necessarily between the owners and the workers. Indeed, if there is a sufficient reduction in profits, the operations of a plant may be curtailed or closed down, with loss of employment. Even if this is not the case, a reduction in profits may lead management to hold costs down by limiting wage increases. Recognition of these factors creates community pressures — particularly if a plant's activity has a major economic impact on the community or region — to resist enforcement of water quality standards or to attain some other solution to the problem, such as increasing the absorptive capacity of the watercourse, easing the standards imposed, and diverting the polluted effluents to the city water-treatment plant for removal of pollutants or other kinds of subsidies. As noted by the Council of Economic Advisers, pollution imposes real "psychic costs" on the community [48], but no less so than the social frictions that may be encountered in forcing pollution abatement measures to be implemented. There would appear to be possibilities for R & D to contribute to the solution of problems of these kinds, thus improving domestic tranquility and serving the basic concepts of social justice upon which our country was founded.

It will be seen that many of the national objectives associated with water resource developments are closely interrelated. Thus, R & D that would improve the deteriorating U.S. balance-of-payments position would also contribute to national economic development. Likewise, R & D that would promote social justice and domestic tranquility would also enhance social well-being. In a somewhat different vein, water resources R & D that would stimulate regional development would generally enhance related national objectives. This is not a certainty, however. Even though the total U.S. economic output is the sum of the output of its respective regions, the gain in output in one region may be at the expense of output in other regions. In this event, the national benefit will be less than the benefit received by the region in which the newly developed technology is applied. Yet there may be other instances where a number of regions benefit from the R & D and few, if any, may stand to lose.

A similar reasoning applies to sub-regional units, such as states or metropolitan areas. The developmental objectives of metropolitan areas which have a somewhat unique set of problems and critical issues were ignored by the Special Task Force Report.

SUMMARY

A crisis of science and technology has resulted from the current widespread mood of social reform which, among other issues, stems importantly from environmental degradation. While research scientists and engineers are not to be blamed for the social misuse of the tools they produce, the technology assessment movement begun in recent years provides a methodology for better protecting and fulfilling society's interests. By evaluating the potentially good and bad effects of technological alternatives, including appropriate institutional measures to control their use, technological assessment promises to be an effective tool for R & D planning in the field of water and related land-use developments.

Technological assessment will most likely reveal that *more*, not less, R & D is needed to protect and enhance environmental quality as well as other national, regional, and urban interests associated with water and land uses. Goal conflicts between different individuals, groups, or political jurisdictions is a major problem associated with water and land resource developments. Research in the social sciences can play an instrumental role in appraising the goal implications of technological alternatives, thus stimulating greater support for R & D in systems technology involving both hardware and software developments. Forecasting technological developments and their potentially good and bad effects is a difficult and complex art. There is a danger that insistence on technological applications without any risk or injury to the ecology or to society may stultify advances in science and technology. While it is desirable to upgrade the quality of technology assessment, the methodology will likely always be subject to limitations of reliably predicting outcomes of social uses of new technology.

The complexity of interrelationships that must be analyzed to improve coordination in R & D planning for water and land-use development invites chaos unless appropriate systematic procedures are pursued. Defining the system's variables and assessing their interrelationships are of crucial importance. Systems analysis related to program planning has certain similarities between the private and public sectors but important differences as well. In the public sector decision-making is *process* oriented, while in the private sector it is *performance* oriented. This distinction is revealed most clearly in the decision criteria for justifying R & D in the private and public sectors, particularly the goal orientations of the appraisal ratio of R & D. This ratio is obtained by dividing net returns, if the R & D is successful, by the cost of the R & D and multiplying by the probability of a successful outcome. These factors cannot be independently estimated because of their functional relationships. In addition, their evaluation is encumbered by a need for a multidisciplinary analysis, which is impaired by professional specialization.

Also, those who pay for the R & D and those who reap social benefits or incur social costs in the implementation of the technology are seldom the same individuals — or at least different parties "pay" and benefit in unequal degrees. Thus technology assessment can aid R & D planning by identifying and evaluating goal conflicts and potential compromises between different groups. This means that R & D planning affecting water and land-use developments needs to evaluate the cooperative R & D roles of the public and private sectors as well as the decision variables of Federally sponsored R & D, as these variously affect national objectives, regional objectives, and sub-regional objectives — particularly those of metropolitan areas.

NOTES AND REFERENCES

1. LESSING, LAWRENCE, "The Senseless War on Science," *Fortune*, 89, March 1971.

2. *Toward Balanced Growth: Quantity With Quality*, Report of the National Goals Research Staff (Washington, D.C.: U.S. Government Printing Office, p. 4, July 4, 1970.

3. For an excellent discussion of the role of the social sciences in technology assessment see, Franklin P. Huddle, "Government Technology Assessment: The Role of the Social Sciences." A paper presented at the Panel-Round Table at the Annual Meeting of the American Political Science Association, Report 70-246SP, The Library of Congress Legislative Reference Service, Washington, D.C., October 2, 1970.

4. *Ibid.*, p. 120.

5. *Technology: Processes of Assessment and Choice*, Report of the National Academy of Sciences to the Committee on Science and Astronautics, U.S. House of Representatives, Washington, D.C.: U.S. Government Printing Office, p. 1, July 1969.

6. See, for example, *Technology Assessment*, House Hearings before the Subcommittee on Science, Research and Development of the Committee on Science and Astronautics, 91 Cong., 1 sess. Washington, D.C.: U.S. Government Printing Office, 1970. Also, *Technical Information for Congress*, Report to the Subcommittee on Science, Research and Development of the House Committee on Science and Astronautics, 92 Cong., 1 sess. Prepared by the Science Policy Research Division, Legislative Reference Service, Library of Congress, Washington, D.C.: U.S. Government Printing Office, 1969. Revised Report April 15, 1971.

7. *Study of Technology Assessment*, Report of the Committee on Public Engineering Policy of the National Academy of Engineering to the Committee on Science

and Astronautics, House of Representatives, Washington, D.C.: U.S. Government Printing Office, p.11, July 1969.

8. MARX, KARL, *Capital: A Critique of Political Economy.* Trans. by Samuel Moore and Edward Aveling from the 3rd ed. New York: The Modern Library, p. 431, 1906.

9. *A Study of Technology Assessment, op. cit.,* 10.

10. *Ibid.,* 5.

11. *Technology: Processes of Assessment and Choice, op. cit.,* 10.

12. *Ibid.,* 14.

13. *Technology Assessment,* House Hearings, *op. cit.,* 222.

14. *A Study of Technology Assessment, op. cit.,* 15-16.

15. Yet, some agencies, such as the AEC and the DOD, which are responsible for sponsoring R & D as well as controlling the implementation of its results, pay inadequate attention to externalities.

16. SHAEFFER, RUTH G., "Systems in Managing Change," *The Conference Board Record,* 48, August 1970.

17. POUNDS, WILLIAM F., "The Nature of Systems," *The Conference Board Record,* 49-50. August 1970.

18. "Schultze Assesses PPBS," *C-E Newsletter,* 5, Operations Research Society of America, March 1970.

19. SHAEFFER, *op. cit.,* 49.

20. POUNDS, *op. cit.,* 51.

21. WILSON, ROBERT E., "The Attitude of Management Toward Industrial Research," *Mechanical Engineering,* p.9, Jan. 1950; and HILL, D.W., *Cooperative Research in Industry*, 188, New York: Hutchinson's Scientific and Technical Publications, 1954.

22. For a fuller discussion of problems and nuances in the use of the appraisal ratio, see SPANGLER, MILLER B., *New Technology and Marine Resource Development: A Study of Government-Business Cooperation*, 62-69. New York: Praeger Publishers, Inc., 1970.

23. DOVE, GRANT A., "Objectives, Strategies, and Tactics in a System" *The Conference Board Record*, 53, August 1970.

24. *Statistical Abstract of the United States*, 520, 1970.

25. *Ibid.*, 522.

26. For a fuller discussion of the logic and problems encountered in the use of this appraisal ratio and its inherent criteria, see SPANGLER, *op. cit.*, 70-81.

27. HATRY, HARRY P., "Reflections From an Outgoing Chairman," *The C-E Newsletter*, 4-5, Operations Research Society of America, October 1967.

28. DADDARIO, EMILIO Q., "Academic Science and the Federal Government," *Science*, 1250, Dec. 13, 1968. Copyright 1968 by the American Association for the Advancement of Science.

29. SCHON, DONALD A., *Technology and Change*, 35, New York: Delacorte Press, 1967.

30. *Technology: Processes of Assessment and Choice*, *op. cit.*, 37-38.

31. As reported in KATZ, MILTON, *The Function of Tort Liability in Technology Assessment*, Reprint No. 9, 597-598. Cambridge, Mass.: Harvard Univ. Program on Technology and Society, 1969.

32. SPANGLER, *op. cit.*, Part VI, 489-586.

33. *Ibid.*, 70-71.

34. *Economic Report of the President*, 114, Washington, D.C.: U.S. Government Printing Office, 1971

35. *Ibid.*, 115.

36. *Environmental Quality*, 2, The First Annual Report of the Council on Environmental Quality, Washington, D.C.: U.S. Government Printing Office, August 1970.

37. *Ibid.*, vii.

38. For the moment we shall ignore that the probability of success is also affected by the quality of the R & D personnel and its management as well as the time period at which the R & D is initiated and the total elapsed time allowed for its conclusion. These matters must be dealt with on a case-by-case basis.

39. For a methodological discussion of subjective probability estimation see, SPANGLER, MILLER B, *New Technology and the Supply of Petroleum: The Treatment of Uncertainty in Resources Planning*, Research Paper No. 2, 10-35, The University of Chicago Program of Education and Research in Planning. Chicago: By the Author, 1956.

40. *Standards for Planning Water and Land Resources*, II-7 to II-19, Report to the Water Resources Council by the Special Task Force, Washington, D.C., July 1970.

41. *Economic Report of the President*, 1971, *op. cit.*, 122.

42. *Ibid.*

43. For a fuller discussion of the long-term outlook for U.S. balance-of-payments problem see, SPANGLER, *New Technology and Marine Resource Development*, *op. cit.*, 24-34.

44. BLOOM, GORDON F., "Productivity: Weak Link in Our Economy," *Harvard Business Review*, 5, January-February 1971.

45. *Annual Report to Congress of the Atomic Energy Commission for 1970*, 211. Washington, D.C.: U.S. Government Printing Office, January 1971.

46. For a discussion of the potentialities for cooperative R & D efforts between the United States and Canada to achieve a fuller utilization of water resources in the Great Lakes Region see BIGLER, ALEXANDER B., *The Role of Marine Sciences in the Multiple Uses of the Coastal Zone of Lake Erie and Lake Superior.* A study prepared for the National Council on Marine Resources and Engineering Development. 282-293. Washington, D.C.: National Planning Association, June 1969. Distributed through the National Technical Information Service, #PB-185-163.

47. It is recognized, however, that in a majority of cases it would seem likely that firms will be able to pass along to the consumer such cost increases in the form of higher prices, thus preserving profit margins. Nevertheless, there will be some inequities in this capability between new and old plants and between river basin systems having unequal assimilative capacities for pollutants.

48. *Economic Report of the President, op. cit.*, 108, 1971.

Note: This chapter is an edited version of Chapter 2 in, Miller B. Spangler, *The Role of Research in Water Resources Planning and Management for Achieving Urban Goals*, a contract study prepared by the National Planning Association for the U.S. Office of Water Resources Research (Springfield, Virginia: National Technical Information Service, May 1972), 26-64. The work upon which this chapter is based was supported in part by funds provided by the United States Department of the Interior as authorized under the Water Resources Research Act of 1964, as amended.

PART II

METHODOLOGICAL CONSIDERATIONS

INTRODUCING SOCIETAL INDICATORS INTO TECHNOLOGY ASSESSMENT

DONALD OVERLY

Senior Professional Staff, International Research and Technology Corporation, Washington, D.C., 20036, U.S.A.

TECHNOLOGICAL ADVANCEMENT — WHO ARE THE BENEFICIARIES AND BENEFACTORS?

In the Western world, growth and development in both industry and government are generally characterized in terms of technological advances which are, in some fashion, causally associated with experienced or proposed economic patterns. This technological bias is well evidenced by the emphasis policy makers and managers place upon the "R & D budget," which somehow evokes images of new products and profits for corporations and new exploits and services for government agencies. This tendency, until quite recently, solicited a minimum amount of serious inquiry and review; after all, technology somehow "produced" computers, teflon, telephones, oxygen furnaces, moon walks, nylon, plastic, polio shots, stereo, electric toothbrushes, window air-conditioners, and jet aircraft. A technological bias to the reporting of an organization's performance and structure was usually accepted as being quite legitimate, or at least was not thought to be worthy of examination. Performance was described in terms of how technology resulted in new products, sales, budgets, revenues, and so on, while structure was often described in terms of how technology influenced staffing, decision processes, new offices, and so on. Implicit may have been the assumption that everyone gains from technology or is a beneficiary, at least to the extent that no one benefits to the detriment of anyone else. This suggests that policy makers and managers in government, universities, and industry believed that technology-oriented programs resulted in, at worst, some crude form of Pareto optimality — that is, no one benefits at the expense of anyone else [1].

Recently, however, policy makers have begun to admit that large scale technology-oriented programs really have a much wider range and diversity of beneficiaries (those who somehow benefit) and benefactors (those who somehow bear the costs) than are considered in conventional program analysis and benefit-cost computations [2].

The wide variation in the unplanned or unidentified impacts of technology, though, results from the wide range of interests which are attempting to utilize technology. Entrepreneurs are looking for new markets, industrial managers want to avoid unnecessary costs, elected officials want to satisfy constituents, and consumers want new products. Up until now, though, major new technological advancements have always resulted in beneficiaries and benefactors who were either not identified or acknowledged in the original program planning and resource allocation processes. The reason for this inability (or refusal) to account for all beneficiaries and benefactors of a program is, in principle, not at all profound.

Every institution or organization must, in some way, establish goals, allocate resources for their achievement, evaluate and report progress, and make necessary adjustments. These decisions and actions are to be made so as to maximize the organization's benefits and minimize costs. The addage, "Maximize benefits and minimize costs" is the acknowledged shibboleth of most industrial and mission-oriented R & D managers, believers in the free market, and other rational (or goal-oriented) thinkers. The ability to rationally maximize benefits and minimize costs seems to be seldom questioned by policy makers and managers in both government and industry due to the belief that available *potential* technology is ubiquitous; that is, any problem can be solved if only there is enough money. A large enough R & D or program budget is often believed to be the major prerequisite for benefit maximization and cost minimization; the actual computation of benefits and costs is assumed to be a rather straightforward accounting exercise.

The mathematics of benefit-cost analysis, however, generally do not acknowledge the issue of selecting the appropriate benefits (beneficiaries) and costs (benefactors) for consideration. Formal policy-level consideration of a program's benefits and costs usually is made for the first time when proposals for budget support are made to the appropriate committee or some equivalent. However, elaborate cost-benefit analysis, by the time they are presented to a corporate budget group, a regional industrial zoning board, a Congressional committee, or a regulatory commission, are seldom in a form which permits brief and intelligent inquiry into the selection and quantification of the benefits and costs used in the analysis [3]. The inability to make the needed intelligent inquiry can, perhaps, be attributed to a combination of two factors: first, the "razzle-dazzle," where the analysis is accompanied by a large computer printout, the math is unnecessarily complicated, and multicolored charts and overlays are used; and second, the "line-up your ducks," where very sophisticated and effective lobbying has been underway during the time the "objective" benefit-cost analyses were underway.

One result of making program decisions in this fashion is that certain costs and benefits which result from an organization's policies and decisions are often not considered as they either were not recognized or were intentionally ignored or disguised. In other words, benefits and costs often are not properly accounted for; proponents of a program tend to exaggerate benefits and ignore costs while opponents, where possible, do the opposite. This phenomnon is suggestive of fraudulent accounting schemes whereby certain costs go unpaid or are shifted to others, and benefits are either substantially overrated or are received without proper payment. Even if there is no attempt to mislead either those who must allocate resources or those who will be affected by a program, there is a good chance that opportunities will not be identified and therefore lost, resources will be spent unnecessarily, benefits will be received which are not paid for, or costs will be incurred which should be paid by others.

The term "fraudulent," while harsh, is not used here in the legal context of fraud of embezzlement. Rather, it is used with reference to a hypothetical market structure (a real-world equivalent, perhaps, of the "purely competitive" market) where *all* benefits and costs are accounted for. An explicit display of all benefits (and beneficiaries) and all costs (and benefactors) probably would result in special interest groups, which represent beneficiaries and benefactors reacting, coalescing, or forming, and finally taking appropriate action.

The question, "What are the avenues for appropriate action?" is one that the special interests themselves — the beneficiaries and benefactors — of proposed and ongoing programs must create and manage through the market legal and political processes. Practices which permit some to receive benefits they have not paid for while forcing others to pay costs for inconveniences incurred through others' actions would, perhaps, be corrected if major proposals and programs were evaluated by an accounting system which is not directly under the control of program proponents and opponents. Such an accounting system would be based upon broad societal indicators which attempt to describe all social, economic, cultural, and political impacts of technology, not just those impacts which a program's proponents and opponents wish to acknowledge. These societal indicators would be designed to measure the impacts of programs and proposals throughout the socio/economic system.

The costs and benefits which are presently not made visible and considered in conventional analysis of technological programs are, broadly speaking, of a societal nature. Programs which have wide ranging social, economic, cultural, and political impacts need to be linked explicitly to program beneficiaries and benefactors if unplanned or unacknowledged expenses and harm are to be avoided.

Both governmental and industrial technology-based programs have resulted in costs which go unpaid or are shifted to others. These include:

- Pollution, where municipal and industrial effluents cause both physical and psychological stresses.

- Transportation, where subsidized highways reduce job and living opportunities for minority groups.

- Unsuccessful R & D and marketing attempts, where extra costs are distributed among profitable goods and services.

- Resource shifts, resulting in unnecessary depletion or discard of capital and human resources.

- Consumer attitudes, where the extra costs of rapidly meeting new demands are distributed among profitable industrial activities.

In the above examples, programs have had or are having a wider range of impacts than was originally understood or acknowledged in early program phases. Beneficiaries and benefactors who were not considered in the original analyses resulted.

These costs are not *a priori* good or bad; they must be considered with reference to some value structure for inclusion in any analysis. For example, the costs of pollution may be considered "bad," and perhaps should be followed by punitive or corrective action. The costs of increased competition, which may drive a firm's costs up and prices down, are discomforting, but they may be "good."

Examples of benefits which are received and often not paid for include:

- Toll-free bridges and freeways, where individual drivers and certain businesses benefit extraordinarily from the public investments.

- Condemnation practices, where owners of property adjacent to the condemned property experience rising values.

- Patent disclaimers, where those who made no investment in the initial discovery use the patent.

- "Piggy-back innovation," where an accessory or

product can be produced by a firm to capitalize on another's already developed and introduced good or service.

As with costs, these benefits are not *a priori* good or bad, and should elicit different responses. Where speculation is inhibiting sound economic development, it may be necessary to impose government controls. Such actions, though, may not be appropriate where a firm can profitably produce a cheap accessory for another's expensive machine.

The concept of beneficiaries and benefactors is useful (again, beneficiaries collect the benefits, benefactors bear the costs), as it serves first, to dampen some of the more hysterical demands to halt immediately any technological advance and second, to ensure that all special interests which might be affected are identified. A more explicit identification of beneficiaries and benefactors may make it possible to develop accounting systems which take into consideration unexpected or hidden benefits and costs and, if necessary, require that those who benefit pay their share of the costs and those who bear a disproportionate share of the costs either have them removed or receive an appropriate compensation. An example of forcing payment for hidden benefits might be to increase the taxes on speculative land which rose in value due to public investments. Compensation for costs incurred might be seen in payments to fishermen whose fishing grounds have been spoiled by municipal sewage. These two cases would probably require government intervention. There may be cases where the question of beneficiaries and benefactors would best be solved in the market place, such as one firm's aggressive marketing resulting in increased costs for its competition.

An accounting system based upon societal indicators could help to force those making demands for arbitrary technological pull-backs to deal with harder facts which describe just who is affected by technology. Benefactors and beneficiaries identified in program proposals, or those who believe they would be detrimentally affected by the program, would have the opportunity to participate directly in various forms of due process, the political process, and the marketplace. In proprietary commercial programs, planners would be able to predict more accurately the acceptance or rejection of specific programs and, perhaps, the plans of their competitors.

This broadened analysis of new programs, requiring that they be evaluated with reference to a set of societal indicators which attempt to represent all beneficiaries and benefactors, is suggestive of a new market mechanism which has two general dimensions. The first dimension considers those costs and benefits which affect the public welfare, requiring the powers of government for any adjustment. The second considers those costs and benefits which affect private competitive positions, requiring adjustment only through the market place.

Although the idea has been suggested that new markets may result from attempts to assess technological impacts [4], probably nothing too disruptive needs to be feared by either policy makers or managers. The concept of technology assessment has only recently been introduced, and current assessment attempts are only beginning to cope with very complex issues.

TECHNOLOGY ASSESSMENT — A NOBLE ENTERPRISE

The previous section suggests a growing need for awareness by policy makers and managers in both government and industry of the wide ranging and disparate social, economic, and cultural impacts of technology. This acknowledged need for improved policy and managerial decision processes necessary for obtaining the full benefits of technology and minimizing unwanted and detrimental consequences has not, however, been matched by the development and testing of specific technology assessment approaches or techniques.

Although there are few reported technology assessment programs to generalize from, one general program definition which may be agreed to is that technology assessment programs are, for the most part, efforts to understand what significant impacts proposed and ongoing technological or applied science programs have upon the social, economic, and cultural systems [5]. There is, however, a distinct difference in interpretation by government and business.

Technology assessment efforts within the government tend to be either historical reviews of technological developments [6] or the compilation of evidence that a new council or board is required [7]. Although these efforts have produced, and are producing, a great deal of needed insight into the problems of technology assessment, only a few specific policy or procedural approaches for technology assessment have been developed and tested [8]. In the federal government, proposals to assess technology usually suggest first the creation of a council (or councils) to provide the legislative and executive branches with the information they need to first, select and manage large-scale technology-based programs and second, to judge large-scale programs proposed by business. The arguments for policy-level technology assessment are, for the most part, reactions to some combination of the following:

- The detrimental effects of technology as seen in unintended (or unreported) impacts such as pollution and food poisoning.

- The risk of allocating huge amounts of resources to areas of endeavor which will prove to be unproductive.

- A growing understanding of the complexity of national economic and social problems.

- A desire to achieve a social and economic state which results in minimum dissatisfactions and perturbations.

Technology assessment in government, then, is perceived as a means to set goals and priorities and rationally allocate resources for their achievement, while minimizing the risk of any detrimental effects. In some cases, these resources would be directly allocated by the government (e.g., pollution abatement and education) while in others government policies would influence the allocation of private resources (e.g., the building of a nuclear power plant and mass transit).

Industry, while acknowledging the need for technology assessment, really emphasizes technology forecasting — that is, trying to determine what technologies will, under certain conditions, be available in the future [9]. This information, while useful in predicting competitors' positions, government R & D policies, and possible market or technological opportunities to exploit, still permits benefits (beneficiaries) and costs (benefactors) to remain unacknowledged.

Technology assessment for the private corporation focuses upon the objective of maximizing benefits and minimizing costs, in a dollar sense, with reference to a statement of existing and possible constraints which are influenced by government action and the competitive market. Technology assessment for the competitive firm is generally a reaction to the following:

- The increasingly high costs of successfully developing and marketing new goods and services.

- A need to understand future government restrictions and competitive patterns so as to avoid the payment of unnecessary costs.

- The growing need to mobilize public and consumer support for beginning new ventures or reducing government restrictions.

The focus here is, as it should be, on the corporate budget and future earnings and markets.

Although the uses of technology assessment differ for both government and business, there are similarities in the rather vague and tentative proposals which today are being offered and discussed and the few programs being implemented. The section below describes a generalized technology assessment procedure without reference to the specialized adaptations required by government or private business.

A Generalized View of Technology Assessment

Although the literature on technology assessment is growing rapidly, few thorough, integrated approaches have been offered or tested. Rather, there seems to be an offering of specific techniques which appear to be required in technology assessment such as Delphi techniques, cross-support analysis, cross-impact analysis, and trend extrapolation [10]; arguments supporting the need for technology assessment [11]; and anecdotes of cases where technology assessment was, in some form, tried [12]. From this melange of insights, critiques, suggestions, and techniques, a generalized (hopefully accurate) list of assessment procedures or requirements is constructed below.

Establish Organizational Goals and Priorities — Although this is understood fairly well in concept, there are great difficulties in practice. Most of these difficulties stem from an inability to trade off possible competing goals (e.g., share of market versus net income and number of officers on Presidential commissions versus introduction of new products). Also, there are many theoretical problems associated with the establishment of priorities. However, some techniques have been developed to overcome these difficulties, such as cross-support analysis, which help to quantify the degree to which single goals support all organizational goals. Even where all difficulties are acknowledged, some statement of goals and priorities can be developed.

Predict Future Developments — Predicting possible future events usually involves rather straightforward statistical analysis (e.g., trend extrapolation) coupled with subjective judgements of those who are experts in fields related to the organization's enterprise. These subjective judgements are often obtained through some form of an iterative voting/judging process, such as a Delphi technique. The real objective here is to determine what broad market opportunities, constraints, and technologies might be available in the future.

Link Organizational Goals and Priorities to Predicted Future Developments — This step is to assure that there will be understanding of the possible future roles the organization may have with reference to changing markets and new mixes of goods and services.

Analyze Organizational Resources — Resources consist of existing and future manpower and capital mixes. The former describe existing capabilities and the latter the ability to mobilize resources. These resources are somehow to be allocated towards the achievement of organizational goals and priorities as they relate to possible future developments. Also, information, whether patents or specialized knowledge, must be considered.

Develop Resource Allocation Strategies — This requires a dynamic understanding of the relationships between corporate goals and priorities and possible future developments. Possible future technological developments may, for example, result in goal modification, either in terms of new opportunities or constraints. With reference to both goals and priorities, it is necessary then to select those allocation strategies which will permit goal achievement at a minimum cost.

The above five generalized steps are often offered by planners as an assurance that future technological advances will have wide ranging benefits and few detrimental impacts. Unfortunately, this is not the case. Some limitations of current technology assessment practices are discussed below.

Some Inherent Limitations in Conventional Technology Assessment

The major weakness in today's form of technology assessment is that it is too technology and technologist-oriented. This is due, principally, to the higher levels of rigor and preciseness in which technological developments can be described relative to social and economic impacts. A picture of future electronically guided roadways, is, for example, much easier to assimilate than some of the impacts which that roadway might have upon people who live near it.

Predicting future developments, whether through trend extrapolation or subjective analysis, takes on a technological bias for a variety of other reasons. First, the question which the analysis (and analyst!) often addresses is, "What can be done?" not "What should be done?"

Questions of feasibility (e.g., Can the SST be built?) rather than useability (e.g., Who will use the first-generation commercial supersonic transport?) predominate and totally bias the projection of future developments. Another, perhaps more fundamental, problem is that this technological bias is seen in the almost automatic discard of any possible future developments which are suggestive of radical social or economic reform. The possibility of major modifications in economic and social institutions is seldom considered. Implicit here is the notion that conflict, bargaining, and negotiation are not legitimate future forces and are not to be acknowledged.

This narrow viewpoint not only avoids proper accounting of beneficiaries and benefactors, it severely reduces recognition of new possibilities which would contribute to new quality-of-life considerations. These considerations — whether defined in terms of housing, education, transportation, consumer products, length of work week, entertainment or sports — could always be translatable by the entrepreneur or manager into terms such as new markets, profits, institutions, taxes, costs, and so on.

One other problem is that future technological developments are often too sponsor-oriented. This, for the most part, is the result of growing government involvement in R & D procurement and the huge amounts of capital required for R & D. Here, for example, the institutional biases are already established: major health and agricultural breakthroughs probably will come from government-supported agencies; advances in telecommunications will come through joint public and private endeavors; aerospace will require huge government subsidies; and mass transit will require public subsidy for at least the required initial investments.

Technological and institutional biases need to be eliminated from program analysis if the wide-ranging impacts of new programs are to be properly assessed. If the scope of technology assessment is not increased, issues of technological feasibility rather than need or desirability will continue to dictate the form in which decisions affecting the future will be made. Institutional biases make it impossible to consider future possibilities without reference to existing institutional structures, even though it may be desirable to advance or modify institutional structures.

Many of the above limitations can be removed through the incorporation of societal indicators into technology assessment efforts.

BROADENING THE ANALYTICAL SCOPE — INCORPORATING SOCIETAL INDICATORS IN TECHNOLOGY ASSESSMENT

The sections above, to summarize, assert the following:

- Technology-based programs often result in benefits and costs which were not identified or acknowledged in the planning and resource allocation processes and, as a result, there are beneficiaries who do not pay for their gains and benefactors who must pay costs they do not incur through their own actions.

- Not accounting properly for beneficiaries and benefactors, whether intentional or unintentional, is the result of inadequate accounting or reporting systems.

- These inadequate accounting systems, at best, consider only those benefits and costs which must legally be acknowledged.

- Current technology assessment programs are limited in their ability to account for all benefits and costs as they are too technology-oriented.

It is suggested here that the incorporation of societal indicators in technology assessment would be a conceptual and methodological advance towards the proper measurement of the impact of technology upon society.

In practical terms, a technology assessment program measures the impacts of specific innovations as described by alternative technological, administrative, and institutional decisions which are or may become available to specific organizations. Usually an innovation can be linked to identifiable proponents or opponents. It must be understood, though, that a single organization attempting to innovate, whether a private firm or a government agency, is not really interested in accounting for the "broad societal impacts" of technology unless exogeneous forces, such as laws or the market place, require that these impacts be acknowledged. In the case of a single firm, a law could require payment of "pollution costs" while the market often requires payment of losses associated with the introduction of an unsuccessful product. In introducing the concept of societal indicators, it is not suggested that all organizations prepare to react with corrective actions to any technological impact. What is suggested is that single organizations become prepared to evaluate their programs with reference to societal indicators going beyond program proponents and opponents, thereby making it possible for private firms to identify new market opportunities, special interest groups which may oppose or support their enterprise, and pending or needed legislation. Certain government agencies would be able to evaluate national science policy, prepare legislation, and exercise regulatory activities in a more knowledgeable way.

The steps of an assessment process are listed below as they might apply to a specific organization; therefore, the term "organization" is used to describe those assessment components and processes which, in reality, would be unique to itself. This technology assessment process has the following components:

(1) A set of organizational goals.

(2) A set of national goal or policy areas.

(3) A set of organizational performance indicators which measure the degree to which organizational goals would be achieved by a given program or innovation.

(4) A set of national societal indicators (technological, social, economic, political, regulatory, etc.) for estimating the degree to which each of the national goal or policy areas would be affected by a given program or innovation.

(5) A set of self-consistent future "worlds" which describe technological, administrative, and institutional innovations and conditions.

(6) Estimates of the impact of each of the future worlds — the innovations and conditions — upon organizational goals, as measured in terms of operating indicators, and national goal areas, as measured by societal indicators.

(7) The desirability of future worlds and individual innovations, measured in terms of their impact, cost, and probability of success in the form of a modified benefit/cost ratio.

Exhibit 1 diagrams one form of technological assessment which incorporates societal indicators in a way which displays the impact of technology upon the social-economic-cultural-technological system. Societal indicators are one of the first developments in the analytical process. The diagram summarizes the flow of information among a number of fairly well accepted analytical components, generally matrix techniques, which are not normally combined. The remainder of this paper will be a discussion of the characteristics of societal indicators and possible implications of their use. A detailed description of the analysis process, as illustrated by Exhibit 1, is discussed by C.A. Ralph in Chapter 31.

The identification of societal indicators (indicators of social, economic, cultural, political, technological, demographic and institutional conditions) is based more upon imagination, common sense, and thoroughness rather than scientific analysis. The objective is to identify those interests and institutions (i.e., beneficiaries and benefactors) upon which programs may impact. Several categories of societal indicators discussed briefly below suggest different components of the larger socio-economic system within which each organization must function. Exhibit 2 is an illustrative list of societal indicators.

Economic — Here the emphasis is upon measures of relative affluence; that is, changes in real income, and changes in patterns of purchasing and consumption.

Exhibit 1. Analysis process.

ASSESSMENT METHOD

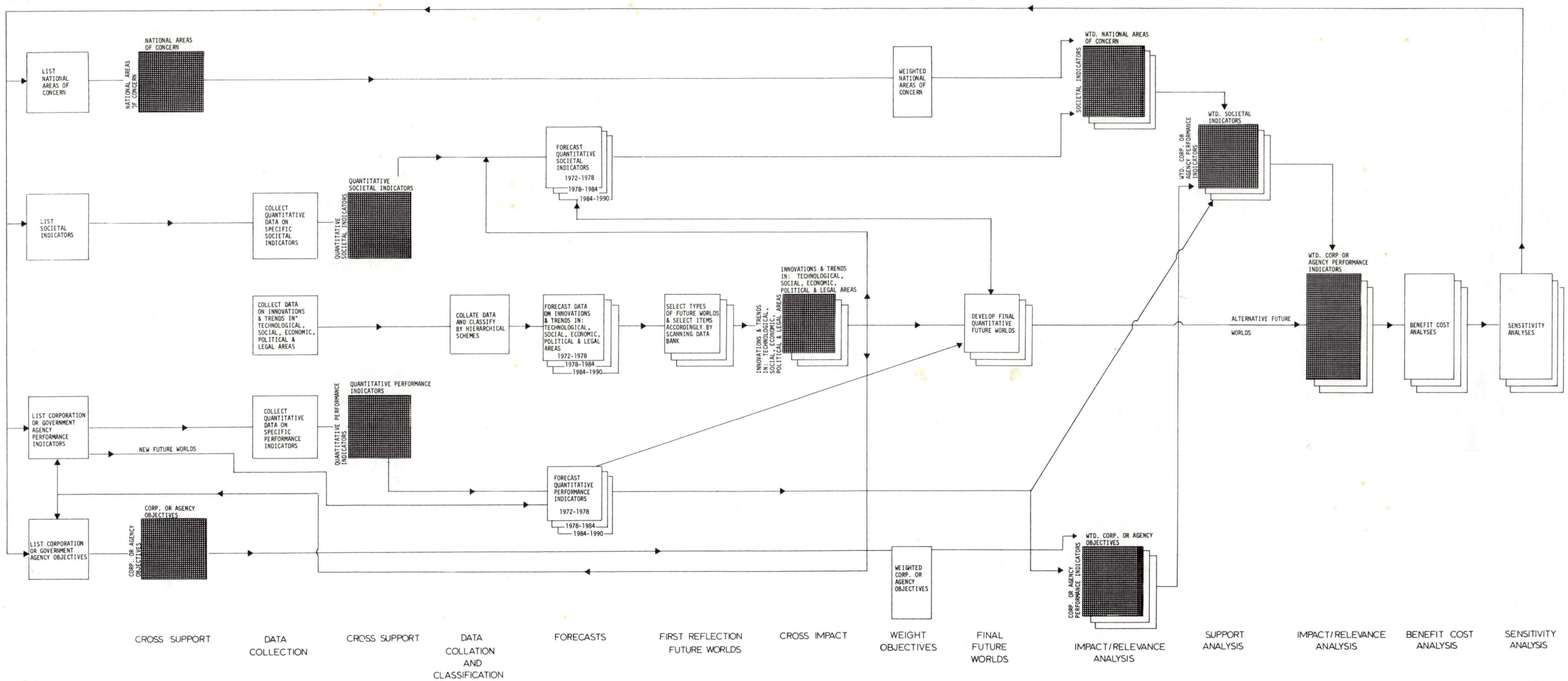

Exhibit 2. Illustration of Outline for Classification of Societal Indicators

1. *Economic*

1.1 Personal income
- 1.1.1 Families
- 1.1.2 Individuals

1.2 Corporate income and profits
- 1.2.1 Manufacturing industries
- 1.2.2 Service industries

1.3 Nonprofit income and budget
- 1.3.1 Universities
- 1.3.2 Hospitals
- 1.3.3 Charitable and private welfare organizations

1.4 Government agency income and budgets
- 1.4.1 Federal
- 1.4.2 State
- 1.4.3 Local
- 1.4.4 Special

1.5 New goods and Services provided
- 1.5.1 Private sector
- 1.5.2 Government

1.6 Capital investments
- 1.6.1 Government
- 1.6.2 Corporate

2. *Demographic*

2.1 Families
- 2.1.1 Size
- 2.1.2 Location
- 2.1.3 Marriage and divorce rates
- 2.1.4 Type of dwelling unit

2.2 Individuals
- 2.2.1 Morbidity rates
- 2.2.2 Mortality rates
- 2.2.3 Years of school
- 2.2.4 Length of work week

Exhibit 2 (continued)

2.3 Manufacturing
- 2.3.1 Size
- 2.3.2 Distribution

2.4 Service sector
- 2.4.1 Size
- 2.4.2 Distribution

2.5 Government agencies
- 2.5.1 Size
- 2.5.2 Distribution

3. *Social*

3.1 Families
- 3.1.1 Children receiving education beyond high school
- 3.1.2 Number and percent of members having contact with police agency
- 3.1.3 Number and percent of families in neighborhood who have incomes 50% lower and higher than the mean

3.2 Government agencies
- 3.2.1 Rehabilitative programs
- 3.2.2 Educational programs
- 3.2.3 Surveillance and control programs
- 3.2.4 Leisure time programs

3.3 For-profit organizations
- 3.3.1 New services and products being offered
- 3.3.2 Personnel policies
- 3.3.3 Public "image"

3.4 Individuals
- 3.4.1 Association memberships
- 3.4.2 Illegal activities participated in
- 3.4.3 Legal actions participated in

4. *Political and legal*

4.1 Federal
- 4.1.1 Domestic economic policy

Exhibit 2 (continued)

4.1.2 Foreign economic policy
4.1.3 Housing
4.1.4 Transportation
4.1.5 Pollution control
4.1.6 Law enforcement
4.1.7 Defense and space policies

4.2 State
4.2.1 Education
4.2.2 Transportation
4.2.3 Welfare

4.3 Local
4.3.1 Education
4.3.2 Transportation
4.3.3 Welfare

4.4 Commissions and special purpose agencies
4.4.1 Transportation
4.4.2 Pollution control
4.4.3 Law enforcement
4.4.4 Communication

5. *Technological*

5.1 Communications
5.1.1 Facsimile reproduction
5.1.2 Picture phones
5.1.3 Cable TV
5.1.4 Other

5.2 Transportation
5.2.1 Guided motor vehicles
5.2.2 Rail mass transit
5.2.3 Aircraft
5.2.4 Other

Analysis focuses upon a variety of economic units, including individuals, families, ethnic and racial groups, for-profit organizations, non-profit private organizations, and all levels of government. Furthermore, careful attention is given to analysis of the changes in the production and consumption of goods and services over time.

Demographic — Analysis of demographic societal indicators, as they are derived from economic, social and political factors, provides insights into the characteristics of human and institutional populations. Of particular interest here is size of units, whether families or corporations; their physical distribution; density and distribution of population; and "vital statistics," such as mortality and morbidity rates, birth rates, and marriage and divorce rates.

Social — Analysis of social indicators would identify the various societal roles of individuals, groups, and institutions. These social units are defined in a variety of ways including economic status, ethnic or racial heritage, type of industry, location, family size, and constituency.

Role is defined as each unit's expectations concerning future types of behavior, that of themselves and that of others, and is to be viewed in terms of each unit's decisions and actions that are designed to bring about the desired future.

Political and Legal — These indicators are defined in rather narrow terms, focusing upon the processes of federal, state, and local government. Consideration is to be given to programs which are now in effect, administrative and legislative proposals, and possible new programs.

Technological — Here, all technological possibilities would be considered. This includes technologies of direct use and competing and complementary technologies.

Ecological — Here, focus is upon physical, chemical, and biological perturbations of the biosphere.

The above examples are illustrative, and would not apply in the same way to every organization.

A review of these general definitions and the illustrations provided in Exhibit 2 should, it is hoped, result in understanding that societal indicators represent those beneficiaries and benefactors which are to be considered in the analysis of technological impacts. Both the existence and absence of indicators *per se* will flavor the analysis. For example, the Exhibit 2 listing identifies certain social characteristics of families, but makes no reference to race or country-of-origin. This means, quite simply, that racial and country-of-origin characteristics *cannot* directly be considered in the process of technology assessment which uses these indicators. They will be considered implicitly only as they relate to other social and economic conditions. No general statement of the relative policy implications of this can be made, as they are dependent upon the nature of the problem or technology being assessed. For example, effective public school programs certainly cannot be developed unless racial and country-of-origin characteristics are considered directly. However, a clothing manufacturer's plans to shift from metal to plastic zippers probably do not have to directly consider minority group composition and values.

Each of the societal indicators represents a specific set of interests — people, organizations, social and economic classes, and special interest groups — who, it is assumed, will either benefit or suffer from a new program. An estimate of what actually will happen is the purpose of the technology assessment process.

POSSIBLE IMPLICATIONS OF TECHNOLOGY ASSESSMENT

The brief description above of incorporating societal indicators into technology assessment ends rather abruptly; a detailed discussion of methodology here would not contribute significantly to a policy or decision maker's decision to utilize some form of technology assessment. Methodology is discussed in detail elsewhere in this book, and does not need to be repeated. Further methodological developments would focus primarily upon development of appropriate feedback loops, the refinement of goals and indicators, and the process of weighting. Consideration here of some of the possible future implications of technology assessment may be of more immediate use to policy and decision makers, assuming that the discussion above gives some insight into the analytical problems associated with technology assessment. An understanding of some of these implications would make it possible to consider rationally what level of commitment to make to the use of technology assessment. Possible commitments include:

- Develop organization-specific technology assessment programs.

- Encourage development of technology assessment programs to which a number of organizations must be responsive.

- Encourage the development of a national technology assessment program.

- Discourage or oppose the development of technology assessment programs.

- Do nothing.

Technology assessment is suggestive of a new market place, or at least new market forces. These new market forces, probably, would be a combination of relatively free price-product competition in an open market and government controls. The open market component could result in a more aggressive offering of new goods and services at different prices in order to capitalize upon new opportunities. Also, measures to prevent external consequences (e.g., pollution) could be instituted in response to predicted future constraints (e.g., fines for polluting).

This open market component, however, could also result in more restrictive practices. Instead of a more aggressive offering of new goods and services with price competition, agreements among potential competitors to restrict these offerings could be made based upon this more thorough knowledge of future possibilities. Agreements could be made to produce only those goods and services which provide a certain profit level with few risks. Another restriction would be seen in early lobbying against possible operating constraints (e.g., pollution controls or higher taxes) before there are any concerted efforts to establish such constraints. So, technology assessment, as practiced by those who operate in the open market, could also be used to establish more restrictive practices.

The use of technology assessment by government would result in new controls, here defined as compensation, direct program support, and punishment. Compensation represents the necessary redistribution mechanisms for, first, removing resources from those who are shifting their costs to others and then, distributing these resources back to those bearing these costs. This compensation mechanism would remove or collect the resources in the form of taxes or fines and redistribute them in the form of direct payments (e.g., compensation to fishermen who have lost their fishing grounds due to pollution) or facilities (e.g., sewage treatment plant). This suggests direct accountability by all organizations and individuals.

Direct program support is not compensation or redistribution. Here, decisions are made concerning new government programs — defense, foreign affairs, domestic affairs, space exploration, medical research, and so on. Taxes are required, but the resources are not allocated back to those who are being "hurt" by those taxed.

Punishment can be thought of in dollar terms (taxes or fines) or criminal sanctions (jail terms).

The degree to which these government programs would increase the effectiveness of technology and minimize its detrimental consequences would be largely contingent upon the visibility of the technology assessment process. If the technology assessment process was a confidential executive or legislative staff operation, there would be no guarantee that final resource allocation decisions would be adapted to established political relationships and bureaucratic processes. If, however, the technology assessment process was visible and appeared to be reliable, entirely new special interest coalitions (e.g., more Ralph Nader-led operations) might form to either oppose or support stated or proposed public policy. The result could be change through the established political processes or, perhaps, more radical forms of change.

A technology assessment program, then, should identify the many impacts of technology, and measure them with reference to both the goals of a specific organization and society and future opportunities and constraints. Any discussion of technology assessment usually implies the development of some kind of an external framework of societal indicators to which all major public and private decisions must be sensitive. The nature of this framework, to be effective, must provide the many beneficiaries and benefactors of technology with the freedom to bargain with the proponents and opponents of technology. As mentioned above, this bargaining process will be some combination of political, legal, and free market processes. (Some might wish to include anarchy and revolution among these processes. Perhaps.) Although such a bargaining process is not to be constructed here, its two most vital characteristics are that it provides an effective forum for complaining and a mechanism for the redress of alleged wrongs.

This discussion of technology assessment could easily bog down in arguments and meanderings concerning the selection of societal indicators — are beneficiaries and benefactors properly accounted for? Rather than meet this problem with maudlin statements about orthogonality and completeness, it is more useful to recognize that all data, for practical purposes, are biased according to the proponents and opponents of technology. This is not inherently bad, neither does it detract from the efficacy of the analysis. However, the existence of biases must be understood, and some attempt made to identify them. Technology assessment involves, for the most part, the quantification of subjectiveness. This means that the analytical results cannot be attributed only to technology assessment as it is a "scientific approach," but also to those who participate in the process. Perhaps some new form of accountability is warranted: technology assessors who guess rightly are rewarded, those who guess wrongly are punished.

Technology assessment is a form of a contemporary rationality which is posited on the belief that human beings have the ability to draw up rational prescriptions for changing society. The failures of the 1960's to apply rationality to areas of public concern can, it is suggested here, be more attributable to the "rational prescribers" than the methodologies they used. More often than not, an elaborate methodology was used to promote predetermined notions and beliefs. If technology assessment, considered here in the broadest context of examining the impacts of any innovation, incorporates a wide range of societal indicators, total program effectiveness should be more accurately measured. (This does not guarantee, however, that it is possible to develop more effective programs. An honest assessment of programs would probably cause much dismay and disbelief.)

The scale of technology assessment which is suggested here almost necessarily leads to consideration of some concept of "social health." The panel of experts which produced *Toward A Social Report* under HEW sponsorship seemed to agree that it was developing something preliminary to a "... procedure for periodic stock taking of the social health of the nation" [13]. Implicit here is the assumption or belief that it is possible for an individual or group to first, understand and second, monitor quality-of-life considerations. If quality-of-life considerations are to be incorporated into any analysis — whether by the government or a large corporation — societal indicators will have to go far beyond measures such as health, rooms per family, pollution, poverty, education, and law enforcement. Generally speaking, these indicators as conventionally measured show general improvement over the past 10 years. It is not at all clear that the quality-of-life or social health is better.

If contemporary rationality is to advance into the 1970's and 1980's, a new form of data is required which has not been incorporated into policy analyses of any form. The notion of expectations must be considered. Social health, or the quality-of-life, consists of some balance between material conditions and expectations. It is necessary to have some understanding of what an individual, group, or organization both has and wants.

Serious attempts to quantify social health or quality-of-life have only recently begun. Here, new variables have to be defined for possible inclusion into a list of societal indicators. Examples of these new quality-of-life indicators which generally deal with expectations might include: sex, sense of participation, sense of power over future events, friendships, access to higher social-economic levels, visibility, and a degree of awe inspired in members of peer groups. How these quality-of-life indicators are to be obtained and incorporated into a list of societal indicators is an open question. The most obvious approach would combine some form of polling, best guessing, and common sense.

The process of technology assessment can, alone, have significant impact upon society. The actual impacts of technology assessment, however, cannot be discussed here in specific terms, as they depend largely upon the form and nature of the assessment process. Possible impacts, however, can be seen as extremes: technology assessment programs could result in more responsive or restrictive corporate practices, increases or decreases in accountability, little or radical change in the form of government, and new coalitions of special interest groups or the *status quo*. Technology assessment has its risks, as does any enterprise. Fortunately, the technology assessment process itself, if applied properly, serves to identify these risks.

REFERENCES

1. BAUMOL, William, J., *Economic Theory and Operations Analysis*,(Englewood Cliffs, New Jersey, Prentice-Hall, 1962), 267-268.

2. U.S. HOUSE OF REPRESENTATIVES, Subcommittee on Science, Research and Development of the Committee on Science and Astronautics, *Hearings, Technology Assessment*, 91st Congress, 1st Session (1969), p. 1.

3. PARKINSON, C. NORTHCOTE, in one of his Studies in Administration, "High Finance or the Point of Vanishing Interest," explains this penhomenon well. Here, he discovers the Law of Triviality which, simply stated, is: the time spent on any agenda item will be in inverse proportion to the sum of the dollars involved.

   This new law of administration is shown by observing that a corporate finance committee agreed to spend $10,000,000 on a reactor in less than 5 minutes. However, the $2,000 bicycle shed for employees required deep deliberation over a period of 45 minutes. See: C. Northcote Parkinson, *Parkinsons Law*, (New York: Ballatine Books, 1957), 39-50.

4. See MESTHENE, EMANUEL, in U.S. House Representatives, *op. cit,* p. 240.

5. U.S. HOUSE OF REPRESENTATIVES, *op. cit.*

6. BARTOCHA, B., NARIN, F. and STONE, C.A., "TRACES-Technology in Retrospect and Critical Events in Science"; *The Science of Managing Organized Technology*, CETRON, M. and GOLDHAR, J., (New York, Gordon and Breach, 1971); U.S. Department of Defense, *Project Hindsight,* October 1966; U.S. Arms Control & Disarmament Agency, *Technological Innovation in Civilian Public Areas*, September 1967; U.S. Department of Defense, *An Exploratory Study of the Structure and Dynamics of the R & D Industry*, June 1964; Organization for Economic Cooperation and Development, *Gaps in Technology* (Paris, 1970).

7. U.S. HOUSE OF REPRESENTATIVES, *op. cit.,* p. 254.

8. Two of the more direct attempts at developing a general methodology with wide-spread applicability are described in: "GNP-T (Goals for National Planning

and Technology)," Joint Institute of Management Sciences and World Futures Society, Symposium on Assessing the Future and Policy Planning, March 9, 1970, Gaithersburg, Maryland, also published as "A Method for Integrating Goals and Technological Forecasting into Planning," in *Technological Forecasting, 2*, No. 1, 1970.

CETRON, M., "A Micro Technology Assessment of the Printing and Publishing Industry," presented at American University, February 1971, and to be published in *Technological Change in Printing and Publishing Management*, Hattery & Bush, eds., Spartan Books.

9. For excellent reviews of forecasting techniques see: AYRES, ROBERT U., *Technological Forecasting and Long-Range Planning*,(New York, McGraw-Hill, 1969); CETRON, M., *Technological Forecasting: A Practical Approach*,(New York, Gordon and Breach, 1969); James R. Bright, ed., *Technological Forecasting for Industry and Government*,(Englewood Cliffs, New Jersey, Prentice-Hall, 1968); and CETRON, M. and RALPH, C., *Industrial Applications of Technological Forecasting: Its Use in R & D Management*,(New York, John Wiley & Sons, Inc., 1970).

10. CETRON, M., *et al.*, *Technological Resource Management, Quantitative Methods*,(Cambridge, Mass., MIT Press, 1970).

CETRON, M. and GOLDHAR, J., *The Science of Managing Organized Technology*,(New York, Gordon and Breach, 1971).

GORDON, T.J. and HAYWARD, H., "Initial Experiments with the Cross-Impact Matrix Method of Forecasting," *Futures, 1*, No. 2, December, 1968.

GORDON, T.J., "Cross-Impact Matrices," *Futures, 1*, No. 6, December, 1968.

DALBY, J., "Practical Refinements to the Cross-Impact Matrix Technique of Technological Forecasting," CETRON, M. and RALPH, C., *Industrial Applications of*

Technological Forecasting: Its Use in R & D Management,(New York, John Wiley & Sons, Inc., 1971).

RALPH, C., "Decision Impact Analysis Applied to the Fishing Industry," (Plessey Radar, Ltd., Report, November 1968).

RALPH, C., "The Beginnings of Cross-Support Analysis," CETRON, M. and RALPH, C., *Industrial Applications of Technological Forecasting: Its Use in R & D Management*,(New York, John Wiley & Sons, Inc., 1971).

RALPH, C., "Resource Allocation Using Cross-Support Analysis," presented at the 27th JSGOMRAM, Institute of Defense Analyses, Arlington, Virginia, April 20-22, 1971.

11. U.S. HOUSE OF REPRESENTATIVES, *op. cit.* Also, National Academy of Sciences, "Report to the Committee on Science and Astronautics — Report to the U.S. House of Representatives," *Applied Science and Technological Progress*,(Washington, D.C., June 1967).

12. MENDELL, JAY S., "Futuristics — An Approach to Thinking About the Future," in *Industrial Applications of Technological FOrecasting: Its Use in R & D Management*, CETRON, M. and RALPH, C., (New York, John Wiley & Sons, Inc., 1971).

13. U.S. DEPARTMENT OF HEALTH, EDUCATION, and WELFARE, *Toward A Social Report*,(Washington, D.C., 1969).

TECHNOLOGY ASSESSMENT; SOME ECONOMIC ASPECTS

The Contribution of Research and Development to Economic Growth in the United States*

EDWIN MANSFIELD

Professor
Wharton School, University of Pennsylvania, Philadelphia, Penna., U.S.A.

I. INTRODUCTION

Technological change is clearly an important factor in economic growth, both in the United States and in other countries, both now and in the past. In recent years — after neglecting the study of technological change for a long time — economists have shown a considerable interest in examining the relationship between research and development (R & D) on the one hand, and the rate of economic growth and productivity increase, on the other. In addition, there have been a number of discussions of whether we as a nation are underinvesting in certain kinds of research and development. Some prominent economists believe that this is the case.

At the outset, two important points should be noted. First, by focusing attention on the economic effects of R & D, we are not implying that only these effects of R & D are important or relevant. On the contrary, increased knowledge is clearly of great importance above and beyond its strictly economic benefits. However, our assignment in this paper was to confine our attention strictly to economic matters. Second, by looking at our Nation's rate of economic growth and productivity increase, we are not assuming explicitly or implicitly that economic growth is,

---

*Reprinted from *A Review of the Relationship Between Research and Development and Economic Growth/Productivity* Published by the Office of Economic and Manpower Studies of the National Science Foundation, February 1971.

in some simple sense, what public policy should attempt to maximize. Clearly, the desirability of a particular growth rate depends on the way it is achieved, how the extra production is distributed, how growth is measured, and many other things. There is, of course, a large literature on the benefits and costs of economic growth.

II. RESEARCH AND DEVELOPMENT AND ECONOMIC GROWTH

To begin with, let's consider the relationship between R & D and economic growth. The pioneering studies in the area — by Robert Solow [1], Moses Abramowitz [2], and Solomon Fabricant [3] — occurred in the mid-fifties. In many respects, Solow's paper was most influential. Assuming that there were constant returns to scale, that capital and labor were paid their marginal products, and that technological change was neutral, Solow attempted to estimate the rate of technological change for the American nonfarm economy during 1909-49. His findings suggested that, for the period as a whole, the average rate of technological change was about 1.5 percent per year. In other words, the amount of output that could be derived from a fixed amount of inputs increased at about 1.5 percent per year.

Based on the findings, he concluded that about 90 percent of the increase during the period in output per capita was attributable to technological change, whereas only a minor proportion of the increase was due to increases in the amount of capital employed per worker. This conclusion received a great deal of attention — and caused some consternation among economists who had focused much more attention on the factors underlying the amount of capital employed per worker than on those underlying the rate of technological change. A flurry of papers followed Solow's seminal article, each modifying his techniques slightly or using a somewhat different data base.

After the first wave of papers in the mid-fifties, investigators began to feel increasingly uneasy about the basic methodology used in these studies. In essence, this methodology was the following: Economists, who view the total output of the economy as being due to various inputs of productive services into the productive process, began by specifying these inputs as labor and capital and by attempting to estimate the contribution of these inputs to the measured growth of output. Then, whatever portion of the measured growth of output that could not be explained by these inputs was attributed to technological change. The crudeness of this procedure is obvious. Since the effect of technological change is equated with whatever increase in output is unexplained by other inputs, the resulting measure of the effect of technological change does not isolate the effects of technological change alone. In addition, it contains the effects of whatever inputs are excluded — which, depending on the study, may be economies of scale, improved allocation of resources, changes in product mix, increases in education, or improved health and nutrition of workers.

To remedy some of these limitations, a number of additional studies were carried out in the early sixties, the most comprehensive and influential being by Edward Denison [4]. Denison attempted to include many inputs — particularly changes in labor quality associated with increases in schooling — that were omitted, largely or completely, in earlier studies. Since it was relatively comprehensive, his study resulted in a relatively low residual increase in output unexplained by inputs included in his study. Specifically, Denison concluded that the "advance of knowledge" — his term for the residual — was responsible for about 40 percent of the total increase in national income per person employed during 1929-57.

Of course, technological change can stem from sources other than organized research and development, as evidenced by John Jewkes's findings [5] concerning the importance of independent inventors as a source of major inventions, and the findings by Hollander [6], and others, concerning the importance of technological changes that depend in no significant way on formal research and development. Denison estimates that about one-fifth of the contribution to economic growth of "advance of knowledge" in 1929-57 can be methodology was the following: Economists, who view the total output of the economy as being due to various inputs of productive services into the productive process, began by specifying these inputs as labor and capital and by attempting to estimate the contribution of these inputs to the measured growth of output. Then whatever portion of the measured growth of output that could not be explained by these inputs was attributed to technological change. The crudeness of this procedure is obvious. Since the effect of technological change is equated with whatever increase in output is unexplained by other inputs, the resulting measure of the effect of technological change does not isolate the effects of technological change alone. In addition, it contains the effects of whatever inputs are excluded — which, depending on the study, may be economies of scale, improved allocation of resources, changes in product mix, increases in education, or improved health and nutrition of workers.

To remedy some of these limitations, a number of additional studies were carried out in the early sixties, the most comprehensive and influential being by Edward Denison [4]. Denison attempted to include many inputs — particularly changes in labor quality associated with increases in schooling — that were omitted, largely or completely, in earlier studies. Since it was relatively comprehensive, his study resulted in a relatively low residual increase in output unexplained by inputs included in his study. Specifically, Denison concluded that the "advance of knowledge" — his term for the residual — was responsible for about 40 percent of the total increase in national income per person employed during 1929-57.

Of course, technological change can stem from sources other than organized research and development, as evidenced by John Jewkes's findings [5] concerning the importance of independent inventors as a source of major inventions, and the findings by Hollander [6], and others, concerning the importance of technological changes that depend in no significant way on formal research and development. Denison estimates that about one-fifth of the contribution to economic growth of "advance of knowledge" in 1929-57 can be attributed to organized research and development. But this is the roughest kind of guess. Denison would be the first to admit that this estimate is based largely on conjecture.

III. FUNDAMENTAL PROBLEMS OF MEASUREMENT

How firmly based is the current state of the art this area? In other words, how reliable are the estimates of the contribution of R & D to economic growth in the United States? I have already indicated some of the difficulties present in these estimates. Unfortunately, there are a number of additional problems of a fundamental nature that must be understood as well. First, the measured rates of growth of output on which these estimates are based suffer from a very important defect, particularly for present purposes, because, to a large extent, they fail to give proper credit and weight to improvements in the quality of goods and services produced, and these improvements are an important result of research and development. For example, the growth rate would have been the same whether antibiotics were developed or not, or whether we devoted the resources used to reach the moon to public works. In general, only those changes in technology that reduce the costs of end products already in existence have an effect on measured economic growth. Unfortunately, the measured growth of national income fails to register or indicate the effects on consumer welfare of the increased spectrum of choice arising from the introduction of new products.

Second, the models on which these estimates are based may not recognize the full complexity of the relationships among the various inputs. In particular, as Nelson, Peck, and Kalachek [7] have pointed out, if the returns to some input are dependent on the rate of technological change and this is not recognized explicitly, some of technology's contribution to economic growth will be attributed incorrectly to other inputs. This may be the case with education, since the returns to education would probably have been less if technological change had occurred at a slower pace. It may also be the case with "the reallocation of resources," a factor sometimes used to explain part of the residual increase in output.

Third, it is not clear how one can get from an estimate of the contribution to economic growth of technological change (or advance of knowledge, in Denison's terms) to an estimate of the contribution to economic growth of research and development. Clearly, there is no reason why these two estimates should be the same; on the contrary, one would expect the latter estimate to be smaller than the former. But the estimate that results from the models discussed in the previous section is the former estimate, not the latter — which is the one we want. As pointed out in the previous section, Denison does make an attempt to derive the latter estimate from the former, and to do so, he is forced to make extremely rough assumptions. To a certain extent, numbers must simply be pulled out of the air.

Fourth, there are difficulties in measuring inputs, the measurement of aggregate capital being a particularly nettlesome problem. Since errors in the measurement of inputs will result in errors in the estimated contribution of these inputs to economic growth, these errors will also be transmitted to, and will affect, the residual unexplained increase in output, which is used to measure the contribution of technological change to economic growth. Also it is difficult to adjust for quality changes in inputs, and there are problems in constructing proper price deflators. According to recent work by Jorgensen and Griliches [8] there are important measurement errors and errors of aggregation in the measures that are ordinarily used; and these errors inflate the residual.

Fifth, there are difficulties caused by the fact that much of the Nation's R & D is devoted to defense and space purposes. For example, some observers note the tremendous increase in R & D expenditures in the postwar period and conclude that, because productivity has not risen much faster in the United States than before the war, the effect of R & D on economic growth must be very small. What these observers forget is that the bulk of the Nation's R & D expenditures has been devoted to defense and space objectives and that the contribution of such expenditures to economic growth may have been limited. Moreover, they fail to realize that improvements in defense and space capability per dollar spent caused by military and space R & D will not show up in our measures because government output is valued at cost. (Also, they fail to recognize the fact that product improvements and new products often fail to register in our output measures and that the effects of R & D often occur with a considerable lag.) More will be said on this score in later sections of this report.

Based on this catalog of problems and limitations, it is clear that the current state-of-the-art in this area is not strong enough to permit us to make very accurate estimates of the contribution of R & D to the economic growth of the United States. At best, the available estimates are rough guidelines. In no sense is this a criticism of the economics profession or of the people working in this area. On the contrary, a great deal of progress has been made since the pioneering ventures into this area a little over a decade ago. Given the small number of people working in this area and the inherent difficulty of the problem, it is difficult to see how much more could have been achieved.

IV. RESEARCH AND DEVELOPMENT AND PRODUCTIVITY INCREASE IN INDIVIDUAL INDUSTRIES

Next, let's turn to the relationship between R & D and the rate of productivity increase in individual industries. During the late fifties, important work was going on at the National Bureau of Economic Research concerning the rate of productivity increase in various industries, this project culminating in John Kendrick's book [9]. As part of this work, N. Terleckyj [10] carried out a study of the relationship between an industry's rate of increase of total factor productivity during 1919-53 and various industry characteristics. According to his results, an industry's rate of growth of total factor productivity was related in a statistically significant way to the amount spent on research and extension. Moreover, the regression coefficient of this variable remains remarkably stable when cross sections are deleted or added, and when the specification of the model is changed somewhat.

The manufacturing study, by Edwin Mansfield [12], was based on data regarding ten large chemical and petroleum firms and ten manufacturing industries in the postwar period. Both for firms and for industries, the measured rate of productivity change was related in a statistically significant way to the rate of growth of cumulated R & D expenditures made by the firm or industry. The specific form of the relationship depends somewhat on whether technological change is assumed to be disembodied (better methods and organization that improve the efficiency of both old capital and new) or capital embodied (innovations that must be embodied in new equipment if they are to be utilized). If technological change was disembodied, the average effect of a 1 percent increase in the rate of growth of cumulated R & D expenditures was a .1 percent increase in the rate of productivity increase. If technological change was capital embodied, it was a .7 percent increase in the rate of productivity increase.

In addition, Jora Minasian [13] studied the relationship between value added, on the one hand, and labor, capital, and cumulated R & D expenditures, on the other, in 17 firms in the chemical industry during 1948-57. (His paper is a continuation of work started in his doctoral dissertation [14].) In all but one of the specifications of the model tried by Minasian, a firm's cumulated R & D expenditures are related in a statistically significant way to the firm's value added, holding its labor and capital inputs constant. Moreover, his estimate of the regression coefficient for cumulated R & D expenditures is strikingly close to the result obtained by Mansfield. Thus, the findings of the two studies trend to reinforce one another.

Finally, Murray Brown and Alfred Conrad [15] carried out a study of the relationship between R & D expenditures (as well as education and other variables) and productivity increase in a number of U.S. manufacturing industries in the postwar period. Their results, published in 1967, indicated that R & D expenditures had a statistically significant effect on the rate of productivity increase. Also, in their judgment, their findings indicate that a given percentage increase in R & D expenditures in durable goods industries produces a substantially larger percentage increase in productivity than does the same percentage increase in R & D expenditures in nondurable goods industries.

V. EVALUATION OF PRODUCTIVITY STUDIES

How reliable are these estimates of the relationship between R & D and productivity increase in individual industries? Clearly, one advantage of these studies over those described in section II. is that the effect of R & D is not derived indirectly as a residual. Instead, an industry's — or a firm's or area's — R & D expenditures are introduced as an explicit input in the productive process. Thus, it is possible to obtain explicit relationships between R & D and productivity increase: We no longer have to attribute to technology or R & D whatever we cannot explain by other factors. This is a real advantage.

But a number of important problems remain. First, too little is known about the characteristics of the activities that firms call "research and development." This lack of information has been a hindrance to progress in this area, since, without a reasonable amount of information on this score, it is difficult to interpret or evaluate models relating R & D expenditures to other economic variables. Clearly, if the figures on "research and development" contain routine technical services and other such activities, the estimates based on these figures will be affected. It is difficult to tell how important this problem is, but for some purposes, I would guess it to be a serious problem.

Second, even if one were sure that the R & D figures were reliable, there would still be the possibility of spurious correlation. Firms and industries that spend relatively large amounts on research and development may tend to have managements that are relatively progressive and forward looking. To what extent is the observed relationship between R & D and productivity increase due to this factor rather than to R & D? Obviously, this is difficult to answer since the quality of management is very difficult to measure. Nonetheless, most investigators seem to feel that only a small part of the observed relationship is due to spurious correlation of this sort.

Third, a large percentage of the R & D carried out by many industries is directed at productivity increase in *other* industries. Consequently, relationships between R & D in an industry or firm and productivity increase in the *same* industry or firm catch only part of the effects of R & D. Unfortunately, too little effort has been directed at introducing interindustry or interfirm flows of technology into the sorts of models that underlie these relationships. Also, the estimates that are obtained depend on the extent of the lag between the time when R & D is carried out and the time when the effects of R & D show up in productivity indexes. Clearly, this lag is often substantial. Unfortunately, the models on which these estimates are based often make very crude assumptions concerning the length of the lag.

Fourth, there are a host of technical problems. To what extent is technological change disembodied, and to what extent is it capital embodied? If R & D is treated as investment in new knowledge — as it is in most of these studies — what depreciation rate should be used? Also, there is the perennial problem of how R & D expenditures should be deflated, as well as the problem of the form of the production function that should be used in particular cases. The answer one gives to these questions can have a significant effect on the estimates one obtains [16]. However, none of these problems is entirely resolved, although some work has been devoted to the deflation problem and to the form of the production function.

Fifth, studies of the relationship between R & D and productivity increase in individual industries suffer, of course, from a number of the same problems that beset studies of the contribution of R & D to economic growth. Some of these problems, discussed in more detail in section III, are inadequacies of the output measures used, poor specification of the relationship among inputs, and difficulties in measuring inputs.

Based on this discussion of the problems in the existing estimates of the relationship between R & D and productivity increase in individual industries, it is clear that the current state-of-the-art in this area is not strong enough to permit definitive estimation of these relationships. Nonetheless, although the results are subject to considerable error, they establish certain broad conclusions. In particular, existing econometric studies do provide reasonably persuasive evidence that R & D has a significant effect on the rate of productivity increase in the industries and time periods that have been studied.

VI. EXTERNALITIES, RISKINESS, AND INVESTMENT IN RESEARCH AND DEVELOPMENT

At this point, let's turn to the question of whether or not, from a purely economic point of view, the United States is underinvesting in research and development. Certain propositions bearing on this question are widely accepted by economists and should be set forth at the beginning of this discussion. The first proposition is that, because the results of research are often of little direct value to the sponsoring firm but of great value to other firms, there is good reason to believe that, left to its own devices, the market would allocate too few resources to R & D — and that the shortfall would be particularly great at the more basic end of the R & D spectrum. The reason for this is fairly obvious: the market operates on the principle that the benefits go to the person bearing the costs, and vice versa. If a firm or individual takes an action that contributes to society's welfare but it cannot appropriate the full gain, it obviously is less likely to take this action than would be socially desirable.

The second proposition is that, because research and development is risky for the individual firm, there is good reason to believe that the market, left to its own devices, would allocate too few resources to R & D. Of course, the risk to the individual investor in R & D is greater than the risk to society, since the results of the R & D may be useful to someone else, not to himself, and he may be unable to obtain from the user the full value of the information. Because the economic system has limited and imperfect ways to shift risks, there would be an underinvestment in R & D. For this reason too, one would expect the underinvestment to be greatest at the more basic end of the R & D spectrum [17].

These defects of the market mechanism in allocating resources to research and development have long been recognized. For example, A.C. Pigou, the well-known English economist, set some of them forth quite clearly in the twenties [18]. Moreover, they have been recognized in the realm of practical affairs and of social organization as well as in the realm of social science. Our society, taking account of these defects of the market mechanism, does not depend exclusively on the market for an investment in R & D. On the contrary, a very large proportion of the Nation's expenditures on R & D stems from government agencies, private foundations, and universities, all of which supplement the R & D supported through the market mechanism. Thus, the relevant question is not whether the market mechanism requires supplementing, but whether the type and extent of supplementary support provided at present is too large or too small, and whether it is allocated properly.

VII. SALIENT CHARACTERISTICS OF THE NATION'S INVESTMENT IN RESEARCH AND DEVELOPMENT

Before discussing various economists' judgments concerning the question posed at the end of the last section, several important characteristics of the nation's investment in research and development must be noted. First, as is well known, the government's investment in R & D is focused very strongly on defense and space technology.

During the early sixties, over 55 percent of the Federal Government's R & D expenditures were for these purposes. With the passage of time, this percentage has decreased, but even in 1970, about 43 percent of the Federal Government's investment in R & D were for these purposes [19]. The relevance to economic growth of much of this huge investment in defense and space R & D has been questioned by many economists.

Numerous groups within the government — an early example being the White House Panel on Civilian Technology — have been interested in the extent of the benefits to civilian technology — the "spillover" or "fallout" — from military and space R & D. Obviously, the extent of this "spillover" has implications regarding the extent to which the investment in defense and space R & D has relevance for economic growth. It is perfectly clear that the value of the "spillover" that has occurred in the past has been substantial, the computer, numerical control, integrated circuits, atomic energy, and many other significant advances having stemmed at least partly from military R & D. However, it is also clear that the contribution of a dollar of military and space R & D to economic growth is considerably less than the contribution of a dollar of civilian R & D. Moreover, in the opinion of some observers, the spillover per dollar of military-space R & D is unlikely to be great as in the past, because the capabilities that are being developed and the environment that is being explored are less closely connected with civilian pursuits than in the past.

Second, just as the government's expenditures on R & D are concentrated largely in a few agencies (DOD, NASA, and AEC) with defense and space missions, so industry's expenditures on R & D are concentrated in a few industries. In 1969, 82 percent of all industrial R & D expenditures took place in only five industries — aerospace, electrical equipment and communication, chemicals (including drugs), machinery, and motor vehicles. Of course, this concentration is due in part to the fact that these industries perform a great deal of R & D for the Federal Government. But if we look only at company-financed R & D, this concentration is nearly the same, these five industries accounting in 1969 for 75 percent of all company-financed R & D expenditures. Moreover, this concentration seems to be increasing [20].

Industry's R & D expenditures are also concentrated largely on products, not processes. For example, according to a survey of business firms carried out in the early sixties, about 47 percent of the firms reported that their main purpose was to develop new products and about 40 percent reported that it was to improve existing products; only 13 percent reported that it was to develop new processes [21]. However, lest there be misunderstanding, it should be recognized that one industry's products may be part of another industry's processes. Thus, when a machinery producer improves its products or when a chemical producer improves its products, this may result in a process improvement in industries that buy and use the machinery or chemicals.

Third, our Nation's investment in R & D is focused very strongly on development, not research. The distinction between research and development, although hazy and indistinct in some cases, is important. Research is aimed primarily at the search for new knowledge, whereas development is aimed at the reduction of research findings to practice. In 1970, according to estimates made by the National Science Foundation, about two-thirds of the Nation's investment in R & D went for development, only about one-third went for research [22]. Much of the development work carried out by industry and government is aimed at very specific objectives and involves large expenditures on prototypes and pilot plants. It is important to avoid the (unfortunately common) mistake of confusing this activity with research.

Moreover, it is important to recognize that much of the R & D carried out by industry is aimed at fairly modest advances in the state-of-the-art. Studies carried out by Hamberg [23], Jewkes [24], and others seem to indicate that the really major inventions seldom stem from industrial laboratories of major firms, which are primarily contributors of minor "improvement" inventions.

Also, surveys indicate an emphasis of firms on relatively short payout periods for R & D, this emphasis being another indication that most R & D carried out by the responding firms is aimed at improvements or minor changes in existing products [25]. In addition, detailed studies of the characteristics of the R & D portfolio of a number of industrial laboratories by Mansfield [26], Meadows [27], and others provide direct evidence that the bulk of the work involves rather small technical risks.

VIII. RECENT JUDGMENTS CONCERNING THE ADEQUACY OF THE NATION'S INVESTMENT IN RESEARCH AND DEVELOPMENT

Let's return now to the question posed at the end of section VI: Is the type and extent of R & D support that society presently uses to supplement the market mechanism adequate from an economic point of view, and is this support allocated properly? In recent years, there have been several prominent discussions of this question, each carried out by people with exceptional qualifications who have devoted considerable time and energy to this task. The purpose of this section is to summarize their views. In the following sections, we shall discuss the evidence underlying their conclusions.

In 1963, the Organization for Economic Cooperation and Development published a report by Christopher Freeman, M. Raymond Poignant, and Ingvar Svennilson which concluded that: "It seems therefore inherently improbable that the scale on which governments supplement civil R & D in sectors other than atomic energy is anything like sufficient to attain the optimum [28]. In their view, "in spite of all the factors which concur to increase the level of R & D activity, there are serious reasons for believing that this level is in many cases inadequate for sustained and rapid economic growth" [29]. In 1964, the U.S. Council for Economic Advisors sounded a similar note when they stated that "in a number of industries the amount of organized private research undertaken is insignificant, and the technology of many of these low-research industries has notably failed to keep pace with advances elsewhere in the economy" [30].

In 1966, the President's Commission on Technology, Automation, and Economic Progress concluded that too little was being spent by the government on R & D in the fields of urban transportation, pollution control, and housing. For example, in housing the Commission stated that: "As it has in agriculture, the Federal Government should actively stimulate research in housing and community development through research grants and through its own building activities. It should also support basic research to establish performance criteria (e.g., moisture resistance, insulation, lighting, etc.) for housing and housing components" [31].

In 1967, Richard Nelson, Merton Peck, and Edward Kalachek, in a book summarizing a two-year study of this question, concluded that there were several important areas where a significant degree of market failure existed which was not remedied adequately by existing government programs. They suggested that a National Institute of Technology be established to provide grants for research and development aimed at placing the technology of various industries on a stronger scientific base and to test the feasibility and desirability of advanced designs. In their view, work of this sort, which falls between basic research and product development, is in need of additional support. Also, in cases where a broad-scale systems view is needed but is prevented by the smallness of firms and fragmentation of markets, the Institute would support work through the middle and later stages of development [32].

At about the same time, William Capron, formerly Assistant Director of the Bureau of the Budget and an economist long interested in this area, stated that: "My own view on this is that we can say nothing with much confidence on theoretical grounds about the social adequacy of our total R & D effort — though my hunch is that as a nation we are underinvesting in R & D over all. However, I think we can say with assurance that the existence of noncompetitive elements and sectors of the economy produces a misallocation of resources within the R & D total" [33]. Specifically, it is Capron's view that too large a fraction of the total R & D effort is spent in oligopolistic industries and on relatively modest improvements, too little being spent in more competitive industries and on more far-reaching work.

IX. NATURE OF THE EVIDENCE

In the previous section, we summarized briefly the conclusions of a number of economists who have been concerned with the question of whether or not the R & D support that society presently uses to supplement the market mechanism is adequate in total and allocated properly. They generally seem to be of the opinion that the Nation's investment in R & D may be too small, but this opinion is often characterized as little more than a hunch. They are much more confident, it appears, that, whether or not the total investment in R & D is too small, this investment is not properly allocated, there being too little R & D devoted to more ambitious attempts to place the technology of various industries on a stronger scientific base (Nelson, Peck, and Kalachek), to urban transportation, pollution control, and housing (Automation Commission), and to more competitive and fragmented industries (Capron). Or, more precisely, this is what they believed at the time they expressed their views in print.

What sorts of evidence are these conclusions based on? First, some of these studies rely largely on judgment combined with economic theory. For example, Nelson, Peck, and Kalachek, who lean heavily on this kind of support, believe that:

> While the present state of knowledge is not strong enough to permit derivation of quantitative rates of return, or optimal allocations of resources, it is strong enough to suggest that for certain kinds of activities there are serious market imperfections... When there are significant external economies, unsupplemented private initiative is unlikely to support work to the extent that is socially optimal. Where government policies already exist which provide added incentive or reduce private costs, or which supplement effort directly, it is difficult to say whether the latent tendency toward under-

> allocation of private effort has been compensated. However, where policies do not exist, where incentive modifications appear minor relative to the gap between private and social returns, or where direct supplements appear small relative to the scope of socially desirable work (clearly a matter of judgment), a presum tion exists that further allocation of resources would yield a higher than average rate of return, and that government policies to achieve such an expansion are in the public interest [34].

Second, these studies rely on the results of several econometric investigations that indicate that, for the industries and fields under investigation, the marginal rate of return from an investment in research and development has been very high. One of the first studies of this kind was Zvi Griliches's study of the returns from agricultural R & D [35]. He found that the rate of return from the investment in agricultural research during 1937-51 in the United States was between 35 and 170 percent. For two very successful projects — hybrid corn and hybrid sorghum — the rate of return on the investment was several hundred percent (but these two projects are obviously far from representative.) In a subsequent study (cited in section IV), Griliches [36] estimated that the gross social rate of return to research (and extension) expenditures was about 300 percent — a figure which he regarded as being quite consistent with his estimate of 35 to 170 percent net social rate of return to agricultural research based on different data and another approach.

Turning to manufacturing, Mansfield [37] and Minasian [38] (in the studies cited in section IV) estimated the marginal rate of return from R & D in the chemical and petroleum industries. Mansfield's results indicated that the marginal rate of return was about 40 percent or more in the petroleum industry, and about 30 percent in the chemical industry if technological change was capital embodied (but much less if it was disembodied). Minasian's results indicated about a 50 percent marginal rate of return on investment in R & D in the chemical industry. In addition, Mansfield provided some evidence that the marginal rate of return seemed relatively high (15 percent or more) in the food, apparel, and furniture industries.

Finally, based on computations for the economy as a whole, Denison [39] concluded that the rate of return from R & D was about the same as the rate of return from investment in capital goods. His estimate of the returns from R & D was lower than the estimates of other investigators, perhaps because he assumed no lag between R & D expenditures and their contribution to economic growth. The calculated rate of return on R & D could be much higher if R & D's contribution occurs only with a lag [40]. More recently, in his 1969 presidential address to the American Economic Association, William Fellner [41] estimated the average social rate of return from technological-progress activities, his conclusion being that it is "substantially in excess" of 13 or 18 percent, depending on the cost base, and that this is much higher than the marginal rate of return from physical investment at a more or less given level of knowledge.

X. EVALUATION OF THE EVIDENCE

How conclusive is the evidence described in the previous section? First, consider the judgmental approach adopted by Nelson, Peck, and Kalachek, among others. Clearly, this approach, although sensible and frequently used in all fields, is limited by the large subjective component that inevitably must enter the calculations. It is very difficult to estimate the extent of the external economies arising from particular types of R & D, or to determine whether incentive modifications are small relative to the gap between private and social returns, or to tell whether supplementary R & D provided by government and nonprofit institutions is small relative to the scope of socially desirable work. The weight one places on this evidence must depend on the confidence one puts in the judgment and objectivity of the investigators [42].

Second, consider the econometric approach adopted by Griliches, Mansfield, and others. This approach is more objective in many respects. Certainly the assumptions underlying the estimates are specified clearly, and one can see how sensitive the results are to changes in these assumptions. But this does not mean that the results can be accepted uncritically. On the contrary, since most of these estimates depend on, and are derived from the studies described in sections IV and II, they are subject to many of the limitations described in sections V and III. The seriousness of these limitations has been stressed earlier, and should be stressed again [43]. In addition, practically all of these econometric studies were carried out several years ago, and the estimates generally pertain to the late fifties or early sixties. It is by no means clear that the results would be different today, but, of course, one cannot rule out the possibility. Also, some of these studies try to measure the social returns from R & D, while others only measure the private returns to R & D (and perhaps part of the social returns not included in the private returns).

Yet, having taken pains to point out the limitations of the individual bits of evidence that have been amassed, we must not lose sight of an impressive fact: No matter which of the available studies one looks at (other than Denison's [44], the conclusions seem to point in the same direction. In the case of those using the judgmental approach, there is considerable agreement that we may be underinvesting in particular types of R & D in the civilian sector of the economy. In the case of the econometric studies, every study of which I am aware indicates that the rate of return from additional R & D in the civilian sector is very high.

XI. NEEDED RESEARCH CONCERNING RESEARCH AND DEVELOPMENT

Previous sections have indicated that, although considerable progress has been made in the last decade in furthering our understanding of the relationships between R & D and economic growth and productivity increase, existing knowledge is too weak to permit very confident or definitive statements concerning these relationships. Although existing knowledge may be of some use in formulating public policy in this area, it is limited by many serious problems and can only be regarded as tentative.

Given that this is the case, what steps might be taken to further our knowledge in this area? The next three sections of this report are addressed to this question, this section being devoted to a description of needed research concerning R & D, the next section being devoted to a description of needed research concerning the process of technological change, and the following section being devoted to a description of needed research concerning economic growth and productivity increase. Needless to say, I shall have to be selective, and my choice of topics may be influenced by my own biases.

With regard to research and development, there are at least six important areas that are in need of considerable additional research. First, we need much more information concerning exactly what is included as R & D in various industries. It is perfectly clear that, without such information, we shall be unable to interpret relationships — or lack of relationship — between measured R & D and other economic variables with any real confidence. What proportion of R & D, as measured by the customary figures, is routine service work? What proportion is aimed at fairly certain and modest design improvements? Detailed studies have been carried out of the characteristics of the R & D portfolios of a sample of firms in the chemical, petroleum, and electrical equipment industries [45]. But this work is only a beginning. Also, work is needed to provide better price indexes for R & D in particular industries, so that it will be possible to compare expenditure data at various points of time more accurately. Some work has been done on this score too [46], but much more needs to be done. In addition, attempts should be made to develop measures of inventive effort which include the work of independent inventors and which are comparable for both large and small firms.

Second, given more detailed breakdowns of R & D in various industries, it is important that we disaggregate R & D in the models used to relate R & D to economic growth and productivity increase. On the bases of existing work, it is perfectly clear that R & D expenditures include outlays on activities of quite different sorts which would be expected to have quite different effects on productivity. Although it was quite reasonable to use total R & D expenditures in earlier studies, we should try now to go beyond this crude beginning. After all, some of the most interesting questions in this area relate to the returns from various kinds of R & D directed at small product improvements, R & D directed at more major investments etc. Unless we disaggregate R & D, such questions cannot be answered.

Third, we need more information concerning the expected profitability and risk attached to the R & D portfolios of particular laboratories and firms, as well as more data concerning the decisionmaking process with regard to project selection and the allocation of R & D funds in various laboratories and firms. Such information would allow us to determine the extent to which firms are risk-averters and to study the social implications of the decision rules employed, explicitly or implicitly, by the firms. A number of studies have been made in this area [47], but these studies pertain to only a small sample of laboratories. Since it has become increasingly clear that, to be useful, studies of this sort must utilize detailed data that can only be derived from intensive work with individual firms, progress in this area depends on researchers being willing to immerse themselves in the operation of individual laboratories.

Fourth, we need more information concerning economies of scale in particular types of research and development. There are numerous reasons for thinking that there are economies of scale in research and development up to some point — "lumpiness" of capital equipment used in R & D, advantages from specialization of labor, reduction of risks due to the law of large numbers, and so forth. However, we know very little — industry by industry — about the extent of these economies of scale for particular kinds of work and about the size of R & D establishment beyond which further increases in size bring little or nothing in the way of further efficiency for the type of work in question. Recent work by Freeman [48] has shed some light on this subject in the electronics industry, but it is still largely unexplored. This is very unfortunate, since the returns derived from a certain expenditure on R & D — and the socially optimal organization of R & D — will depend on these economies of scale.

Fifth, we need more information concerning the conditions and mechanisms leading to the application of basic science and its translation into new products and processes. According to recent studies [49], the United States has been more adept than Western Europe at the application and translation of the findings of basic science into economically significant innovations. What are the reasons for this superiority, if indeed it exists? What is the mechanism in various areas leading to the translation of new basic science into technology? These are important questions, and ones about which little is known — although the TRACES study [50] provides some relevant and significant information. To appreciate the importance of these questions, it should be noted that the country that does the basic scientific work in a particular area may not be the one that reaps the greatest economic benefits from the technological innovations in the area. The extent of the economic benefits from fundamental research depends on the facility and efficiency with which the results of fundamental research are applied. Fortunately, according to the OECD studies, the United States seems to have been relatively adept at the application of fundamental research, the consequence being that the economic returns from basic research have probably been relatively high in the United States. But why has this been the case, and can we be sure it will continue to be the case in the future?

Sixth, we need more information concerning the coupling of industrial R & D with marketing and production. Industrial R & D can have little economic impact unless it is applied. And the difficulties in bridging the gap between R & D, on the one hand, and marketing and production, on the other hand, are greater than typically recognized. Systematic studies in depth of the problems in this area — and the ways that industry has attempted to solve these problems — would be of considerable use.

It is high time that we begin to build this aspect of the R & D process into models relating R & D expenditures to productivity increase and economic growth. Also, it should be recognized that a large part of the riskiness of industrial R & D is due to commercial, not technical, uncertainty. That is, it appears that the probability is much greater that a firm can solve the technical problems involved in the typical R & D project than that it will turn out to be economically justified in having gone to the trouble and expense of solving them [51]. If this is indeed the case, it raises questions concerning the extent to which there is proper coordination between the R & D people, on the one hand, and the marketing and production people, on the other. Detailed and intensive studies should be carried out to shed light on this question, which has received limited — and often superficial — treatment in the past.

XII. NEEDED RESEARCH CONCERNING THE PROCESS OF TECHNOLOGICAL CHANGE

In addition, considerable research is needed to promote a fuller understanding of the process of technological change. First, we need to know much more concerning the role of R & D in the entire process of technological innovation. Until a few years ago, there was a tendency to equate R & D with innovation, the consequence being that the non-R & D activities associated with innovation were neglected. Recent work [52] indicates that the non-R & D aspects of innovation — tooling and construction of plant, manufacturing start-up, marketing start-up, etc. — frequently account for as large a proportion of the total costs of a successful product innovation as does R & D. It is important that we recognize the importance of these non-R & D prerequisites to innovation in our models. Also, it is important that we learn more about the areas where — and conditions under which — little or no formal R & D is required for innovation. Studies by Myers and Marquis [53], as well as others, show quite clearly that many innovations require little in the way of formal R & D.

We need to know more about the origin, type, significance, and frequency of such innovations in a variety of industries.

Second, studies are needed of the conditions that promote or thwart the rapid conversion of an invention into an innovation, given that market and technical factors make such a conversion socially desirable. According to recent OECD studies [54], American firms are much more adept at achieving such a conversion than Western European firms. In the United States, there is some evidence [55], albeit crude, that firms achieve such a conversion more rapidly than in the past. But existing information tells us far too little. To what extent does an industry's market structure determine the average rate of conversion? To what extent do problems of interindustry and interfirm coordination lessen the average rate of conversion? Does the use of various management techniques that are currently in vogue have a perceptible or demonstrable effect? What are the characteristics of the managers and managements that seem to perform best in this regard?

Third, we need to know much more about the sources of invention and innovation in various industries. With regard to invention, what has been the relative importance of independent inventors, small firms, large firms, universities, etc., as sources of significant inventions in particular industries? The studies by Jewkes, Sawers, and Stillerman [56], Hamberg [57], Enos [58], and others are valuable, but there is a need for much more empirical work of this sort. Moreover, with regard to innovation, what has been the relative importance as innovators of firms of various kinds — large, small, conglomerate, single product, etc. — in particular industries? At present, we have reasonably useful data for only a handful of industries — petroleum, steel, coal, ethical drugs [59]. Attempts should be made to identify the firms that pioneered in the introduction of important new processes and products in other industries. Such information is needed if we are to obtain a better understanding of the factors and conditions conducive to invention and innovation and the relative efficiency and creativity of various kinds of organizations.

Fourth, studies are needed of the effects of market structure on an industry's rate of technological change. What is the effect of market structure on the amount spent by an industry on R & D and other innovative activities, the sort of R & D and other innovative activities carried out, the productivity of the industry's R & D, the quickness of the firms to innovate, and the rate of acceptance of new techniques and products (both those arising within and those arising outside the industry)? To what extent are giant firms required in various industries to ensure a rapid rate of technological change? These are extremely important questions. Yet, despite considerable progress in this area in the last decade, we are far from having definitive answers to these questions.

Fifth, we need to know much more about the factors influencing the rate of diffusion of innovations. Mansfield [60] and Griliches [61] have formulated models of the diffusion process and obtained detailed data concerning the diffusion of a number of important innovations in manufacturing and agriculture. These models have been used by Mansfield [62] and others for technological forecasting. However, existing data pertain to only a handful of industries, and much more work of theoretical and econometric sorts is needed. Also, we need to know much more about the mechanism and costs of transferring technology from organization to organization and from country to country. Although a 1966 conference [63] sponsored jointly by NSF and the National Planning Association helped to clarify some aspects of this topic, a great deal of work remains to be done. Theoretical work — like that begun by Kenneth Arrow [64] — and detailed empirical work — like Hall and Johnson's study [65] of the transfer of the production of the F-104 to Japan — is needed.

Sixth, it seems to me that much more information and work is needed to measure more accurately the "spillover" to civilian technology from military and space R & D. Of course, I realize that an enormous amount of verbiage and papers of dubious distinction have been produced on this topic. But, so far as I know, the amount of penetrating, quantitative, objective analysis has been surprisingly limited [66]. In view of the great importance of this question, more should be done. Also attempts should be made to study and evaluate various approaches designed to increase such "spillover." For example, NASA has adopted a number of approaches in its work with Midwest Research Institute, the Aerospace Research Applications Center at Indiana University, the University of Maryland, and other places. It would be extremely valuable to find out what the experience of these groups can teach us about the relative cost and effectiveness of various approaches.

XIII. NEEDED RESEARCH CONCERNING ECONOMIC GROWTH AND PRODUCTIVITY INCREASE

In addition, considerable research is needed to promote a fuller understanding of the process of economic growth and productivity increase. First, we need to improve the measures of output on which the productivity statistics depend. As noted previously, existing output measures do not record the effects of the introduction of new or improved products. This is a very important limitation. Some very competent observers, e.g., the Price Statistics Review Committee, have recommended that the government might experiment with price series that allow, even roughly, for product improvement. More might be done along this line. Also, measures like gross national product do not recognize the social costs — pollution, accidents, etc. — arising from technological changes. To the extent that these costs are borne by government, not industry, these costs are counted as end product of a positive sort. Economists are becoming increasingly aware of this problem, but more should be done.

Second, better information of other kinds is needed as input to studies of the rate of productivity increase and technological change. For one thing, the price indexes used to deflate construction expenditures are questionable; indeed they seem to be cost, not price indexes [67]. Also, there are problems in adjusting capital inputs for the extent of capacity utilization. (And better estimates are needed of the elasticity of substitution.) Obviously, errors from these sources can result in errors in the estimated rates of growth of capital and labor, which in turn can result in errors in productivity estimates. Also, turning to a different, but related topic, we need studies of the rate of technological change in various sectors of the economy that make greater use of engineering data and experience. For various reasons, economists, with some notable exceptions, have tended to avoid using engineering estimates. Here is place where interdisciplinary work is badly needed.

Third, we need a better understanding of the complex interrelationships among R & D, education, management, and capital formation in the process of economic growth. For example, consider the relationship between R & D and education. In most economic models, the contribution of education is perfectly straightforward: More educated workers are able to produce more than less educated workers. As far as it goes, no one can take issue with this hypothesis. Nonetheless, the hypothesis may result in a misspecification of these models because it oversimplifies the relationship between education and economic growth [68]. In part, the effect of education on economic growth depends on the rate of technological change since an important effect of more education is to make managers and workers more adaptable to change and quicker to adopt innovations. Moreover, an investment in education of certain kinds is likely to increase the rate of technological change. For these and other reasons, the relationships among these variables are richer and more complex than pictured in most contemporary economic models. We must learn more about the nature of these relationships and formulate our models accordingly.

Fourth, taking up a point that is related to the previous one, we should learn more about the extent to which technological change in various industries has been capital embodied or disembodied. That is, attempts should be made to estimate the extent to which new techniques and products in various industries in recent years have required new plant and equipment, the extent to which they could be "grafted" onto old plant and equipment and the cost of doing so, and the extent to which they can be accommodated or used without altering existing plant and equipment [69]. Also, much more should be learned about the process of "learning by doing" [70]. These questions are important because the effects of capital formation on economic growth — and the extent to which the utilization of new technology requires capital formation — depend on them.

Fifth, in models designed to relate R & D to productivity increase in particular sectors of the economy, we must take better account of interindustry — and in some cases, interfirm — flows of technology. Specifically, we must take more realistic account of the fact that R & D in one firm or industrial sector often increases productivity in another firm or industrial sector. Some progress has been made in incorporating this fact into economic models, the study by Brown and Conrad [71] being a beginning. But much more must be done. Unless we learn how to do this more effectively, our estimates of the effects of R & D cannot help but be very crude.

Sixth, we need to extend many kinds of studies of productivity increase to a larger number of countries. In many areas, we lack reliable data for countries other than the United States. For example, according to a recent OECD study [72], the rate of economic growth in the member countries is closely correlated with their performance in the diffusion of technological innovations. Yet very little is known about the diffusion process in countries other than the United States. A study is now underway to compare the rate of diffusion of selected innovations in various countries, but this study is only a beginning [73]. Although there are many pitfalls in international comparisons, they obviously can be helpful, if carried out carefully, in disclosing the factors influencing diffusion rates.

In passing, it may be worthwhile to mention a few of the common pitfalls in international comparisons. For present purposes, perhaps the most important thing to note is that one cannot conclude that the returns from R & D in the United States are small because some countries that have invested a fair amount in R & D have relatively low growth rates, and some countries that have invested little in R & D have relatively high growth rates. Many factors other than R & D affect economic growth and should be taken into account in any such comparisons [74]. Moreover, the fact that some countries spend much more than others on defense and space R & D should be taken into account — together with the fact that many important effects of R & D on true economic growth are not reflected in measured growth rates. Further, some countries that have not invested heavily in R & D have nonetheless achieved high rates of technological change by importing technology and by imitating and "catching up" with the technological leaders. Obviously, the technological leaders cannot achieve high rates of technological change in this way. Also, some countries that have invested relatively large amounts in R & D have not been as efficient in converting the results of their R & D into innovation as the United States has. Thus, the returns from R & D in these countries may be a poor guide to those in the United States.

XIV. CONCLUSIONS

In conclusion, technological change has certainly contributed in a very important way to economic growth in the United States. Although existing studies have not been able to estimate this contribution with great accuracy, they have certainly indicated that this contribution has been large. Moreover, although econometric studies of the relationship between R & D and productivity increase have been subject to many limitations, they provide reasonably persuasive evidence that R & D has an important effect on productivity increase in the industries and time periods that have been studied. Turning to the adequacy of the Nation's investment in research and development, there is too little evidence to support a very confident judgment as to whether or not we are underinvesting in certain types of research and development. However, practically all of the studies addressed to this question seem to conclude, with varying degrees of confidence, that we may be underinvesting in particular types of R & D in the civilian sector of the economy, and the estimated marginal rates of return from certain types of civilian research and development seem very high.

Additional research is badly needed to determine more adequately the relationship of research and development to economic growth. A number of specific areas where work is needed have been indicated in this paper. In recent years, economists have been encouraged by government agencies and the foundations to turn their attention to urban problems, environmental problems, health problems, educational problems, and other areas. Without denying for a moment the importance of these areas, I think that there is some danger that research on technological change may be receiving less encouragement than it warrants.

NOTES AND REFERENCES

1. SOLOW, ROBERT, "Technical Change and the Aggregate Production Function," *Review of Economics and Statistics*, August, 1957.

2. ABRAMOWITZ, MOSES, "Resource and Output Trends in the United States Since 1870," *American Economic Review*, May, 1956.

3. FABRICANT, SOL, "Economic Progress and Economic Change," *34th Annual Report of the National Bureau of Economic Research*, New York, 1954.

4. DENISON, EDWARD F., *The Sources of Economic Growth in the United States*, New York: Committee for Economic Development, 1962.

5. JEWKES, JOHN, SAWERS, DAVID and STILLERMAN, RICHARD, *The Sources of Invention*, revised ed. New York: W.W. Norton & Co., 1970.

6. HOLLANDER, S., *The Sources of Increased Efficiency*, Cambridge: M.I.T. Press, 1965.

7. NELSON, RICHARD R., PECK, MERTON J., and KALACHEK, EDWARD D., *Technology, Economic Growth, and Public*

Policy. Washington: The Brookings Institution, 1967.

8. JORGENSEN, DALE W., and GRILICHES, ZVI, "The Explanation of Productivity Change." *Review of Economic Studies*, July, 1967. It is important to note that quality change in inputs often is a consequence of technological change.

9. KENDRICK, JOHN, *Productivity Trends in the United States*, National Bureau of Economic Research, Princeton: Princeton University Press, 1961.

10. TERLECKYJ, NESTOR, "Sources of Productivity Change." Unpublished *Ph.D. thesis*. New York: Columbia University, 1959.

11. GRILICHES, ZVI, "Research Expenditures, Education and the Aggregate Agricultural Production Function," *American Economic Review*, December 1964.

12. MANSFIELD, EDWIN, "Rates of Return from Industrial Research and Development," *American Economic Review*, May, 1965. This material also appears, with some modifications, in Edwin Mansfield, *Industrial Research and Technological Innovation*, W.W. Norton for the Cowles Foundation for Research in Economics at Yale University, 1968. When we speak of productivity increase in the paragraph of the text, we really refer to rates of movement of the production function based on labor and capital alone.

13. MINASIAN, JORA, "Research and Development, Production Functions, and Rates of Return," *American Economic Review*, May, 1969.

14. MINASIAN, JORA, "The Economics of Research and Development" in *The Rate and Direction of Inventive Activity*, National Bureau of Economic Research, Princeton: Princeton University Press, 1962.

15. BROWN, MURRAY and CONRAD, ALFRED, "The Influence of Research and Education in CES Production Relations"

In: *The Theory and Empirical Analysis of Production*. New York: Columbia University Press, 1967.

16. MANSFIELD, EDWIN, "Comment" In: *The Theory and Empirical Analysis of Production, ibid.*

17. See ARROW, KENNETH J., "Economic Welfare and the Allocation of Resources for Invention," *The Rate and Direction of Inventive Activity, op. cit.*; and NELSON, RICHARD R., "The Simple Economics of Basic Scientific Research," *Journal of Political Economy*, June, 1959.

18. PIGOU, A.C., *The Economics of Welfare*, third ed., London: Macmillan & Co., Limited, 1928.

19. National Science Foundation, *National Patterns of R & D Resources, 1953-70*. p. 36. Washington: Supt. of Documents, U.S. Government Printing Office, 1969.

20. National Science Foundation, *Research and Development in Industry, 1968*. p. 9. Washington: Supt. of Documents, U.S. Government Printing Office, 1970.

21. McGraw-Hill, *Business Plans for Expenditures on Plant and Equipment*. Annual. New York.

22. National Science Foundation, *National Patterns of R & D Resources, 1953-70, op. cit.*, p. 7.

23. HAMBERG, DANIEL, "Invention in the Industrial Laboratory," *Journal of Political Economy*, April, 1963.

24. JEWKES, SAWERS, and STILLERMAN, *op. cit.*

25. McGraw-Hill, *op. cit.*

26. MANSFIELD, EDWIN, "Industrial Research and Development: Characteristics, Costs, and Diffusion of Results," *American Economic Review* May 1969, as well as my books cited in footnotes 12 and 44.

27. MEADOWS, E., "Estimate Accuracy and Project Selection Models in Industrial Research," *Industrial Management Review*, Spring 1968.

28. FREEMAN, CHRISTOPHER, POIGNANT, M. RAYMOND and SVENNILSON, INGVAR, *Science, Economic Growth, and Government Policy*. p. 43. Paris: Organisation for Economic Co-operation and Development, 1963.

29. *Ibid.*, 35.

30. U.S. President, *Economic Report of the President*. Together with the *Annual Report* of the Council of Economic Advisers. Transmitted to Congress, January 1964. Washington: Supt. of Documents, U.S. Government Printing Office, 1964.

31. National Commission on Technology, Automation, and Economic Progress, *Technology and the American Economy*. p. 87. Washington: Supt. of Documents, U.S. Government Printing Office, February 1966.

32. NELSON, PECK, and KALACHEK, *op. cit.*

33. CAPRON, WILLIAM, "Discussion," *American Economic Review*, 508, May 1966.

34. NELSON, PECK, and KALACHEK, *op. cit.*, 172-73.

35. GRILICHES, ZVI, "Research Costs and Social Returns: Hybrid Corn and Related Innovations," *Journal of Political Economy*, October 1958.

36. GRILICHES, ZVI, "Research Expenditures, Education, and the Aggregate Agricultural Production Function," *op. cit.* Also, in a subsequent paper, Willis L. P Peterson estimated that the rate of return to poultry research has been about 20 to 30 percent. See PETERSON, WILLIS L., "Return to Poultry Research in the United States," *Journal of Farm Economics*, August 1967.

37. MANSFIELD, EDWIN, "Rates of Return from Industrial Research and Development," *op. cit.*

38. MINASIAN, JORA, "Research and Development, Production Function, and Rates of Return," *op. cit.*

39. DENISON, *op. cit.*

40. ABRAMOWITZ, MOSES, "Economic Growth in the United States," *American Economic Review*, September 1962.

41. FELLNER, WILLIAM, "Trends in the Activities Generating Technological Progress," *American Economic Review*, March 1970.

42. Also, some of the studies were made several years ago and the authors may feel that some of the problems existing then have been ameliorated considerably since their studies were made. However, this is certainly not the case for NELSON, PECK, and KALACHEK, and CAPRON; and it is very doubtful for the others.

43. For a discussion of these limitations, see MANSFIELD, EDWIN, *The Economics of Technological Change*. New York: W.W. Norton & Co., 1968.

44. Also, the paper by JORGENSEN and GRILICHES suggests that the social rate of return from R & D is "comparable" to that on other forms of investment; but the authors are careful to point out that they "have made no attempt to isolate the effects of expenditures on research and development from expenditures on other types of current inputs or investment goods ..."

45. MANSFIELD, EDWIN, RAPAPORT, J., SCHNEE, J., WAGNER, S., and HAMBURGER, M., *Research and Innovation in the Modern Corporation*. New York: W.W. Norton & Co., 1971.

46. MILTON, H., "Cost-of-Research Index, 1920-65," *Operations Research*, November 1966. Also, work in this area has gone on at the RAND Corporation and at the National Science Foundation.

47. MANSFIELD, EDWIN and BRANDENBURG, RICHARD G., "The Allocation Characteristics, and Outcome of the Firm's R & D Portfolio: A Case Study," *Journal of Business*, October 1966; and MANSFIELD, EDWIN,

RAPAPORT, J., SCHNEE, J., WAGNER, S., and HAMBURGER, *op. cit.*

48. FREEMAN, CHRISTOPHER, "Research and Development in Electronic Capital Goods," *National Institute Economic Review*, November 1965.

49. BEN DAVID, J., *Fundamental Research and the Universities*. Paris: Organisation for Economic Co-operation and Development, 1968.

50. IIT Research Institute, *Technology in Retrospect and Critical Events in Science*. Prepared for the National Science Foundation. Chicago, 1968. Other recent studies are the Defense Department's Project Hindsight, which, of course, has been the subject of considerable controversy, and National Academy of Sciences, Materials Advisory Board, *Report on Principles of Research Engineering Interaction*. Washington, 1966.

51. MANSFIELD, EDWIN, RAPAPORT, J., SCHNEE, J., WAGNER, S., and HAMBURGER, M., *op. cit.* Also, for some discussion of the importance of this topic, see National Academy of Sciences, *Applied Science and Technological Progress*. Washington, 1967.

52. See U.S. Department of Commerce, *Technological Innovation: Its Environment and Management*, Washington: Supt. of Documents, U.S. Government Printing Office, January 1967; and MANSFIELD, EDWIN, RAPAPORT, J., SCHNEE, J., WAGNER, S., and HAMBURGER, M., *op. cit.*

53. National Science Foundation, *Successful Industrial Innovations* by MYERS, SUMNER and MARQUIS, DONALD. Washington: Supt. of Documents, U.S. Government Printing Office, 1969.

54. Organisation for Economic Co-operation and Development, *Gaps in Technology: General Report*. Paris, 1968.

55. LYNN, F., "An Investigation of the Rate of Development and Diffusion of Technology in Our Modern Industrial Society," In: *The Employment Impact of Technological Change*, Appendix Volume II, of *Technology and the American Economy*. The Report of the National Commission on Technology, Automation, and Economic Progress. Washington: Supt. of Documents, U.S. Government Printing Office, February 1966.

56. JEWKES, SAWERS, and STILLERMAN, *op. cit.*

57. HAMBERG, *op. cit.*

58. ENOS, JOHN, *Petroleum Progress and Profits*. Cambridge: M.I.T. Press, 1962.

59. MANSFIELD, EDWIN, *Industrial Research and Technological Innovation, op. cit.*; and MANSFIELD, RAPAPORT, SCHNEE, WAGNER and HAMBURGER, *op. cit.*

60. MANSFIELD, EDWIN, "Technical Change and the Rate of Imitation," *Econometrica*, October 1961; and MANSFIELD, *Industrial Research and Technological Innovation, op. cit.*

61. GRILICHES, ZVI, "Hybrid Corn: An Exploration in the Economics of Technological Change," *Econometrica*, October, 1957.

62. MANSFIELD, EDWIN, *Numerical Control: Diffusion and Impact in the Tool and Die Industry*, In press.

63. National Science Foundation, *Proceedings on a Conference on Technology Transfer and Innovation*. Washington: Supt. of Documents, U.S. Government Printing Office, 1967.

64. ARROW, KENNETH, J., "Classificatory Notes on the Production and Transmission of Technological Knowledge," *American Economic Review*, May 1969.

65. HALL, GEORGE R., and JOHNSON, ROBERT E., "Transfers of United States Aerospace Technology to Japan," *The Technology Factor in International Trade*, New York: Columbia University Press, 1970.

66. For an interesting study, see Denver Research Institute, *The Commercial Application of Missile-Space Technology*. Denver, 1963.

67. GRILICHES, ZVI, "Production Functions in Manufacturing: Some Preliminary Results," In: *The Theory and Empirical Analysis of Production, op. cit.*

68. NELSON, PECK, and KALACHEK, *op. cit.*

69. DENISON believes that the importance of this question has been exaggerated. See his comments in the March 1964 *American Economic Review*.

70. ARROW, KENNETH J., "The Economic Implications of Learning by Doing," *Review of Economic Studies*, June 1962.

71. BROWN and CONRAD, *op. cit.*

72. Organisation for Economic Co-operation and Development, *Gaps in Technology: General Report, op. cit.*

73. RAY, G., "The Diffusion of New Technology," *National Institute Economic Review*, May 1969.

74. For some estimates of residual growth rates — after the effects of labor, capital, education, and other non-R & D factors have been taken into account — in various European countries, see DENISON, *Why Growth Rates Differ*, Washington: The Brookings Institution, 1967.

Editor's Note: Obviously, one of the extremely important considerations in any technology assessment is the economic benefit. Much has been said about how good R & D is, and how it has helped economic development. However, there are many questions as to the relationship between R & D and economic growth and this is exactly what is addressed by Ed Mansfield. (M.J.C.)

THE ROLE OF LAW AND LAWYERS IN TECHNOLOGY ASSESSMENT

HAROLD P. GREEN

Professor
The George Washington University National Law Center, Washington, D.C., U.S.A.

Any discussion of the role of the law in technology assessment must proceed from the premise that technology assessment is nothing more than a procedure underlying a public policy decision that a particular aspect of technology should be brought under some measure and form of social control. Neither the problem of social control nor the process of technology assessment is new or novel. Technological advance has always produced social problems of one type or another, and public policy decisions as to social control of technology have always been based on the kind of assessment of benefits versus risks/costs contemplated by the concept of "technology assessment." When Congressman Daddario originated the term "technology assessment" in 1967, he did so not in terms of proposing a new function or operation but rather solely in terms of creating a new mechanism to assist Congress in performing the technology assessment function which is an inherent part of the legislative process.

It is in this sense that I am discussing technology assessment in this paper. No consideration is given to technology assessment (i.e., forecasting) by business entities or to the manner in which assessment functions are performed in the policy-making activities of the Executive Branch. Moreover, technology assessment should not be regarded as a means for producing "correct" policy decisions. In a democratic society, public policy decisions, even with respect to science and technology, should ideally reflect the desires and values of the public whether these are right or wrong, rational or irrational. Our public policy decisions are made in the crucible of the political process, and irrational or incorrect decisions are frequently made by Congress with respect to such issues as foreign policy, economic policy, tax policy, and confirmation of nominees to the Supreme Court. Indeed, it is a rare piece of major legislation which does not to some extent reflect political compromises. There is no reason why we cannot tolerate equally irrational or incorrect decisions in the area of public policy for technology.

But, whatever the concept of technology assessment, it is clear that the concept does not supersede, but rather is built upon and supplements, existing institutions. There are a variety of basic institutions in our society which perform a technology assessment function. The most basic of these is the economic market place where the public votes with its dollars for the products and the services it desires. These votes are counted by businessmen and translated into demand, which in turn is weighed against the costs of producing what the public wants. Businessmen invest in technological innovation because they sense a public demand, and hence a market, for the innovations. Unfortunately, however, the demand usually does not reflect a realistic appraisal of the hazards and other social costs of the technology.

Risks associated with the technology may, however, be reflected in costs if the law imposes liability with respect to the risks. Liability may be imposed on the manufacturer or the seller, in which event his costs may be increased, thereby making use of the technology more expensive and decreasing demand, so that his investment in the technology is deterred. Alternatively, liability may be imposed on the user of the technology, thereby increasing his potential costs and discouraging his use of the technology. The process of associating potential liability with the manufacture, sale or use of technology is termed "internalization of costs" by economists. If liability is not imposed, costs related to the hazards of the technology are regarded as "external costs." To the extent that costs attributable to hazards are internalized, the market place operates to deter the development and use of hazardous technologies. Putting this another way, the technology will be developed and used only if the anticipated benefits outweigh the costs, including those attributable to potential liability. The deterrent effect of potential liability is somewhat minimized by the availability of insurance, which permits the firm to translate potential liability of an uncertain magnitude into periodic payments of known insurance premiums. It is apparent, however, that insurance itself is an institution for technological assessment. The availability of insurance reflects an assessment by the insurance companies that premiums paid for insurance against specified risks will compensate for potential liability on account of such risks. Premium rates, moreover, reflect the insurance companies' experience with the risks, so that technologies which have proved to be hazardous will be saddled with higher premium rates than those which are relatively innocuous. Thus, if insurance is unavailable, the threat of liability may deter hazardous activities, and even if it is available, the level of premiums for the insurance may operate to encourage or discourage such activities. It is worth noting in this connection that insurance companies are the real professionals in assessment of risk and the quantification of risk in terms of dollars.

The liability and insurance mechanisms are useful instruments for technology assessment, but they are clearly inadequate. Insurance is based on liability, and principles of liability applicable to new technologies evolve slowly and uncertainly only after cases arise, and cases arise only after the technology produces actual harm. Given today's revolutionary rate of technological advance, coupled with the enormous potentialities for harm in modern technology, practice of a new technology may result in unacceptably severe and widespread harm before effective rules of liability have evolved. Moreover, modern technology produces insidious forms of hazards which do not produce instant, dramatic injury, but which operate slowly and cumulatively, so that the existence of the hazards and the concomitant injuries may not become manifest for many years or perhaps generations. Consider, for example, the problem of pollution of the atmosphere caused by automobile exhaust fumes. Automobiles have been part of our environment for more than half a century, but only recently have we become aware that their exhaust fumes imperil life. Steps are now being taken to deal with the problem, but no one can know, or even estimate, how much injury has been sustained in the past.

There are other reasons why the liability and insurance mechanisms do not operative effectively with respect to such insidious hazards. The injuries produced are generally non-specific, and it is frequently extremely difficult, if not virtually impossible, to prove, in a legally adequate manner, a causal relationship between such hazards and an injury. And even if the causal relationship can be demonstrated, who can be held liable when the causative factor was produced by numerous, nameless, faceless sources?

Looking at the overall operation of the market-liability-insurance mechanisms for technology assessment, it is obvious that they are not adequate from the standpoint of appropriate and effective social control of technology. Aside from the intrinsic difficulties involved in imposing liability for hazards associated with technology in a timely manner, it is clear that our legal system has not yet succeeded in internalizing all relevant costs. For example, costs which fall upon the public generally, or on specific groups other than those who sell or use the fruits of the technology, as in the case of industrial pollution of air or water, are not internalized. Moreover, the public persists in demanding products it recognizes as hazardous and usually is unwilling to pay a higher price to obtain the products in safe form. It is clear, nevertheless, that these private insitutions — market, liability, and insurance — are a first line of technology assessment, and that any other technology assessment institution must be built upon and supplement their operation.

It is obvious that the legal system plays a major role in the operation of these private assessment institutions since it is the legal system itself which, through the imposition of liability in specific cases, creates a rule of law which in effect translates potential technological hazards into costs which must be considered in connection with private decisions as to the development, manufacture, sale, and use of the technology's products. This in turn comes about as lawyers represent adversary clients in litigation arising out of injuries allegedly resulting out of hazards associated with the technology. Although lawyers in performing this role are merely representing clients and do not percieve of their role as part of any grand technology assessment function, it is clear that they are in fact vital elements in the assessment process.

The existence of social problems notwithstanding, the operation of these private assessment institutions frequently calls for legislative action. Legislatures are the ultimate technology assessment agencies, and they implement the conclusions reached in the assessment process through the enactment of laws. Frequently they establish statutory rules of liability in particular areas of technology. They can prohibit the practice of hazardous technologies, or they can prescribe regulatory controls to govern the manner in which technologies are practiced. All such legislative actions are predicated on a weighing of benefits against risks — i.e., on technology assessment. In many instances, the legislature merely establishes a general regulatory policy and assigns to an administrative agency the responsibility for implementing the regulation. For example, the Food and Drug Administration is clearly an agency for technology assessment. Through a process of weighing benefits against risks under a general statutory standard, it determines what foods and drugs may be sold.

It should be noted, however, that the legislatures do not usually perform a technology assessment function until an obvious social problem exists and is thrust upon them. They do not seek out potential hazards of technology; rather they react to hazards which are shown to exist. Parenthetically, it should also be noted that the known existence of a technological hazard does not mean that the legislature will act. There is typically a period of gestation until the need for the legislation sinks in to overcome natural inertia and the confidence of the promoters of the technology, and then a further period of struggle against vested interests before final legislative action is taken. There is no button which can be pushed to produce a certain and timely societal response to technological advance.

The present pressures for a new technology assessment institution are based on the belief that society should direct, and not merely react to, technological advance. The concept of technology assessment involves basically the continuing analysis of the potential impacts of a technological development on society, including analysis of alternative approaches to achievement of the technological goal. This exercise would place in the hands of the policy makers a vast body of data underlying the carefully analysed pros and cons of various alternative policies. The policy makers will be able, therefore, to apply the accelerator or the brake, and to choose the direction in which technological advance should proceed. It will do this, presumably, through two sets of decisions. If the technology is dependent upon government funding support, there will be a decision to spend or not spend; otherwise, the decision will be as to the role of law — should law encourage or discourage certain relevant activities, or should it be neutral?

Most of the discussion of technology assessment has ignored the fundamental distinction between assessment of technologies which evolve wholly or primarily within the private sector and assessment of technologies which evolve wholly or primarily through government support.

In the former case, assessments and consequent governmental actions are superimposed upon the incentives and deterrents inherent in the market system, and the effect will be to channel technological development and use along lines regarded as optimum by government rather than along lines indicated by market forces. In the latter case the only incentives and deterrents operating with respect to the technology are those created by the government itself as it makes decisions as to funding the development of the technology. Putting this another way, government funding is a powerful stimulus which forces technological advance, and the only deterrents are the government's own self-restraint.

Government support and sponsorship of technological advance has not been limited to areas of traditional governmental concern, but extends as well to areas in which technological advance has traditionally been a matter of purely private concern and investment. We may accept as axiomatic that government has a legitimate interest in research and development to improve and develop military technology. We may also accept government's role in development of technology determined to be important for purposes of national prestige which will be developed, if at all, only through government investment. The national space program is an example of this. Similarly, we can understand government's support of medical technology in the interest of promoting the public health. But government has also intervened actively, with enormous financial support, to accelerate development of such technologies as nuclear power, saline water conversion and the supersonic transport plane, to mention only a few examples, all of which are intended for primarily commercial application by private enterprise.

The massive government commitment to forcing technological advance has a number of implications from the standpoint of technology assessment. First, and most important, it represents the principal reason why technology is presently being developed at so revolutionarily rapid a rate, thereby placing immense strains on the present conventional mechanisms for technology assessment. This is so because these conventional mechanisms operate largely on the basis of experience derived from molecular advance in technology. Government sponsorship of technology results in extremely rapid advance which has the effect of leapfrogging experience, affecting more people, more substantially, and on a much more rapid time scale. Second, government support for technologies which are esssentially commercial constitutes a form of subsidy, disrupting the normal operation of the market forces and making government, rather than the private market place, the principal determinant of resource allocation. Third, when government makes a decision to develop a technology on a more or less predetermined time scale, there is a natural tendency for those who have a vested interest in the program — in government (both in the executive and legislative branches) and in industry — to become obsessed with their programmatic objectives and to minimize the social hazards or problems which may be inherent in use of the technology after it has been developed.

All such government programs in support of technological advance have obvious merit from the standpoint of national policy. They will produce technologies which will give the American people a better and more secure life. But they also involve hazards and social costs. Since government sponsorship of such programs in effect pre-empts the normal operation of the market place and the liability mechanism, we must rely solely on government for technology assessment in these areas. It is apparent, however, that government is usually more interested in achieving its technological goals than in protecting the public against hazards. In seeking further support for these programs, the hazards are consistently minimized. To the extent that they are articulated, we are promised that research and development to eliminate, minimize or control the hazards will be conducted. And we are assured that, after the technology is developed, government will see to it that the technology will not be practiced in a manner which will subject the public to unreasonable hazards.

It is obvious that government programs for development of technology do not proceed in callous disregard of the public welfare. It is equally obvious, however, that in the process of balancing benefits against risks, the government sponsors of technological programs are making decisions as to whether the public will be asked to assume certain risks or burdens which are determined to be "reasonable" or not "undue." Since such determinations are presently made within small closed circles of experts who have a vested interest in the technology, the basic question is whether the public itself, if it had comprehensible access to all relevant facts, would be willing to assume these risks and burdens for the sake of obtaining the promised benefits.

The proposal for a new technology assessment institution, at least as originally articulated by Congressman Daddario, would force the process of balancing benefits against risks into an open forum, in which all relevant information is freely available and in which sound decisions can be made before, and not long after, technology gives rise to serious social problems.

At the same time, it should be recognized that the type of technology assessment institution which is proposed will perform its function primarily on the basis of sophisticated prediction of the consequences of technological development. In many instances, the prediction will be based on only the fragmentary evidence available in the early stages of development. The process can undoubtedly determine the effects which are already desirable or clearly undesirable, and also those which are uncertain. It is in the area of uncertain effects that the greatest problem exists. Will obvious and important expected benefits induce a decision to proceed with development of a technology in the face of uncertain effects? Presumably the technology assessors will carefully monitor further developments which would contribute to a resolution of the uncertainties, so that if uncertain effects become undesirable effects a new decision can be made. But, especially in light of the fact that there will rarely be unanimity of opinion on the nature of the effects which, in any event, are rarely all black or all white, is it realistic to assume that technological development, once begun, can be turned off in the face of the vested interests which will have come into being?

It should also be noted that reliance on prediction runs counter to the traditions and practices of our system. Although prediction is usually an element of decision-making, it seems obvious that major commitments are usually made only when there is a sound basis in experience. One aspect of the massive government commitment to technological advance in recent years has been that major technological programs have proceeded in a manner leapfrogging experience, based on the predictive judgment that undertain effects will prove not to be undesirable in the light of future experience. We all know, however, how frequently the most carefully based predictions of scientists and technologists prove to be inaccurate, sometimes with disastrous consequences.

The basic problem in technology assessment is to assure an adequate articulation of potential costs and risks. It is axiomatic, I believe, that in any technology assessment the benefits are usually obvious and relatively immediate, and will always be more than adequately articulated and pressed by those with vested interests in the technology. Costs and risks are, on the other hand, typically much more remote and speculative and there rarely are authoritative interests which can be relied upon to articulate and press the costs and risks upon the assessors and decision-makers. It is obvious, particularly where the technology evolves through government financial support, that striking an appropriate balance between benefits and costs/risks requires that some mechanism be found to ensure that costs and risks are given equal time, attention, and dignity.

The law and lawyers have an important role to play in this process. Much of the discussion of technology assessment has proceeded on the assumption that this function is to be performed largely or wholly by scientific and technological experts. It is obvious, however, when we consider that benefits and risks involve what benefits the public wants and what risks the public will find acceptable, that the more broadly constituted the assessment body is , the more valid its conclusions will be. In particular, what is required is the active participation of generalists — men of practical affairs — who can bring to bear a skepticism based upon broad experience. Lawyers, by reason of their training and experience, may well be the consummate generalists in our society.

The law and lawyers are not concerned, as is often believed, solely with the processes of litigation. Their interest and their role extends to the entire process of law-making by judges, by legislatures, and by administrative bodies. Law is the cement which holds society together, and lawyers have traditionally played a major role as architects of public policy. Their daily work involves the application of legal principles to complex and esoteric problems over which they must have a complete mastery. In a single day, the lawyer may have to deal with the medical facts involved in a whiplash injury, the engineering detail and scientific principle involved in a complex piece of machinery, and the economic implications of a particular course of action. The good lawyer has the special competence to separate the relevant from the irrelevant and to ask the right questions. Beyond this, any assessment of benefits and risks necessarily must turn upon consideration of the social environment in which the technology will be used. The existing legal system will always have some bearing on whether, the extent to which, and the manner in which, benefits can and will be enjoyed; and the legal system will also have some bearing on whether and how the risks can be minimized or controlled. It may well be that enjoyment of benefits or control of risks will necessitate some change in relevant law, and such a change in law may in itself involve undesirable, and very important, first, second, or third order effects which should be taken into account in the technology assessment.

The social consequences of technology, whether desirable or undesirable, cannot be considered in a vacuum. They must necessarily be assessed within the context of the legal system viewed as a body of existing or potential rules which have or can have a directing, deterring, stimulating, or neutral effect on the technology. They must also be assessed with comprehension that introduction of a new technology may in itself result in changes in relevant portions of the legal system. In a real sense, therefore, there is a reciprocal relationship between law and technology which must be taken into account in the assessment process. Beyond this, it must be recognized that, whatever the nature and outcome of a form of technology assessment, as the ultimate decision is implemented in one way or another, private rights will be affected and vested interests disrupted, and affected parties will take to the courts to assert their interests. The formal technology assessment is, then, only one stage in a continuous progress of assessment. The legal system will provide inputs, and will operate to shape the course of the technology, before and after the formal assessment takes place, and lawyers will play an integral role in the process of providing these inputs. It is not clear, however, whether lawyers will play a substantial role in formal technology assessment procedures. Whereas the private assessment mechanisms cannot function without lawyers, there is no imperative that lawyers be part of a formal technology assessment institution. That such a role for lawyers exists is self-evident, but that whether this role will in fact exist is dependent on a recognition by those who control the assessment mechanism that lawyers can contribute something of value to the process.

MEASUREMENT AND TECHNOLOGY ASSESSMENT

MARVIN J. CETRON and DONALD N. DICK

President and Vice-President
Forecasting International Ltd., Fairfax, Va., U.S.A.

"Man does not live by bread alone." This age-old saying, if we contemplate its ramifications, gets at the crux of this chapter. If our savant merely had food in mind, as the strictest construction might imply, then we are led to believe that man has interest in meats, vegetables, dairy products, etc. We could make a very meaningful, quantified (numerically measured) assessment of his diet, given his weekly or daily menu. Through calorimetrics it is possible to measure the energy value of various menu items and, hence, the value of a specific diet. Our unit of measurement (also called an interval) would be the familiar calorie. At this point we would then be able to rank order menus by food energy value content, interesting and useful information, but by no means "conclusive" measurements. What is lacking is a point of reference, i.e., the daily, minimum caloric intake requirement. With this reference in hand we can make a positive assessment of food value: whether the minimum daily requirement is met, exceeded or fallen short of. Through the use of other dietetic measurements and standards the consequences thereof can then be stated and an assessment completed.

The last sentence of the foregoing paragraph makes a critical point that must be observed in any assessment and most certainly technology assessment. *All* parameters must be assessed, otherwise an assessment might not only be incorrect and incomplete, but also misleading. Our dietary assessment in the preceding paragraph would not have been a complete one because other nutritional factors had not been addressed for measurement. You, of course, realize that we had not touched upon the necessity for evaluating mineral and vitamin values and the characteristics of the fat content. One must always, methodically and at the outset, set down the independent and interdependent attributes of that which is to be measured and assessed.

There is a recent example of the havoc that can ensue if this is neglected. A medical center, at great expense in time and effort, assessed the efficacy of two different surgical techniques in terms of the length of the convalescence period experienced with each. Intuitively one method was felt to be better, but an examination of the resulting data failed to confirm the expected result. As it turned out, one important data category had not been included in the "measurements" taken: Age of the Patient. That surgical technique which had been considered the better intuitively, had been applied to a group of patients whose average age was considerably higher than the second group under the alternative technique. Convalescence data was obscured by this improper loading of the experiment.

Let us return to our dietary example. We find that man does not live by bread alone, nor by calories alone. To this we add the meaning which our savant had in mind: He does not live by food alone. Now we transit into the complexities to be faced in technology assessment. If he does not live by food alone, we may surmise that man attaches importance and value to other things: Love, honor, patriotism, creature comforts, career, purpose, fulfillment, security, peace, aggressiveness, and so on. At best life is a compromise of all of these factors; trade-offs are involved and a common value scheme is *implied*. However, this universal value scheme is most elusive and we are forced to use other logic in its place, methodology. Guns and butter cannot both be assessed in terms of calories, and added and subtracted under the usual rules of mathematics. The same problem exists in assessing technology as we will presently describe. As we observed from the example in the foregoing paragraphs, the problems of technology assessment are complex. To be comprehensive and useful, technology assessment must give careful attention to the interaction that obtains between government and industry. It must measure accurately the impact of technology utilization on social, political and economic patterns. It must establish priorities among the multitude of problems on the one hand and opportunities on the other that confront contemporary society. It must study methodologies of forecasting and planning as well as the structure of hierarchical aspects of independent assessments. And it must evaluate alternative solutions to problems in one area that may complement, or conflict with, proposed solutions to problems in another area.

Various methods are employed in technology assessment to meet these stringent demands, but, for the most part, current approaches are based on *system analysis* techniques. In general, these approaches consist of conducting an analysis of a given problem, proposing alternative solutions, and, guided by numerous inter-related measures of cost and benefit, selecting the optimum solution. In the past, most of these system analyses gauged cost solely in dollars and benefit strictly in economic terms. To meet contemporary needs, however, technology assessment requires the expansion of this approach. The concept of cost must be expanded to include social as well as economic costs, and the concept of benefit must be broadened to include society's collective goals and the aims and aspirations of individuals within the society. In short, we must expand beyond the calories, minerals and vitamins of our original example and consider the less tangible and less easily quantified factors that make for improved quality of life.

This expansion leads us into the area of "soft" data. Data required for the comparisons involved in technology assessment may be labeled as "hard" or "soft" (or somewhere in between) depending upon the degree of universal acceptance of the manner in which the data was generated. *Hard data* would be data from established fields of the physical sciences or accepted economic indicators. *Soft data* would be data from some social indicators or data based totally on judgment. The more easily data can be demonstrated and quantified, the harder it may be considered to be. In a technology assessment methodology involving the measurement and comparison of both types of data, the utmost caution obviously must be used in assigning numerical values to the softer data.

New analysis approaches such as cross-support, cross-impact and relevance matrices attempt to combine the use of "judgment" (soft) data with hard. These techniques, which require the determination of quantitative interrelationships between technological, social and economic factors, are beginning to be widely applied [1]. However, quantitative considerations can be meaningfully related to an overall perspective in an assessment methodology only to the degree permitted by the fundamental properites of measurement.

Since technology assessment requires measurement of the technological, economic and social sectors, it would be appropriate at this point to make a cursory excursion into the principles of measurement itself as a background to our treatment of the main theme.

The role of *measurement* in establishing any particular value system is to assign numerals to objects or events according to some set of rules [2]. This is simply the process of associating the characteristics of the objects or events with the rules of a number system. The system used is that of real numbers, and the rules are those normally associated with arithmetic operations. The rules are based upon three fundamental properties of numbers: identity, rank order, and additivity. The numerical properties, in turn, are used to distinguish the various levels of measurement. Each set of rules that assigns numerals to objects or events generates a level of measurement — a measurement scale. Four levels of measurement are generally used, namely: nominal, ordinal, interval and ratio scales. Statistical measures, the conditions for measurement, and the arithmetic operations permitted vary with the type of scale employed. The properties of each scale are summarized in Table 1. Let us briefly examine each of them in ascending order.

Nominal scales are the lowest order — the weakest — of all measurement scales. Numerals are used only as labels to distinguish among the items within a group, a function that could be performed by letters or other symbols. An example of the nominal scale in common use is the number on the jersey of a football player; it serves only to distinguish him from all the others on the field. We cannot safely say that player #40 is twice as good as #20, or that players #10 and #20 together are as effectual as player #30. The assignment of numerals to each item in a nominal scale is arbitrary, and the numbers have no meaning other than identification among the items. Only the statistical measures of frequency and mode have value with a nominal scale.

The next level in strengthening a measuring system is the *ordinal scale*. In this scale the numerical property of rank order is incorporated. Rank ordering requires that each item be compared to a single common attribute. This comparison results in the identification (or definition) of an order of preference among the items. If each item can be ranked relative to all other items, the ranking is called a simple order. If more than one item is placed in each ranking position (class) the ranking is called a weak order. Numerical designators can be used to distinguish between classes, but additive properties cannot be employed within the measuring system. As an example, consider the first-,second- and third-class buses of Mexico. It may not be argued that a third-class bus is the equal in comfort or efficiency of a second- and a first-class bus together. Numerical comparison between the classes is meaningless. The statistical measures one can use with ordinal scales include not only frequency and mode, but also median and percentiles. Although these latter measures can be applied, they cannot be extended to include interpolating between class intervals. The type of distribution between intervals is unknown in an ordinal scale.

TABLE 1. Measurement Scales

Scale	Measurement Conditions	Major Characteristic	Example	Statistical Measures	Numerical Properties
Nominal	Generate a collection of distinguishable names for each item.	Lowest level of measurement. Identify (or define) distinguishable characteristics of items.	Football player uniform number	Frequency, mode	Identity: Either $A = B$ or $A \neq B$ If $A = B$, then $B = A$ If $A = B$ and $B = C$, then $A = C$
Ordinal	Compare each item to a *single common* attribute.	First level effort to identify (or define) an order of preference.	Simple order — one item in each class — President of US, VP of US. Weak order — more than one item in each class — college seniors, juniors, sophomores, freshmen.	Frequency, mode, median, centiles, rank order, coefficient of correlation.	Identity plus rank order: If $A > B$, then $B \not> A$. If $A > B$ and $B > C$, then $A > C$.

TABLE 1. Continued

Interval	Standardize a unit of preference — arbitrary reference.	Intervals between points on scale are ordered. Ordered set consists of real numbers.	Centigrade (Celsius) and Fahrenheit temperature scales	All ordinal measures plus standard deviation, skewness, product-movements.	Identity, rank order plus: With an arbitrary zero, difference may be treated as absolute amounts and the differences assumed to have additivity property.
Ratio	Establish that measurement scale is referenced to an absolute or "normal" zero.	The ratio of one measurement to another remains the same when the scale is transformed.	Absolute — Kelvin temperature scale "normal" — Gregorian calendar	All interval measures plus coefficient of variance.	Identity rank order, plus additivity: If $A = P$ and $B > 0$, then $A + B > P$; $A + B = B + A$; If $A = P$ and $B = Q$, then $A + B = P + Q$; $(A + B) + C = A + (B + C)$

Interval scales use the numerical properties of identity and rank order, as did the ordinal scale, with the additional qualification that an order between points on the scale be established. This is accomplished by standardizing a unit of preference among the items for the common attribute. The units employed may be arbitrary but must be internally consistent when applied to each item. One important characteristic of an interval scale is that the zero point is a matter of convention or convenience. This limits the applicability of the numerical manipulation one can use. However, differences on the scale may be treated as absolute amounts and these differences may be assumed to have additive properties. Thus, normal arithmetic operations can be performed with the differences. A widely used example of an interval scale is the Fahrenheit temperature scale. Each scale division is equal in terms of a single measure (temperature); the difference between 190° and 200° is the same as the difference between 40° and 50°F. The statistical measures applicable to interval scales include all ordinal scale measures plus standard deviation, skewness, and product moments. The interval scale is the lowest order scale where quantitative comparisons among items can be made. Their credibility is restricted to the scale acceptance and are only as meaningful as the units of preference used.

The highest of the four levels of measurement is the *ratio scale*. It is referenced to either an absolute or a "normal" zero, the former having been established by scientific fact and the latter by convention. An example of an absolute scale is the Kelvin temperature scale. An example of a "normal" scale (with zero set by convention) is the Gregorian calendar, which is in reality an interval scale. The major characteristic of the ratio scale is the constancy of the ratio of one measurement to another when the scale is transformed. It is with this scale, and only this scale, that one is permitted to employ all the arithmetic operations of addition, subtraction, multiplication and division.

As Stevens notes, "It is an interesting fact that the measurement of many physical quantities has progressed from scale to scale. When men knew temperature only by sensation, when things were only "warmer" or "colder" than other things, temperature belonged to the ordinal class of scales. It became an interval scale with the development of thermometry, and after thermodynamics had used the expansion ratio of gases to extrapolate zero, it became a ratio scale" [3].

These four measurement scales have specific applications in technology assessment. Any methodology with the fundamentals of measurement must attempt to insure that any data, particularly soft data, is referenced to the appropriate measurement scale and placed in a credible perspective.

S.S. Stevens once said, "When description gives way to measurement calculation replaces debate" [4]. *Measurement provides the basis for reproducibility, objectivity and credibility.* But how do we measure — or assess — subjective judgments of social costs and benefits? What numerical value do we assign to costs like the loss of self-respect or the desecration of beauty? What value do we give to benefits such as peace of mind and intellectual enlightenment? The critical element in technology assessment is the quantification of subjective judgments in order to get reproducible outputs. *There are no standards for checking validity but one should expect reproducible outputs and objectivity so that one can see visibly what has taken place in each step of the assessment.*

Proceeding from the weakest measurement scale (nominal) to the highest (ratio), each higher scale incorporates all of the conditions of the previous one; i.e., the conditions are cumulative. The very nature of the conditions under which data is generated for the social, economic, and technological factors of technology assessment requires that any assessment method incorporate the conditions of the lowest level of measurement and subsequently meet the conditions for successive scales of measurement, up to the highest scale conditions permit.

Specifically, it should begin with the nominal scale, which requires that a collection of distinguishable names for each item be generated. For technology assessment purposes, the factors to be incorporated must be classified and placed in an overall structure that is meaningful to the assessment. This is such an obvious first step that its significance may be overlooked. Establishing an adequate structure will, in general, be an attempt to generate a set of mutually exclusive, collectively exhaustive descriptors. In many cases this is not achieved. One approach which has considerable merit is to develop a "functional" hierarchy that is based on the problem to be solved. A functional hierarchy is a structure that describes the problem area, rather than propose solutions or prescribe techniques. Chapter 37, Ted Taylor's, on agricultural assessment is an example of such a structure [5].

Proceeding to the ordinal scale requires the establishment of, and comparison to, a *single common attribute*. This is the first level of measurement where an attempt is made to identify or define an order of preference. The goal is to isolate common attributes that are meaningful to the assessment problem. A separate ordinal scale must be generated for each attribute to be measured. When judgment is used to establish the order among items, it must be exercised with regard to one common attribute and the judgment procedure must be explicit in this respect.

Interval scales require the standardization of a *unit of preference* for a common measurement attribute. This unit of preference and a suitable *scale reference point* are used to distinguish quantitative differences between positions on the scale for each item under consideration. Although the unit of preference and reference point are arbitrary, they must be derived in a manner that supports the desired utility of the assessment and must reflect the procedures employed to collect the supporting data.

When dealing with data from the physical sciences, the problem is minimal; suitable physical standards and references exist, and data collection procedures are fairly uniform. The softer sciences, however, must operate with a more nebulous body of knowledge and procedures.

An example is the collection of subjective judgments that incorporate social implications or establish preferences as well as values. Such judgments, incidentally, are a central issue in technology assessments.

Efforts in the field of subjective judgment collection are viewed with a great deal of skepticism, if not outright distrust, and not without some cause. This is especially true when system analysts and futurists wander into the field and solicit quantitative estimates of broad issues without using adequate procedures and conditions to insure credible estimates. The results are then used as if a valid ratio scale had been developed.

Even in the generation of a valid *interval scale* the data collected should:

(a) be arranged within a problem-structured classification system,

(b) demonstrate a single common attribute,

(c) include a standardized unit of preference, and

(d) be related to a meaningful scale reference point.

In the collection of subjective judgments for such a scale, the participants should be interrogated by directing their judgments.

Ratio scales of measurement can be used only if it can be stablished that the measurement scale is referenced to either an absolute or normal zero. The primary factors involved in technology assessment — social indicators, alternative future, etc. — as well as the nature of the forecasting effort preclude using such a scale. Thus, the highest scale one could hope to achieve would be an *interval* scale. It is important to note that selection of an accepted unit of measurement does not in itself provide a valid ratio or interval scale. Manipulation of quantitative values is always limited by data collection procedures and their associated measurement references.

Any technology assessment approach incorporates the fundamentals of measurement to some extent, whether those fundamentals are explicitly stated or not. Credible approaches involve the successive steps of systematic arrangements for (1) problem definition, (2) measurement of the items concerned, and (3) analytic comparisons. By examining the role of measurement in the assessment, the inherent problems become visible and permit the analyst to effectively demonstrate the areas of applicability of the study as well as its limitations.

At present, most technology assessments do not consider, or, at least, do not make visible these issues. This may be due in part to lack of knowledge of their existence, but is primarily due to failure to use rigorous techniques in attacking such problems. There are, however, several cross-correlation techniques which are beginning to be applied in technology assessment which have a dual relationship to measurement. They can assist in attacking problems such as structuring and interdependence, but their applicability and interpretation are limited by the level of measurement scale developed. Three of these cross-correlation techniques are *cross-support, cross-impact,* and *relevance matrices.* A brief examination of the data collection framework used in these techniques and of its relationship to measurement scales can serve as an example of the role of measurement.

Cross-support is used to determine the support effect of each item of a field on all other items. It is used to clarify complex relationships. For example: If Item A is developed, what will be the support effect on Items B, C, and D? It is used to define the extent of the support inter-relationship; the resulting information can serve to rank order each item from the point of view of cross-support. The cross-correlation is displayed as a matrix. It is a square array with the item-to-item effect described by the matrix elements.

Generation of the matrix elements is accomplished by soliciting subjective judgments. Hard data generally does not exist for such relationships. As we saw earlier, the manner in which the judgments are gained determines the level of measurement scale one can use.

One method used to generate the matrix elements uses a group of experts to associate numeric values with an adjectival scale as follows:

Major contribution	= 8
Considerable contribution	= 4
Some contribution	= 2
Negligible contribution	= 1

The effect one item of the field may have on another can be either beneficial (positive) or detrimental (negative). The judgments are made using varied comparisons.

Using such an adjectival-numeric scale will generate an ordinal ranking. The scale is ordinal, and although numbers are used to designate the scale, one cannot simply add numbers across the matrix as in simple addition. A *unit of preference* has not been established. A sum across the matrix which lists the sum of 8's, the sum of 4's, the sum of 2's and the sum of 1's is valid, and would indicate the relative total *cross-support* of each item as defined for levels of contribution. Note that changing the 8, 4, 2, 1 to a 4, 3, 2, 1 scale still does not permit a conventional arithmetic sum without implying that one "major contribution" is equivalent to four "negligible contributions." If a unit of preference is defined and used to generate an interval scale, the matrix rows can be summed and a numeric rating given to an item's total cross-support (Fig. 1 is an example of such a matrix). Differences in numbers would then have meaning - meaning in terms of the unit of preference. There have been many varied uses of cross-support recently [6].

Another cross-correlation method used is called *cross-impact*. Cross-impact is used to establish the *change in the probability* of individual events with the occurrence or non-occurrence of related events. A matrix of future events is formed as shown in Fig. 2. The first column indicates the best estimate that an event will occur, independent of the other events. The elements of the matrix are completed by a panel

Goals	g_1	g_2	g_3	g_4	—	—	—	g_n	Total cross support
Housing for 90% population by 1979 (g_1)		1	1	2					27
90% literacy rate for 10-25 yr age group by 1979 (g_2)	1		1	4					21
Adequate nutrition for 100% population by 1980 (g_3)	1	4		1					24
8% increase in GNP per yr (g_4)	8	2	4						16

Fig. 1. Goal cross-support matrix (illustrative)

By 1985 original probabilities	If this innovation occurred	The probabilities of these innovations would be affected as follows:					
		T_1	T_2	T_3	T_4		T_n
.75	CATV in 80% homes (T_1)	–	9	0	0		
.4	80% communications by satellite (T_2)	8	–	4	0		
.8	News by remote facsimile transmission (T_3)	5	0	–	0		
.2	Manufacture of drugs in space (T_4)	0	0	0	–		

Fig. 2. Cross-impact matrix (illustrative)

of experts using a scale which can be either positive or negative, depending on the enhancement or inhibition of the innovation concerned. The question that can be posed is: Given that Event 1 will occur in Time 1, how would the probabilities of each of the other innovations be affected? Any manipulation of the numerics here is once again limited by the choice of scales used for the estimates. This technique has been expounded in great detail by Gordon and Becker in Chapter 29 and is in wide use [7].

Of the three types of matrix techniques, the relevance matrices are probably the oldest [8]. These matrices are used to establish the relationship between two fields by considering each item within the fields. (Figure 3 is an example.) There are many examples of this technique, including PATTERN [9], PROFILE [10], QUEST [11], and the TRI-MATRIX Technique, as explained by M.J. Cetron in Chapter 30. The key point of these matrix techniques is to develop a means of assigning a "value" to technology. Admittedly there is subjective judgment involved in putting some kind of numerical value on projects, but properly used, such numbers can help a manager make decisions without causing him to ignore his experience and judgment.

When we begin the formal process of quantifying judgments by matrix techniques, a group working through the matrix benefits by numerous built-in advantages of this approach. We will list some of the important ones.

(1) More individuals' views and expertise are brought to bear on various aspects of the problem.

(2) The matrix approach obliges the participants to structure their thinking and may well give them a new perspective of the problem.

(3) Because the numerics must be written it gives great visibility to the process and has the potential of opening up a more extensive dialogue among the participants.

(4) It is an excellent way to gain the involvement of the participants since it becomes *their* program.

Weights	W_1	W_2	W_3	W_4	—·—·—·—	W_n	Total relevance index
Indic. \ Goals	g_1	g_2	g_3	g_4		g_n	
e_1	β_{11}	β_{12}	β_{13}	β_{14}		β_{1n}	$\sum_{j=1}^{n} \beta_{1j}W_j = u_1$
e_2	β_{21}	β_{22}	β_{23}	β_{24}		β_{2n}	$\sum_{j=1}^{n} \beta_{2j}W_j = u_2$
e_3	β_{31}	β_{32}					
e_4	β_{41}	β_{42}					
1 1 1							
e_m	β_{m1}	β_{m2}				β_{mn}	$\sum_{j=1}^{n} \beta_{mj}W_j = u_m$

Fig. 3. Indicators/goal impact/relevance matrix

It should be noted that older forms of decision-making still in vogue today involve subjective judgments. With these matrix techniques we employ them too, but they force us to do so explicitly rather than implicitly, thereby reducing the likelihood of poor decision-making due to unconscious loading.

The role of measurement in technology assessment is a fundamental one. Quantitative relationships are required to adequately describe the effects that technology utilization can have on all sectors of society. In many assessments this role is masked by the overall approach and the techniques used. Nevertheless, basic measurement considerations must be satisfied in order to obtain a level of credibility. The newer matrix techniques, touched on here, make this role obvious. The visibility afforded by making measurement conditions explicit permits a reviewer to quickly discern the limits of the assessment and qualify in his mind an appropriate perspective for utilization. As such, understanding the role of measurement can be a contribution to the arsenal of techniques for analysis.

There are still omissions and weaknesses in quantitative technology assessment techniques based on the type of analysis illustrated here. It should be stressed again that the data generated does not yield decisions, rather information that will facilitate decision-making. Indeed, these techniques are merely thinking structures to force methodical, meticulous consideration of all the factors involved in the assessment. In other words, data subjected to analysis yields information, and information subjected to judgment yields decisions.

Those of us who are working to build these assessment systems are firmly convinced that if we had to choose between any system with miles of computer print-out and the human brain, we would take the brain. The brain has a marvelous way of learning from experience and has an uncanny way of pulling out salient factors and rejecting useless information. It is wrong to say, however, that one must choose intuitive experience over analysis, or minds over machines. The two are not alternatives but partners that complement each other. Used together properly, they can be counted on to reduce the most complex problem to an answerable question.

A close look at a few quantitative technology assessment methods will show that most of these approaches are still experimental management techniques. The fact that a computer or an adding machine may be used to facilitate data handling should in no way distract from the basic fact that human subjective inputs are the foundation of these systems. Accurate human calculation, as opposed to use of a computer for the calculation of all the inter-relationships considered, would not alter the basic principles of these management tools in any respect.

These approaches represent the latest thinking on how to use the collective judgments of technical staffs and decision-makers in such a manner that the most logical and sound decisions evolve. The goal is the greatest pay-off possible for the investment of resources, whether in terms of men, money or facilities. To make an incorrect decision is understandable, but to make a decision without a conscientious and thoroughgoing effort to get as much usable information as possible is unforgivable. The managers who design and work with technology assessment systems, however, must realize that the correlation matrices, technological forecasts, quantitative estimates of project value, and other aids to technology assessment are simply planning tools.

Even this caveat, however, does not quell the cries of the critics and there are some very vocal critics around, both in government and in business. Some of the criticism is a reaction to the fear of "mechanization" of a task felt to be rightfully in the province of human evaluation. Other critics claim that building up a logical system, computerizing the output, and quantifying what are essentially intuitive and judgment decisions may insulate some managers with a false sense of security. In some way, they fear, management responsibility will be hard to pin down.

Systematic analysis, however, tends to force managers to consider their technology assessment tasks more comprehensively. It also highlights problem areas that might easily be overlooked by more traditional approaches. At the same time, quantitative matrices are considered threatening by some managers because such techniques may tend to expose their value judgments to critical analysis by others. As the high degree of sophistication of these planning devices improves, however, managers will be won over, especially when they find that they will be able to spend more, not less, time doing real decision-making.

REFERENCES

1. CETRON, M. et al., *Technological Resource Management, Quantitative Methods*, Cambridge, Mass.: MIT Press, 1970, and CETRON, M. and RALPH, C., *Industrial Applications of Technological Forecasting*, New York: John Wiley & Sons, Inc., May, 1971.

2. HALL, A.D., *A Methodology for System Engineering*, Princeton, N.J.: Nostrand Company, Inc., 244, 1962.

3. STEVENS, S.S., *Handbook of Experimental Psychology Mathematics, Measurement, and Psychophysics*, New York: John Wiley & Sons, Inc., 25, 1951.

4. *Op. cit.*, 1.

5. TAYLOR, THEODORE, "Food Production Wastes - Summary of a Case Study of a Systems Approach to Research Program Planning," *Paper presented at the Technology Assessment Symposium*, Stockholm, Sweden, October 11-12, 1971, and as Chapter 37 in this volume.

6. RALPH, C.A., "The Beginnings of Cross-Support Analysis (DIANA) as Applied to the Fishing Industry," CETRON, M. and RALPH, C., *Industrial Applications of Technological Forecasting: Its Use in R & D Management*, New York: John Wiley & Sons, Inc., 1971. Also, the research laboratory of the Mead-

Johnson Corporation in Evanston, Illinois and UNESCO are applying the techniques. The way in which a modified cross-support is used in the developing countries in Latin America, Southeast Asia, and Africa is included in Chapter 12 entitled, "A Method for Planning and Assessing Technology Against Relevant National Goals in Developing Countries," by M.J. Cetron and L. Connor.

7. Two other illustrations of the utilization of cross-impacts are Jim Dalby's article entitled, "Practical Refinements to the Cross-Impact Matrix Technique of Technological Forecasting," in CETRON, M. and RALPH's, C. book, *Industrial Applications of Technological Forecasting: Its Uses in R & D Management*, New York: John Wiley & Sons, Inc., 1971 and the Monsanto Corporation's program which is explained in some detail by Kenneth Craver on pp. 97-100 in the same book.

8. CETRON, M.J. et al., *Op. cit.*

9. ESCH, MAURICE E., "Planning Assistance Through Technical Evaluation, PATTERN," *Paper presented at the 17th National Aerospace Electronics Conference*, Dayton, Ohio, May 11-12, 1965.

10. CETRON, M.J., "Programmed Functional Indices for Laboratory Evaluation, 'PROFILE,'" *Paper presented at the 16th Military Operations Research Symposium (MORS)*, Seattle, Washington, October 10-14, 1965.

11. CETRON, M.J., "Quantitative Utility Estimates for Science & Technology, 'QUEST,'" *Paper presented at the 18th MORS*, Fort Bragg, N.C., October 19-21, 1966.

UTILIZATION OF CROSS-IMPACT IN TECHNOLOGY ASSESSMENT

T.J. GORDON and H.S. BECKER

President and Treasurer and Director of Programs, respectively, The Futures Group, Glastonbury, Conn., U.S.A.

INTRODUCTION

Technology assessment involves anticipating the likely impact of a technology on a variety of hard-to-quantify environmental, social, political, and value factors, as well as the effect of potential changes in these areas on the acceptability of the technology. The SST debate furnishes an example of a technology assessment issue. Would operation of the new aircraft adversely affect the global environment? Would travelers value very short trips in Europe? How would the Soviet or the French-English aircraft fare if ours were to be cancelled? What of jobs in Seattle and elsewhere? Typically, technology assessment problems are multidimensional, multidisciplinary, socially-oriented, and, in the broadest sense, political. Analytic models are very useful in studying technical parts of such problems (e.g., the output of contaminants from the SST's exhaust and their effect on the atmosphere), but the social, political, and value components of technology assessment problems rarely yield to modeling. Cost-benefit approaches, in which attempts to interpret non-economic costs and benefits (such as impact on public wealth) in pseudo-economic terms, are possible in some situations, but are generally difficult, sometimes unsatisfying and often quite dependent on the analyst's subjective interpretation of the value structure of concerned groups.

Cross-impact methods overcome some of these problems.

They furnish a means of exploring the effects of interactions between different kinds of potential events associated with technology assessment (for example, between a proposed new technology and institutional changes). Judgment is still an essential ingredient in this process, but, unlike most other approaches, it is dealt with explicitly, and its contribution to the solution can be evaluated. The aim of the cross-impact approach in technology assessment is to form a wholly explicit model which is useful in testing policies designed to improve or diminish the probability of occurrence of events associated with the technology being considered.

Cross-impact analysis is based on the concept that the occurrence or non-occurrence of a possible event or the enactment of a particular policy may affect the probability of occurrence of a host of other events and policies. The method requires that such interactions be defined and their strength estimated. In the absence of hard data about these future interactions, opinion and judgment are used. With these judgments about interactions in hand, a large number of scenarios can be analyzed (usually by computer) to identify privotal events and policy consequences.

HISTORY OF DEVELOPMENT

The cross-impact method had its origin in *Future*, a simple simulation game, designed in 1966 by Olaf Helmer and T.J. Gordon for the Kaiser Aluminum and Chemical Corporation. In this game, sixty potential future events, technological as well as societal, were described on 5" x 5" cards. The cards also showed the designers' assessment of the probability of occurrence of these events by 1985. Since event probabilities changed during the play of the game, a scoring system was provided to record the probability of each event. A number of the cards were shuffled and "played" as follows. The players were first given the opportunity to adjust the probability of the card at the top of the stack by making "investments." Then through the use of a randomizing device (an icosahadron die with 10 percent of its faces marked with a "10", 20 percent with a "20", 30 percent with a "30", and so on), the event was "decided" by applying the usual rule that if the number on the die equalled or exceeded the event's stated probability, it was said to have occurred. If the event occurred, it was taken to be part of the world of 1985, and the card was flipped over to reveal the other events in the set that would be affected by this outcome. The probability of each of these events was shifted up or down by a predetermined amount (and so recorded by the scoring system) so that later, when its turn came to be decided, its new probability was used. In this way, all of the selected items were "decided", thus forming one possible "world" of 1985.*

Of course, the feature of interest here was the inclusion of the event interactions. The game *Future* stimulated further work. At the University of Illinois, the process was automated so that a player sitting at a CRT console could view the events and judge whether they were a desirable component of the future. At UCLA, Syracuse University, and elsewhere, the game *Future* formed the basis of seminars on which new forecasted events central to the theme of the seminar replaced those used in the original game.

At UCLA in 1968, T.J. Gordon and H. Hayward developed the first computer-based approach to dealing with cross-impact analysis as a technique in its own right [1]. In this approach, significant events judged relevant to a particular aspect of the future and their probability were first described. These events were displayed in the form of an orthogonal matrix with the diagonal cells excluded. At each intersection the question was posed: "If the event in the row were to occur, how would it affect the probability of occurrence of the event in the column?" These interactions were judged to be enhancing, inhibiting, or unlinked (no influence). The judgments were recorded in the matrix cells, together with the suspected strenth of the interaction. The computer "solution" involved:

---

*In the sense used here, a "world" is a listing of events which is assumed to occur by a particular time in the future.

1. Selecting an item at random from the list of events.

2. "Deciding" it on the basis of the item's initial probability.

3. Adjusting the initial probabilities of the remaining items according to the information about interactions furnished in the matrix.

4. Continuing the process until each item had been "decided" as having happened or not happened and appropriately adjusting the probabilities of the remaining events.

5. Repeating this process a large number of times until enough trials were available to determine new probabilities for the items by recording the number of times each item occurred during these trials.

This process suffered from many faults, including the arbitrary nature of the algorithm used to convert the estimated strengths of the interactions into probabilities, the difficulty in supplying the original independent probabilities, the limitation of the process to event pairs only, and the omission of information about the consequences of non-occurrence (i.e., if an event didn't happen, how were the probabilities of other events likely to be affected?).

Nevertheless, even in this form the process had several interesting applications:

1. It was sometimes useful as a heuristic device to identify which events were likely to be most important to some aspects of the future and how these events were likely to interact. Making such judgments explicit raised to visibility many decision elements often glossed over.

2. The approach made it possible to integrate the opinions of experts of different disciplines by asking them to furnish judgments about interactions meaningful to them. These judgments were recorded in the appropriate matrix cell.

3. The matrix, once completed, served as a sort of model against which prospective policies could be tested.

For example, if a new item, representing a policy, were added to an existing matrix, the differences in outcome between an earlier "policy-less" matrix and the one with policy, could be attributed to the affect of the policy. Thus, if one entry in the matrix were a goal (e.g., likelihood of achieving a desirable return on investment) the worthiness of the policy under test could be assessed by the degree to which it shifted the probability of the goal [2].

In its early form the approach was used by the Education Policy Research Center at Syracuse University in the analysis of educational policy, and by the Monsanto Corporation in the construction of a more sophisticated, dynamic version of the process. Cross-impact approaches now form an important part of the long-range planning process at Monsanto. The approach has also been used in corporate planning by other companies, such as General Electric.

At the Institute for the Future, the method was gradually improved by T.J. Gordon, S. Enzer, R. Rochberg, O. Helmer, and others. An intermediate step was reported in "Research on Cross-Impact Techniques with Applications to Selected Problems in Economics, Political Science, and Technology Assessment "[3].

Subsequent work has solved some of the earlier problems and has produced a technique for assessing the impact of policies. Contributing to these efforts have been projects undertaken by Otto Sulc in Czechoslovakia, Norman Dalkey at The Rand Corporation, Howard Johnson at the University of Texas, Teruo Shito and others at the Science and Technology Agency of the Japanese Government (who are exploring the use of cross-impact technologies to evaluate policy alternatives suggested by the Agency's major Delphi study published in June 1971), and William Brauers from the Center for Research in Defense in Belgium (who is working to introduce cross-impact analysis into the planning-programming-budgeting system in the Center), and analysts in various agencies of the Canadian Government.

One aspect of these advances was the application by S. Enzer and O. Helmer, of statistical formulae which set limits to the assessed probabilities of the dependent events. In addition, input probabilities are taken to be mutually dependent estimates, the data in the matrix is in the form of revised probabilities rather than hard-to-translate impact strength, and a non-occurrence matrix is computed, rather than estimated [4]. Among the earlier criticisms still not satisfied is that the method is constrained to event pairs. This limitation, however, arises only because of the amount of work that would be imposed on the persons providing the estimates if they were asked to deal with groups of events; it is not a theoretical weakness of the approach.

An essential point: since the process permits the selection of prospective events of almost any sort--technological, institutional, goal-oriented, social, political, economic, value-derived--it is, in concept, a technology assessment tool.

EXAMPLE OF METHOD

Perhaps the most direct way to illustrate the technique is through an example. Suppose we were concerned about the impact of the electric automobile [5]. Some events we might include in the matrix are:

	Potential Event	1985 Probability
1.	Manufacture and scale of 8 million electric autos/yr	.5
2.	Depression & unemployment in the petroleum industry	.3
3.	Great expansion of the battery industry	.5
4.	Significant improvements in pollution levels	.3
5.	Introduction by utility industry of SO2 & NOx removal or control systems	.7
6.	40 percent increase in power demand	.5

The occurrence matrix might appear as follows:

Probability	Event	1	2	3	4	5	6
.5	1	X	.60	.70	.60	.70	.90
.3	2	.60	X	.50	.30	.70	.50
.5	3	.60	.30	X	.20	.70	.70
.3	4	.20	.30	.20	X	.40	.50
.7	5	.60	.30	.50	.40	X	.50
.5	6	.50	.30	.50	.10	.70	X

The decimal figures in the matrix represent the authors' assessment of the probability of the event in the column if the event in the row were certain to occur. The information contained in the matrix is a model, subjective to be sure, which depicts the interrelationship between potential future events of the sort normally included in technology assessments.

With this information, the non-occurrence matrix can then be calculated using standard statistic definitions. In this case it would be as follows:

Event	1	2	3	4	5	6
1	X	0	.30	0	.70	.10
2	.46	X	.50	.30	.70	.50
3	.40	.30	X	.40	.70	.30
4	.63	.30	.63	X	.83	.50
5	.27	.30	.50	.07	X	.50
6	.50	.30	.50	.50	.70	X

To illustrate the analysis of this matrix, we can follow the computer process step-by-step.

1. From among the items 1-6, one is selected at random, say item 3.

2. A second random number (from 0 to 1) is generated and compared with the stated probability of item 3. Say the number is .6. Since the probability of item 3 is less than the selected number, the event is recorded as having not occurred.

3. The probabilities of the remaining items are adjusted as indicated by the odds ratios from the non-occurrence matrix.

4. A second item is selected at random from these remaining items, say item 6.

5. It is decided on the basis of its new probability.

6. Now, the probabilities of the remaining items are computed using either the occurrence or non-occurence matrix, depending upon whether item 6 "occurred" or "did not occur".

7. All of these items are decided in this manner. At the end of the sequence a "happen" or "didn't happen" notation is associated with each item. This is a single scenario.

8. The process is repeated many times, typically 1,000, to form new scenarios, and new probability estimates are computed for each item based on the number of times each appeared in the scenarios.

9. These final probabilities and the initial probabilities will essentially coincide if the matrix is balanced; that is, if the entries in the matrix do not violate rational statistical limits.

Now suppose a proposed policy action were to be tested. For this case, let us assume that we decide to make it certain that 8 million electrical automobiles are manufactured; in other words, the probability of the first item of the matrix goes to unity. The policy is shown at unity probability since it is an action which, for the test at least, is certain to occur. The computational sequence is repeated. The difference in probabilities between the first and second sets of runs can be attributed to the policy under test. In this example, the impact of a decision to manufacture the 8 million electric cars has the following effects:

	Probability	
	Original	With Electrical Automobile Decision
1. Manufacture of 8 million electric autos/yr.	.5	1.0
2. Depression and unemployment in the petroleum industry	.3	.43
3. Great expansion of the battery industry	.5	.57
4. Significant improvements in pollution levels	.3	.39
5. Introduction by utility industry of SO2 & NOx control systems	.7	.69
6. 40 percent increased in power demand	.5	.69

Clearly it would be advantageous to test how other additional policies (in sequence or in parallel) might affect the world depicted by the cross-impact model.

SUMMARY

At its root, the process of technology asessment is one of anticipating primary and secondary interactions. The most important of these interactions cross disciplinary lines and are therefore difficult to handle analytically. Cross-impact techniques are by no means a panacea for all of the difficulties of technology assessment, but they may be a step in the right direction.

1. GORDON, T.J. and HAYWARD, H., "Initial Experiments with the Cross-Impact Matrix Method of Forecasting," *Futures, 1*, No. 2, 100-116, December 1968.

2. GORDON, T.H., "Cross-Impact Matrices: An Illustration of Their Use for Policy Analysis," *Futures, 1*, No. 6, 528-531, December 1969.

3. GORDON, T.J., ROCHBERG, R., and ENZER, S., *Report R-12*, The Institute for the Future, August 1970.

4. A more detailed description of the techniques as of this writing will be found in: ENZER, S., BOUCHER, W.I., and LAZAR, F.D., "Futures Research as an Aid to Government Planning in Canada: Four Workshop Demonstrations", *Report R-22*, The Institute for the Future, July 1971.

5. The events used in this example are drawn from NETSCHERT, BRUCE, "The Economic Impact of Electric Vehicles," *Bulletin of Atomic Scientists*, 2935, May 1970. The 1985 probabilities shown are merely illustrative and were estimated by the authors.

THE TRIMATRIX--AN INTEGRATION TECHNIQUE FOR TECHNOLOGY ASSESSMENT

MARVIN J. CETRON

President
Forecasting International Ltd., Fairfax, Va., U.S.A.

A NOTE OF CAUTION

There was some consideration as to not using this chapter on the trimatrix method since the method failed to take into consideration three aspects needed in the systematic integration. However, after discussions with various practitioners in the field, it was felt that presenting the method, together with its drawbacks, could serve a useful purpose. Therefore, the trimatrix technique, with its three omissions, and the ways to overcome the deficiencies are presented with the assumption that others in the field would learn by not making the same mistakes I made. This is the practical application of the concept that, "Experience is the best thing you can buy, if you're smart enough to get it secondhand." Therefore, I present this chapter essentially as it was developed in 1969, and at the end of the chapter, present the test findings, and explain how the three omissions from the method could be corrected and incorporated into a more acceptable model, which is presented in the next chapter by Christine Ralph.

BACKGROUND

Diminishing natural resources, need for environmental restoration, and economic difficulties make it mandatory to seek new and more effective methods for long-range planning and decision-making.

There is also the need to better understand the future challenges brought about by current trends in technology. Effective projection of scientific trends is required as early as possible in order to efficiently allocate available resources. Obvious planning deficiencies in current environmental, economic, social, and defense activities can be found in every sector. Consequently, there is an urgent requirement for an integrated system of technological forecasting and planning.

Traditionally, many corporations are preoccupied with the day-to-day task of conducting business, and consider technological planning only as it applies to their specific realm of interest. Furthermore, corporate managers tend to react to technological developments and changes in their operating environment only when they occur. This is not enough; indications are that corporations must evaluate their position and effect on the total environment in which they operate. This provides the corporate manager the opportunity to act--not just react--in determining the corporation's future.

The role of technology in everyday life has come under increased scrutiny, not only from corporate managers, but from all sectors of society. A considerable portion of this interest is due to the realization that technological exploitation can result in undesirable long-term effects. Focus at the national level is currently directed at assessing technology, but the capability required for performing a comprehensive assessment of present and emerging technologies has not been developed. Using Gabor Strasser's (White House Staff, Office of Science and Technology) definition: "Technology assessment is nothing more than a systematic planning and forecasting process, delineating options and costs, encompassing economic as well as environmental and social considerations, that are both external and internal, with special focus on technology-related 'bad' as well as 'good' effects."

During the last two years, "technology assessment" has become a household word, being used as frequently as "environment," "ecology," and "social indicators." However, the term "technology assessment" is frequently misused and even more frequently misunderstood. It is not to be construed as being synonymous with the protection and improvement of Man's environment. If, in overreaction to these problems, resources are allocated predominantly for environmental or ecological restoration, instead of in a more balanced and comprehensive program to improve the quality of life, society will surely suffer from imbalances and secondary side effects. In other words, we cannot afford a reduction in our gross national product or a stagnation of technological innovation, for the sake of an improved ecology. We will simply have to face the problem of priorities in an earnest way in order to reach an acceptable compromise, because all our problems, such as transportation, power, the environment, crime, etc., cannot be solved simultaneously. This chapter provides a method of systematically reaching such an acceptable compromise.

Technology assessment, a term coined by Philip Yaeger, Counsel for the House Committee on Science and Astronautics, is an attempt to establish an early-warning system to detect, control, and direct technological changes so as to maximize the public good while minimizing the public risks. It is a relatively new approach to allocating scientific resources, establishing technological priorities and seeking relevant alternatives to current technology.

It is easy to conclude that solutions to problems involving technology are technological. This simplistic attitude ignores the human factors which may have initially created the problems, or which may inhibit solutions.

The recognition of new techniques and technologies has provided sufficient incentive to pursue these opportunities. Where industry has not taken the initiative, government has provided it through economic incentives.

Decision-aiding models have embraced every segment of the R&D spectrum from the most fundamental research through operational development. In every instance, the major elements have fallen into three general categories: utility, feasibility, and cost. Sometimes one of these factors has been omitted, and consequently the models were considered deficient by other practitioners. [1].

There is also the growing opinion that a greater GNP is not the most important goal for a country [2]. Such factors as quality of life and ecology may even be more important *once a reasonable GNP has been achieved.* The same is true for a person. Many studies have shown that salaries are not the most important factor to an individual's well-being (gestalt). Factors such as professional development, opportunity to publish, the freedom to "do your thing" are often more important to an individual after some minimum threshhold salary is met. (There is always some *minimum* salary.) An analogous situation applies to a firm. Profit is not a firm's most important product--long-term growth and continuity are more important-- but only *after a minimum profit is generated.*

One can see then in all three areas, the individual, the corporation, and the government, that after the *survival* (financial) goals and the *long-term growth and continuity goals* are achieved, the individual, corporation or government begins to consider *societal goals*. For the individual these may include the Parent-Teachers Association (PTA), the Youngs Mens Christian Association (YMCA), Red Cross, or community organizations. For a company they may include sponsoring educational programs such as "Sesame Street", day care centers to help working mothers, centers for the performing arts, csholarships fro needy students, etc. Countries on the other hand will work toward international cooperation on the enviornment, health, safety, adn possibly even non-nuclear proliferation treaties and cooperation on scientific projects, such as oceanography or space.

It is interesting to notice that the output of the societal goals of each of the similar groups in the hierarchy is the input to the next larger group in the hierarchy. The societal goals of the individual are inputs to organizational units and the outputs of corporations at the societal level are the inputs into the local or federal government. The outputs of the various societal goals of countries are inputs into the goals of international organizations.

A heirarchical list of the societal units or components is attached as Exhibit A at the end of this chapter. This illustrative list, which was prepared by Forecasting International (FI) as part of a consultation for the Swedish Board for Technical Development, is much more comprehensive than the Maslow heirarchy [3], which is only applicable to individuals, not groups (institutions and organizations). The FI list of societal components is also not nearly as existential or as esoteric as the Graves heirarchy [4]. Graves, in his "level of existence theory," does consider 8 levels of societal components, and takes into account that growth and development are endless, open and continuous, but he does not consider that each group has survival, growth and societal goals.

An informal conglomeration of specific technology assessments does not necessarily address all the desired features of technology assessments on a national scale. What remains to be accomplished is the effective integration of: national priorities, the sensitivity of the national priorities to specific technology assessments, and a method to show relationship between specific technologies and areas of national interest. An overview of the synergistic effect of various technologies is necessary in order to provide a basis for informed decision-making at the national policy-making level.

AN APPROACH

It is an established fact that it takes from 5-10 years to develop a system from an operational concept through the research, development, engineering, testing, and implementation phases in most moderate-to-large companies, or any other similar organizational entities (public, private, or government). If such a system is intended to be an integral part of the company's long-term continuity, growth, and profit, then significant resources are usually expended in assuring its creation. This means that, within the 0-5 year time frame at least, barring unforeseen radical changes in the operating environment, it would be uneconomical to consider major changes in plans for that period. Hence, in the 5-10 year time frame, there is greater opportunity for a company to consider changes in its operational plan, though there are still some limitations on the magnitude and scope of changes which can be considered. This limitation is based on many factors (e.g., amount of resources previously committed, importance of the program or system to the survivability or profit posture of the company, etc.). In the 10-20 year time frame, however, there is freedom in nearly every industry to consider and specify major changes in operational plans.

Most corporations operate under some short range (3-7 year) business plan, which is updated annually, and which defines the direction to be taken during that period of time. Inclusion of a 0-5 year forecast and assessment is desirable because (1) it provides continuity to and serves as a baseline for the 5-10 and 10-20 year forecasts and assessments; and (2) it can be compared with the existing short-range plan, and the differences, if any, evaluated to identify elements of the plan which the corporation may want to redirect to achieve a greater continuity with long-range goals. In conjunction with his colleagues [5], the author has developed one approach to technology assessment which considers this process, while also taking into account near- and long-term national and/or corporate priorities and their sensitivities to technological advances. The method provides a framework for technology assessment macro-analysis, which, in turn, provides the overview necessary for strategic policy decisions. The following illustrates the feasibility and decision-making utility of this method when applied to any organizational entity--public, private or government.

The method has five principal elements. These include:

1. A set of national goals or areas of concern are

explicitly stated--transportation, health, and national defense are examples of areas where national goals would be stated. (These can be generated separately and concurrently for later use, to measure the effectiveness of the assessment.) This same assessment can be formed using corporate goals.

2. The national technology base is defined as where organized knowledge and scientific advances are applied to solve problems. Atomic energy conversion and organic materials are examples of technological areas that form the national technology base.

3. The transformation of the advances of sciences to meet national needs requires that the socio-economic organizations create systems to meet national goals. Examples of such systems of national importance include electric power distribution, ground transportation, and telephone communication.

4. The impact (progressive and regressive) of using technology to meet national goals, by various systems, is measured by social and economic indicators*.

5. The social and economic consequences of technological alternatives are referenced to a common data base, and these data arrayed to provide an overview of all alternatives.

The information flow for the above and the method used in assimilating and processing the information is shown in Fig. 1. Each area describes the nature of the effort and the type of activity to be performed.

The specific objectives of such an approach include:

1. To establish a relevant technology base and the timing for new technologies

2. To define the most probable alternative future environments ("new worlds") in which the country or corporation may be required to operate

---

*Two examples of social and economic indicators are shown elsewhere in this chapter.

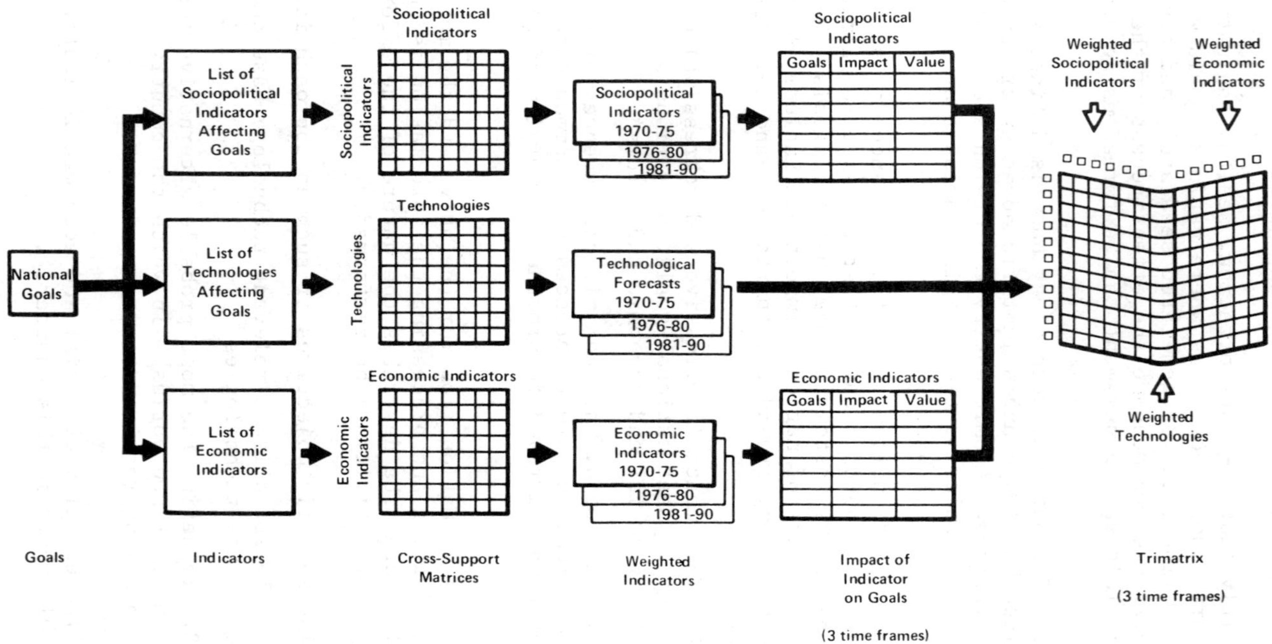

Fig. 1. Technology assessment study

3. To assess the impact of technology in creating these new environments, both inside and outside the corporation (or country)

4. To determine the relative advantages and disadvantages of various technological courses of action--opportunities vs. threats--open to the corporation, and to porvide guidance in the evaluation of such optional courses of action

5. To identify and assess major operational, new business and service opportunities and constraints to the corporation in the near and long-term future

6. To identify and assess current and new business opportunities which may be of particular interest, and of economic and/or social value to the corporation

7. To provide the basis for evaluating the existing goals and operational policies of the corporation; and, perhaps, assess how such goals impact on segments of the "external" national environment with which the corporation interacts or could interact strongly

8. To develop a working management planning tool which can be used on a continuing basis.

EXPLORATORY FORECASTS

Technological forecasting during the last ten years has emerged as one technique which can aid in the decision-making process for resource allocation and priority selection. The technique is broad enough to cover, for instance, such items as the planning of a research program to fight lukemia, or evaluating technical alternatives to provide sufficient electrical power to the states along the eastern seaboard.

Existing information on the future of various technologies (e.g., nuclear power, materials, etc.) can be transformed, where necessary, into the functional areas of the systems. However, recent forecasts of technology do not appear to be sufficiently oriented to reflect both the beneficial and detrimental impacts they may have on national goals especially the social and economic sectors.

Deficiencies in information are accommodated by initiating supplementary forecasts using such forecasting techniques as trend extrapolation or Delphi. Exploratory forecasts used in technology assessment studies are usually referenced to at least three base points of time: 1971--the current state, 1976--five-year plans, and 2000--the focal point of a number of extensive long-range projections. Other dates of particular interest can be included. The major aim in the initial stages of such studies is to filter and supplement available forecast data to meet the study needs.

NATIONAL AND/OR CORPORATE GOALS/TECHNOLOGY BASE

The selection of a set of credible quantitative national and corporate goals is based upon a review of existing suggestions, updating and expanding where necessary.

Selection criteria emphasizes completeness of coverage, realism in terms of current trends, and adequate data backup. The set of goals proposed by Leonard Lecht was presented to several Congressional Committees, and is considered by many to be a good initial reference basis for such studies [6].

The listing of the goal classification and an abstract of standards was proposed in 1968. Each of the goals represents a quantitative standard based upon the 1968 trend of informed opinion. Each goal is supported by historical data. Also, the costs of the goals have been estimated. (They are attached as Appendix B to this chapter.) Additional discussion on national goals is included in Lecht's chapter in this book entitled, "Shifts in National Priorities, Economic Grants and Requirements for Scientists and Engineers." (A new set is currently being prepared by Lecht at the National Planning Association.)

The national technology base can be used to develop

---

*LECHT, Leonard, *Goals, Priorities, & Dollars: The Next Decade*, New York, The Free Press, 1966.

a technology goal framework of functional and support technologies based upon the systems necessary to meet the goals. Goal areas can then be used to develop a functional system structure compatible with technology, social, and economic considerations. This permits the relationship of supporting systems to the technological systems to be investigated and, if desired, subsequently assessed.

The goals themselves are not independent, and so, to help in their quantification, it is first necessary to determine the magnitude of their interrelationships. To do this a cross-support matrix is formed, as described in Fig. 2. From this matrix, the total contribution the achievement of one goal makes to the achievement of all the other goals can be determined. This is done by considering the support one goal has on each of the others, separately, and then adding these separate contributions to form a total. The "contribution" one goal may have on another can be either a beneficial (or positive), or a detrimental (or negative) contribution. (Cross-support analysis is discussed in more detail in Chapter 28 on measurement and scaling by Cetron and Dick).

The development of a standard hierarchical system of social and economic indicators is under consideration. Two examples are shown in Figs. 3 and 4. (Details of social and economic indicators are given in the Chapter 25 by Overly.) The sets of social and economic indicators are analyzed to determine the extent of interrelationships by conducting cross-support analyses. The result of this step in the exercise is to obtain cross-support indices for each indicator in terms of the other indicators; these are arrived at in the same manner as those in the goal cross-support matrix. Again, the support can be positive or negative, depending upon the beneficial or detrimental effects. Fig. 5 provides an illustrative example of this process.

Once a cross-support index has been established for each indicator, the next step is to relate the indicators to the weighted goals, by obtaining a sensitivity value for the degree to which a given indicator (modified by its cross-support index) reflects in the achievement of the

	g_1	g_2	g_3	g_4				g_n
g_1		a_{12}	a_{13}	a_{14}				a_{1n}
g_2	a_{21}		a_{23}	a_{24}				a_{2n}
g_3	a_{31}	a_{32}						a_{3n}
g_4								
g_n	a_{n1}							a_{nn}

g_1 = goal 1

a_{12} = magnitude of contribution of goal 1 to goal 2

Fig. 2. Cross-Support Matrix

various goals [7].

Since the procedures for treating economic and socio-political indicators are similar, Fig. 6 illustrates the method and format for treating economic indicators. These can then be combined for all indicators in the matrix shown

---

*See, CETRON, Marvin J., "Goals for National Planning--Technology (GNP-T)," Presented March 9, 1970, at the joint Institute of Management Sciences-World Future Society Symposium on Assessing the Future and Policy Planning, and under the title, "A Method for Integrating Goals and Technological Forecasts into Planning," in *Technological Forecasting and Social Change*, 2, No. 1, 1970.

FIG. 3. Selected Relevant Socio-Political Indicators
(Illustrative Example)

1. Population characteristics
   - •Current level and growth
   - •Distribution (density) by:
     - geographic area
     - age
     - profession
     - race
     - religion
     - income level
     - educational level
     - special interests
   - •Migration patterns (changes in distribution)
   - •Family size
   - •Number of persons living in houses/apartments
2. Employment characteristics
   - •Types of jobs (manufacturing, sales, management, etc.)
   - •Number of total jobs by type and industry
   - •Number of welfare jobs
   - •Number of average job changes per worker by job type and industry
   - •Creation of new professions as a result of the evolution from a manufacturing to a services-oriented economy
3. Land usage policies
4. Number of publications by type (newspapers, general interest magazines, professional journals, catalogs, textbooks, etc.)
5. Circulation volume of publications by type
6. Professional organizations
7. FCC policies
8. Labor strike activities
9. Influence groups (industrial, community, state, local, federal, etc.)
10. Federal administrative political philosophy (liberal, conservative, moderate)

FIG. 3. (Continued)

11. Foreign policies
12. Domestic policies
13. Manpower education and retraining policies
14. Leisure time
15. Leisure time activities
16. Number of dwelling starts
17. Number of houses and apartments owned/rented
18. Number of telephones
19. Telephone usage (number of calls, local and long distance, for business or personal use, etc.)
20. Number of radio and TV stations
21. Number of radios and TV's

Goals--Area of National Concern	Possible Economic Indicators	Data Sources
Agriculture	Wheat (acreage, yield, production value, and foreign trade) Rye - Rice - Oats, etc. Fertilizers Farm machines Farm employment Migratory farm workers	Agricultural Statistics, Statistical Abstracts, Bureau of Census Machine Age
Housing	New housing starts (types, location, etc.) Construction materials (brick, lumber, concrete, glass) Employment of masons, carpenters, plumbers Mortgages (number, interest rates, types, area) Mobile homes Size of family unit Number of phones per unit Number of bathrooms	Housing and Urban Development Department Publications Department of Commerce Statistical Abstracts Wall Street Journal National Materials Advisory Board

FIG. 4. Economic Indicators

	SELECTED RELEVANT SOCIO-POLITICAL INDICATORS*	1 Population Characteristics	2 Employment Characteristics	3 Land Usage Policies	4 # Publications	5 Circulation Vol. of Publ.	6 Professional Organizations	7 FCC Policies
1	Population Characteristics	X	2	8	1	4	1	1
2	Employment Characteristics	8	X	4	1	2	4	-
3	Land Usage Policies	4	1	X	-	1	1	1
4	# Publications	4	2	2	X	4	4	4
5	Circulation Vol. of Publ.	8	4	4	8	X	4	8
6	Professional Organizations	2	8	-	-	1	X	-
7	FCC Policies	8	1	1	1	2	-	X
8	Labor Strike Activities	4	8	1	1	2	4	-
	⋮							
23	Programming Format/Content	8	1	2	2	4	1	8
	WEIGHTED CROSS-SUPPORT VALUES, $\sum S$ (NORMALIZED)	17	10	8	5	8	8	9

Fig. 5. Cross-Support Matrix of Selected Relevant Socio-Political Indicators (Illustrative Example)

Indicator...E_i
Sensitivity index of E_i...e_i

	Goals	Degree of contribution of indicator to achievement of goal (z_{ij}) 8	7	6	5	4	3	2	1	Value
v_1	g_1		✓							$v_1 z_{i1} e_i$
v_2	g_2					✓				$v_2 z_{i2} e_i$
v_3	g_3								✓	. . . . .
v_4	g_4			✓						
										. . . . .
v_{n-1}	g_{n-1}									
v_n	g_n									

Scale for contribution:
8 Critical
4 Very Important
2 Important
1 Insignificant

Total Value U_i

Fig. 6. Goals for National Planning - Economic

WEIGHTS	W_1	W_2	W_3	W_4	– – – – –	W_n	TOTAL RELEVANCE INDEX	
INDIC. \ GOALS	g_1	g_2	g_3	g_4		g_n		
e_1	β_{11}	β_{12}	β_{13}	β_{14}		β_{1n}	$\sum_{j=1}^{n} \beta_{1j} W_j$	$= u_1$
e_2	β_{21}	β_{22}	β_{23}	β_{24}		β_{2n}	$\sum_{j=1}^{n} \beta_{2j} W_j$	$= u_2$
e_3	β_{31}	β_{32}						
e_4	β_{41}	β_{42}						
1 1 1								
e_m	β_{m1}	β_{m2}				β_{mn}	$\sum_{j=1}^{n} \beta_{mj} W_j$	$= u_m$

Fig. 7. Goal/Indicator Matrix

CORPORATE GOALS (WEIGHTED) \\ SELECTED RELEVANT SOCIO-POLITICAL INDICATORS (WEIGHTED)		DESCRIPTION	1 Population Characteristics	2 Employment Characteristics	3 Land Usage Policies	4 # Publications	5 Circulation Vol. of Publ.	6 Professional Organizations
	DESCRIPTION	∑G \\ ∑S	17	10	8	5	8	8
1	Low Volume, High Margin	18	8 / 2448	2 / 360	8 / 1152	-4 / -360	8 / 1152	1 / 144
2	Cash Flow 6% of Investment	26	-1 / -442	4 / 1040	-8 / -1164	-1 / -130	-2 / -416	-1 / -208
3	12-15% Annual Growth Rate	15	-1 / -225	-8 / -1200	1 / 120	1 / 0	2 / 240	8 / 960
4	5 Years to Reach Maximum Profits	22	-4 / -1496	1 / 220	1 / 176	1 / 110	-2 / -352	-2 / -352
5	Not more than 25% Gov't Business	18	1 / 306	2 / 360	- / 0	1 / 90	1 / 144	- / 0
6	Corporate Image	1	2 / -34	-1 / 10	1 / 8	-2 / -10	-2 / -16	1 / 8
WEIGHTED CROSS-IMPACT VALUES, (NORMALIZED)		∑S(G)	16	21	-2	-6	20	19

Fig. 8. Impact Matrix of Weighted Corporate Goals & Weighted Selected Relevant Socio-Political Indicators

(Illustrated Example)

Weights	u_1	u_2	u_3		u_l		w_1	w_2	W_3	w_k	Weights
Economic Indicator	E_1	E_2	E_3		E_l	techs.	S_1	S_2	S_3	S_k	Social Indicator
	p_{11}	p_{12}	p_{13}		p_{1l}	T_1	q_{11}	q_{12}	q_{13}	q_{1k}	
	p_{21}	p_{22}	p_{23}		p_{2l}	T_2	q_{21}	q_{22}	q_{23}	q_{2k}	
						T_3					
	p_{n1}	p_{n2}	p_{n3}		p_{nl}	T_n	q_{n1}	q_{n2}	q_{n3}	q_{nk}	

Fig. 9. Tri-Matrix

In Fig. 7, a practical example of which is given in Fig.8, for socio-political indicators.

The overall sensitivity value is obtained by multiplying the "contribution" of the indicator to the measureable degree of achievement of each goal by the indicator cross-support index for each, and also by the weight for each goal. The scale for the "contribution" (z_{ij}) has been extended to allow interpolation from the logarithmic scale used for the cross-support matrices to increase the sensitivity.

This relationship can be represented mathematically as follows:

$$u_i = e_i \sum_{j=1}^{n} z_{ij} \quad v_j.$$

where

u_i is the total weight associated with economic indicator E_i

e_i is the cross-support index for economic indicator E_i

z_{ij} is the contribution of the $_i$th indicator to the goal g_i (one of n goals)

v_j is the weight of the goal g_j

Similar sensitivity values W are obtained for the socio-political indicators. All these values are used as weights in the subsequent operation.

The final stage in the process of the assessment of technologies is the "tri-matrix" shown in Fig. 9. This presupposes a set of identified technologies, T_1, T_2... T_m, which are related according to their impacts on all the indicators. The scale is that which has been used in the previous cross-support matrices. The relationships are determined in a similar manner, by adding the impacts of the technologies on each weighted indicator, to give two final ratings for each project--one in terms of socio-political indicators, the other in terms of economic indicators.

Mathematically, these relationships can be defined for the socio-political indicator as:

$$t_{si} = \sum_{j=1}^{k} w_j \, q_{ij}$$

where

t_{si} is the rating of the technology T_i with respect to the socio-political indicators

w_j is the weight of the socio-political indicator S_j

q_{ij} is the impact of the technology T_i on the socio-political indicators

and, for the economic indicators

$$t_{ei} = \sum_{j=1}^{l} u_j \, p_{ij}$$

where

t_{ei} is the rating of the technology T_i with respect to the economic indicators

u_j is the weight of the economic indicator E_j

p_{ij} is the impact of technology T_i on the economic indicator E_j.

Because of the possible effects of technologies on each other, or because the technologies support or inhibit each other, the degree of interdependency should be taken into account in the final decision-making process. To aid this determination, a cross-support analysis for technologies can be performed in the same manner as for the other cross-support matrices, and a value is obtained for each technology by summing across the rows:

$$r_i = \sum_{j=1}^{m} b_{ij}$$

where

r_i is the total support for technology T_i, and

b_{ij} is the support of T_i for T_j.

However, we also need to know the extent to which technologies support or depend on the others, and so we sum down the columns, giving:

$$r'_i = \sum_{j=1}^{m} b_{ji}$$

where

r'_i is the total dependency value of technology T_i on the other technologies, and

b_{ji} is the dependency of T_i on T_j.

Thus we have four ratings for each project, t_{si}, t_{ei}, r_i and r'_i. These terms allow the decision-maker to take into account both the socio-political and the economic effects on the goals and the degree of interdependency of technologies before their final selection. We could combine all of these ratings into one function, but in doing so, we would lose some of the effectiveness of the method.

The overall objective of the Goal/Technology coupling effort is to provide visibility of the goal/technology relationship, either beneficial or detrimental. Multi-level structures of information are derived to assist in fully utilizing the relationships established by the analysis. The information is then collated in a benefit/cost function to give an overall desirability ranking for each technology.

RESULTS

This trimatrix method was tested [8]. The following positive results were obtained:

a. It was possible to prepare a macro-technology assessment methodology which provided visibility of policy alternatives at various decision levels

b. It was possible to generate a listing of possible consequences of changing priorities of the national goals on the chosen system (e.g., power on transportation) and the technologies supporting the system

c. It was possible to develop a hierarchical set of social and economic indicators for the system studies and

d. It was possible to use the exploratory forecasts of a group of technologies supporting the system with associated social and economic effects.

Now that we have "walked through" this tri-matrix approach and discussed the positive aspects, let us discuss the three problems which were not adequately addressed.

One problem with this method grew out of the fact that the socio-political indicators were never cross-supported with the economic indicators. This oversight was corrected in the next chapter by Christine Ralph by broadening the concept of societal indicators to include social, demographic, economic, legal, political, etc., and then performing the cross-support on all the important indicators.

A second problem dealt with the absence of a procedure for determining the timing and internal consistency in the assessment methodology except for the 0-5, 5-10, and 10-20 year time frames. By incorporating the cross-impact techniques, discussed by Ted Gordon and Hal Becker in the previous chapter and as used by Chris Ralph in her article, internal consistency and timing was incorporated into the model.

The third missing element in the tri-matrix method was the lack of "future worlds" or "mini-scenarios" which could be used for in-depth analysis. This innovative procedure of "future worlds" is also incorporated in the next chapter.

The manner in which the three missing elements were incorporated in the new flow is shown in Fig. 10 and is explained in detail in Chapter 31.

Continuing, Fig. 11 illustrates the detailed analysis process obtained by extending the basic "analysis process" shown in Fig. 10 through the addition of corporate goals and corporate indicators. Also Fig. 11 incorporates an economic analysis (input/output analysis) that is useful in comparing the technology assessment results with a purely economic analysis.

UTILITY AND FUTURE EFFORTS

The demonstration of the feasibility of this approach to macro-technology assessment has provided a potentially powerful tool to systematically place into perspective the major considerations of technology assessment. These are: the national and corporate priorities, relative measures of the national and/or corporate value of technologies, and beneficial and detrimental features of technological exploitation. Assessment can be accomplished in a manner that recognizes (and uses) the qualitative and quantitative aspects of a problem. This method organizes relevant information into an objective data base which allows assessment of the consequences of alternatives within a procedural framework which accommodates the information of independent assessments of technology.

Implementation of the macro-view assessment is recommended to derive the necessary refinement benefit from iterative processes. Following this, a natural evolution would be an assessment of another major system affecting

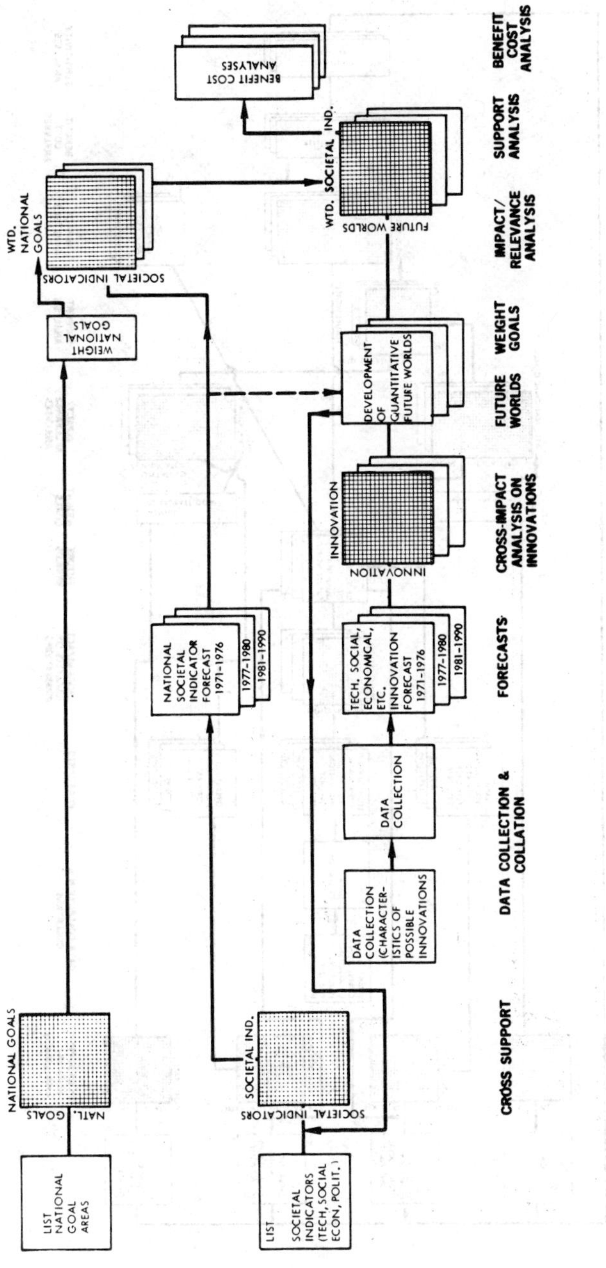

Fig. 10. Analysis Process

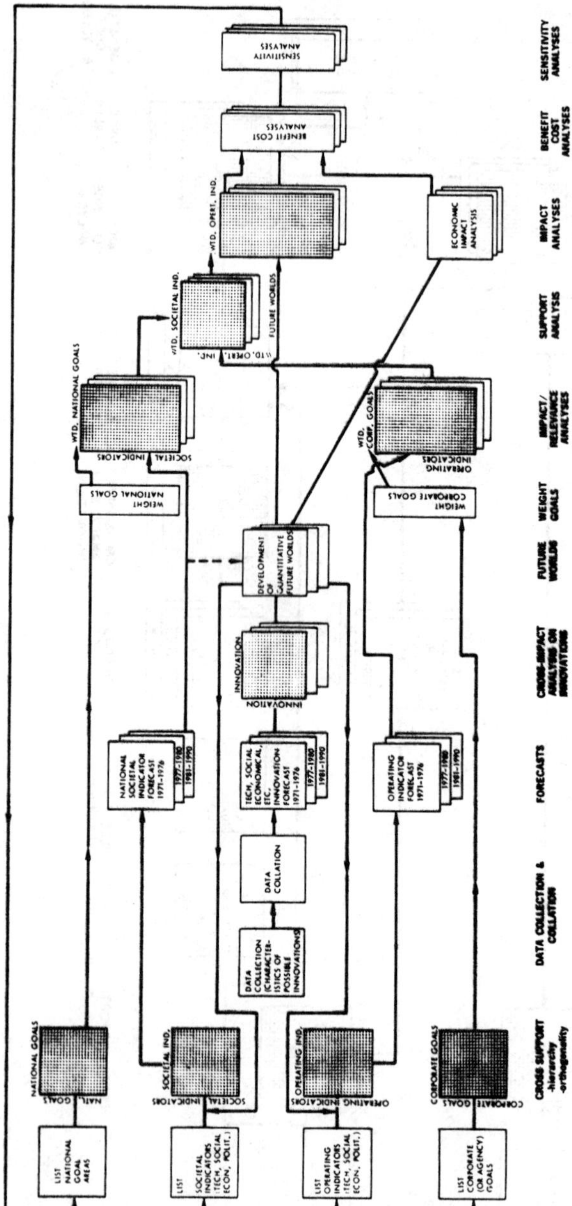

Fig. 11. Analysis Process

a moderate number of goal areas and technologies. Full-scale implementation should be attempted only after gaining operating experience with the technological and procedural framework, and, as above, marginal acceptance of a set of national goals.

We spend a lot of time talking about technology assessment, but very few assessments are being made of the methodology and tools that exist. We should be making these assessments. Perhaps the real problem is not technology assessment, but "planning" in general. On the other hand, there are very intelligent vocal critics of planning who want freedom to "do their thing" and live decent, free lives, and on the other hand, another vociferous minority who call for more planning and rational management of the world's resources. These schools of thought must be brought closer together, if only by using the technology assessment and planning studies as an indication of what could be and then letting an informed management or the legally constituted and elected governing bodies make decisions.

NOTES AND REFERENCES

1. There is really a simple cost-benefit formula: utility x feasibility, divided by dollars, which provides a "desirability index." These same factors can be applied to technology assessment utility (based on social, political, and economic indicators), feasibility (based on technological forecasts), and cost (based on technological forecasts and the value measured against the relative contribution of each alternative to national goals.)

2. MEADOWS, DONNELLA H., *et al.*, *The Limits to Growth*, A reporter for the Club of Rome's Project on the Predicament of Mankind (New York: Universe Books) 1972. Even the President of the Commission of the European Communities, Dr. Sicco L. Mansholt, says the Common Market should start optimizing on Gross National Happiness rather than GNP.

3. MASLOW, ABRAHAM H., *Motivation and Personality*, New York: Harper & Brothers, 1954.

4. GRAVES, CLARE W., "Levels of Existence: An Open System Theory of Values," *Journal of Humanistic Psychology, 10*, No. 2, 136, Fall 1970.

5. DICK, DONALD N., Forecasting International, Ltd., and RALPH, CHRISTINE of International Research and Technology Corporation.

6. LECHT, LEONARD, *Goals, Priorities, & Dollars: The Next Decade*, New York: The Free Press, 1966.

7. See, CETRON, MARVIN J., "Goals for National Planning — Technology (GNP-T)," Presented March 9, 1970, at the joint Institute of Management Sciences-World Future Society Symposium on Assessing the Future and Policy Planning, and under the title, "A Method for Integrating Goals and Technological Forecasts into Planning," in *Technological Forecasting and Social Change, 2*, No. 1, 1970.

8. The tri-matrix technique was tested on a micro-scale in the U.S. Postal Service and was presented as a paper under the title, "A Micro-Technology Assessment of the Printing and Publishing Industry," at the Institute of Technological Change at the Center of Technology and Administration of the American University on February 8, 1971, by Marvin J. Cetron. It is also being included in *Technological Change in Printing and Publishing Management*, edited by Drs. Hattery and Bush, Spartan Books, New York, 1972.

APPENDIX A

ILLUSTRATIVE HIERARCHY OF SOCIETAL UNITS

I. Individual or Family

1. *Survival*

a. Food
b. Clothing
c. Shelter
d. Health
e. Other

2. *Long-Term Growth and Continuity*

a. Professional development
b. Education and training
c. Work conditions - standards
d. Savings
e. Other

3. *Societal*

a. Church work
b. Parent-Teachers association
c. Consumer organizations
d. Social or charitable work
e. Other

II. Organizations (Labor, Religious, Societies, etc.)

1. *Survival*

a. Common goals
b. Income
c. Influence
d. Standards of behavior

2. *Long Term Growth and Continuity*

a. Meetings
b. Communications (publications)
c. Education
d. Other

3. *Societal*

   a. Picnics
   b. Old-age homes
   c. Retreats
   d. Other

III. Corporation or Firm

1. *Survival* (Financial)

   a. Minimum profit
   b. Discounted cash flow
   c. Share of market
   d. Stockholder equity

2. *Long-Term Growth and Continuity*

   a. Product with significant market for 3 to 5 years
   b. R & D program
   c. Produce products that have same total profit but longer life
   d. Career opportunities for development and advancement

3. *Societal*

   a. Support day care centers
   b. Hire minority groups
   c. Provide employee time off from work to work on civic and social committees
   d. Athletic facilities

V. National

1. *Survival*

   a. GNP
   b. New housing starts
   c. Number of machine tools
   d. Productivity
   e. Balance of payment
   f. National security

2. *Long-Term Growth*

   a. Resources
   b. Work force utilization
   c. Administration
   d. Health programs
   e. Conservation programs
   f. Food and drug inspection
   g. Welfare programs

3. *Societal*

   a. Non-nuclear proliferation treaties
   b. Joint international scientific cooperation
   c. Tourist exchange
   d. National prestige
   e. Aid to lesser developed countries

VI. International

Ecological, monetary policies, trade policies, coexistence policies, etc.

APPENDIX B

Classification of Goals and the Abstract Standards, National Goals Project. Source: *Manpower Requirements for National Objectives in the 1970's* by Leonard A. Lecht, Center for Priority Analysis, National Planning Association, 1968.

1. *Agriculture*

Includes cost of programs to raise income of commercial farm families to a close approximation of income of non-farm families plus the cost of programs to encourage movement of 150,000 low-income members of the farm labor force into non-farm employment each year.

2. *Area Redevelopment*

Cost of expanded redevelopment programs following the objectives of the Area Redevelopment Act.

3. *Consumer Expenditures*

Living standards rise to limit set by savings rate of approximately 8% of disposable personal income. Includes additional increases in consumer expenditures from other goals such as health, education, and transportation, plus family allowance system to increase incomes of families below $3,300 to this level.

4. *Education*

Assumes an increase in proportion of students from eligible age group receiving high school and higher education amounting to a 50% increase in proportion for college group. Allows for doubling of faculty salaries over decade, increased teacher supporting staff, expansion of adult education and vocational training role of junior colleges, and increased plant and equipment at all levels of education.

5. *Health*

Stresses programs to enlarge access to modern health technologies by providing families with a level of health care equal to that enjoyed currently by families with most comprehensive health insurance plus expanded provisions for dental and psychiatric care. For persons over 65 includes level of medical care costing 50% more than HEW estimate of cost of adequate medical care for the aged in early 1960's with two-thirds of costs financed by public funds. Also includes increase in ratio of hospital beds to population following Hill-Burton Act standards, together with increase in health research expenditures sufficient to employ the 77,000 health research professionals NIH estimates will be needed in early 1970's.

6. *Housing*

Includes elimination of all remaining substandard housing between 1966 and 1975, plus increase in number of housing starts from 1.5 million in 1962 to 2.5 million by 1975 to provide for new family formation, adequate housing for nonwhites, higher income levels, greater emphasis on special housing for the aged, and for vacation "second" housing. Also includes cost of R&D program to develop synthetic building materials, mass production of housing components, and building codes geared to potentialities of modern technology.

7. *International Act*

Goal includes cost to United States of UN Decade of Development target that each industrialized nation contribute 1% of GNP from public and private sources to supply capital to developing nations. Also includes military support to developing nations at early 1960 levels, plus support for international nonfinancial organizations rising to $1.5 billion in 1975, primarily for WHO, UNESCO, FAO, and for an expanded UN Peace Force.

8. *Manpower Retraining*

Cost of programs for retraining 1% of the labor force a year following the outlines of the Manpower Development and Training Act passed in 1962, and providing basic literacy training as provided by 1963 amendments in MDTA.

9. *National Defense*

Expenditures for an adequate national defense taking into account applications of technological advances such as antimissile missiles, nuclear aircraft carriers, space vehicles, and requirements to maintain conventional armed forces.

10. *Natural Resources*

Cost of programs for increasing and economizing the supply of natural resources in an urbanized and affluent society. Largest expansion in expenditures projected for water purification and storage with programs derived from studies of Senate Select Committee on Water Resources.

11. *Private Plant and Equipment*

Expenditures for plant and equipment needed to sustain economic growth in the 1970's plus the additional private plant and equipment expenditures projected for specific goals: utilities in urban development, transportation equipment in transportation, private and non-profit facilities for schools, hospitals, welfare, etc.

12. *Research and Development*

Standard stresses increases in "civilian economy" R&D with total R&D expenditures projected increase from 3% of GNP in early 1960's to 4% of trillion-dollar GNP in mid-1970's. Includes substantial increases in expenditures for basic research, for water desalineation and oceanography, for health and social science research, for R&D information systems, and for an R&D extension service for the private economy with objectives similar to State Technical Services Act of 1965.

13. *Social Welfare*

Includes expenditures from public and private sources for providing typical pension to retired couple covering cost of "modest but adequate" standard of living for an elderly couple in American cities in early 1960's plus allowance for increases in earnings levels and standards of living. Also includes family allowance system to establish income maintenance floor for families with poverty incomes in the 1970's and provisions for incorporating nationwide protection against income loss from illness as part of OASDI, expansion of unemployment compensation and benefits similar to proposals of recent Administration.

14. *Space*

Includes expenditures for sustained space research and development program involving manned lunar landing about 1970 followed by exploration of moon, earth orbiting laboratories, and development of technology leading to human landing on Mars before year 2000. Also stresses expansion of research in basic space sciences and in application of space technology in such areas as weather observation satellites and in long-distance telecommunications.

15. *Transportation*

Expenditures for transportation equipment and R&D allowing for projected increase in automobile stock and for changes in transportation resource use following line of President Kennedy's 1962 Transportation Message to Congress. Also includes cost of R&D and advances such as supersonic planes, nuclear ships, hydrofoils, gas-turbine engines, etc.

16. *Urban Development*

Includes expenditures, generally derived from other goals, which are attributable to programs for providing adequate transportation, housing, cultural and recreation facilities, schools, hospitals, and industrial, commercial, and governmental buildings for over three-fourth of the population who are expected to be living in urban areas in 1975. Also includes expenditures for new mass transit technologies and for equipping automobiles with devices for controlling air pollution. Involves overall increase in spending for urban facilities rising from 11% GNP in 1962 to 13% in 1975.

A MACRO-ASSESSMENT METHODOLOGY

CHRISTINE A. RALPH

Senior Professional Staff
International Research and Technology Corporation,
Washington, D.C., U.S.A.

During the last few years, technology assessment has become the "in-topic" in government, academe and, although to a lesser extent, in industry. However, just what is meant by technology assessment is not at all clear, and its purpose and usefulness even less so. Certainly there is a need for assessment of some sort, for determining the direct and indirect impacts of various actions but how that is to be accomplished still seems to be open for question. One group of would-be technology assessors looks at the subject in terms of a micro-view, namely taking one specific technology and trying to assess all its potentially detrimental effects. Another group looks at it from a macro-view, where the assessment is to determine the relative worths of several technologies in terms of their beneficial and detrimental effects. This type of assessment appears to be similar to project selection and resource allocation techniques that have been used by industry and government for some years, and because of this, many people claim that they have already performed technology assessments. This is, however, not the case as the assessments being made for purposes of project selection of resource allocation were only considering economic costs and benefits, in terms of R&D budgets, profits, sales, company growth and so on. Few impacts which did not directly affect the company were considered.

Technologies do not exist in isolation — they are usually developed for some specific purposes, and therefore have a known intended and direct impact, and a known cost (in dollar terms). However, the effects do not stop there; technologies, and the actions associated with them, often have much broader effects than are first realized. We need only look at the development of the transistor or the automobile to see the differences that they have made to our way of life. Some of these changes are for the better — improved medical equipment, closed-circuit television for security purposes, improved communications, improved transportation, and so on. Other changes resulting from technology are generally worse — increased noise pollution, decreasing inclination for self-entertainment, air pollution, and traffic congestion. We are all familiar with the despairing cries of the ecologists and environmentalists, and for these people the micro-view of technology assessment is a boon. The linking of water pollution to steel mills and air pollution to the automobile is a relatively straightforward task. This micro-view is usually prejudiced, as the beneficial effects of technology are ignored. What must be understood, however, is that any assessment is only as reliable as the information it uses, and that the criteria chosen as the basis for the assessment can echo the biases of the assessor. This is true wherever subjective judgement is used for decision-aiding — the well-used techniques in operations research and systems analysis can be subject to the same problems.

It is not the purpose of the author to denigrate assessment techniques, or any other management aids, but to place them in proper perspective. Technology assessment *can* be a very useful and important tool for management — any management — whether of a nation, an agency or a corporation, but it is essential that it is kept as a decision-aiding technique, not a decision-making one. We can devise a set of criteria, against which technologies and science and technology programs can be compared in some fashion, but what comes out at the end of such an assessment, possibly in terms of a ranked set of numbers, should not be treated as gospel. The actual function of performing the assessment, of developing criteria and of comparing the technologies against these criteria, is equally as important if not more so than the final rankings. We also believe that it is not sufficient to assess merely technologies and technology programs, but that they should be viewed in terms of operations and administrative and regulatory functions. Therefore when we perform an assessment we should also be assessing all innovations, government and corporate actions, changes in society and the economy, and the potential introduction of new regulations and legislation.

The technique described in this chapter is an outgrowth of the very early attempt described in the previous chapter by Marvin Cetron, and is currently undergoing its first major testing. The method used can take two forms; the first, shown in Exhibit 1, is the generalized form which can be used for any government agency or corporation in planning its technology and R&D programs; the second, shown in Exhibit 2, is for use in planning national science and technology policy. As the second method is actually a sub-set of the first, the first will be described in detail, with explanations given later as to the structuring of the second.

The basic premise of the method is that it is possible to develop a relatively unbiased set of criteria against which the technologies, innovations and actions can be assessed. The criteria chosen in our model are two sets of goals: a set of national goals, or areas of concern, and a set of agency of corporate goals. The reason for these choices is that in establishing national science and technology policy, the purpose is, or should be, to meet the national goals, or help satisfy areas of concern. For a government agency, the agency's goals should be compatible with the nation's goals. From the corporation's point of view, national goals, or areas of concern can affect regulations and legislation concerning quality of goods, permissible effluents, trade agreements, taxes and so on, and therefore have a direct effect on corporate policy.

Before the assessment can take place, therefore, it is necessary to develop two sets of goals: one for the nation (which can better be called a set of areas of national concern); and one for the agency or corporation. Several attempts have been made at developing a set of

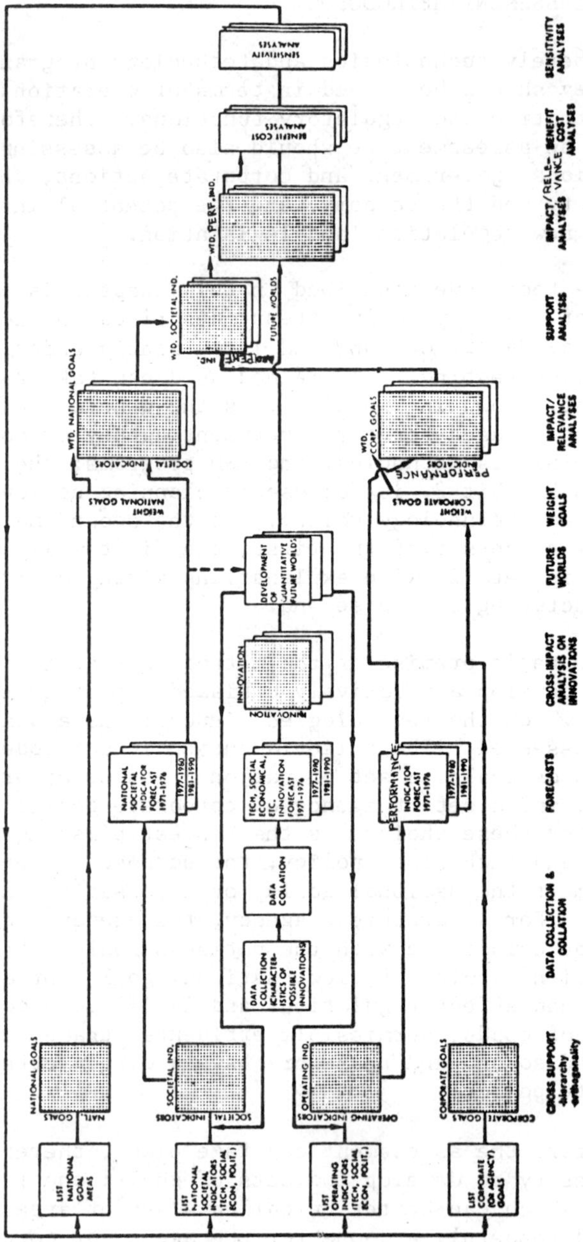

Exhibit 1. Technology Assessment Method

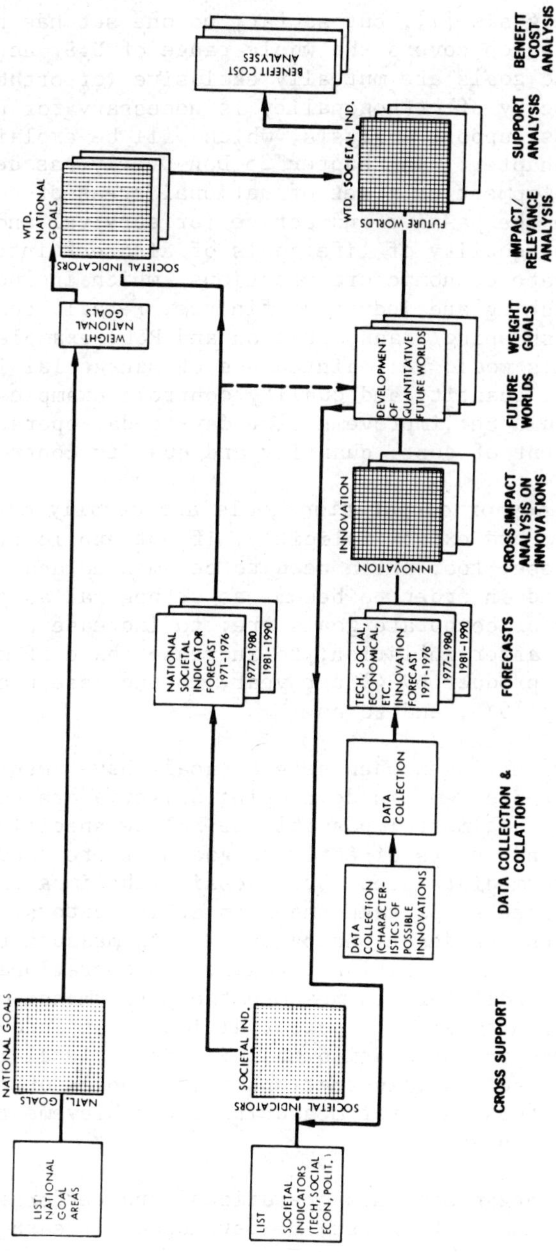

Exhibit 2. Technology Assessment Method for National Science and Technology Policy

national goals [1], but so far, no one set has been developed which covers the whole range of U.S. activity where the goals are mutually exclusive (or orthogonal). The property of orthogonality is necessary for performing the cross-support analysis, which will be explained later in the chapter. In Chapter 25 Don Overly has described the development of a set of national areas of concern based on the passive and active (or survival and growth) goals and quality of life goals of discreet interest groups. For private economic organizations, which include: heavy manufacturing and industry, finance, retail, communications, energy, shipping/transportation and R&D, examples of passive goals would be: maintenance of managerial integrity and cost, quantity and quality control; examples of active goals would be: improvement of day-to-day operations and improvement of cost, quantity and quality control.

Agency or corporation goals are usually much more specific, and exist implicitly, if not explicitly — however, these, too, often need to be amended and sometimes quantified in order to become as orthogonal as possible. Examples of corporate goals are: to increase profit by 5% per year after inflation; to increase share of market of a specific product by 4% per year; to increase productivity by 30% by 1973, and so on.

However, even when sets of goals have been developed, our problems in developing criteria are not over, since it is almost impossible to relate specific innovations and events directly to goals, there needs to be some intermediate step for assessing the impacts. These intermediate steps take the form of indicators: societal indicators, as described by Overly, to measure the degree of achievement of national goals, and operational and performance indicators to measure the performance of the technological programs, or potential impact of innovations upon the agency or corporation goals. The societal indicators which he used as illustrations are examples of the types of indicators used for measuring the achievement of the areas of concern.

The organization's operational and performance indicators must, of course, be developed for each particular agency or corporation, although the same structure can be used. We can essentially use four hierarchies, as shown in Exhibit 3, to describe the total operations of the agency or corporation. This description has been chosen for two reasons: the first, is that it provides a useful classification scheme for categorizing the data needed in the assessment method; the second, and more important reason, is that it provides an interface between innovations and actions and corporate or agency goals, and also between the organization's performance indicators and societal indicators.

The first hierarchy is of services or of products. If we were considering the transportation industry, we would consider the industry as providing transportation for goods or people, which could be described, at a lower level of the hierarchy, in terms of providing inter-city air transportation for people, or intra-city transportation of small packaged goods by pedestrian delivery men. This service or product hierarchy can be directly related to the organization's goals in terms of the revenues and costs of the service, its efficiency, reliability, volume, and so on.

The second hierarchy describes the processes involved in providing the service, or manufacturing the product. It is in this hierarchy that new technologies can be implemented to substitute for existing processes, or to completely transform the function while still providing the same service. Using the previous example of intra-city delivery of small packages, a substitution would be the provision to all delivery men of small, economic electric cars; a complete replacement would be to develop a system of penumatic tubes between office buildings and apartment houses, and so on.

Each process can be described in terms of the indicator types which form the third hierarchy. For example, the inventory for a process takes two forms: the inventory of the organization which is needed to that process — capital, equipment, facilities, materials, personnel, etc.; and the inventory of whatever is being processed — small packages, people, and so on.

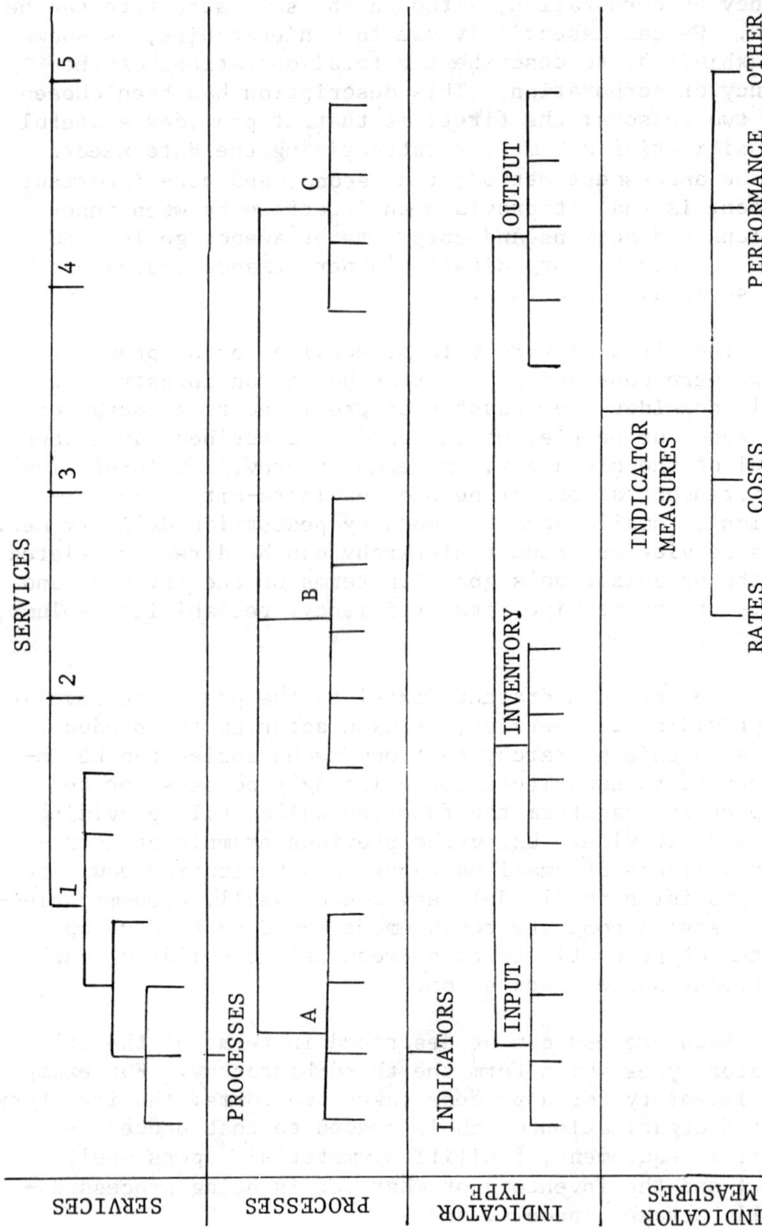

Exhibit 3. Four hierarchies for developing performance indicators

This particular quantity or quality of each of the indicator types is listed in the fourth hierarchy of indicator measures. These describe, for example, the number of people needed to perform a process with a particular piece of equipment; the capacity of the equipment in terms of number of items being processed per unit time; the cost of processing a number of items per unit time; the quantity of effluents generated by a process per unit time . . . and so on. These indicator measures can be related to societal indicators in terms of employment for example, or permissible effluents. Let us assume a corporation employs x number of people to work y number of machines to perform a given process with a given level of reliability. A new machine is developed which has 5% greater reliability, but requires only x/2 people to work y machines. The cost per new machine is greater than that of the old one, but laying off x/2 people cuts direct labor costs. In terms of performance indicators related to the organization's goals, it is very likely that the installation of such new machines would benefit the corporation, but to lay off half the work force on those machines would detract from the nation's goal of increasing employment, and could cause major problems with labor unions.

Once all these hierarchies have been developed then the assessment method can take place. It is possible to perform "surprise-free" projections of, for example, the achievement of an organization's goal against time, and then measure against that curve the effects caused by implementing some new piece of technology, or some government regulation. However, this is a very lengthy process, and frequently several innovations and actions tend to be compatible, and can be implemented within a short time of each other. The process can be facilitated by the development of several "future worlds" for the organization, for a series of different time frames. These future worlds comprise self-consistent lists of possible technological innovations, social, political, economic and regulatory occurrences and various organizational programs, which may be included within each of the desired time frames, say, for example, 1971-76, 1976-80, 1980-90. These future worlds will cover a "least suprise" future, an extrapolated future, and then several others which include potential new goods and services; and innovations of a more "suprise" nature. Cross-impact analysis is used to ensure self-consistency of these future worlds. If we assume a set of innovations $T_i (i = 1,2...n)$ with associated probabilities of occurrence, P_i, we wish to measure the impact on the other innovations if T_m, say, actually does occur, that is, if $P_m = 1$. If T_m occurs, it could have an enhancing or an inhibiting effect on the other T_i. Cross-impact analysis [2,3,4] provides a means of measuring the effect by considering the connections between the innovations.

A matrix is set up, as in Exhibit 4. The elements of the matrix S_{ij}, are completed by a panel of experts, using a scale (possibly 0-9) which can be either positive or negative, dependent upon the enhancement or inhibition of the innovation concerned. Thus the probability of T_i, if T_m occurs, can be represented by P_i, where

$$P_i' = f(P_i, k, S, t_m, t)$$

where P_i is the probability of T_i before t_m occurred
- k is a function of the connection mode denoting positive or negative
- S is a measure of the strength of the connection (i.e., S_{ij})
- t_m is the time of occurence of T_m
- t is the time in the future for which probabilities are being estimated.

The choice of functions is dependent, to some extent, on the desired rate of convergence, but the one which will most probably be used is

$$P_i' = kS \frac{t-t_m}{t} P_i^2 + \left[1-kS\left(\frac{t-t_m}{t}\right)\right] P_i$$

where $k = \pm 1$, depending upon the mode, and the function is bounded by the appropriate constraints.

Now, having developed our lists of goals, indicators

ORIGINAL PROBABILITIES	IF THIS INNOVATION OCCURRED	THE PROBABILITIES OF THESE INNOVATIONS WOULD BE AFFECTED AS FOLLOWS:					
		T_1	T_2	T_3	T_4		T_n
P_1	T_1	--	S_{12}	S_{13}	S_{14}		
P_2	T_2	S_{21}	--	S_{23}			
P_3	T_3	S_{31}	S_{32}	--	S_{34}		
P_4	T_4	S_{41}	S_{42}	S_{43}			
P_n	T_n	S_{n1}	S_{n2}	S_{n3}	S_{n4}		

Exhibit 4. Cross-Impact Matrix

and future worlds, the assessment method, as shown in Exhibit 1, can commence.

The first step in the analysis is to weight both sets of goals. The technique in both cases is identical, namely first, a cross-support matrix, [5,6,7] followed by paired comparisons rank ordering.

Cross-support analysis is a method for determining the degree to which one variable of a set contributes to another variable of the same set. Examples of this can be the degree to which the achievement of one goal supports the achievement of another, or the support one technology program has for another. The final product of such an analysis is a value number for each variable, based upon its contribution to the rest. In many cases, these value numbers can be used directly as weighting factors, but in others they require modification. The reason for this is that the weighting of any set of variables is dependent upon the criteria chosen for their weighting. If the only criterion is degree of support, then the variables should be as independent as possible in all dimensions except that of support, so that the cross-support analysis can take place. If support is not an appropriate criterion, then some other means for weighting must be found. In this case, where the purpose of the assessment is to relate innovations and actions to an organization's goals and national areas of concern, it would seem likely that an innovation which directly increases the chance for achieving a particular goal, which in turn supports several other goals, should have more value than an innovation which helps achieve one goal which does not contribute to another.

The cross-support matrices for goals (organization and national) are formed as in Exhibit 5, to determine the total contribution the achievement of one goal makes to the achievement of all the other goals. This is done by considering the effect one goal has on each of the others, separately, and then adding those separate contributions to form a total. The "contribution" one goal may have on another can be either a beneficial (or posi tive) contribution, or a detrimental (or negative)

Exhibit 5. Goal Cross-Support Matrix
g_1 = goal 1
α_{12} = magnitude of contribution of goal 1 to goal 2.

	g_1	g_2	g_3	g_4				g_n
g_1		a_{12}	a_{13}	a_{14}				a_{1n}
g_2	a_{21}		a_{23}	a_{24}				a_{2n}
g_3	a_{31}	a_{32}						a_{3n}
g_4								
g_n	a_{n1}							a_{nn}

g_1 = goal 1

a_{12} = magnitude of contribution of goal 1 to goal 2

contribution. The degree to which there is a contribution can be assessed by associating numeric values to an adjectival scale, as follows:

Major contribution	(H) = 8
Considerable contribution	(M) = 4
Some contribution	(L) = 2
Negligible contribution	(O) = 1

If the effects are negative, then the same scale can be used, but with negative signs (i.e., -H, -M, -L, -O).

In this cross-support matrix, and in all the other matrices of impact/relevance, or support, the scale is chosen to be logarithmic [8,9]. In the chapter on measurement and scaling Cetron and Dick provide much more information on the problems involved in scoring and scaling, and the choice of scoring techniques.

In each case detailed descriptions of the numerical values are provided, and care is taken to ensure that the scales, and the arithmetic manipulation of them, conform to the laws of scaling [10].

In looking at this cross-support matrix for goals, we have that mathematically, the value (G) for each of n goals (g) can be given by

$$G_i = \sum_{j=1}^{n} \alpha_{ij} \qquad i \neq j.$$

where the α_{ij} are the assigned contribution (H, M, L or O). This means that for each goal (g_i) there is an associated number (G_i) which indicates the degree to which that goal affects the others.

This is an important input to the ranking of the goals, in order of priority, and their subsequent weighting, which is the next stage in the process. The method for weighting the goals is, first, to list them in order of importance, and then to divide 100 points between them, using paired comparisons [11]. Where one criterion is the cross-support, and another is the intrinsic importance of the goal this is based upon subjective judgement, with the cross-support numbers used to aid this judgement.

The next step is to measure the impact or relevance of the indicators to the goals. This is performed for performance indicators to organization goals and for societal indicators to national goals. A matrix is set up as in Exhibit 6 relating goals (g_i) to indicators (e_i), where the elements of the matrix, β_{ij}, conform to the previous scale, and are then summed to provide a total support index u_i for each of the indicators e_i. Where

Exhibit 6. Indicators/Goal Impact/Relevance Matrix

WEIGHTS	W_1	W_2	W_3	W_4	— — — —	W_n	TOTAL RELEVANCE INDEX	
INDIC. \ GOALS	g_1	g_2	g_3	g_4		g_n		
e_1	β_{11}	β_{12}	β_{13}	β_{14}		β_{1n}	$\sum_{j=1}^{n} \beta_{1j} W_j$	$= u_1$
e_2	β_{21}	β_{22}	β_{23}	β_{24}		β_{2n}	$\sum_{j=1}^{n} \beta_{2j} W_j$	$= u_2$
e_3	β_{31}	β_{32}						
e_4	β_{41}	β_{42}						
1 1 1								
e_m	β_{m1}	β_{m2}				β_{mn}	$\sum_{j=1}^{n} \beta_{mj} W_j$	$= u_m$

$$u_i = \sum_{j=1}^{n} \beta_{ij} \quad w_j$$

Those relevance indicators which describe the relevance of the indicators to the goals are then used as weights for the appropriate indicators, in the next part of the analysis.

In order to relate the agency or corporation to its environment, a support matrix is set up, as shown in Exhibit 7, to determine the support of performance indicators to societal indicators. If, as in the previous step, we have determined the weights, u_j, of the societal indicators, e_j, and also the weights, v_i, of performance indicators, P_i, we can obtain the relationship

$$x_i = v_i \sum_{j=1}^{m} \gamma_{ij} \; u_j$$

where x_i is the total support index of performance indicator P_i, and the γ_{ij} are scaled in a manner similar to that used previously. These support indices which describe the magnitude of support between the performance and societal indicators are then used as weights in the final matrix which relates performance indicators to future worlds.

The relationships of impact between the weighted performance indicators and the future worlds are shown in the form of a support matrix, as in Exhibit 8. The matrix is completed in the same manner as before, and the elements summed to provide total value coefficients for each innovation, action and program comprising each future world and for each future world as a whole.

Generally, the value coefficient for each innovation is given by:

$$y_{pq} = \sum_{j=1}^{r} \delta_{pqj} \; x_j$$

Exhibit 7. Performance Indicator/Societal Indicator Support Matrix

	Weights	u_1	u_2	u_3	u_4	— — —	u_m	Total Support Index	
Wts	Performance Ind. \ Societal Ind.	e_1	e_2	e_3	e_4		e_m		
v_1	p_1	γ_{11}	γ_{12}	γ_{13}	γ_{14}		γ_{1m}	$v_1 \sum_{j=1}^{m} \gamma_{1j} u_j$	$= x_1$
v_2	p_2	γ_{21}	γ_{22}	γ_{23}	γ_{24}		γ_{2m}	$v_2 \sum_{j=1}^{m} \gamma_{2j} u_j$	$= x_2$
v_3	p_3	γ_{31}	γ_{32}	γ_{33}	γ_{34}		γ_{3m}	$v_3 \sum_{j=1}^{m} \gamma_{3j} u_j$	$= x_3$
⋮									
v_r	p_r	γ_{r1}	γ_{r2}	γ_{r3}	γ_{r4}		γ_{rm}	$v_r \sum_{j=1}^{m} \gamma_{rj} u_j$	$= x_r$

Exhibit 8. Future World/Performance Indicator Impact Matrix

Future Worlds	Wts. / Indicators / Innovations	x_1 p_1	x_2 p_2	x_3 p_3	x_4 p_4	– – – –	x_r p_r	Total Value Index		
FW_1	T_{11}	δ_{111}	δ_{112}	δ_{113}	δ_{114}		δ_{11r}	$\sum_{j=1}^{r} \delta_{11j} x_j$	$= y_{11}$	Y_1
	T_{12}	δ_{121}	δ_{122}	δ_{123}	δ_{124}		δ_{12r}	$\sum_{j=1}^{r} \delta_{12j} x_j$	$= y_{12}$	
	T_{13}	δ_{131}	δ_{132}	δ_{133}	δ_{134}		δ_{13r}	$\sum_{j=1}^{r} \delta_{13j} x_j$	$= y_{13}$	
	T_{14}									
	⋮									
	T_{1k}	δ_{1k1}	δ_{1k2}	δ_{1k3}	δ_{1k4}		δ_{1kr}	$\sum_{j=1} \delta_{1kj} x_j$	$= y_{1k}$	
FW_2	T_{21}	δ_{211}	δ_{212}	δ_{213}	δ_{214}		δ_{2kr}	$\sum_{j=1}^{r} \delta_{21j} x_j$	$= y_{21}$	Y_2
	T_{22}	δ_{221}	δ_{222}							
	⋮									
	$T_{2\ell}$	$\delta_{2\ell 1}$	$\delta_{2\ell 2}$				$\delta_{2\ell r}$	$\sum_{j=1}^{r} \delta_{2\ell j} x_j$	$= y_{2\ell}$	
⋮										⋮
FW_m	T_{m1}	δ_{m11}	δ_{m12}	δ_{m13}	δ_{m14}		δ_{m1r}	$\sum_{j=1}^{r} \delta_{m1j} x_j$	$= y_{m1}$	Y_m
	T_{m2}									
	⋮									
	T_{ms}	δ_{ms1}	δ_{ms2}	δ_{ms3}	δ_{ms4}		δ_{msr}	$\sum_{j=1}^{r} \delta_{msj} x_j$	$= y_{ms}$	

where δ_{pqj} are the elements of the matrix.

The value coefficient for each future world is given by:

$$Y_p = \sum_{q=1}^{m} y_{pq}$$

These value coefficients represent the value of the impact of the future worlds and of individual innovations, actions and programs upon the agency or corporation goals.

Once the values of the future worlds and their components have been assessed, the next step is to determine the desirability of each future world, and of each innovation, action and program. For each individual component, there is an associated probability of success or achievement, ps, and associated cost, f. The probability of success is obtained from the forecasts, and the cost is obtained from the detailed economic analysis performed on each innovation, and includes R & D costs, operating costs, and capital expenses. These costs are then discounted over the time periods involved using a discounted cash flow technique [12].

Using the benefit/cost ratio,

$$\text{Desirability} = \frac{\text{Value} \times \text{Probability of Success}}{\text{Cost}}$$

We can obtain desirability indices for the three time frames, for *each* component, if desirability is d_{ij} of innovation, action or program T_{ij} and cost is f_{ij}

$$d_{ij} = \frac{y_{ij} \times ps_{ij}}{f_{ij}}$$

For desirability of future worlds, we can combine the costs (f_{ij}) of the components (T_{ij}) where

$$F_i = \sum_{j=1}^{t} f_{ij}$$

and the probabilities (ps_{ij}) can be combined in the usual statistical manner to provide a total probability Ps_i so that the desirability of each future world can be given by:

$$D_i = \frac{Y_i \times Ps_i}{F_i}$$

for each time frame.

We also consider desirability indices for combinations of innovations and programs to provide inputs to the comparative analysis.

These desirability indices then provide the basis for any decision-making, that can then take place, concerning the organization and its future role.

However, as a check on the sensitivity of the method, a sensitivity analysis is performed. This is accomplished by varying goal weights and scaling factors and measuring the differences in impacts at various points throughout the technique. Also measured are changes in impacts, by altering innovations and programs within the future worlds.

The assessment method shown in Exhibit 2, and mentioned earlier as being for assessing national science and technology policy follows exactly the same form. The only differences are that there are no organizational goals or operational performance indicators to be included. This is because this particular assessment is not organization specific. The policy decisions for science and technology emphasis can be conducted through the societal indicators in terms of national goals, so that the science and technologies are related directly to national goals. Apart from this, the matrices are completed in the same fashion as before, as are the benefit/cost and sensitivity analyses.

There is nothing sophisticated about this technique — it has not been designed to be an elegant piece of theory — its sole purpose is to provide sufficient information for the decision-makers to act upon, whenever decisions must be made concerning the development and implementation of technologies. Also, and it is important to remember, this is not an exercise to be performed once, and the results accepted for an indefinite future period. The method should be performed at least once a year, and more frequently if breakthroughs or new developments occur in any field, or society or the economy and so on. If assessment techniques are treated in this manner, and claims are not made as to their panacean properties, then the future of assessments as management aids looks fairly bright, but until these pretentious claims decrease, the skeptics and cynics are likely to continue to prevent widespread acceptance and use of such methods.

REFERENCES

1. LECHT, L.A., *Manpower Requirements for National Objectives in the 1970's*, National Planning Association, Washington, D.C. 1968.

2. GORDON, T.J. and HAYWARD, H., "Initial Experiments with the Cross-Impact Matrix Method of Forecasting," *Futures, 1,* No. 2, December, 1968.

3. GORDON, T.J., "Cross-Impact Matrices," *Futures, 1,* No. 6, December, 1968.

4. DALBY, J., "Practical Refinements to the Cross-Impact Impact Matrix Technique of Technological Forecasting," CETRON and RALPH, *Industrial Applications of Technological Forecasting: Its Use in R & D Management,* New York; John Wiley & Sons, Inc., 1971

5. RALPH, C.A., "Decision Impact Analysis Applied to the Fishing Industry," Plessey Radar, Ltd. Report, November, 1968.

6. RALPH, C.A., "The Beginnings of Cross-Support Analysis," CETRON and RALPH, *Industrial Applications of Technological Forecasting: Its Use in R & D Management*, New York; John Wiley & Sons, Inc., 1971.

7. RALPH, C.A., "Resource Allocation Using Cross-Support Analysis," presented at the 27th JSGOMRAM, Institute of Defense Analyses, Arlington, Virginia, April 20-22, 1971.

8, DEAN, B.V. and ROEPCKE, L.A., "Cost-Effectiveness in R & D Organizational Resource Allocation," *IEEE Transactions on Engineering Management*, EM-16 No. 4, 222-242, November, 1969.

9. RALPH, C.A., "Decision Impact Analysis", *op. cit.*

10. HALL, A.D., *A Methodology for Systems Engineering*, p. 244, Princeton, N.J., Van Nostrand Co., 1962.

11. CHURCHMAN, C.W., ACKOFF, R.L., *Introduction to Operations Research*, New York, John Wiley & Sons, Inc., 1957.

12. DISMAN, S., "Selecting R & D Projects for Profit," *Chemical Engineering*, *69*, December, 1962.

PART III

CASE STUDIES

TECHNOLOGY ASSESSMENT AND POLICY-MAKING IN THE UNITED STATES*

ROBERT U. AYRES and SUSAN C. SIMON

Vice President and Senior Staff Member
International Research and Technology Corporation,
Washington D.C., U.S.A.

with

JACK W. CARLSON

Assistant to the Deputy Director
of the Office of Management and Budget,
Washington, D.C., U.S.A.

INTRODUCTION

The rapid and increasing pace of technological change has been so much remarked on in decades past that the phenomenon has almost lost the capacity to capture and hold our wandering attention. Public concern is presently focussed on campus disorders, crime in the streets, the apparent disintegration of traditional society and values, and environmental deterioration. Yet, these are in some degree attributable to, and symptomatic of, the mismatch between rates of technological and institutional change.

In the 19th and earlier centuries, the typical length of time between initial conception of a new idea and its

---

*Prepared originally for the Seminar on Technological Forecasting, United Nations Economic Commission for Europe December 7-12, 1970, Warsaw, Poland.

widespread implementation was often more than 100 years. Technological progress consisted of creeping forward blindly but slowly enough so that the damage done by an unexpected hazard would not be lethal. During these long gestation periods, people and institutions had time to adjust.

As the 20th Century has unwound, the time lag from invention to widespread use has progressively decreased [1]. The world entered the atomic age less than ten years after the possibility of a self-sustaining nuclear reaction was first deduced theoretically. It was also seven years after fission was first observed, and barely four years after the first self-sustaining reaction. (Of course, the exigencies of World War II greatly accelerated the process. However, the rapid progress in electronics, computers, and space technology especially, has proceeded without benefit of wartime urgency — though not without military funding.) Nylon required only 13 years from first discovery to commercial use. What is significant, however, is that though we are speeding ahead — technologically speaking — at a much faster pace, we are still blindfolded to the full impact of our actions, and when we come to grief, the results are likely to be much more painful and quick.

Technological change, we feel, is primarily stimulated by public and private demand for goods and services. For example, the demand for more time-saving and higher performance transportation relative to other goods has lead to a significant commitment of R&D resources to improve transportation techniques and subsequently has resulted in several technological breakthroughs, including jet engines and new stress-resistant metals. The relative increase in labor costs has resulted in many labor saving techniques either built-into new equipment or facilities or reflected in new procedures of "recipes" [2].

Both the greater investment in R&D and the resulting proliferation of new techniques and their more rapid diffusion has contributed to seemingly uncontrolled technological side effects. However, the fact that the American economy and society and indeed the world economy and society is much more integrated, and has created an environment where the spillover of individual decisions cause uncompensated damage or benefits to other people.

The problem today is how to deal with the risks involved; how to avoid the pitfalls of technological progress without forfeiting the benefits. In essence, this has come to be an institutional question, requiring institutional answers, since some "institution" or method or approach has to be established before individuals, communities and nations can control their destinies. And hopefully, the institutional answers which evolve will have global significance, and will eventually permit an international solution to our common problems.

The fundamental question to be dealt with now, it seems to us, is to what extent the future development of technology can be guided into desirable directions without stifling innovation. Assessment is desirable, to be sure, but coercive constraints on innovation are all too likely to become coercive restraints on progress. The issue, then, is can there be a way of institutionalizing the technology assessment process without destroying that which we merely wish to improve? And, if we can institutionalize the process, will the "system" then respond to rational persuasion and deal effectively with the risks that are implied?

TECHNOLOGY ASSESSMENT AND POLICY CHOICE

In the United States, there is a growing awareness of this need to monitor technological progress systematically, and we are in the early stages of attempting to define an institutional solution to the problem.

Some assessment of technology has occurred through the traditional decision-making channels in the public sector. For example, discussion of annual Federal budgets and the development of new government initiatives (legislation) has included some assessment of the resulting technological change. This has been true to an extent with the decision to proceed with the Second and Third Generation of Nuclear Reactors (e.g. Liquid Metal Fast Breeder Reactor). Some assessment occurred before the decision was made to invest in improving rapid rail transportation (e.g. Metro-Liner). Some assessment is being made concerning the impact of new techniques for rehabilitating criminals before further investment is made. However, these efforts in the past have been incomplete, using only the information that was available, and most of them occurred concurrent with a commitment to invest significant resources in implementation of the new technology.

Over the past four years, there has been a small but significant movement within the Federal government to set up permanent institutional mechanisms to provide an ongoing technology assessment capability to aid decision-makers. At this point, the movement includes a number of interesting proposals: one calls for an independent commission to provide this capability [3]; another would set up an Office of Technology Assessment as an arm of Congress [4]; a third would create a joint congressional committee [5]; whereas a fourth prefers a select congressional committee [6]. And within the executive branch, there are those who see the need for a separate unit, probably within the Office of Science and Technology, to be available to the Administration and to provide technology assessments on demand [7].

This movement is of interest to us primarily because it illuminates the nature and problems of decision-making in our complex society. Never before have our policy-makers been forced to decide not merely on issues beyond their own formal competance but on controversial technical issues on which everyone has personal views, and on which there is no established body of "expert" opinion. Many of the so-called ecological issues are of this kind. In the past two or three years both the Executive Branch and the Congress have been faced with the task of considering all the available information and all the opinions concerning that information on such "hot topics" as DDT, the SST, radioactive wastes, birth control, and pollution control. From this they have had to decide upon some course of action, or inaction, as the case may be, and the task has never been simple. Thus, the impulse to set up a permanent mechanism for technology assessment expresses the hope, not only that such a unit will make the decision process simpler and more objective, but will by its very existence help to forestall activities or projects that would later produce problems requiring corrective measures on the part of government. Whatever mechanism is finally selected will have to be an integral part of the major decision-making channels such as budgeting and planning.

The use of the term "technology assessment" has become widespread lately, and it is important to keep its various definitions in mind. As the term is predominantly used, it refers to a form of policy research to evaluate both the short-term and long-term impact on society of a technology or a technological program [8]. This is not technological forecasting, as such, but an analysis of the social consequences of technological change. Ideally, an assessment would be carried out *before* a new technology was applied, thus elucidating not only the expected benefits but also the potential unwanted side-effects that might occur. However, most technology assessment, as it has occurred thus far in the United States, has dealt retrospectively with a developed technology whose ill effects are already in evidence or at least strongly anticipated. For example, with DDT, the full effect of the pesticide on the natural environment was not examined by any government body until its bad effects were documented and publicized by a private individual — nearly twenty years after it became widely used [9]. With the SST, the program was nearing the point of no return before its environmental impact became a subject of heated public debates. Therefore, most technology assessments have tended to focus on technological counter-measures to a recognized problem such as solid wastes or automotive air pollution. Less frequently, there have been examinations of Federally-sponsored technological programs in progress including an evaluation of the drawbacks as well as the benefits to be derived from its continuation (e.g., the SST, the space program).

Up to the present time, there have been no noteworthy *prospective assessments* covering the entire spectrum of social, economic and environmental impacts from a technological development originating or significantly supported by the private sector of the economy.

To facilitate a deeper examination of the role of technology assessment in policy-making, it seems worthwhile to explore a specific case history in some detail. The example we have chosen illustrates virtually the entire spectrum of ways in which technical information or judgements are or can be, used to influence government policy. It is the issue of air pollution from automobiles, which touches all levels and all branches of government in the U.S., both local and national, and also involves a direct challenge to the economic interests of some of the most prominent corporations in the United States. It is a good illustration of the potential impact of rational persuasion, alone, on the policy-making process in the United States.

THE CALIFORNIA EXPERIENCE

AUTO AIR POLLUTION

The improvement in city conditions by the general adoption of the motor car can hardly be overestimated. Streets clean, dustless and odorless, with light rubber-tired vehicles moving swiftly and noiselessly over their smooth expanse, would eliminate a greater part of the nervousness, distraction, and strain of modern metropolitan life.

Scientific American, 1899

As the quote from *Scientific American* illustrates, prospective technology assessment of automotive technology has not been notably successful. The existence of a major problem was first recognized by the residents of Los Angeles basin in Southern California in the 1940's. It's sympton was "smog," a brownish, irritating haze which seemed (but was not) like a combination of smoke and fog.

In the early 1950's, a professor of biochemistry, Dr. Ari J. Haagen-Smith, at the California Institute of Technology in Pasadena, diagnosed the chemical nature of smog in Los Angeles, and demonstrated its origin from photochemical reactions among several constituents of automobile exhaust, namely unburned hydrocarbons — especially olefins — and oxides of nitrogen (NO_x).

It was not until 1960 that attempts were made to control this source of pollution, with establishment of a Motor Vehicle Pollution Control Board (MVPCB) to regulate auto emissions. In cooperation with the State Department of Public Health, emission standards were set and auto modifications were required. New cars sold in California (beginning in 1963) were required to have crankcase emission controls, and cars (beginning in 1966) were required to come equipped with exhaust controls limiting HC and CO emissions to specified levels.

During this early phase, the emphasis among those looking into the issue centered around demonstration of of the detrimental effects of air pollution on the health of California citizens, plus an investigation of the internal combustion engine (ICE) to provide pollution reduction through modifications. A pattern of on-going technology assessment, under the auspices of the MVPCB, was being established, whereby sources both inside and outside the industry were consulted concerning automobile technology and air pollution abatement.

In 1967, the MVPCB was replaced by the California Air Resources Board (CARB) through legislation that directed the Board to establish statewide ambient air quality standards, to set automobile emission standards, and to exercise authority over pollution reduction devices. Since its creation, the CARB has literally pioneered, for the state and the nation, in creating a technical assistance - and assessment - aid to the California legislature for air pollution abatement.

During this period, the California state legislature, and primarily the Assembly Transportation Committee under the chairmanship of John F. Foran (Democrat of San

Francisco) became very active, establishing a pattern of legislative initiative and responsibility in this area. Part of the reason for this was the continuing crisis in air pollution in California, and the increasing public demand for action. Also, the program initially set up was not working according to expectations. Once emission standards had been set up by the Board and the Department of Public Health, it was assumed that they would be implemented with control devices for new and used cars. This was not the case, at least for used cars. In the beginning there had been widespread interest in device development, but when several prototype units were certified by the CARB, the auto makers promptly introduced devices of their own, refusing to adopt those developed outside the industry and causing the smaller companies to withdraw from the market. The net result was that the auto manufacturers were left alone in the field and had little stimulus for further device development. The state legislature decided to take additional steps.

Early in March 1967, the Assembly Transportation Committee held hearings throughout the state on the strength and weaknesses of existing pollution control laws, on the health aspects of air pollution, on other economic aspects of pollution, and, for the first time, on alternative propulsion systems. The committee brought together a panel of witnesses who challenged the public position of the auto industry, arguing that drastic reductions of emissions were, in fact, technically feasible. In this, the committee was aided by the Assembly Office of Research (the only research staff attached to any legislative body in the world, as far as we know) and by the Washington representative of the Legislature.

The principal results of this activity were the passage, in 1968, of the Pure Air Act in California, which gave to state agencies even greater control over air pollution activities. In Assembly Bill 356, a companion piece to the Act, the state agreed to a mandatory purchase of up to 25% of its yearly vehicle stock additions to be low pollution vehicles, at 200% regular cost. A low pollution vehicle was defined in terms of stringent emission levels, adopted partly to exclude ICE cars and to stimulate a market for development of alternative systems. Also in Assembly Bill 357, California passed even more strict control standards up through 1974, which, for the first time, included a limitation on the oxides of nitrogen emissions.

Other results of the hearings in 1967 and of the continuing efforts of the California Assembly Office of Research, was a steam car test program, initiated in early 1968, to be conducted by the California State Highway Patrol. (This effort subsequently died, due to the fact that no steam powered vehicles were offered for tests.) Late in 1968, the legislature also applied for, and received a grant from the Urban Mass Transit Administration of the U.S. Department of Transportation to conduct experiments in San Francisco and Alameda-Contra Costa Counties in California with steam buses. In addition, California has been conducting tests of natural gas-powered vehicles on state cars. In 1969, hearings were held on the subject of leaded fuels, and in 1970 the California Air Resources Board proposed even more stringent emission standards, which, by this time, has been surpassed by standards proposed on the Federal level.

During this period, the courts were brought into the picture via a suit, filed in the Federal District Court in Los Angeles, charging that the major automakers conspired to delay development and installation of pollution control devices. Nine months later, in September 1969, the Justice Department handed down a consent decree, settling for a promise of good behavior. Several state and city governments have continued the suit, partly for the purpose of forcing the evidence collected by the Department of Justice to be made public.

The California experience is important to examine for a number of reasons. It was here that the impact of air pollution was greatest and public demand for abatement loudest. It was here that the first attempts were made to "control" some of the ill effects of prevailing automotive technology via enforcing emission standards. And it was California that the first attempts were made to initiate or sponsor both the development of control technology and the development of technology for alternative propulsion systems. The decision to adopt these approaches was the product, largely, of official, state-sponsored technology assessment, and, as will be seen, the California approach was adopted in essence by the Federal government, which, though it arrived later on the scene, has matched, and even surpassed the California program in some ways.

FEDERAL EXPERIENCE

The assessment of automotive technology, from the standpoint of pollution did not really become a major concern of Congress until 1966-67. This is not to say that Congress had not been involved in automobile air pollution control. In the mid-sixties, a number of bills had been passed to provide R&D funds for all levels of governments for research and technical assistance in auto air pollution control. In 1965, the Secretary of Health, Education and Welfare was authorized to determine and set exhaust emission standards, and specifically, standards for CO and HC emissions, as were determined, were authorized to go into effect. (They went into effect on 1968 model cars, and were identical to existing California standards.)

However, the suspicion that the ICE might never achieve desired cleanliness goals was gaining ground, both in California and across the nation. Therefore, in the spring of 1967, a set of hearings were sponsored by the Senate Commerce Committee on Public Works. The hearings were on "Electric Vehicles and Other Alternatives to the Internal Combustion Engine," [10] and were triggered by proposed legislation to sponsor R&D demonstration projects for research into electric vehicles and other propulsion systems. Testimony and statements were taken from nearly 100 people, including engineers, scientists, government agency officials, auto industry representatives, electric car developers, utility industry spokesmen, and virtually every expert available who was in some way concerned with this issue.

The hearings had some impact on the policy process, if only to the extent that they spotlighted the auto air pollution problem and demonstrated the minimal extent of the auto industry's technological competence in this area. Public interest in the electric car, as such, was considerably abated as a result of the hearings, due to several drawbacks — the most important being the lack of adequate rechargeable battery technology. The hearings also showed the need for further investigation into other propulsion systems.

Another response partially attributable to the submission to Congress of legislation to promote electric vehicles occurred a few months prior to the hearings. Pursuant to an inter-departmental agreement, the Department of Commerce/Technical Advisory Board (CTAB) set up a Panel on Electrically Powered Vehicles to examine the feasibility of electric cars for the U.S. The fifteen member panel was chaired by Richard S. Morse of MIT, and was composed of six university professors, representatives of the three big auto companies, and representatives from the coal, oil, chemical, and electric utility industries. Beginning in January 1967, the panel first set up six subpanels in important technical areas, while the main panel held briefings from outside experts, collected and examined available published literature, and took field trips to major laboratories where relevant work was underway. Their assigned task was to assess the likelihood that electric powered vehicles would be a viable alternative to the internal combustion engine, in terms of both technical feasibility and economic practicality. In October, 1967, they issued a summary of the results of their work, including sixteen recommendations to the Federal Government, called a "Program for Progress" [11]. Among these were recommendations that standards should be set to provide no increase in the level of lead emitted to the atmosphere from gasoline combustion; a call for the creation of a five-year, $60-million research program in alternative vehicular power systems; and a recommendation that Federal, state and local governments include low emission criteria as factors when purchasing vehicles.

Critics of the report point out that the panel did not confine its efforts to the alternatives, as was its stated task. The report, however, was a very broad one and represented the first major attempt to assess the internal combustion engine in the light of alternative technologies. Work in this direction was also initiated by HEW's National Center for Air Pollution Control in early 1967. Arthur D. Little, Inc., was commissioned to study the feasibility of electric vehicles [12] while Battelle Institute was asked to examine other types of unconventional power systems including nuclear, thermoelectric, thermionic and so-called external combustion engines [13]. In addition, the Center sponsored a state-of-the-art symposium on electric vehicle research at Columbia University, in the spring of 1967. The contributed papers were later published in book form [14].

Undoubtedly some of the thinking in the above investigations had an impact on the Air Quality Act of 1967, [15] a watershed piece of legislation that not only established the national approach to air pollution control in general, but also specifically called for increased R&D into fuels and vehicles. However, no Federally sponsored R&D program on alternatives to the ICE was established at that time.

The next major step in the assessment of automotive technology was a second set of hearings, again jointly sponsored by the Senate Commerce Committee and the Air and Water Pollution Subcommittee of the Public Works Committee. These hearings were initiated by Senator Warren Magnuson, Chairman of the Senate Commerce Committee, at the urging of Ralph Nader, a private citizen who lobbies for consumer welfare legislation. The subject was the "Automobile Steam Engine and Other External Combustion Engines," held in late May, 1968 [16].

On this occasion, the committee staff invited testimony from twelve witnesses and over forty other sources of information were included in the record. These hearings were not held to consider specific legislative proposals, but were undertaken primarily to explore the potential of external combustion engines (ECE's) as a possible satisfactory alternative to the present automobile engine. Apart from Ford and GM and the Chairman of the electric vehicle panel, Richard Morse, the principal testimony was offered by one of the authors (RUA), based on an independent study of low pollution technology sponsored by Resources for the Future, Inc., a non-profit research organization [17]. The physical presence of two operational steam powered automobiles, built in the 1950's, also played a significant role in the hearings.

The result of these congressional investigations culminated several months later in a report prepared by the staff of the Senate Commerce Committee entitled, "The Search for a Low-Emission Vehicle," March, 1969 [18]. Here, in 24 pages, the Committee staff summarized the results of the hearings, plus a considerable amount of supplementary investigation. It concluded that the Rankine cycle (steam) engine is likely to be a satisfactory alternative to the ICE, both in terms of performance and emissions. It also recommended that state, local and the Federal government adopted procurement policies for low-emission vehicles; that the Department of Transportation finance demonstration tests in these areas; that the Department of Health, Education and Welfare spend more of its research funds on low-emission systems, and that economic incentives for private ownership of such vehicles should be considered.

By early 1970, the "environment" had become the big issue of the decade, and on the executive level, President Nixon, in his state of the union message, pointed to the automobile as the worst polluter to be dealt with. In an Environmental Message, delivered February 10, he further proposed a comprehensive approach to auto pollution, recommending a Federal purchase program for low-emission vehicles and emphasizing that since the ICE may never meet standards, it was necessary to look into alternate systems.

Before and after this open commitment to the environment, a number of special entities were created, both temporarily and permanently, to assist the President and his staff in determining policy concerning air pollution and the automobile. The Air Quality Act of 1967 had created a Presidential Air Quality Advisory Board to provide a link between the White House, the Department of Health, Education and Welfare and the private sector. In May, 1969, the President set up a cabinet-level Environmental Quality Council (to be distinguished from the subsequent permanent Council on Environmental Quality), composed of the heads of the various concerned agencies who were assigned to "examine the full range of variables which effect environmental quality." At a White House meeting in November 1969, the Council reported to the President, recommending tighter emissions standards for 1975 (comparable to those already established for California) and a 5-year, $45 million Federal program to develop a pollution free automobile. However, White House approval remained in abeyance pending a report by an Ad Hoc Panel on Unconventional Vehicles, reporting via the President's Office of Science and Technology. The chairman was Dr. Ernest Starkman, an eminent engineering professor at the University of California (Berkeley), who is also Chairman of the Technical Advisory Panel of the California Air Resources Board (CARB). Sitting *ex officio* for the Office of Science Technology was Dr. S. William Gouse, formerly of M.I.T., and a contributor to the Morse Panel. While setting up the cabinet-level Council, the President also created a 15 member Citizens Advisory Committee on Environmental Quality, to be chaired by Laurence S. Rockefeller.

On November 18, 1969, the President announced the creation of a Task Force on Air Pollution, to be chaired by Dr. Ari J. Haagen-Smit of California Institute of Technology to evaluate the effectiveness of present efforts to curtail air pollution and to recommend further action that may be needed. On January 29, 1970, he named the members of his recently created three-member Council on Environmental Quality, established officially January 1, to function in a manner similar to the Council of Economic Advisors. The aforementioned Ad Hoc panel submitted its report to the President on February 1, 1970, calling for urgent Federal action on unconventional vehicle development. The program was officially announced July 16, 1970, along with the appointment of an eight-member Advisory Committee on Advanced Power Systems, again chaired by Professor Ernest Starkman of the University of California, to assist the overworked staff of the Council on Environmental Quality in overseeing the program. The plan called for development of two or more virtually non-polluting vehicles to be produced by 1975 with the Department of Health, Education and Welfare (specifically, the National Air Pollution Control Administration), the lead agency in the program. At this time, the advisory committee concluded that the most promising alternatives to the ICE are (1) gas turbines, (2) hybrids (heat engine + battery) and (3) Rankine cycle ("steam") engines, in that order. The panel was apparently impressed by progress in developing inexpensive ceramic materials for turbine parts (by the Ford Motor Company) and advanced R&D on hybrid systems, by TRW, Inc. (No comparable recent programs on Rankine cycle systems was shown to the panel.)

In response to the announcement of proposed 1975 emissions standards, plus revised testing procedures, the auto manufacturers separately announced in early 1970 that clean air would require the elimination of lead from gasoline, and divulged plans to build a lower compression engine capable of burning unleaded 91 octane fuel. The reason given for this step was that lead additives gradually interfere with the effectiveness of catalytic exhaust control devices. Some oil companies immediately announced plans to market such a gasoline, while others attacked the idea.

In the early stages of the debate on lead additives, the National Air Pollution Control Administration (NAPCA) and the President's Science Advisor, Lee A. Dubridge, asked the (Department of) Commerce Technical Advisory Board (CTAB) to look into this issue. A CTAB Panel on Automotive Fuels and Air Pollution was selected in December, 1969, under the chairmanship of Professor David V. Ragone, an Associate Dean at Carnegie-Mellon University, formerly of Gulf-General Atomic Corporation and also a prominent member of the earlier Morse panel. It included five representatives from industry (no direct representative from an oil or auto company), and three government agency staffers. Three subpanels were set up and 32 different organizations made presentations. The panel worked toward a 90-day deadline, and issued its first interim report in June, 1970 (The final report is due in December, 1970) [19]. It recommended that the government have authority to regulate fuel additives (i.e., lead) and proposed that price incentives be created to favor the purchase of unleaded fuels.

This problem of leaded fuels is an interesting sidelight to the main debate, but it is less one of "technological" choice for policy makers and more one of economics and timing. It is a fact that most control devices available for testing *do* lose effectiveness when leaded fuels are used, and several oil companies have agreed to produce lead-free fuels in recognition of this fact. The problem is how to encourage consumers to buy the fuel, which will be more expensive than leaded fuel of the same octane, and the debates center, at this time, on proposals for taxes on leaded fuels. Curiously, many Senators who have led in the fight for tougher emissions standards have opposed such a tax. Their reasons in doing so are not clear. However, it may well be attributable to a misunderstanding of the purposes of the tax — which the Administration presented as a "revenue" bill and never adequately explained to the House Ways and Means Committee.

Meanwhile, the Senate Commerce Committee held another round of hearings in late January, 1970, on a "Low Emission Vehicle Act," introduced in the House and Senate October 27, 1969. By means of the Hearings, another survey was made of the state-of-the-art in advanced automotive technology, to demonstrate that a procurement plan for preferential purchase of "low emission vehicles" by the government would be feasible.

Also, Congress was preparing for another round of "Clean Air" Legislation, with hearings held before the House Subcommittee on Public Health and Welfare (Interstate and Foreign Commerce Committee) beginning in December, 1969, and running intermittently through April, 1970. A house version of the bill was passed in June, only to be overwhelmed by much tougher legislation proposed by the Senate and passed by a vote of 73-0 on September 22, 1970.

A glance at the Senate version of the bill, linked mostly to Senator Edmund Muskie (D-Me), shows the near revolutionary nature of its measures, at least with respect to previous legislation. Members of Congress are not only confronting industry with an incentive to stimulate potential competition in the form of the preferential low-emission vehicle procurement plan, but they have overridden all objections by the industry and have decreed that autos (by 1975 with a one year maximum grace period) will have to meet proposed 1980 emission standards, which the industry claims it does not have the technology to do. The bill would also phase out the use of leaded fuel by new cars, a factor that will directly affect engine design. Unfortunately for the auto makers they suffer from a severe "credibility gap" due to their historical intransigence and Congressmen are now evidently assuming that they *will* be able to meet legal specifications, even if they protest, because they have invariably resisted earlier air quality legislation, even while apparently possessing the capability to do *better*. Should this bill pass as it stands and it will likely do so by December 1970, Congress will have, in effect, dictated the objectives of future automotive technology (although not the technological specifics), a role never before assumed.

CONCLUSIONS AND RECOMMENDATIONS

The Process and the Participants

All the above activities add up to an impressive amount of seemingly responsible investigation into an important technological issue. And, in fact, it is impressive, if only for the fact that the automobile air pollution problem has been studied more extensively, by more separate groups, than any other single environmental issue.

As is also evident from the foregoing, the predominant approach to retrospective technology assessment in this and other "environmental" areas has been to canvass or consult all the "experts" in the field, and from there decide upon a palliative or ameliorative program. This has happened time and again, and the same organizations and often the same individuals have testified at hearings or have been asked to serve on a board, a panel, a task force, or a committee. The procedure is cumbersome and redundant, but it is somewhat effective, insofar as Congress and the administration cannot plead ignorance concerning this issue.

In many ways, the policy-making "system" has worked, at least in this case, though much fault can be found. The relevant information has been assembled and made available to decision-makers, though belatedly and rather inefficiently, and a body of legislation has been generated to deal with the problem, vigorously, if not with maximum finesse.

Since we are concerned with the processes of technology assessment and the ways in which they can be improved, we should understand why the system worked — to the extent that it did — in this case. First of all, the *nature of the problem* and its causal agents was never seriously in doubt. There has been relatively little effort wasted in argument concerning these fundamentals. Urban air pollution demanded abatement measure, and since the automobile was shown to be responsible for over half of the problem, it was inevitable that attempts would be made to correct this situation. Also, the automobile has already been subject to a number of controls on the state and federal level, be they licensing and registration requirements or performance and safety regulations. Thus, it was possible to extend various existing control mechanisms to facilitate air pollution abatement, eliminating the need to create entirely new institutions.

Second, *technological alternatives and experts were available*. Again, it was fortunate that some pertinent technology had already been developed *outside* the auto industry, and that alternative low-emission power plants had been built or were under development. The fact that steam and electric cars were physically available to be brought to Washington for Congressmen to see was evidently quite important in influencing policy. Equally important were the several independent experts who were fortuitously in a position to provide a contrasting position to that enunciated in the public statements of the automobile industry. Their testimony, concerning the feasibility of large-scale, low emission alternatives, both in California and later in Washington, was crucial in forcing the industry to commit itself to building a clean ICE, saying over and over that the internal combustion engine could be cleaned up to the same degree as any likely alternative [20].

Third, *some form of precedent had been established.* Here we refer to the fact that the California experience functioned as a prototype example for the Federal government. The lead-role played by a particular committee and its chairman, the occurrence of hearings and public meetings, and the collection and dissemination of valuable information were all repeated on a larger scale in the Federal approach. The willingness and ability of a state to confront and prevail over a giant industry on the issue of pollution was an experience not ignored by national political leaders.

Fourth, *pressure and public opinion played a role.* As the "Environment" and the "Quality of Life" came to be popular national issues, the pressure of public opinion was able to maintain and increase the momentum of abatement activity in this area. Pressure brought to bear by special interest associations, and independent spokesmen such as Ralph Nader, as well as the communications media themselves, served to reinforce and complement the impact of expert testimony and existing technology. And the activities of Congressmen and certain committees producing legislation, hearings, recommendations and reports also provided political incentives for the administration to respond in kind, and to assume some form of leadership in the effort. All of this, of course, produced a great duplication of efforts at technology assessment, though this may have been an unavoidable concomitant of the educational process which was undeniably taking place. (Each participant in the decision-making process has a different perspective and must have basic information in order for him to formulate his own conclusions.)

The major weakness of the "system" is that there was not enough authorization of the problem and that it took persistant public outcry concerning the automobile and air pollution to bring about an appropriate response by Congress and the Administration. As a result of this public concern, independent experts and witnesses made their services available at no cost to the government. (Far more often, at hearings of state legislatures or regulatory bodies, no independent witnesses are available, for the simple but devastating economic reason that such witnesses must usually testify at their own expense — and often in a hostile environment.) Without this essential collection of outside expertise, there would have been nothing to contradict the familiar arguments of the auto industry. Ralph Nader played an important, though little-known, role in the auto air pollution problem, and it is highly likely that without his skillful use of the media and effective intervention on behalf of the need for public hearings, there might never have been a thorough examination of "steam" and other alternative technologies, at least at the time it was needed. One cannot help but conclude that the retrospective technology assessment of automotive technology took place largely as a result of a series of fortuitous occurrences. For this reason, the existing technology assessment "system" is basically inefficient.

INSTITUTIONALIZING TECHNOLOGY ASSESSMENT

The enormous cost and difficulty of undoing the damage resulting from an entrenched technology (such as the automotive internal combustion engine), have convinced a number of persons both inside and outside Congress, that an adequate "early warning" system is essential to permit ameliorative action at an earlier stage of development. The effort has taken the form of a variety of proposals, mentioned earlier, to set up some specialized mechanism to provide a technology assessment capability that will both help prevent applications of harmful technologies and will perform assessments of current technologies as an aid to decision makers.

The pioneering effort in this direction has been made by Congressman Emilio Daddario, chairman of the Subcommittee on Science, Research and Development of the House Committee on Science and Astronautics [21]. Following extensive hearings over a period of several years, his committee has proposed the establishment of an Office of Technology Assessment (OTA), with a 13-member Board composed of two House and two Senate members, the Comptroller General, the Director of the Legislative Reference Service, the Director of the Office of Technology Assessment (as distinguished from the Board), and six members from the "public". Their purpose would be to consider existing and probable impacts of technological applications, and to provide a supplementary information system to congressional hearings. The OTA would cooperate closely with the National Science Foundation (NSF) to prevent duplication of efforts; and assessments would be initiated by any committee chairman, by the Board, or by the Director. The majority of these assessments would be performed by contract with outside organizations.

A possible weakness that some critics have discerned in this proposal is that the OTA would be too closely linked with Congress, both through budget dependency and in subject matter to be assessed. The status quo with the identifiable beneficiaries who have loyal commitments from legislators would work against a process that is intended to arouse attention about developing problems. As a working unit of Congress, also, it could conceivably be overwhelmed by short-term activities that committees and their staffs have been performing to date, leaving little time, energy and resources to investigate new technologies.

Another major suggestion, proposed by Senators Warren Magnuson and Philip Hart, would establish an independent unit, to be called the Independent Technology Assessment and Environmental Data Collection Commission [22]. The hope behind this proposal is to have a group with no innate orientation toward either technological advancement or environmental protection. Though the Commission would, in theory, be independent of all branches of government, the initial members of it would be designated by Congress, and would, in turn, determine the composition of the final Commission. They would not, however, be subject to the usual (political) appointment process. The Commission would make assessments, without regulatory force, of proposed new technologies and of technologies in use or planned; plus collect, store, and make available all information to all branches and agencies of the Federal and state governments and to private persons and entities. Also, it is designed to be especially helpful to the Courts, who must rule on technological issues.

The major criticism of this plan not only concerns the degree to which its theoretical independence is borne out in practice, especially with those it is supposed to regulate,(which appears somewhat questionable at this stage,) but also the fact that, as an independent commission, it may become more "regulatory" in impact than might be the case with other institutions. Though its assessments would not be regulatory in purpose, the very existence of the organization and the assessments it performs could stifle innovation with the threat of governmental disapproval. On the other hand, there is some fear that its indepencence would remove it from the seat of power and reduce its influence.

Two other proposals call for establishment of congressional committees to perform this function. A Senate Resolution sponsored by Edward Muskie, would set up a Select Committee on Technology and the Human Environment [23]. It would have a limited life (three years) and would have no jurisdiction over legislation. At the end of its duration, it would render a comprehensive report on the future impact of scientific and technological change. A proposal by Senator Gordon Allott and others would set up a more permanent mechanism, by establishing a joint congressional committee on science and technology, whose functions would be similar to the proposed select committee, but would involve an annual report, and possibly a longer lifespan [24].

Both of these suggestions, as with the OTA, would be highly "congressional" in outlook, and would help Congress in deciding policy, but would be insufficient for a "system" including the executive functions of government. They would probably need to be balanced by a similar unit for the executive branch, and possibly for various agencies — a prospect that may be inevitable — but seems unnecessarily cumbersome at this point.

Several other proposals have been offered, some of which prefer not to place the technology assessment capability in a single unit. The National Academy of Public Administration recommends that "assessment should be made a continuing responsibility of each Executive Department or Agency, and should be integrated into program formulation, review, decision, and execution through the planning-programming-budgeting system. "Central responsibility for government-wide policy recommendations, coordination, and review of technology assessment should be vested in the Council on Environmental Quality" [25]. Others have suggested that either a "fourth branch of government" be created for this purpose [26] or even an independent agency, such as a Department of Science and Technology [27] or an Agency for Technological Development [28]. However, a panel for the National Academy of Sciences has urged "the creation of a constellation of organizations, with components located strategically within both political branches that can create a focus and a forum for responsible technological assessment activities throughout government and the private sector [29].

RECOMMENDATIONS

All the above proposals have some merit, if only because they recognize the need for prospective assessments and provide some mechanism for bringing together the necessary skills and disciplines. The approaches differ mainly in terms of where in the government structure the primary impact is designed to occur. They all attempt to balance between the opposite dangers of being too close to the decision-making process or being subjected to pressures of political expediency and being too remote from the hurley-burley and consequently lacking in influence. They differ, too, in the value set on technological progress per se and the degree to which they would lean over backward to prevent technology assessment from becoming technology "harassment" [30]. These goals are difficult to accomplish, at best, and may in fact be contradictory. That is, any technology assessment unit may in fact have some tendency to inhibit technological progress in certain fields.

We have noted, also, that in these various proposals, there has been no explicit attempt to deal with one critical issue, namely ensuring that the available information reaches the responsible decision-makers. If there is one thing the auto air pollution case history has shown, it has been the importance of having the necessary information at the right places at critical times. If a number of independent witnesses had not been willing to take the time and energy to testify — without payment — at the various hearings, there would probably have been no noticeable results from these efforts. Also, various agencies, such as NAPCA and DOT, might have escaped some criticism for lack of concern if their program examiners had been more familiar with the importance of some of their projects. And recently, if certain members of Congress had not been forearmed with technical arguments concerning both the prospects for reducing emissions from ICE vehicles, and the feasibility of alternative systems, the Congress would not have been able to confront the auto industry so directly. Plural channels and sources are obviously seeded for plural decision-makers.

It seems to us that, in addition to a technological assessment capability *per se*, what is needed most is the institutionalization of the role Ralph Nader has played in recent years in the policy process. There needs to be some unit to perform the crucial function of collecting data and injecting it directly into the system via direct face-to-face contact with Congressmen, examiners and administrative and other decision-makers. There needs to be a continuous, explicit attempt to present, not only the other side, but all sides of various issues; and the process *must* be continuing, since a single exposure rarely has a lasting impact.

An essential link in the technology assessment mechanism, in our estimation, would be a clearinghouse that would function as an active liason between those individuals and institutions performing, or capable of performing, on-going technology assessments, and critical persons or committees in the policy-making process. This clearinghouse would be a funnel for requests for "expert" opinion, and would (hopefully) have sufficient independent resources of its own to invite and *pay for* expert witnesses. It would be responsible for tracking of major issues, and potential issues, and would provide guidance to foundations and other sources of funding for studies to fill gaps in current knowledge. It would also be a channel through which experts outside government could raise technological questions concerning programs that may have inherent, but unforseen drawbacks. It would, in short, attempt to see that the full range of opinions were presented when Congress investigated such an issue. Preferably, such a unit would be independent of any single branch of government, with funding from both the Congress and the executive branches, as well as private foundations.

The function of this hypothetical clearinghouse would not be to create new information, but to channel and focus it. It is a function carried out in our broader society by the media: newspapers, magazines, radio and TV. However, the news media cannot adequately provide the kind of specific, issue-related, technical information needed by decision-makers. They are oriented to a mass audience. Their primary mission may be to entertain even more than to inform. The news-media in general are not themselves outstandingly competent to assimilate information of a technical nature and translate it into terms familiar to the public. Although many special purpose publications with limited audiences can do this, they are not the normal reading matter of the decision-makers who may need the information. And even if the material reaches its target — by accident — there is no mechanism for answering questions, filling in gaps, or exploring further implications.

Thus, there is no substitute for personal face-to-face contact with experts, whether informal hearings or informal sessions. There are no interesting methodological uncertainties or other corrollary risks to attract the interests of academics — which may perversely explain why the literature on technology assessment has never touched on this key area. However, it is worth pointing out that a mechanism for locating *and paying for the time* of expert witnesses who already exist (in many instances) would inexpensively achieve a great deal to improve the operations of the decision-making "system", including its ability to hedge against future technological disasters While an improved capability to forecast the social implications of technological change is undoubtedly desirable, it seems to us that there is no particular need for the work to be done by some special group with direct institutional links to the Congress or the White House. What is gained in terms of face-to-face access to the seat of power may well be lost by compromising its independence of judgement. We suggest that face-to-face access to decision-makers is indeed important, but that it can be achieved — perhaps more easily — by any reputable outside group, without the necessity of having a private launching pad within the governmental apparatus.

NOTES AND REFERENCES

1. According to GILFILLAN, S. COLUM, "The Prediction of Technical Change", *The Review of Electronics and Statistics, 34,* 368-385, Nov. 1952, of 19 inventions "voted most useful" and introduced between 1888 and 1913, a span of 176 years occurred between the first thought of invention to first working model or patent and 24 years from the latter to practical use.

2. For a description of the primary role of derived demand, in directing R&D investment and the subsequent use of new techniques, see CARLSON, JACK W. "Aspects of the Diffusion of Technology in the United States", paper for the Fifth Meeting of Senior Economic Advisors, Economic Commission for Europe, United Nations, November 1967.

3. S. 4044, 91st Congress, 2nd Session, June 30, 1970. Bill introduced by Senator Warren Magnuson for himself and for Senator Philip Hart.

4. H.R. 6698, 90th Congress, 1st Session, March 7, 1967. Bill introduced by Congressman Emilio Q. Daddario.

5. S. Resolution 68, 90th Congress, 1st Session, January 25, 1967, Senator Edmund Muskie and others.

6. S. 1305, 90th Congress, 1st Session, March 15, 1967, Senator Allott and others.

7. Report to the President, April, 1969, by the President's Task Force on Science Policy.

8. See statement by Emilio Q. Daddario, Chairman, Subcommittee on Science, Research and Development of the Committee on Science and Astronautics, U.S. House of Representatives. 90th Congress, 1st Session, Serial 1 (Revised August 1968). Title: "Technology Assessment". According to Daddario, "Technology Assessment" is a form of policy research which provides a balanced appraisal to the policymaker. Ideally, it is a system to ask the right questions and obtain correct and timely answers. It identifies policy issues, assesses the impact of alternative courses of action and presents findings. It is a method of analysis that systematically appraises the nature, significance, status, and merit of a technological program.

9. CARSON, RACHEL, *The Silent Spring*, Boston, Houghton-Miflin, 1962.

10. "Electric Vehicles and Other Alternatives to the Internal Combustion Engine." Joint Hearings before the Committee on Commerce and the Subcommittee on Air and Water Pollution of the Committee on Public Works, United States Senate, 90th Congress, 1st Session, March 14, 15, 16, 17 and April 10, 1967.

11. "The Automobile and Air Pollution: A Program for Progress." Part 1, Report of the Panel on Electrically Powered Vehicles to the Commerce Technical Advisory Board, John F. Kincaid, Chairman, Assistant Secretary for Science and Technology, U.S. Dept. of Commerce, Alexander B. Trowbridge, Secretary, October, 1967.

12. Arthur D. Little, Inc., "Prospects for Electric Vehicles: A Study of Low-Pollution-Potential Vehicles - Electric," GEORGE, J.B.H., STRATTON, L.J. and ACTON, R.G., for the Department of Health, Education and Welfare, National Center for Air Pollution Control, May 15, 1968.

13. Battelle Memorial Institute, "Study of Unconventional Thermal, Mechanical, and Nuclear Low-Pollution-Potential Power Sources for Urban Vehicles," to the U.S. Department of Health, Education and Welfare, National Center for Air Pollution Control, Cincinnati, Ohio, March 15, 1968.

14. See the work of the Task Force on Environmental Quality, Council of Economic Advisors, U.S.A., 1967 (Jack W. Carlson, editor).

15. "Power Systems for Electric Vehicles," a symposium sponsored by the U.S. Department of Health, Education and Welfare, Columbia University and the Polytechnic Institute of Brooklyn, April 6-8, 1967, HEW, Public Health Service, Bureau of Disease Prevention and Environmental Control, National Center for Air Pollution Control, Cincinnati, Ohio.

16. "Automobile Steam Engine and Other External Combustion Engines," Joint Hearings before the Committee on Commerce and the Subcommittee on Air and Water Pollution of the Committee on Public Works, United States Senate, 90th Congress, 2nd Session, May 27 and 28, 1968.

17. AYRES, ROBERT U. and McKENNA, RICHARD, *Alternatives to the Internal Combustion Engine*, Resources for the Future, Inc., to be published by Johns Hopkins University Press, 1972.

18. "The Search for a Low-Emission Vehicle," Staff Report prepared for the Committee on Commerce, United States Senate, March, 1969.

19. "The Implications of Lead Removal from Automotive Fuel," an interim report of the Commerce Technical Advisory Board Panel on Automotive Fuels and Air Pollution, U.S. Department of Commerce, June, 1970.

20. "Automobile Steam Engine and Other External Engines," Joint Hearings . . . See above. One example of this argument occurred in the dialog between Senator Howard H. Baker, Jr. and Lawrence R. Hafstad, then Vice President, General Motors Research Laboratories, General Motors Corp.:

Sen. Baker: Last question. In the judgment of either of you or all of you, is it reasonably probable that in the foreseeable future, the internal combustion engine in its present format and configuration can ever be designed to perform realistically with hydrocarbon emissions of 35 parts per million or less?

Mr. Hafstad: I would answer that by saying yes as far as laboratory devices are concerned.

Sen Baker: Of course my question is whether it is reasonable and practical to assume that this can be done for the mass production of automobiles, that is, produce an internal combustion engine in production runs for general operation, general maintenance, for the general public that will emit unburned hydrocarbons of 35 parts per million or less.

Mr. Hafstad: I think it is probably going to be possible, but it will take a lot of development work to make it reliable enough and I would say this is the reason we are in the horserace here between the gasoline engine and some of these other three to see which comes out the best.

Sen. Baker: Would you judge it will be any more difficult to perfect and produce a reliable production run of internal combustion engines that would produce 35 parts per million or less?

Mr. Hafstad: We think the operating problems and problems of cost and the bulk of the steam engine will be so high that probably the gasoline will come out ahead. This remains to be seen.

21. Mr. Daddario originally introduced a bill, H.R. 6698, March 7, 1967, primarily to stimulate discussion. It called for establishment of a Technology Assessment Board. Following a Seminar on Technology Assessment, held by the subcommittee, and a set of hearings on the subject, a second draft of the bill was introduced by Mr. Daddario and Congressman Mosher, April 16, 1970, as H.R. 17046. Another set of hearings was held, and a new bill, based on the hearings, was introduced July 15, 1970, as H.R. 18469.

22. S. 4044, "Commercial Technology Assessment Act of 1970", introduced June 30, 1970.

23. S. Resolution 69, 90th Congress, 1st Session, January 25, 1967; S. Resolution 78, 91st Congress, 1st Session, January 29, 1969.

24. S. 1305, 90th Congress, 1st Session, March 15, 1967.

25. "A Technology Assessment System for the Executive Branch," Report of the National Academy of Public Administration, Committee on Science and Astronautics, U.S. House of Representatives, 11, July, 1970,

26. GOLOVIN, N.E.,"The Evaluative Function in Government," Octover 25, 1968, 46 pages, "A fourth branch of government - the evaluative branch - is proposed as a mechanism for coping with complex, long-range socio-economic problems."

27. ROBACK, H.,"Do We Need a Department of Science and Technology?" *Science, 165,* 36-43, No. 3888, July 4, 1969.

28. ETZIONI, A.,"Agency for Technological Development for Domestic Programs," *Science, 164,* 43-50, No. 3875, April 4, 1969.

29. U.S. Congress, House Committee on Science and Astronautics, "Technology: Processes of Assessment and Choice," Report of the National Academy of Sciences, 117, Washington, D.C., April 4, 1969.

30. Morris Tanenbaum, Manager of Engineering, Western Electric, and member of the National Academy of Sciences Panel on Technology Assessment voiced a concern with "potential effects of new technology - assessment mechanisms on the funding and direction of research and development in the private sector." He states: "To the extent that assessment can define the social requirements of new technology at an early stage, this could encourage private investment. However, to the degree than new assessment mechanisms create new uncertainties, this could discourage private investment in areas of technology which are at the focus of assessment activities. These may be the areas where innovation is most important and private participation most desirable." Appendix B, pages 148-150, in the above noted NAS report.

A PRELIMINARY TECHNOLOGY ASSESSMENT OF SOLID STATE MICROWAVE DEVICES* †

RAYMOND BOWERS

Program on Science, Technology and Society and Department of Physics

and

JEFFREY FREY

Department of Electrical Engineering

Cornell University, Ithaca, New York, 14850, U.S.A.

INTRODUCTION

Most technological developments have brought in the wake of their primary intended effect, a series of unforeseen secondary effects, some adverse and some beneficial. It is characteristic of much technological development that those concerned with the development of a new device are so preoccupied with the primary effect

---

*Supported by funds provided to the Cornell University Program on Science, Technology and Society by the National Science Foundation and the General Electric Foundation.

†Reprinted with permission from *Scientific American*, where it appeared under the title "Technology Assessment and Microwave Diodes" in the Feb. 1972.

that they give inadequate attention to possible secondary consequences.

The possibility of anticipating secondary consequences which may result from the introduction of new technologies has received much attention in recent years and has led to the concept of "Technology Assessment" [1]. Technology assessment goes beyond the anticipation of secondary technological effects; it includes the attempt to foresee some of the potential social impact of a new technology. For example, the introduction of devices such as the automobile and the telephone have had far-reaching effects on the character of our society, surpassing anything that was foreseen at the time of their introduction and perhaps surpassing what it would have been possible to foresee at the time of introduction.

While it is generally agreed that the ability to anticipate the consequences of technological change is a desirable goal, some skeptics doubt that we have the methodology, the imagination and the understanding to achieve successful or worthwhile analysis in view of the very great changes that can be brought about by new technology. An example which might support this view is the following: in 1948 there were 100,000 television sets, the next year there were a million, a decade later there were 50 million. Faced with this kind of extraordinary growth and consequent impact, one can understand the view that analyzing potential societal impacts of devices such as television is beyond our present capacities, no matter how desirable. It is indeed possible to find many examples, especially in the field of biology (e.g. the possibility of increasing longevity, predetermination of sex of children, human cloning) of technological changes that could have such a profound effect possible adverse effects, but also to suggest areas of investigation that will increase the potential for benefit that can come from the application of these new devices.

on society that it is difficult to know where to begin the process of analysis. However, the fact that one can specify problems of such a magnitude should not deter us from trying to do a better job of anticipating consequences of technological development in areas where the social consequences are less profound, where the new technology is almost at our doorstep and where it does not seem difficult to foresee some possible secondary and tertiary consequences. Experience gained in analyzing these simpler situations may be helpful in future attempts to deal with more important areas.

The recent development of new and potentially very cheap sources of low power microwave devices [2] is an example of a new technology which seems a suitable subject for technology assessment. The physical characteristics of the new devices are now fairly well outlined and it is possible to foresee at this stage a number of technological options that will become available. The devices should result in a major extension in the application of microwaves. The developments in this area are evolutionary in the sense that the new devices will enable us to do things we already know how to do, but much more simply and cheaply. Thus past experience with microwaves will be useful in projecting future uses. In this respect, we are faced with a simpler problem than one encounters with a revolutionary device such as the telegraph, which, in making possible instantaneous communication between very distant points, made it feasible to perform many functions that could not be carried out previously.

This paper constitutes a preliminary attempt at a technology assessment of solid state microwave devices. While devoted primarily to the problem of the regulation of microwave devices so as to ensure proper *use of the electromagnetic spectrum*, the assessment will also touch briefly on the issues of *potential health dangers*, and invasion of *privacy*. Our purpose is not only to identify

Microwave devices have been developed for more than three decades and are now in widespread use for communications, navigation and in industrial electronics. Present microwave sources are expensive and microwave systems are largely the property of military and industrial organizations; they are rarely privately owned and used in the home* or in the private auto. Electron tube sources, the basis of past microwave technology, cost many hundreds and even thousands of dollars. But this situation is likely to change radically within the next decade. Reliable and cheap microwave sources whose ultimate cost in mass production may be only a few dollars are now being developed and seem likely to produce a most dramatic proliferation in the application of microwave technology. Devices will soon be on the market with prices within reach of many householders and it is likely that within a decade these devices will be used in the home, in the private auto and in the private boat. They may even be incorporated into toys. One can also expect commercial organizations to use them on a massive scale for transmission of information. It is already clear that a number of microwave component manufacturers hope that this will come to pass. The proliferation of microwave devices could conceivably match that of television sets. In order to assess the potential growth and consequences that may result from that growth, a number of questions need to be considered:

a) What are likely uses for the new devices?

b) What are the estimated costs of complete microwave systems?

---

*Our concern in this paper is with relatively low power microwave sources. We are not considering the use of microwaves for heating since it does not seem likely at this time that recent developments in the solid state field are likely to have much impact on very large average microwave power production. Hence we are excluding, for example, microwave cooking ovens from most of our discussion.

c) What impact will they have on the use of the electromagnetic spectrum considered as a consumable or saturatable resource?

d) What special problems will there be in reducing interference between systems?

e) If segments of the frequency spectrum are too congested to allow for uncontrolled proliferation of new devices, what are the principles that should determine allocation and priority within the frequency spectrum?

f) Are there potential health hazards resulting from increased likelihood of exposure to microwave radiation?

g) Will the new devices present special problems with respect to privacy and the protection of confidential information?

These questions represent only a fraction of those that should be considered in a comprehensive technology assessment. Most of them deal with the use of an important resource, the electromagnetic spectrum. Our list does not include more important non-technological issues such as the impact of the new devices on social processes. We know that the introduction of devices such as the telephone and the automobile have influenced where and how people live. We are not suggesting that the introduction of new cheap microwave sources will have comparable social effects; nevertheless, it would be extremely nearsighted to assume that these new sources will not have substantial social impact eventually since they will affect both communication and transportation systems. However, sophisticated considerations of sociological consequences are beyond the competence and qualifications of the present authors and we hope that these first steps towards a technology assessment will encourage social scientists to collaborate with technologists in order to investigate the potential social implications.

SOLID STATE MICROWAVE SOURCES

It is little more than a decade since the theoretical ideas leading to the microwave solid state generators we are discussing here were published [3]; it is even less than a decade since the first experiments demonstrating these devices indicated their enormous potential [4]. In the years following the first demonstration of high frequency oscillations in gallium arsenide by Gunn in 1963, and of the Read and IMPATT diodes by Lee, et al., and DeLoach and Scharfetter in 1965, an armory of devices has been developed which in certain ranges can now compete in output and efficiency with the older and well-established electron tube technologies. At the same time, the performance of transistors in the microwave range has been greatly improved. There is a bewildering array of microwave-oscillatory phenomena to be found in the technical journals, many of which might lead to important devices in the future. Apart from transistors, which probably will be used in applications requiring relatively large CW power in the low microwave range, and the new thermionic emission diodes, which promise moderate output power but very low noise, four particular devices have been reasonably well developed and will be of major importance in the future. These devices act, when used in the proper circuits, as negative conductances: a microwave voltage applied to their terminals causes a current to flow that is 180° out of phase with the voltage. Unlike positive conductances, in which voltage and current flow are in phase, and which absorb energy, negative conductances generate energy — or, more correctly, can transform the DC energy supplied by the device's "bias" supply into microwave energy. The four devices are:

(1) *The Transferred Electron, or Gunn Oscillator* [5]

This "bulk negative differential conductance" device depends on the fact that above a certain value of electric field, increasing the electric field in certain materials results in a decreased current through the material. The origin of this effect is as follows: in certain materials, such as GaAs, it is possible to excite electrons, using moderate electric fields, from their normal high mobility states which carry large currents into higher energy states that have much lower mobility and can consequently transport less current. When small electric fields are applied, the electrons increase their velocity (and hence the current they carry) as the electric field is increased. In this range the conductivity is normal or "positive". However, as the electric field is further increased, electrons are excited to the higher energy states of lower mobility and lower current. When this condition is reached, increasing electric fields result in decreasing current and the effective differential conductivity becomes "negative". The existence of a negative conductivity leads to oscillations, as in other electron devices. The Gunn oscillator is operated under conditions that lead to the formation of domains of space charge that traverse the sample in fractions of a microsecond; it is the motion of these domains that is responsible for the oscillatory microwave currents.

(2) *The Limited Space Charge Accumulation (L.S.A.) Diode* [6]

Invented by John Copeland of the Bell Telephone Laboratories, his device is related to the Gunn oscillator in that it is also based on the existence of negative conductance at certain electric fields; however, the mode of operation is quite different from that of the Gunn Oscillator. The device is operated so as to prevent the full formation and transit of domains across the solid; the bias voltage is chosen so that the space charge would dissipate before it had time to travel across the solid. This leads to a mode in which the high frequency electric energy generation is distributed across the sample and consequently the L.S.A. device is closer to a bulk negative resistance oscillator than is the Gunn Oscillator. This generation of high frequency energy throughout the bulk material makes possible larger power outputs than can be achieved by Gunn oscillators.

(3) *Read and IMPATT Diodes* [7]

These are basically similar devices, based on principles first developed by W.T. Read, Jr., of Bell Telephone Laboratories. In Read's original proposal, an intense radio frequency field across a semiconductor p-n junction results, during one-half cycle, in the creation of a large concentration of mobile charge carriers in the vicinity of the high electric field region at the junction, due to ionization resulting from the collision of the accelerated electrons with the atoms of the solid. If this cloud of charge is allowed to drift from its site of creation through an intrinsic semiconductor region to a collector electrode, the current flow associated with this drift of charge will be out of phase with the applied field when it arrives at the collector. The phase difference will be determined by the time required for the avalanche to build up, plus the time required for the charge to travel to the collector electrode. While the former time lapse is not easily adjustable, the latter can be fixed in fabrication. Consequently it is possible to arrange for the current collected to be one-half cycle out of phase with the applied field. Under such conditions, the device acts as a negative resistance and will oscillate. The IMPATT diode is similar in principle to the Read diode, but the avalanching and drift regions are combined in a single zone of resistivity.

(4) *The TRAPATT (for Trapped Plasma Avalanche Triggered Transit)* [8]

This device is related to the IMPATT diode, but the doping of the semiconductor is adjusted so that the peak of the electric field which produces the avalanche moves through the diode, rather like a shock wave, instead of being stationary. After the high-field peak has passed through the diode, the drift region is filled with a large number of mobile charge carriers, produced in the avalanche. These carriers can now be extracted from the diode to produce current in the external circuit. Since there is little current during the high-voltage (high-field) portion of the cycle, when the avalanching occurs, and the high-current carrier extraction takes place during a low-voltage portion of the cycle, the efficiency can be quite high. The carrier extraction occurs under the influence of smaller fields than in the IMPATT diode so that it takes longer for the carriers to be extracted, and the frequency of the device is lower than that of an IMPATT diode.

The devices described above do not necessarily compete with each other for each has its own advantages and disadvantages in terms of optimum frequency, power, CW or pulsed performance, noise, tunability, cost, etc. Many factors must be considered in picking the device that will give the optimum performance for a particular application. For example, a receiver local oscillator requires only small amounts of CW power but must be relatively free of FM noise, while a simple radar requires a source which produces a large amount of power in brief pulses and noise performance is relatively unimportant. Since the devices should not be compared as if they were directly competitive, we will state the best current performance levels we know of, and indicate the relevance of such performance to applications.

Figure 1 indicates current best performance for CW devices. Such devices are useful in radars that give information on both range and closure rate, in communications systems, for telemetry, and for intrusion alarms. For short-range alarms or radars, only a few tens of milliwatts average power is necessary; this is readily achieved today. For local oscillators, under a milliwatt of virtually noise-free power is required; the quietest device on the chart is the Gunn diode, and it can easily achieve such power levels, at least up to ten to twenty GHz. For communications links, power levels up to watts are required, depending upon distance and other factors; both gallium arsenide and silicon IMPATT diodes, and CW TRAPATT diodes already have achieved sufficient power levels for this application.

Figure 2 gives data on pulsed devices, which are also useful in radar systems, and as transponders. Very high peak power levels have already been achieved — kilowatts in the lower microwave ranges, and hundreds of

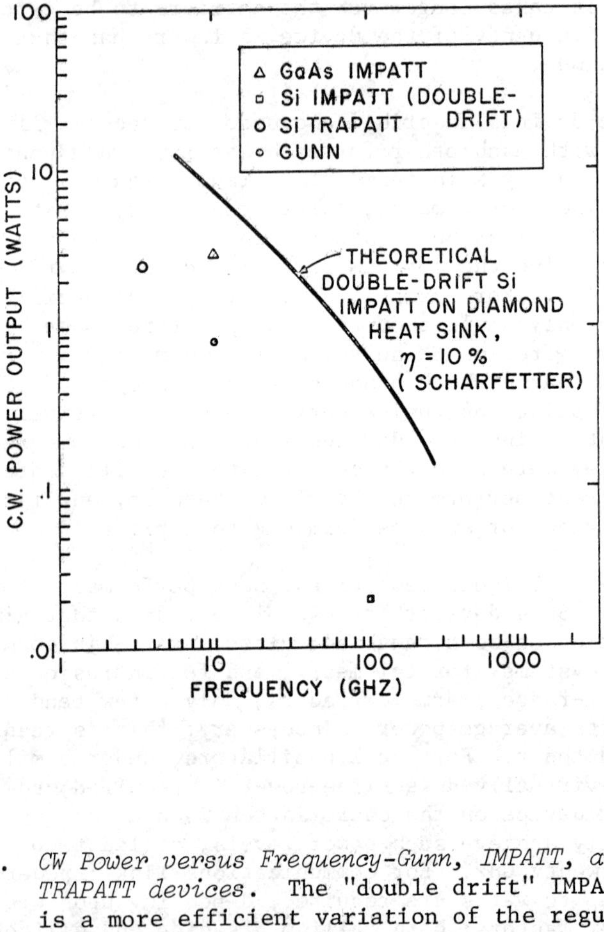

Fig. 1. *CW Power versus Frequency-Gunn, IMPATT, and TRAPATT devices*. The "double drift" IMPATT is a more efficient variation of the regular IMPATT. The curve of theoretical performance is based on the reasonable assumption of 10% efficiency for silicon devices. Greater efficiencies can be achieved with gallium arsenide (GaAs) devices but difficulties with that material have hindered their widespread application. In terms of noise, the Gunn device is the "quietest", the TRAPATT the most "noisy".

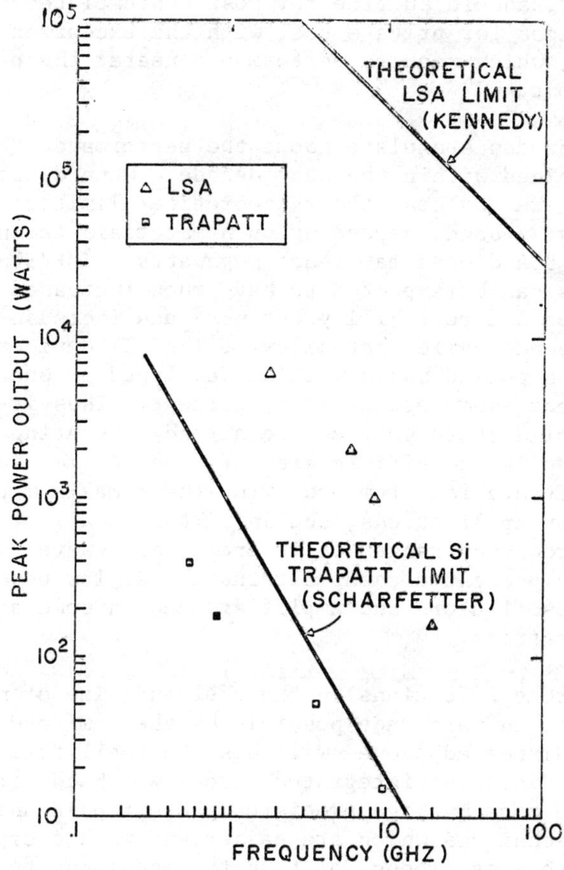

Fig. 2. *Pulsed Power versus Frequency-LSA and TRAPATT Devices*. TRAPATT devices, while producing less peak power than LSA devices, are considerably more efficient. Efficiencies up to thirty percent can be expected to be achieved with LSA devices (18% has been reached) but TRAPATTs have already operated in excess of sixty percent efficiency.

watts above 10 GHz — and the potential of the pulsed devices has not yet been reached. Current power levels, in fact, should suffice for most contemplated radar applications for private use, with the exception of weather radars, which require performance nearer the predicted LSA limit.

One can speculate about the performance that will be attained within the next decade. Gunn diodes operating CW are now near their theoretical limits, a few watts at X-band, depending upon materials technology. Pulsed LSA diodes may reach megawatts. IMPATT and Read devices can be expected to have much increased efficiencies and reliability but not much increase in power. It is conceivable that kilowatt TRAPATT devices operating on a pulsed basis will be developed to operate at efficiencies exceeding fifty percent. Thus it is clear that solid state sources are already operating at powers, frequencies and efficiencies of great commercial interest, and they are far from achieving their maximum performance. For many applications, and in virtually all of the microwave frequency ranges of interest, microwave solid state devices present a cheap alternative at low powers to klystron oscillators and amplifiers and in some applications, to magnetrons.

Great reductions in the cost and size of microwave systems, in part made possible by the semiconductor devices discussed above, will also be facilitated by the use of microwave integrated circuits. Such circuits will be made by high precision photographic and thin-film techniques which are easier and much cheaper to adapt to mass production than the machining techniques required for waveguide and coaxial microwave circuits. For example, a typical microwave receiver, in the past, had to be assembled from precisely machined metal waveguide parts, occupied considerable volume and was expensive. Such construction was acceptable when microwave systems were assembled in small quantities for specialized uses, but the clumsiness and expense of sophisticated waveguide circuits would, if they were necessary, certainly hinder the popularization of microwave applications. In microwave integrated circuits, waveguides are replaced by strips of gold a few thousandths of an inch thick and less than .050" wide, placed in the desired pattern on a substrate of glass, ceramic, sapphire, or other materials, by techniques that can be as simple as silk screening. A microwave IC receiver need occupy no more than a few cubic inches, and clearly silk-screening or photolithographic delineation of patterns followed by etching is cheaper than machining of fine parts.

APPLICATIONS OF MICROWAVE SOLID STATE DEVICES

The microwave part of the spectrum (especially the range 1 to 10 GHz) has been in use for some time, and a wide variety of applications has evolved. Probably the most widespread current civilian use of microwaves is in the field of communications. Current and proposed long-distance communications links (A.T.&T., Western Union, Microwave Communications, Inc., Datran, etc.) operate in, or are proposed for, the 3.7-4.2 GHz and 5.925-6.425 GHz bands; as these bands become saturated, new links will appear in the 11.7-12.2 GHz band.

Microwaves do not bend with the curvature of the earth, so that for long links, repeaters which receive, amplify, and re-transmit the signal must be used between terminals of the link. The spacing between repeaters in the lower two bands is determined by the curvature of the earth, features of the terrain, and acceptable antenna heights, but above 10 GHz atmospheric attenuation becomes a major factor so that the repeaters have to be more closely spaced. In the lower two bands, thirty miles is a normal spacing between repeaters. The total electronics cost (exclusive of antennas) of a typical repeater can be under ten percent of its total cost [9]. In addition, at each terminal, expensive equipment must be provided to switch incoming and outgoing calls to the proper circuits. When the cost of this switching equipment is included for a link below 10GHz, the fraction of the total system cost attributable to microwave components becomes very small. Consequently, there is little cost benefit to be gained by using solid state devices below 10 GHz. However, as the 4 and 6 GHz bands become saturated — as is already the case in some urban areas — use of higher frequency bands becomes necessary and here the cost situation is significantly different. At 12 GHz, as the result of atmospheric attenuation, maximum practical repeater spacing is about four miles; at 18 GHz, 2.3 miles; and at 30 GHz, 1.3 miles. For a long link, 23 times as many repeaters are required at 30 GHz as at 6 GHz, and microwave costs can be a sizable fraction of total system cost. Low cost and high reliability are essential for many closely spaced repeaters. This consideration, combined with the possibility of using only a few tens of milliwatts of power because of the ease of construction of millimeter wave high-gain antennas, has led the Bell System to conclude that for high frequencies "our best hope of building a repeater of adequate reliability at a competitive cost is based on solid-state devices and integrated microwave circuits" [10]. These devices, in effect, open up the spectrum above 10 GHz to long-haul microwave links, and could therefore have a considerable effect on spectrum use.

Another field of application for microwave solid state devices will certainly be in direct satellite-to-earth receiver communications, using the devices as local oscillators. NASA's ATS-F satellite, to be launched in 1973, might ultimately carry direct transmission to 600,000 village receivers in India. In addition, directed links are planned to provide educational programs to 21 remote Alaskan villages. While the satellite-to-earth transmission frequency of the Indian system is in the UHF band, where transistor local oscillators can be used, systems in more well-developed countries — particularly in countries with widespread land-mobile services already established in the UHF and lower microwave regions — can be expected to operate at microwave frequencies. Here the most economical local oscillator might be a Gunn or IMPATT device. At the recent World Administrative Radio Conference in Geneva, both UHF frequencies (620-790 MHz) and the bands 22.5-23.0, 41-43, and 84-86 GHz were set aside for broadcast satellite purposes.

The current congestion of the UHF bands, as well as the possibility that direct satellite broadcast to earth will necessitate the construction of microwave TV receivers anyway, has led to consideration of the use of microwaves for omnidirectional land-based broadcast TV transmission [11]. Such a system might be more efficient for broadcast use over small areas than direct satellite transmission, and would provide a large area of application for microwave solid state devices as receiver local oscillators.

The new developments in solid state sources have the potential for a major improvement in land mobile communications systems, and in fact could make automobile telephones very common. Widespread installation of automobile telephones is limited by the narrow band now assigned to this service; if the number of automobile phones is to be increased, some microwave or millimeter wave bands are clearly needed. The directionality of short wavelength systems would probably necessitate a large number of local transmitting/receiving terminals, with the phone signal being passed along from terminal to terminal as the car moves along. A system for locating the car, probably also involving microwaves, would be required.

In the field of guidance and control — radar, radio location, etc., — the availability of microwave solid state devices is, once again, not expected to result in new types of applications, but, rather, to vastly increase the number of systems in use. It will soon be possible for light aircraft to carry both altimeter and collision-avoidance radars at costs comparable to other general aviation electronic equipment. Weather radar is also a possibility but will require higher peak powers than are likely in the immediate future. Collision avoidance radar might within a decade be considered necessary on aircraft equipped for instrument flying. Radars for small boats may also be practical devices, particularly for operation at night or in areas where fog is frequent. This market is potentially much larger than that for aircraft radars.

Use of radar in automobiles, for automatic headway control, clearlane indication, backup obstacle warning, and triggering of passive restraint devices, may lead to an entirely new, and very large, radar market. Such radars, probably operating in an FM-CW or two-frequency pulsed mode to provide information on both range and closure rate, could be used either as simple threshold devices or as continuously operating devices to regulate automobile speed. The requirements of reliability and low cost suggest that here is another area for application of microwaves which could not be pursued without solid state devices.

Simple Doppler systems, or systems utilizing changes in sensed antenna impedances, can be used for intrusion alarms. While ultrasonic devices are now in widespread use, microwaves may offer greater directionality and be less sensitive to disturbance by, say, extraneous sounds. They also have the potential for development as fire alarms or as combination intrusion and fire warning devices.

Proximity sensors — automatic door openers, automatic crocodile roarers [12], etc., can work in the same way as intrusion alarms.

Microwaves have always been important for telemetry. Sangamo Electric Company is testing an automated meter-reading system in which a slowly moving truck, using a microwave transceiver and magnetic tape units, interrogates small transponders on each house to obtain the meter reading. The scheme being tested works at 915 and 1830 MHz using transistors, but if this telemetry is used at higher frequencies, simple amplitude-modulated Gunn or IMPATT devices may be more suitable. Some argue that broadest meter reading is a wasteful use of the spectrum and claim that more rational alternatives have been put forward by cable TV companies and the common carriers.

Vehicle location may be facilitated by microwaves. Buses, service trucks, police cars, or cars equipped with mobile telephones could be equipped with coded low-power transmitters, sensed by simple receivers on buildings or utility poles, which in turn relay location information to base via land-line. Alternatively, vehicles may be equipped with passive identifiers which respond to microwave interrogation in a unique way — say, by reflecting a harmonic of the transmitted frequency.

Various applications in process control, e.g., thickness monitoring, counting, etc., can also be imagined and one can expect use of solid state devices for these purposes.

Medical and biological applications are also likely. Changes in physiologically significant parameters of the circulatory and respiratory systems could be detected by simple transmission/reflection measurements [13]. Power levels can be very low, well within the range of solid state devices. The frequency will probably be in the low microwave range (below 3 GHz) to permit penetration of the skin.

Cheap microwave sources might also be incorporated into some toys. (In this context, we mean toys for children and are not referring to the more sophisticated toys for grownups which will surely emerge in such areas as private boat radars, etc.) The technology is here to provide children with a primitive radar to watch cars going by, and it takes little imagination to invent a new generation of military games that have a primitive radar component. This is a potentially troublesome area because of the implications of uncontrollable use of cheap systems with poor frequency stability, which might interfere with other microwave systems. One can also ask whether it is desirable for children to learn about first and second strike through the medium of toys!

BENEFITS, PROBLEMS AND NUMBERS

The benefits that could result from the development of new microwave sources are manifold. The entire communication system could be improved by opening up the frequency range above 10 GHz and relieving the congestion at lower frequencies. The new sources also provide a potentially economic means of communication for places where "wired" systems are unavailable or impractical. They have the potential to improve the technologies associated with transportation (including traffic control and reduction of injury by collision), reduction of damage by fire, crime detection and prevention, and health care. These benefits will be accompanied by many problems which need attention soon if we are to maximize the benefits and minimize the problems. In this section, we shall discuss these problems in general terms, leaving more detailed discussion to the next section.

Most of the projected problems ultimately result from the possibility of a massive proliferation of microwave devices resulting in heavy demands on part of the electromagnetic spectrum and the likelihood of mutual interference. Also if such widespread use should come to pass, we shall have to devote much more attention to possible health hazards that might result from exposure to microwave beams.

In the microwave range, we have past experience and institutional arrangements for controlling systems (largely military and industrial) whose numbers do not exceed a few tens of thousands and which cost anywhere from 10^4 to 10^7 dollars per system. But if unit prices fall to the range of a thousand dollars and there is widespread use in such things as light aircraft and large private boats, the numbers could mount to hundreds of thousands or perhaps a million. We are surely unprepared to deal with the numbers of units that will be in use when the unit price of microwave systems falls below one hundred dollars and the systems are widely used in auto-hundred dollars and the number of units rises to 10 or 100 million.

We have made no sophisticated analysis of the number of devices that are likely to be produced as a function of unit price; a number of manufacturers of both systems and devices that we have contacted have indicated they are currently pursuing such studies. We have, however, made some rough guesses of the orders of magnitude involved. On the axes of Fig. 3 we have designated the orders of magnitude of the price per unit and the number of systems that may result. These estimates are given for particular classes of potential users, ranging from the very expensive common carrier and military systems, through the intermediate cost systems for light aircraft or small boats, to the very cheap systems that one might expect to see on private autos. The chart is meant to illustrate nothing more than the fact that even if only a small percentage of potential users are equipped with radar or communications systems, the numbers will be very large indeed. We shall now consider some of the potential problems in detail.

SOME SPECIFIC PROBLEMS

1) *Impact on the Use of the Electromagnetic Spectrum*

In a study published in 1965 the Joint Technical Advisory Committee noted [14] that "the spectrum space above 10 Gc/s is unique at this point in history, in that there are relatively few services implanted in the band..." Today, however, new common carrier land transmitters are assigned to the band 11.7-12.2 GHz; Community Antenna Relays are at 12.7-12.95; and most satellite-to-earth links may well be above 12 GHz. These services, with a potential for great growth, are being forced into the band above 10 GHz by pressure on the spectrum from below, not because it is technologically advantageous.

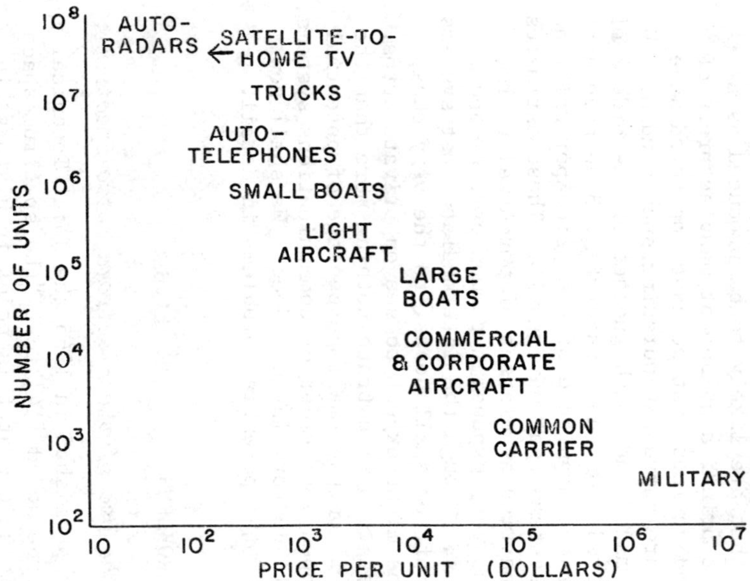

Fig. 3. *A very rough estimate of the number of microwave systems as a function of price.* This figure was assembled by estimating the price at which the applications considered might be realized, and drawing an area for the application representing both the uncertainties in acceptable price and potential numbers of units.

While the spectrum in this country is allocated, in blocks, for specific uses up to 90 GHz — and recent proposals would extend block allocations to 300 GHz — not all of this spectrum is in use. One of the problems in dealing with spectrum use is that "virtually none of the statistics explicitly required in the examination of the spectrum usage is available in the existing data base maintained by the FCC" [15], so that utilization information must be inferred from a variety of other sources such as number of applications, number of stations authorized, etc. From the data that are available, however, one can conclude that the spectrum below 10 GHz is filling rapidly to the point where future growth of microwave services may be affected. This saturation, moreover, arises only from the more "classical" communications services, such as common carrier communications, safety and special radio services (business, police, etc.), TV broadcast auxiliary, and current satellite communications, all using the older thermionic microwave sources. Safety and special radio services, for example, have been growing at the rate of ten percent per year. Moreover, the recent decision by the FCC in favor of competition in data transmission services will result in major increases in the number of point-to-point links; almost two thousand station applications for this service have been received already by the FCC.

Spectrum congestion is a local problem, and studies of selected areas show that congestion is already severe and in some cases some bands are saturated, in areas as diverse as New York City, Houston, Fargo (N.D.), and Venice, (La.) [16]. The nature of the locality determines the type of congestion; the commercial and entertainment center, New York, is afflicted with complete saturation in the 3.7-4.2 GHz common carrier band, and severe congestion in the other two common carrier bands below 12 GHz, while the petroleum area around Venice has nearly full safety and special service bands. The FCC's block allocations system does not allow the transfer of common-carrier spectrum to safety and special services, and vice versa. As a result of this policy and the growth of commercial microwave systems, specific microwave bands in certain locations are already saturated.

Thus we must assume that there is going to be widespread use of the spectrum above 10 GHz once economic and reliable systems are available. Indeed, if the very large numbers that are implicit in the potential application of microwave techniques are to be accommodated, the only logical place for these services is in the frequency range above 10 GHz. Since solid state devices already span a range of frequencies up to 100 GHz, it is tempting to assume that the new increase by a factor of 10 in the microwave frequency range available should accommodate all expected uses without difficulty. But such an optimistic assumption may not be justified.

We can make three suggestions which might better prepare us for the proliferation of microwave systems for communications, guidance, and other uses: 1. Calculations of estimated use of spectrum should be made; 2. An adequate data base for such calculations, and for correlating such calculations with the real situation should be established; and 3. Consideration should be given to new frequency assignment principles and procedures.

It is obviously desirable to attempt model calculations for communities of varying population densities, equipped with all of the foreseeable microwave systems of the future including fixed and mobile communications systems, car radars, etc., taking into account the frequency requirements and narrow beaming possible at the higher frequencies in order to assess the magnitude of the expected congestion. In the absence of such calculations, prudence based on past experience would surely lead us to assume that congestion is likely to arise, and in the future spectrum allocation must take into account the potential for such congestion.

One of the requirements for minimizing spectral congestion is for complete information on spectral use: a computerized data base containing information on location, frequency, radiated power, and power contour for each transmitter, both operating and proposed. Until very recently, the organization that compiled much of this information is the Western Electric Company; its data base covers only the common carrier bands. This information is made available to others outside the Bell system. We believe that a more comprehensive data system should be developed and maintained by the appropriate government agency. It should be noted that lack of an adequate data base must result in application of excessive safety margins on spectral overlap, so that the spectrum may not be optimally utilized.

Finally, the principle of block allocation of frequencies in the microwave spectrum needs to be reconsidered. In addition, performance requirements, differing for different uses in different parts of the spectrum and applicable to both transmitting and receiving units must be established. Simple block allocation is excessively rigid, as was seen in the cases of New York City and Venice (La.), and flexibility should be expressly built into the system. Account should be taken of beam directionality and polarization, and multiple use of the same frequency might well be allowed, even in the same area. Some standard unit of utilization of spectrum space, such as the PODAF (Power Density exceeding a specified level over an Area within an assigned Frequency band) [17] — might be used in assigning spectral space, rather than just field intensity and contour. Special bands analogous to the Citizen's Band in the lower frequency region might be set aside for very low power uses, which will be hard to regulate, with certain general performance criteria enforced for these systems.

We wish to emphasize a special danger resulting from the existence of very cheap sources. In the past, microwave sources have been very expensive and this has led naturally to development of sophisticated and costly systems to obtain maximum benefit from the sources However, the new sources now make possible "cheap and dirty" systems and it is possible that some manufacturers will sacrifice narrowness of beam and precision of frequency control in order to achieve lower costs. This is especially likely for devices whose power level is below that which normally requires licensing and regulation. In this area there is obviously a heavy responsibility on the engineering profession, if government regulation of these very low power devices is to be avoided, to ensure that they are "non-polluting" from the electromagnetic point of view. It is worth examining whether the profession can establish standards which ensure that the minimum bandwidth, beam width and power are used to accomplish objectives. This is a principle of conservation of a natural resource which should be applied whether or not a problem of congestion is immediately foreseen. If this much responsibility is not exercized, then detailed regulation and control will surely be necessary.

One area to which some engineering and manufacturing attention should be given in order to facilitate conservation of the spectrum is that of antenna design. Techniques for the design of inexpensive narrow-beam antennas, perhaps fiberglass paraboloids or dielectric molded antennas, do not seem to have kept up with advances in source development. Active antennas, which employ some amplification on the receiving end, might be developed in order to allow use of lower transmitter powers than would otherwise be required. In addition, frequencies for specific applications should be chosen wherever possible to take advantage of the natural attenuation of the atmosphere at certain frequencies. With proper engineering, many of the low power microwave devices can be no more troubling than a flashlight.

There are some who feel that even at this stage the use of microwaves in toys, both children's and adults', and in systems for which wired transmission can do the job equally well should be prohibited. The argument is that since the electromagnetic spectrum in free space is a saturatable resource, it should not be used for trivial purposes or used when alternative wired systems are available. It seems to us to be wholly impracticable *ab initio* to prevent the development of such devices and indeed involves a restraint on the use of the spectrum that impinges on the rights of some developers. It seems much more realistic to assume that such systems will be developed and to allocate them frequency ranges well separated from the more sensitive and vital functions such as common carrier communications, data transmission, and safety and special services, to prevent serious and even disastrous interference. Of course, the time may come when the frequency spectrum is so heavily congested that an embargo becomes necessary on all new systems transmitting microwaves through free space if the function can be performed via cables.

2) *The Potential of Microwaves as a Health Hazard*

The expectation of very large numbers of microwave systems in the relatively unsupervised control of private individuals makes it imperative that the biological effects of microwave radiation be fully explored. Standards that were established in a period when microwave systems were fairly uncommon, and when the average citizen was unlikely to be irradiated by a microwave beam, may be inadequate when microwave beams are emitted from every car, traffic light, or utility pole. A measure of the magnitude of the problem can be obtained by considering radar-equipped cars. A reasonable average power output for a collision avoidance radar on a car is 50 mW; if this power were transmitted within a beam angle of 2°, the power density at a distance of five meters from the car would be of the order of ten microwatts/cm^2. Admittedly the beam would be scanned and the car may be moving, so the time-average power density at any distance from this one car may be considerably below the calculated figure — but a given individual would be irradiated by many cars. We shall see later that such power levels are inconsequential in terms of the current American safety standards but may be of consequence in terms of the safety standards adopted in Eastern Europe.

Microwave radiation can be harmful to living organisms. Documented cases exist of cataracts and testicular damage in man, and death in animals exposed to microwave radiation in controlled laboratory experiments. These effects were probably caused by heating due to absorption of microwave energy at power levels much in excess of those we would expect to be produced by the kinds of systems considered here. But other, more subtle, effects have been reported at much lower power levels. They are designated "athermal" because they do not seem to be directly attributable to heating. For example "Garlic root tips have exhibited mutations when grown in a high-frequency field; micro-organisms have swum in preferred directions associated with the field direction; snails and planaria have shown the ability to sense the presence and polarity of very weak electric and magnetic fields; several people have heard sounds which correspond to the frequency of modulation of an incident microwave beam; some fish are able to navigate by detecting distortions in their self-generated electric fields; visual stimuli have resulted in humans from fields too weak to depolarize visual neurons in the usually understood way..." [18]

The frequency and hence wavelength of the radiation relative to the size of the absorbing object and that object's dielectric properties will determine the microwave energy absorbed. For humans, radiation begins to be significantly absorbed at frequencies in excess of about fifteen MHz. The absorptivity of microwaves varies over parts of the body and also varies with time. The depth of penetration of microwave radiation in fat can be expected to be of the order of ten times as great as that in muscle and this is presumably reflected in the absorption. Certain organs, notably the eye and the testes, are particularly sensitive to heating effects. The minimum power density required to produce cataracts in rabbits at a particular frequency is known [19] but the data is not readily available for humans. Experiments correlating exposure to microwave radiation with testicular damage in animals have also been performed, and at least one example, of a radar technician who performed maintenance on a high-power weather radar while it was in operation and was rendered infertile, is in the literature [20].

Athermal effects were not included when the current American radiation safety level was set at an average of 10 mW/cm^2 for long-time exposures; but these effects were apparently influential in establishment of the minimum standard of 10 $microwatts/cm^2$/working day [26] in the USSR. The U.S. standard of 10 mW/cm^2 average long-term rf power as a safe maximum was confirmed by the Tri-Service Program of research on the biomedical aspects of microwave radiation, administered at Rome Air Development Center from 1957-1961 [21]. However, some workers feel that the Tri-Service research is "largely irrelevant" [22] to the subject of low-power microwave radiation hazards because of the disregard for athermal effects and particularly because of a tendency in the Tri-Service work to reject East European data on athermal affects. These data have included evidence of disturbed rhythm of heart contractions, hypertension, and decreases in sensitivity of various sense organs. Average power levels at which these effects were noted range upward from 30 microwatts/cm^2, and frequencies were usually in the UHF to low microwave range where absorption by the skin and bone of the skull is small. Recent American experiments in this area have demonstrated that the metabolic activity of embryonic chick hearts is disturbed by 24 GHz radiation [23] and that the development of insect pupae could be adversely affected by 10 GHz radiation [24]. Both of these experiments involved power levels too low to cause significant heating.

Another factor not included in the setting of the current U.S. safety standard is that of duty cycle of the applied radiation. In an experiment involving two groups of rabbits, no members of one group that received 80 mW/cm^2 CW radiation for an hour developed cataracts; but all of the second group, which received pulsed radiation of 400 mW/cm^2 (with a duty cycle of 20% and hence the same average power) developed cataracts. The body temperature was, in both instances, the same [25]. Hence a radiation standard based solely on average power may not be adequate.

Having considered both the U.S. and USSR experimental results, as well as their own experiments, Czechoslovakia has established the following standards for mean daily radiation dosage in the 0.3-300 GHz frequency range for the general population [26]: 2.5 microwatts/cm^2 for CW and 1.0 microwatts/cm^2 for pulsed radiation. People who work with microwave generating devices and who have specific periodic health examinations are allowed 10 times the above limits.

It is clear that our understanding of microwave effects on biological systems is poor and this has led to different countries adopting safety standards that differ by factors of thousands. It is obvious that we must greatly increase research in this area, emphasizing low power effects, and re-examine safety standards before microwave devices proliferate. We should be concerned not only with effects on humans, but also other biological systems in our environment. A national study is needed which brings together and weighs all the relevant information in order to avoid yet another "after the fact" controversy.

3) *Invasion of Privacy and Interception of Data Transmission*

Many people are concerned with the possibility of new developments in electronics being used as means for the invasion of privacy. Also, as more information is transmitted through microwave beams, industrial organizations and banks may also be concerned about the possibility of interception of transmitted data. The problem is complex and one must consider not only the communication technique used but also the interplay between communications systems and computer data banks.

We have examined the privacy question in a preliminary way and have come to the tentative conclusion that the new sources do not represent a *special* new problem in the privacy area in the sense of adding a new dimension to the privacy problem. In some respects, the new microwave systems seem to have some advantages over telephone lines with respect to the maintenance of privacy. First of all, one must gain access to the beam whose position may not be physically apparent. Secondly, moderately secure transmission seems somewhat easier to achieve in the microwave range than in the standard telephone system. One technique uses double frequency transmission: the first frequency carries the coded message and the second frequency transmits the code. Anybody trying to intercept the information will have to find both frequencies and the signal which transmits the code can occupy an exceedingly narrow band. Such techniques can also be used on telephone wires, but we believe they can be accomplished more simply with microwave links.

Obviously, it is almost impossible to prevent somebody who is really determined to do so from intercepting information. Our concern has been with how to make interception difficult enough to discourage frequent or casual interception. If large amounts of information, whether thought to be sensitive or not, are to be transmitted by microwave systems, it should be normal practice for them to be coded at least in some simple fashion so that access to this information would require the possession of a simple "key." Since such coding is not hard to do, its implementation as standard procedure would seem prudent.

Our conclusions with respect to privacy are highly tentative. Our purpose is to bring the issue of privacy before the technical community and to stimulate a public discussion of it. We are not concerned here with the issue of the ethics of information interception; we are implying that the citizen should be informed of what techniques are possible so that he can decide for himself what steps to take if he wishes to ensure privacy.

CONCLUDING REMARKS

We have discussed some of the prospects, benefits and problems that may be associated with the development of cheap solid state microwave sources. Our discussion has been limited to problems associated with use of the spectrum, potential hazards to health and the issue of privacy. We have not attempted to assess the impact on social processes; they are likely to be substantial in those areas where wired communication systems do not exist (e.g. the Indian and Alaskan projects). Our purpose was not to present firm conclusions but rather to initiate debate on these issues and to indicate areas where more detailed analysis is necessary. We especially hope that the technical community will devote some of its resources at meetings and in its publications to the continued discussion of these problems, inviting contributions from social scientists as well as from physical scientists and technologists. This is part of the public responsibility of the research and development community.

ACKNOWLEDGEMENTS

The authors are indebted to Harvie H. Branscomb for his effective research assistance, and to our many colleagues at Cornell for suggestions and critical comments. We would also like to thank members of the staff of the Federal Communications Commission and the Office of Telecommunications Policy, and members of the industrial organizations who provided us with helpful information.

REFERENCES

1. "Technology: Processes of Assessment and Choice", *Report of the National Academy of Sciences*, Committee on Science and Astronautics, U.S. House of Representatives, July, 1969.

"A Study of Technology Assessment", *Report of the Committee on Public Engineering Policy*, National Academy of Engineering, Committee on Science and Astronautics, U.S. House of Representatives, July, 1969.

BROOKS, HARVEY and BOWERS, RAYMOND, "The Assessment of Technology", *Scientific American*, *222*, 13, February, 1970.

2. BOWERS, RAYMOND, "A Solid-State Source of Microwaves", *Scientific American, 215*, 22, August 1966.

3. READ, W.T., "A Proposed High-Frequency Negative Resistance Diode", *Bell Syst. Tech. J., 37*, 401, 1958.

RIDLEY, B.K. and WATKINS, T.B., "The Possibility of Negative Resistance Effects in Semiconductors", *Proc. Phys. Soc., 78*, 293, 1961.

HILSUM, C., "Transferred-Electron Amplifiers and Oscillators", *Proc. IRE, 50*, 185, 1962.

4. GUNN, J.B., "Microwave Oscillations of Current in III-V Semiconductors", *Solid State Electronics, 1*, 88, 1963.

LEE, C.A. et al., "The Read Diode-An Avalanching Transit-Time Negative-Resistance Oscillator", *Int'l. Solid State Circuits Conference Technical Digest*, 28, 1965.

JOHNSTON, R.L. et al., "A Silicon Diode Microwave Oscillator", *Bell Syst. Tech. J., 44*, 369, 1965.

5. CARROLL, J.E., *Hot Electron Microwave Generators*, Edward Arnold (Publishers) Ltd., London, Ch. 3, 1970.

6. COPELAND, J.A., "LSA Oscillator Diode Theory", *J. Appl. Phys., 38*, 3096, 1967.

7. DELOACH, B.C., Jr., "Modes of Avalanche Diodes and Their Associated Circuits", *IEEE J., SC4*, 101, 1969.

8. DELOACH, B.C., Jr., and SCHARFETTER, D.L., "Device Physics of TRAPATT Oscillators", *IEEE Trans., ED-17*, 9, 1970.

9. COSGROVE, T. and CHIPP, R.D., "Economic Considerations for Communication Systems", *IEEE Trans. Comm. Tech., 16*, 513, 1968.

10. TILLOTSON, LEROY C., "Use of Frequencies Above 10 GHz for Common Carrier Applications", *Bell Syst. Tech. J., 48*, 1563, 1969.

11. FELDMAN, J. et al., "A Study on the Technical Feasibility of Terrestrial Omnidirectional Television Transmissions in the 12 GHz Band", *IEEE Trans. Comm. Tech., 17*, 475, 1969.

12. A reliable source reports that a large amusement park in Florida is contemplating the use of a microwave link to trigger the roaring of plastic crocodiles at passing tourist boats!

13. SCHWAN, HERMAN P. and VOGELHUT, P.O., "Scientific Uses", In: *Microwave Power Engineering*, E.C. O'Kress, ed., Academic Press, N.Y., 1968.

14. Joint Technical Advisory Committee of the Institute of Electrical and Electronic Industries Association (JTAC), "Radio Spectrum Utilization", *IEEE*, New York, 1965.

15. JTAC, "Spectrum Engineering - The Key to Progress", *IEEE*, New York, 1968; Supplement 3.

16. FCC Microwave Study, Communications and Systems Inc., February 1971.

17. JTAC, *op. cit.*, Supplement 8.

18. SHER, LAWRENCE D., "Mechanical Effects of AC Fields on Particles Dispersed in a Liquid: Biological Implications". Office of Naval Research TR #37, 1963.

19. CARPENTER, R.L. et al., "Opacities in the Lens of the Eye Experimentally Induced by Exposure to Microwave Radiation", *IRE Trans., ME-7*, 152, 1960.

20. ROSENTHAL, DAVIS S. and BEERING, STEVEN,C., "Hypogonadism After Microwave Radiation", *J. AMA, 205*, 105, 1968.

21. MICHAELSON, S.M., "The Tri-Service Program...", *IEEE Trans., MTT-19*, 131, 1971.

22. FREY, A.H., "Biological Function as Influenced by Low-Power Modulated RF Energy", *IEEE Trans., MTT-19*, 153, 1971.

23. PAFF, G.H. et al., "The Effects of Microwave Irradiation on the Embryonic Chick Heart as Revealed by Electro-Cardiographic Studies", *Anat. Rec., 142*, 264, 1962.

24. CARPENTER, R.L. and LIVSTONE, E.M., "Evidence for Nonthermal Effects of Microwave Radiation: Abnormal Development of Irradiated Insect Pupae", *IEEE Trans., MTT-19*, 173, 1971.

25. CARPENTER, R.L. et al., *op. cit.*, 152.

26. MARHA, KAREL, "Microwave Radiation Safety Standards in Eastern Europe", *IEEE Trans., MTTT-19*, 165, 1971.

BENEFIT-COST STUDIES IN SOCIO-TECHNICAL SYSTEMS

CHAUNCEY STARR

Dean,
School of Engineering and Applied Science,
University of California, Los Angeles, U.S.A.

INTRODUCTION

The general problem of balancing utility against total societal costs should be a part of all planning in our society. Traditionally, technical performance as a function of monetary cost is always involved in engineering design decisions. What we are discussing however is inclusion of all societal costs, indirect as well as direct, and all measures of utility, direct as well as indirect. Clearly, existent social-technical systems, over a period of many years, have developed an empirically acceptable balance between utility and social cost. In addition, we have examples of national decision making involving future social technical systems which contain implicit predictive trade-offs of societal benefits versus societal costs.

As the number of our socio-technical systems increase and their impact on the individual becomes more apparent, concern with achieving a planned balance of the utility of these systems against their societal costs has also increased. It has become apparent, therefore, that greater insight, analysis and predictive planning are essential for the future development of new or larger socio-technical systems.

These problems originate from certain general assumptions inherent in the operations of our society. First, it is traditionally accepted that everyone should have the opportunity for a natural death, that is, a death from old age, from a natural wearing out of the human body. Second, it is now commonly accepted that every individual should have the opportunity to use and enjoy the fruits of our centuries of technological development. Third, recently we have emphasized our responsibility to assure for succeeding generations the best environmental and genetic inheritance that we can provide. And fourth, it is the philosophy of an egalitarian society that where the activities of an individual infringes on others in an undesirable way, the society may intervene to control individual activities in order to achieve a balance between group well-being and the privileges of the individual.

It is evident that these inherent basic assumptions are not fully compatible. Technology's contribution to enlarging the range of personal powers of the individual unfortunately also provides the individual with the power to damage others. In fact, the usual situation is that the benefits of technology are concentrated on the user, but the penalties are diffusely spread to many. Under these circumstances, the governing agencies of our society intervene and impose controls on both the individual and the technological system. Unfortunately, the criteria used by our society to achieve this balance of group and individual interests are rarely explicit, and are most often hidden in the complex empirical adjustments of our social, political, and economic subsystems.

We now face a general situation in which widespread use of a new technological development may occur before its social impact can be properly assessed, and before any empirical adjustment of the benefit-versus-cost relation is obviously indicated. It has been clear for some time that predictive technology assessments are a pressing societal need. However, even if such assessments become available, obtaining maximum social benefit at minimum cost also requires the establishment of a relative value system for the basic parameters in our objective of improved "quality of life." The empirical approach implicitly involves an intuitive societal balancing of such values. A predictive analytical approach will require an explicit scale of relative social values.

For example, if technology assessment of a new development predicts an increased per capita annual income of x percent but also predicts an associated annual per capita accident probability of y fatalities, then how are these to be compared in their effect on the "quality of life?" Because the penalties or risks to the public arising from a new development can be reduced by applying constraints, there will usually be a functional relationship (or trade-off) between utility and risk, the x and y of our example.

In order to provide insight to this problem of societal criteria for evaluating benefits versus costs, I undertook a study of one specific social value - fatalities arising from socio-technical systems. A quantitative analysis has been made of accidental deaths arising from technological developments in public use. This analysis was used to develop some generalized understanding of the public's approach to this social balance. The initial results were published in *Science, 165,* 1232-38, 19 September, 1969.

The analysis makes two assumptions. The first is that historical national accident records do contain consistent patterns of fatalities in the public use of technology. The second assumption is that such historically revealed social preferences and values are sufficiently enduring to permit their use for predictive purposes. We have evidence that these assumptions may not be adequate in all cases. Quite clearly in the technological trade-offs involving environmental pollution, historical data is not available. Nevertheless, in those areas where information has been available for a considerable time, the study is revealing.

There are many historical illustrations of such trade-off relationships that were empirically determined. For example, automobile and airplane safety have been continuously weighed by society against economic costs and operating performance. In these and other cases, the real trade-off process is actually one of dynamic adjustment, with the behavior of many portions of our social systems out of phase, due to the many separate "time constants" involved.

Readily available historical data on accidents and health, for a variety of public activities, provide an enticing stepping-stone to quantitative evaluation of this particular type of social cost. The social benefits arising from some of these activities can be roughly determined. On the assumption that in such historical situations a socially acceptable and essentially optimum trade-off of values is being approached or has been achieved, we could say that any generalizations developed might then be used for predictive purposes. This approach could give a rough answer to the seemingly simple question "How safe is safe enough?" Because this methodology is based on historical data, it does not serve to distinguish what is "best" for society from what is "traditionally acceptable." It also does not establish cause-effect relationships - rather, only the observable results of the cumulative operations of our social system.

VOLUNTARY AND INVOLUNTARY ACTIVITIES

Societal activities fall into two general categories - those in which the individual participates on a "voluntary" basis and those in which the participation is "involuntary," imposed by the society in which the individual lives. The process of empirical optimization of benefits and costs is fundamentally similar in the two cases - namely, a reversible exploration of available options - but the time required for empirical adjustments (the time constants of the system) and the criteria for optimization are quite different in the two situations.

In the case of "voluntary" activities, the individual uses his own value system to evaluate his experiences. Although his eventual trade-off may not be consciously or analytically determined, or based upon objective knowledge, it nevertheless is likely to represent for that individual, a crude optimization appropriate to his value system. For example, an urban dweller may move to the suburbs because of a lower crime rate and better schools, at the cost of more time spent traveling on highways and a higher probability of accidents. If, subsequently, the traffic density increases, he may decide that the penalties are too great and move back to the city. Such an individual optimization process can be comparatively rapid (because the feedback of experience to the individual is rapid), so the statistical pattern for a large social group may be an important "real-time" indicator of societal trade-offs and values.

"Involuntary" activities differ in that the criteria and options are determined not by the individuals affected but by a controlling body. Such control may be in the hands of a government agency, a political entity, a leadership group, an assembly of authorities or "opinion-makers," or a combination of such bodies. Because of the complexity of large societies, only the control group is likely to be fully aware of all the criteria and options involved in their decision process. Further, the time required for feedback of the experience that results from the controlling decisions is likely to be very long. The feedback of cumulative individual experiences into societal communications channels (usually political or economic) is a slow process, as is the process of altering the planning of the control group. We have many examples of such "involuntary" activities, war being perhaps the most extreme case of the operational separation of the decision-making group from those most affected. Thus, the real-time pattern of societal trade-offs on "involuntary" activities must be considered in terms of the particular dynamics of approach to an acceptable balance of social values and costs. The historical trends in such activities may therefore be more significant indicators of social acceptability than the existent trade-offs are.

In examining the historical benefit-risk relationships for "involuntary" activities, it is important to recognize the perturbing role of public psychological acceptance of risk arising from the influence of authorities or dogma. Because in this situation the decision-making is separated from the affected individual, society has generally clothed many of its controlling groups in an almost impenetrable mantle of authority and of imputed wisdom. The public generally assumes that the decision-making process is based on a rational analysis of social benefit and social risk. While it often is, we have all seen after-the-fact examples of irrationality. It is important to omit such "witch doctor" situations in selecting examples of optimized "involuntary" activities, because in fact these situations typify only the initial stages of exploration of options.

THE CONTINUUM OF RISK EXPOSURE

Fatalities may arise from technological systems as either the result of discrete, statistical events, i.e., the true accident - or as a result of a continuous exposure to a cumulative hazard, i.e., environmental pollution. Although the distinction usually may be unimportant for an annual actuarial analysis - as in the case of motor vehicle deaths - both the societal and individual reaction to the risk is generally different. This is best illustrated by considering the extreme case of infrequent discrete catastrophes.

A tabulation of such catastrophes, both man-made and natural, are shown in Table 1. The frequency of events is very low, with a thousand-fold range of fatalities per event. If one excludes the people directly affected, the societal response to such a range of catastrophes is roughly uniform. The dramatic impact

TABLE 1. Single Event Major Catastrophes

TYPE OF EVENT	TIME PERIOD	MAGNITUDE (DEATHS/EVENT)			FREQUENCY EVENTS/ YEAR
		Mode	Maximum	Average	
Major air crashes	1965-1969	82	155	78	6.0
Major auto crashes	1966	10	40	19	11.0
Earthquakes	1920-1970	30,000	180,000	25,000	0.50
Explosions	1950-1968	10	100	26	2.0
Major fires	1960-1968	12	322	35	0.67
Flood, tidal waves	1887-1969	2,000	900,000	28,000	0.54
Hurricanes	1888-1969	400	11,000	1,105	0.41
Major railroad crashes	1950-1966	19	79	30	1.0
Major marine accidents	1965-1969	32	300	61	6.0
U.S. tornados	1900-1969	72	689	78	0.74
Typhoons, cyclones, blizzards	1880-1970	10,000	300,000	37,240	0.18

communicated by the news media (newspapers, radio, TV) is about the same for the crash of a large commercial airliner, the hurricane damage over a large territory, or the massive loss of life in an extreme earthquake.

It is likely that the very low frequency of such events tends to reduce their significance to the observer. The typical individual reaction is that these events are so rare both in time and location, that the consequent individual risk is negligible. So, the concept of "it will never happen to me" is easily accepted.

However, these infrequent catastrophes are generally the end-points of a continuous spectrum of like events, whose frequency increases exponentially as their individual severity decreases. The typical logarithmic relationship is shown in Fig. 1, where the frequency and magnitudes of earthquakes are compared. The rapidly increasing frequency with diminishing severity of earthquakes begins to approach a steady series of tremors. The public awareness of this exposure is, of course, limited to the severe, infrequent earthquakes that do visible harm.

A similar continuum of frequency logarithmically inverse to severity can be shown in an analysis of motor vehicle accidents. In Fig. 2 is shown the annual number of motor vehicle accidents in the State of California for the years 1968 and 1969 as a function of the number of vehicles involved in the accident. Because the vehicle population in California is very large (more than 9 million registered) and the accidents are roughly stochastically random, the frequency versus severity curve may be expected to approximate a Poisson probability law - similar to that shown in Fig. 2.

The continuous spectrum of damaging events is also illustrated by theoretical studies of hypothetical nuclear power plant accidents. A recent analysis by Otway [1].

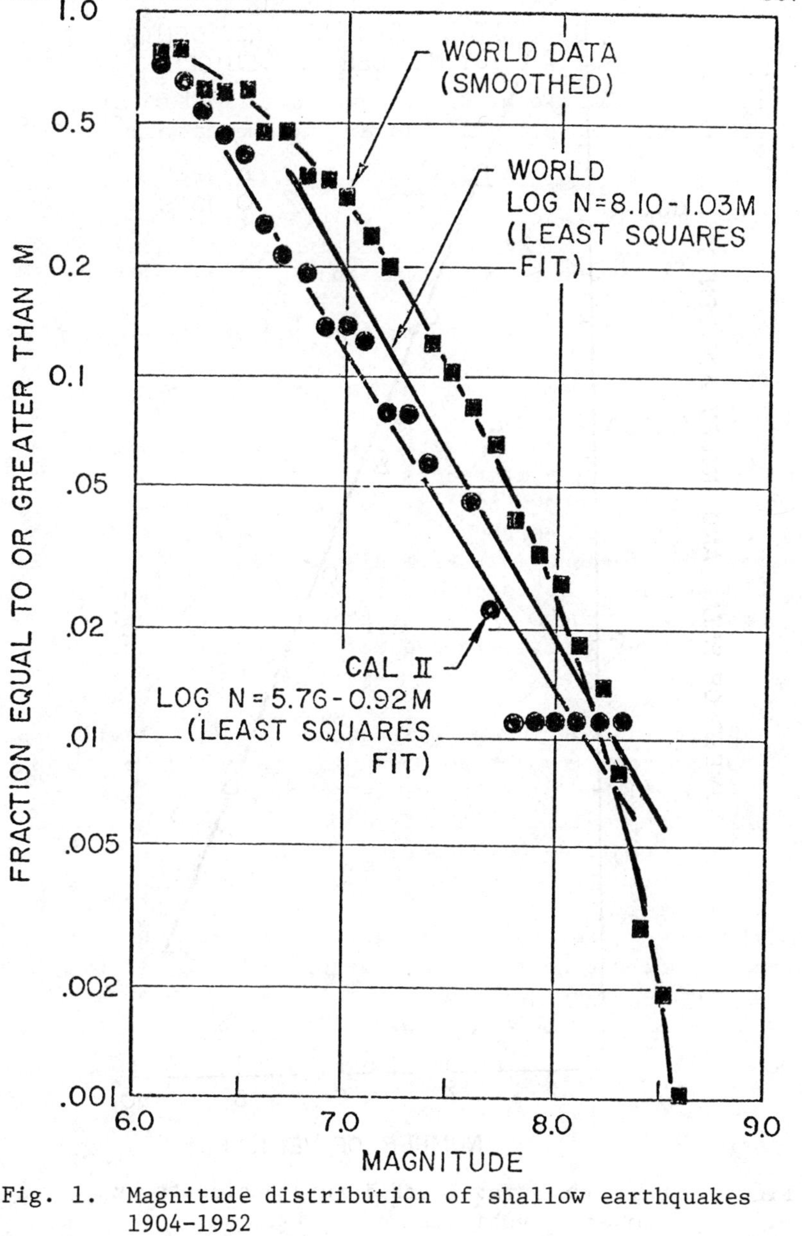

Fig. 1. Magnitude distribution of shallow earthquakes 1904-1952

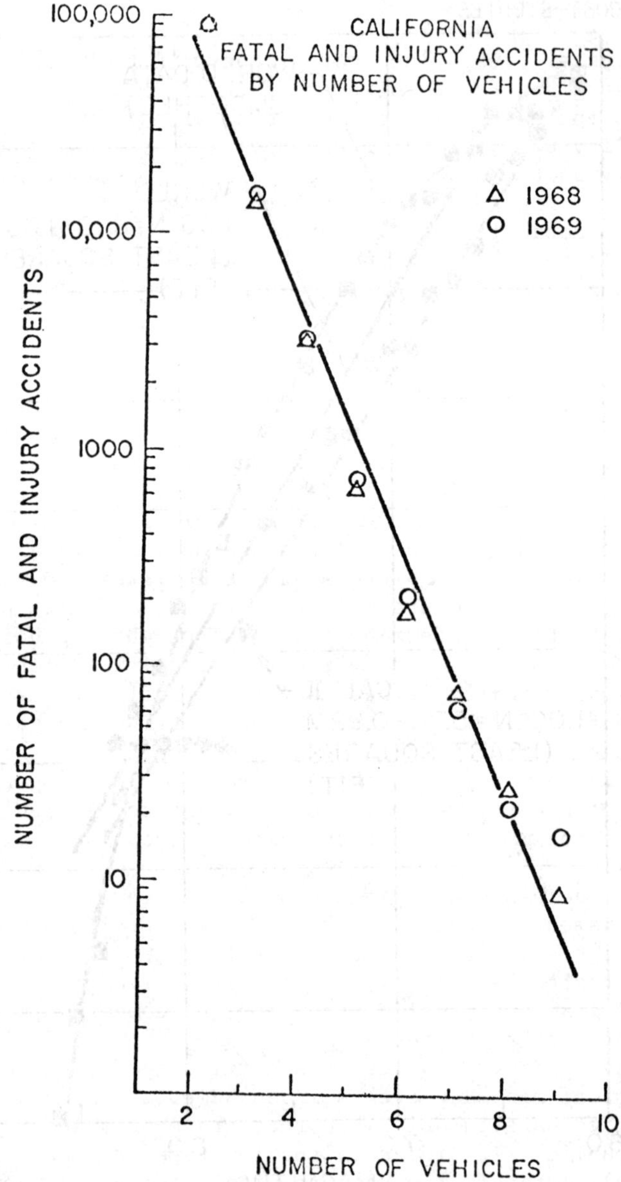

Fig. 2. California fatal and injury accidents by number of vehicles

for a typical pressurized water reactor 1000 MWe power plant has given the results shown in Fig. 3. The similarity to the earthquake curves is apparent. Here the probability per year (or frequency) approximately varies inversely as the square of the fission product release per event. Figure 3 also shows that there is a characteristic probability curve for each type of accident chain. This would be expected, as major differences in initiating events produce largely different accident sequences.

The spectrum of relationships between accident frequency and their severity may be generalized by the three-dimensional sketch shown in Fig. 4. The range of disability goes from minor inconvenience to death. At each point on this disability scale, is an associated curve of frequency versus the number of people involved. In this paper, I have chosen to discuss only the fatal end point. However, a complete study would integrate the effects of all accidents, minor to fatal, and with proper cost factors, determine the total weighted public risk and the social cost. Even for fatalities alone, the weighting factor of age is very significant - as shown in Fig. 5. This type of comprehensive analysis should be undertaken for a meaningful benefit-cost comparison for a specific technical system. This paper is not intended to be comprehensive, and reviews only the scope and magnitude of the basic methodology.

The nuclear power plant example mentioned raises some fundamental questions about our societal approach to public risk. The dramatic aspect of the unfamiliar large catastrophe - but with exceedingly low probability - tends to draw both public attention and concern. Unlike the natural catastrophes - earthquakes, typhoons, floods, tidal waves, etc. - society has not learned to place such hypothetical man-made events in an acceptable comparative perspective particularly when they are poorly understood by the public. If, in fact, the low release events are of more importance, the concern with imaginary large catastrophes may seriously distort both societal policy decisions and, as a result, engineering design emphasis.

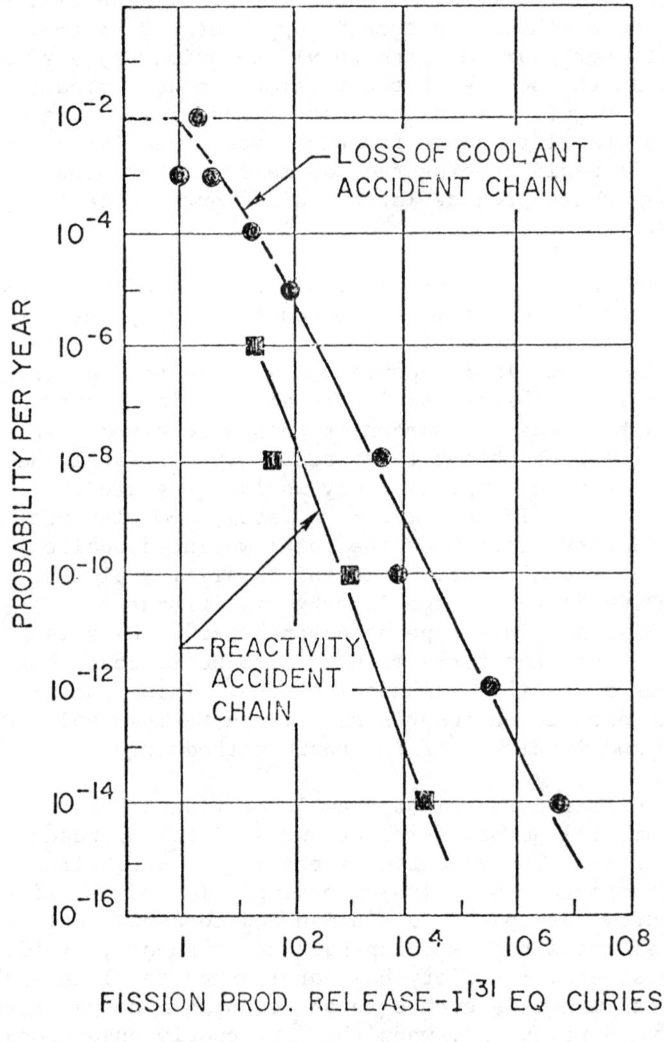

Fig. 3. Fission product release vs probability

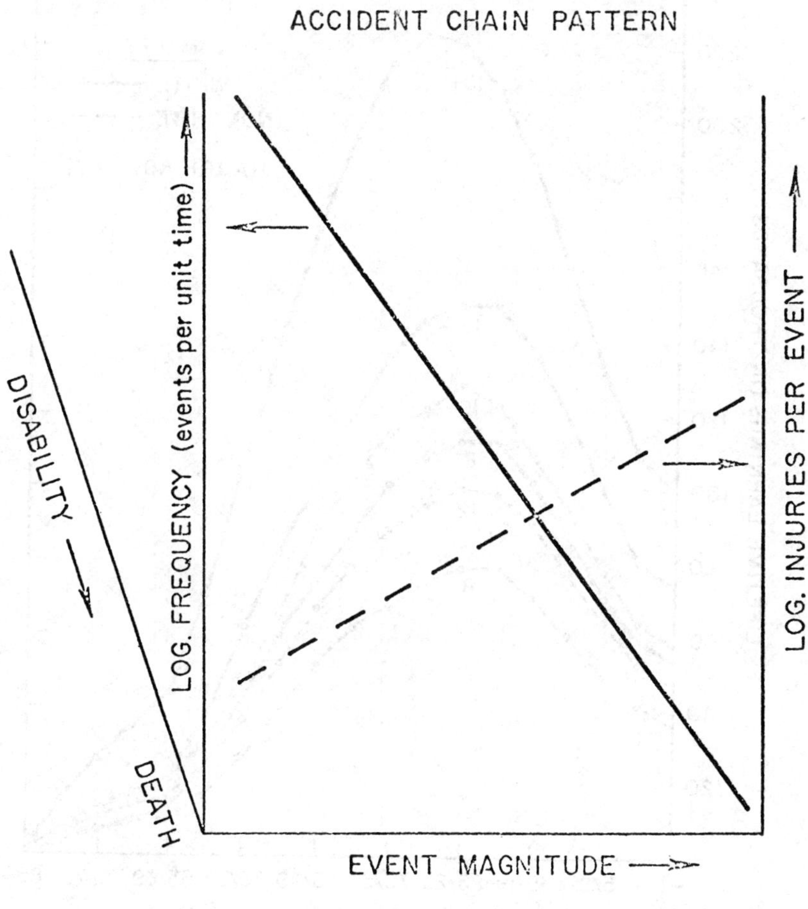

Fig. 4. Accident chain pattern

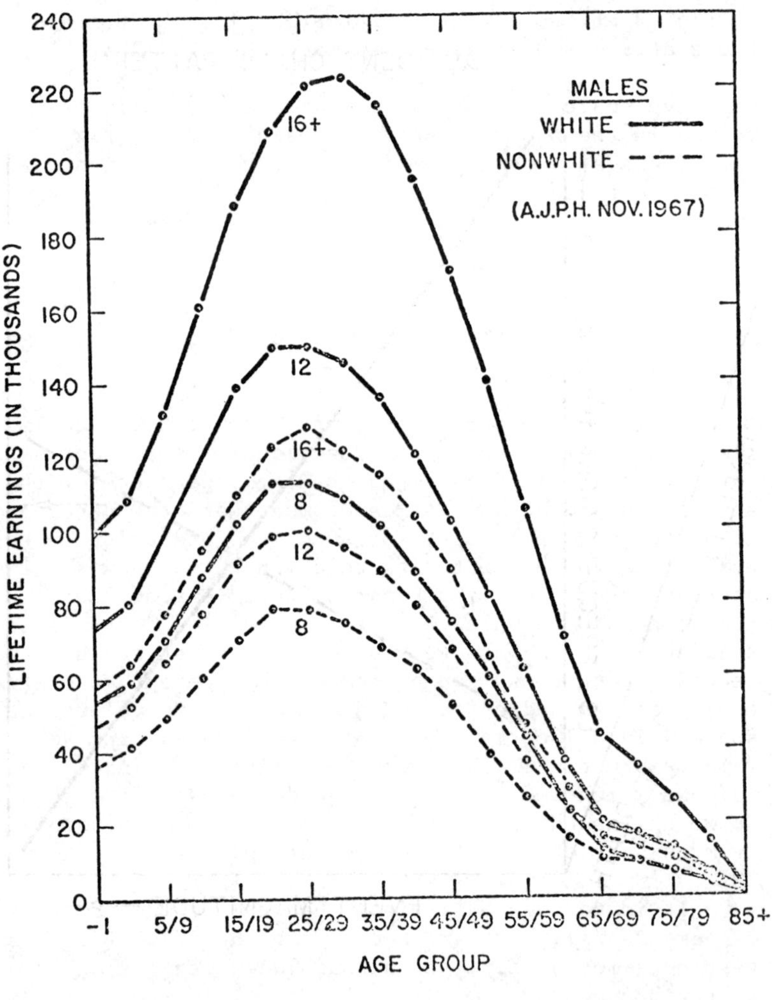

Fig. 5. Present value of lifetime earnings of males, by years of school completed, discounted at 4%, 1964

In the particular case of nuclear power, it is unlikely that this issue will be resolved by practical experience. The probability of the large events plotted in Fig. 3 are so low, that actuarial insight based on the statistics of real nuclear accidents will probably never exist. It is important to recognize the comparative scale of the extremely small probabilities involved in the nuclear accident chain shown in Fig. 3. Let us use as a comparative measure natural disasters - floods, earthquakes, tornadoes, and storms. Averaged over the U.S., these cause five to ten deaths per million population per year. The probability of causing this same level of fatalities by the nuclear accident chain of Fig. 3 is less than 10 per year - that is, once in a million years of operation of a nuclear power station.

It is, of course, very comforting that the public risk from nuclear power plant accidents is in fact so low. However, it does emphasize the importance of the analytic approach to risk assessment, and the development of a societal philosophy, perspective, and criteria for evaluating the benefit-risk relationships. How should we approach the "once in a hundred years" earthquake, or the somewhat equivalent "once in a million years" nuclear accident? Do we passively accept the first and not the second? Both risks can be reduced by preventive engineering design and added costs. What is the value to our society in allocating resources now for reducing such time-remote risks? What is a "negligible" risk - that is, one which we consciously decide to neglect? When do we philosophically shift the burden of responsibility from an "act of man" to an "act of God?"

It is to these questions that the remainder of this discussion will be addressed - with the objective of suggesting a methodology for a rough quantitative approach to their answers.

QUANTITATIVE CORRELATIONS

With this definition of the problem, and the associated caveats, we are in a position to discuss the quantitative correlations. For the sake of simplicity, the measure of the risk to the individual has been restricted to the fatalities (deaths) associated with each type of activity. Although it clearly would be useful to include all injuries (which are 100 to 1000 times as numerous as deaths), the difficulty in obtaining data and the unequal penalties of varying disabilities would introduce inconvenient complexity for this study. So the risk measure used here is the statistical probability of fatalities per hour of exposure of the individual to the activity considered.

The hour-of-exposure unit was chosen because it was deemed more closely related to the individual's intuitive process in choosing an activity than a year of exposure would be, and gave substantially similar results. Another possible alternative, the risk per activity, involved a comparison of too many dissimilar units of measure; thus, in comparing the risk for various modes of transportation, one could use risk per hour, per mile, or per trip. As this study was directed toward exploring a methodology for determining social acceptance of risk, rather than the safest mode of transportation for a particular trip, the simplest common unit - that of risk per exposure hour - was chosen.

The social benefit derived from each activity was converted into a dollar equivalent, as a measure of integrated value to the individual. This is perhaps the most uncertain aspect of the correlations because it reduced the "quality-of-life" benefits of an activity to an overly simplistic measure. Nevertheless, the correlations seemed useful, and no better measure was available. In the case of the "voluntary" activities, the amount of money spent on the activity by the average involved individual was assumed proportional to its benefit to him. In the case of the "involuntary" activities, the contribution of the activity to the individual's annual income (or the equivalent) was assumed proportional to its benefit. This assumption of roughly constant relationship between benefits and monies, for each class of activities, is clearly an approximation. However, because we are dealing in orders of magnitude, the distortions likely to be introduced by this approximation are relatively small.

In the case of transportation modes, the benefits were equated with the sum of the monetary cost to the passenger and the value of the time saved by that particular mode relative to a slower, competitive mode. Thus, airplanes were compared with automobiles, and automobiles were compared with public transportation or walking. Benefits of public transportation were equated with their cost. In all cases, the benefits were assessed on an annual dollar basis because this seemed to be most relevant to the individual's intuitive process. For example, most luxury sports require an investment and upkeep only partially dependent upon usage. The associated risks, of course, exist only during the hours of exposure.

Probably the use of electricity provides the best example of the analysis of an "involuntary" activity. In this case the fatalities include those arising from electrocution, electrically caused fires, the operation of power plants, and the mining of the required fossil fuel. The benefits were estimated from a United Nations study of the relationship between energy consumption and national income. The contributions of the home use of electric power to our "quality of life" - more subtle than the contributions of electricity in industry - are omitted. The availability of refrigeration has certainly improved our national health and the quality of dining. The electric light has certainly provided great flexibility in patterns of living, and television is a positive element. Perhaps, however, the gross-income measure used in the study is sufficient for present purposes.

Information on acceptance of "voluntary" risk by individuals as a function of income benefits is not easily available, although we know that such a relationship must exist. Of particular interest, therefore, is the special case of miners exposed to high occupational risks. In Fig. 6, the accident rate and the severity rate of mining injuries are plotted against the hourly wage. The acceptance of individual risk is an exponential function of the wage, and can be roughly approximated by a third-power relationship in this range.

RISK COMPARISONS

As a reference level for comparing risks, it seemed appropriate to take the risk of death due to disease. This is shown in Fig. 7 for the U.S. population, both for disease and accidents. The average accident fraction is about one-tenth of the disease average, and its presence is not significant. However, if one considers the younger population (10-30), the disease level is a magnitude less than the average, and the accident level is actually larger. The reverse is obviously true for the oldest group. In a more complete study, the varying value of human life with age - from a societal viewpoint - would require special emphasis on the age group from 20-50 years, whose social value may be several times that of the younger and older. Aside from the complex humanistic judgments intrinsic to this approach, the present study is too general to justify this refinement.

The results for the societal activities studied, both "voluntary" and "involuntary," are assembled in Fig. 8. Also shown in Fig. 8 is the third-power relationship between risk and benefit characteristic of Fig. 6. The average risk of death from disease is indicated for comparison.

Several major features of the benefit-risk relations are apparent, the most obvious being the difference by several orders of magnitude in society's willingness to accept "voluntary" and "involuntary" risk. As one would expect, we are loathe to let others do unto us what we happily do to ourselves.

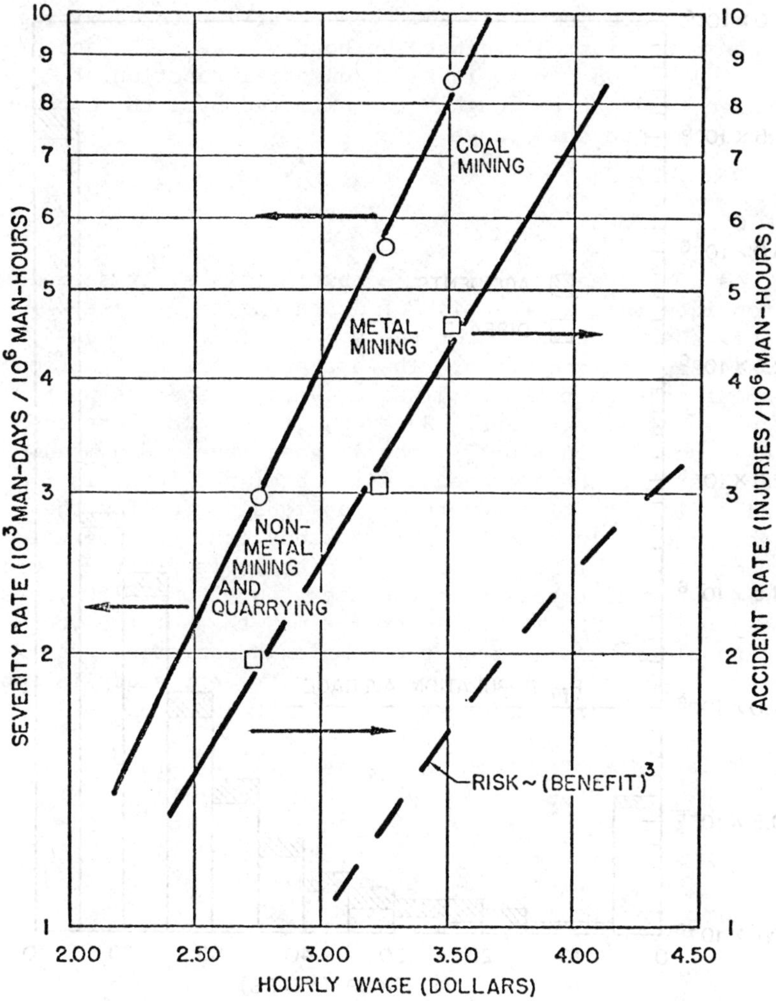

Fig. 6. Mining accident rates vs incentive

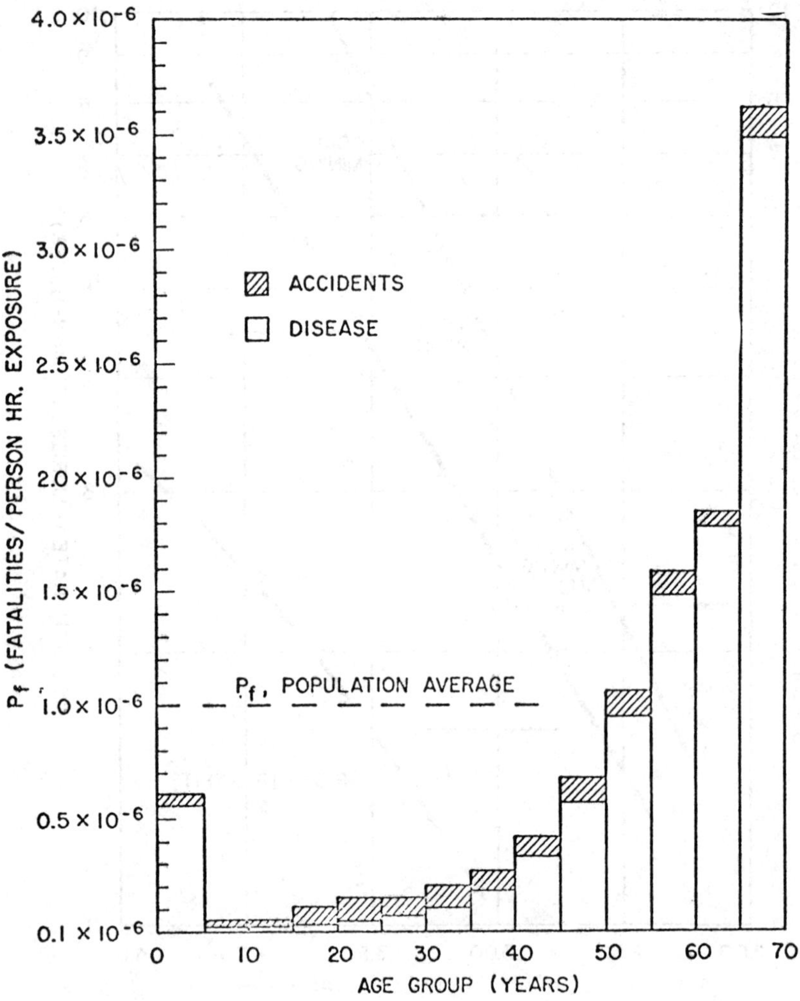

Fig. 7. Risk vs age group for accidents and disease

The rate of death from disease appears to play, psychologically, a yardstick role in determining the acceptability of risk on a voluntary basis. The risk of death in most sporting activities is surprisingly close to the risk of death from disease - almost as though, in sports, the individual's subconscious computer adjusted his courage and made him take risks associated with a fatality level equaling but not exceeding the statistical mortality due to involuntary exposure to disease. Perhaps this defines the demarcation between boldness and foolhardiness.

The simple ratio of benefit to risk is conventionally used as an index of societal acceptability. This is not consistent with the hypothesized logarithmic relationship of Fig. 8. As one moves toward higher benefits along either suggested band, the benefit/risk ratio drops rapidly due to the $R \sim B^3$ relation. Thus, the simple ratio is apt to be quite misleading in considering social value and acceptability of an activity.

The risks associated with general aviation, commercial aviation, and travel by motor vehicle deserve special comment. The latter originated as a "voluntary" sport, but in the past half-century the motor vehicle has become an essential utility. General aviation is still a highly voluntary activity. Commercial aviation is partly voluntary and partly essential and, additionally, is subject to government administration as a transportation utility.

Travel by motor vehicle has now reached a benefit-risk balance, as shown in Fig. 9. It is interesting to note that the present risk level is only slightly below the basic level of risk from disease. In view of the high percentage of the population involved, this probably represents a true societal judgment on the acceptability of risk in relation to benefit. It also appears from Figure 9 that future reductions in the risk level will be slow in coming, even if the historical trend of improvement can be maintained.

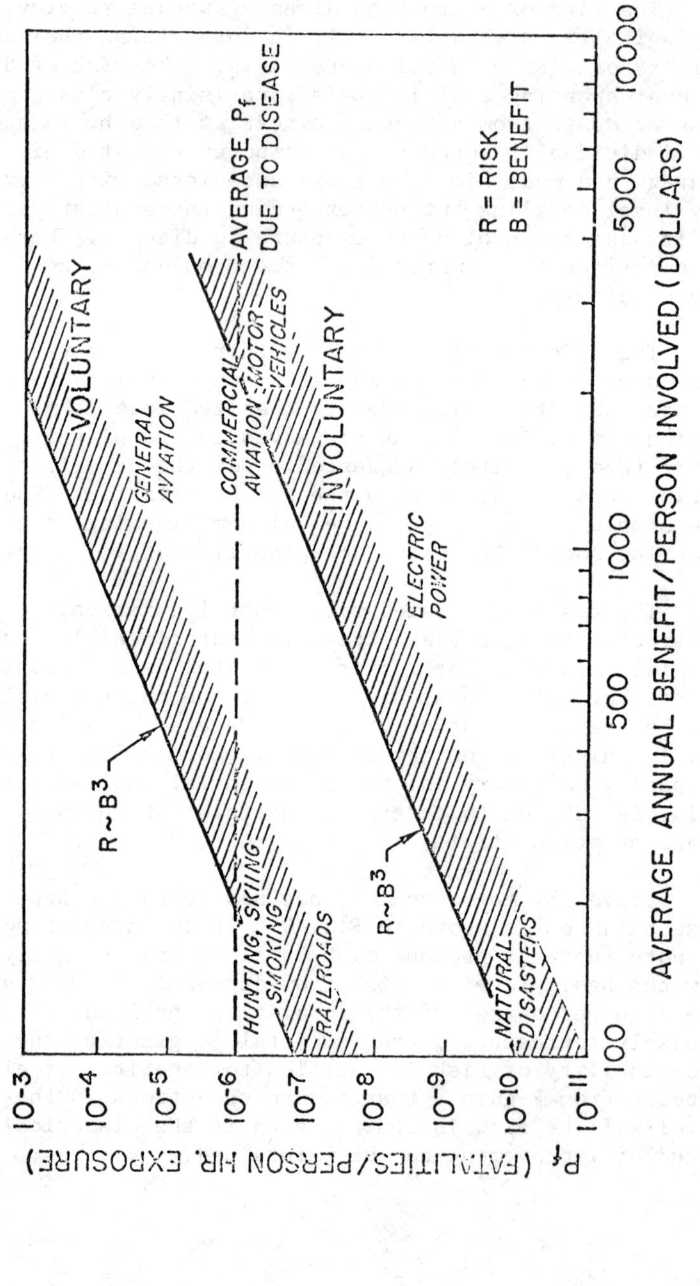

Fig. 8. Risk vs benefit - voluntary and involuntary exposure

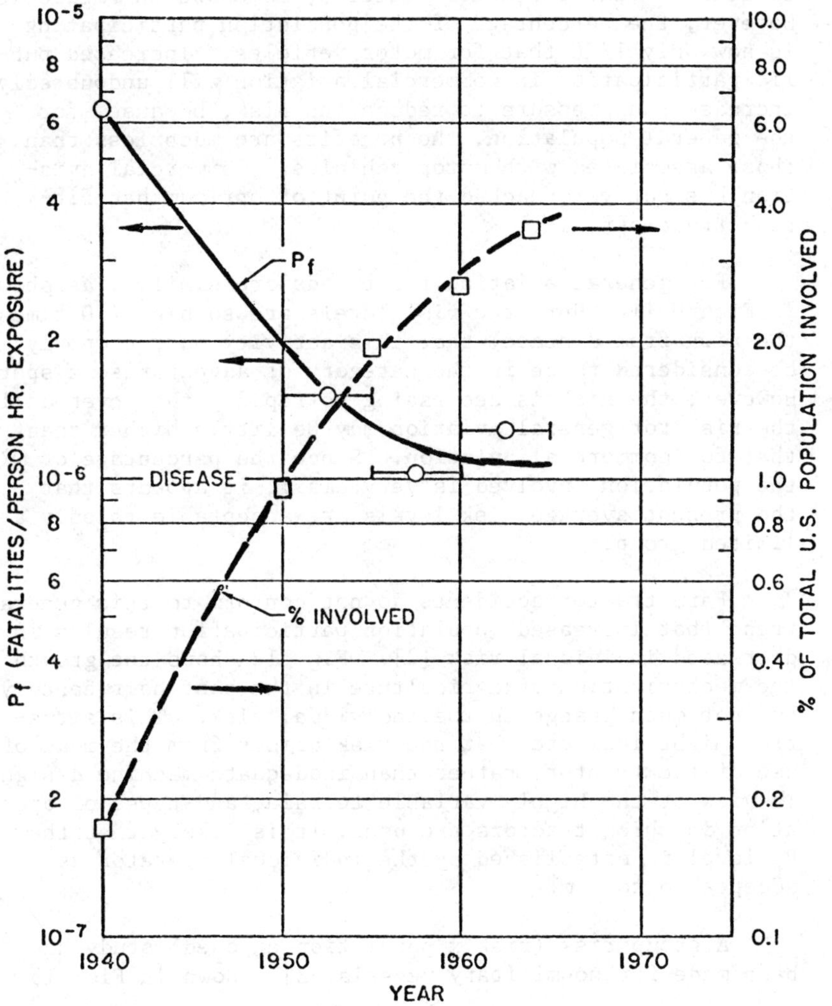

Fig. 9. Risk and participation trends for motor vehicles

Commercial aviation has barely approached a risk level comparable to that set by disease. The trend is similar to that for motor vehicles, as shown in Figure 10. However, the percentage of the population participating is now only 1/20 that for motor vehicles. Increased public participation in commercial aviation will undoubtedly increase the pressure to reduce the risk, because, for the general population, the benefits are much less than those associated with motor vehicles. Commercial aviation has not yet reached the point of optimum benefit-risk trade-off.

For general aviation the trends are similar, as shown in Figure 11. Here the risk levels are so high (20 times the risk from disease) that this activity must properly be considered to be in the category of adventuresome sport. However, the risk is decreasing so rapidly that eventually the risk for general aviation may be little higher than that for commercial aviation. Since the percentage of the population involved is very small, it appears that the present average risk levels are acceptable to only a limited group.

Farm tractor accidents do not conform to this general trend that increased population participation results in decreased individual risk [2]. Fig. 12 shows the growth in the mechanization of agriculture in the past half century, but not much change in the individual risk. This situation might indicate that the risk arises from the mode of use of the tractor, rather than inadequate machine design. In view of the highly variable terrain, and types of operation in which tractors are used, it is likely that the P_f level is established by the individual operator as acceptable to him.

A group risk (risk x population exposed) study has been made for nonmilitary vessels [3], shown in Fig. 13.

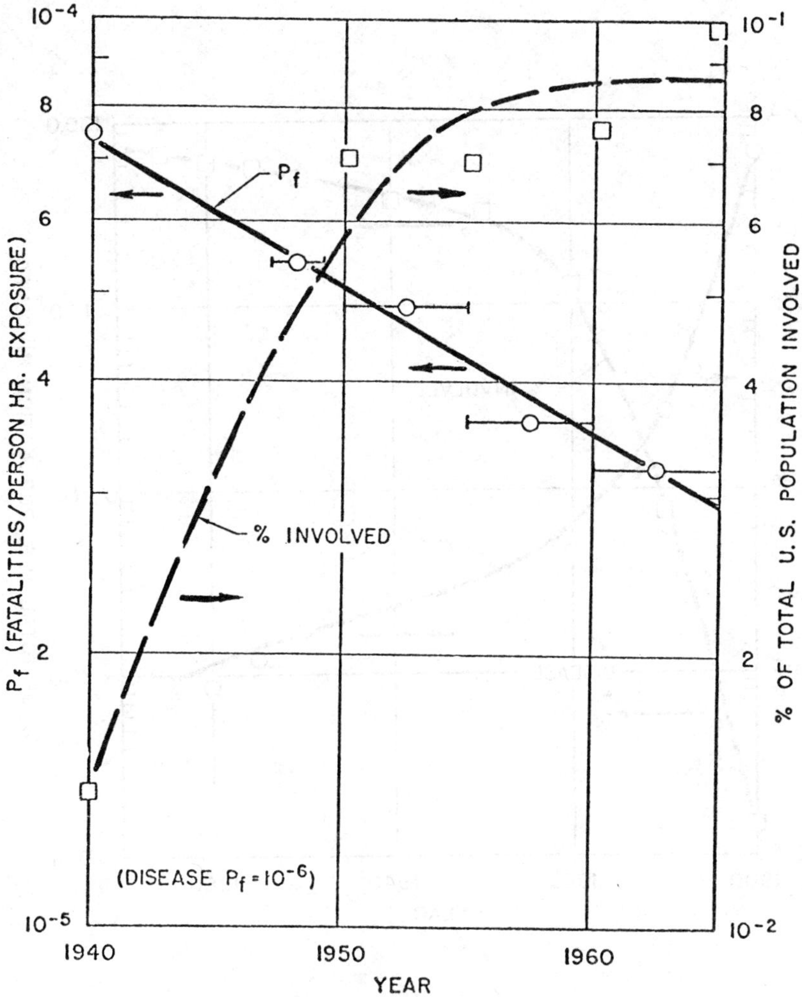

Fig. 10. Risk and participation trends for certified air carriers

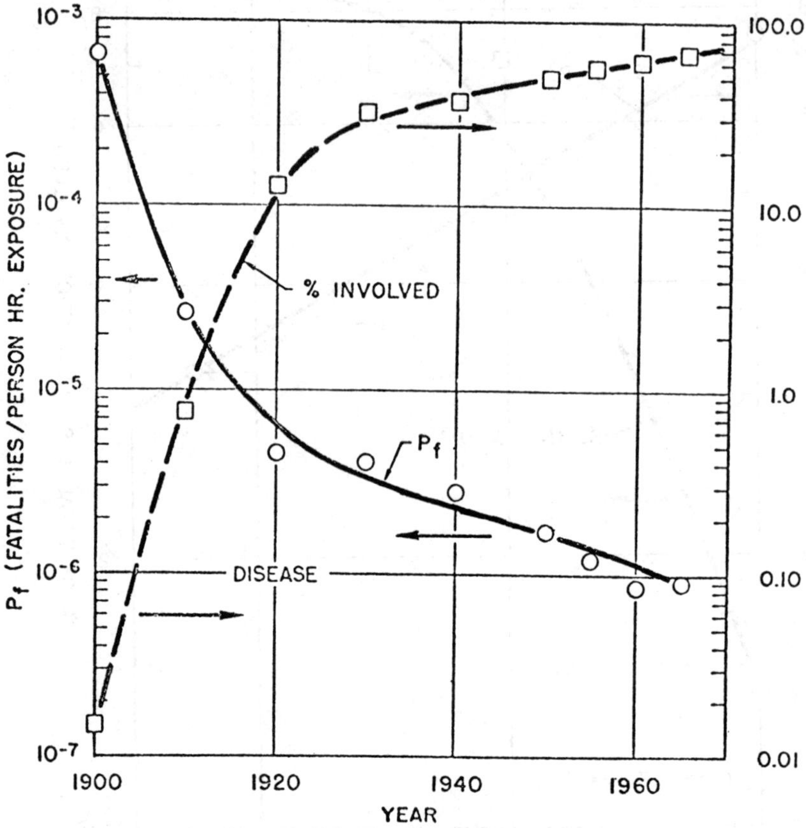

Fig. 11. Risk and participation trends for general aviation

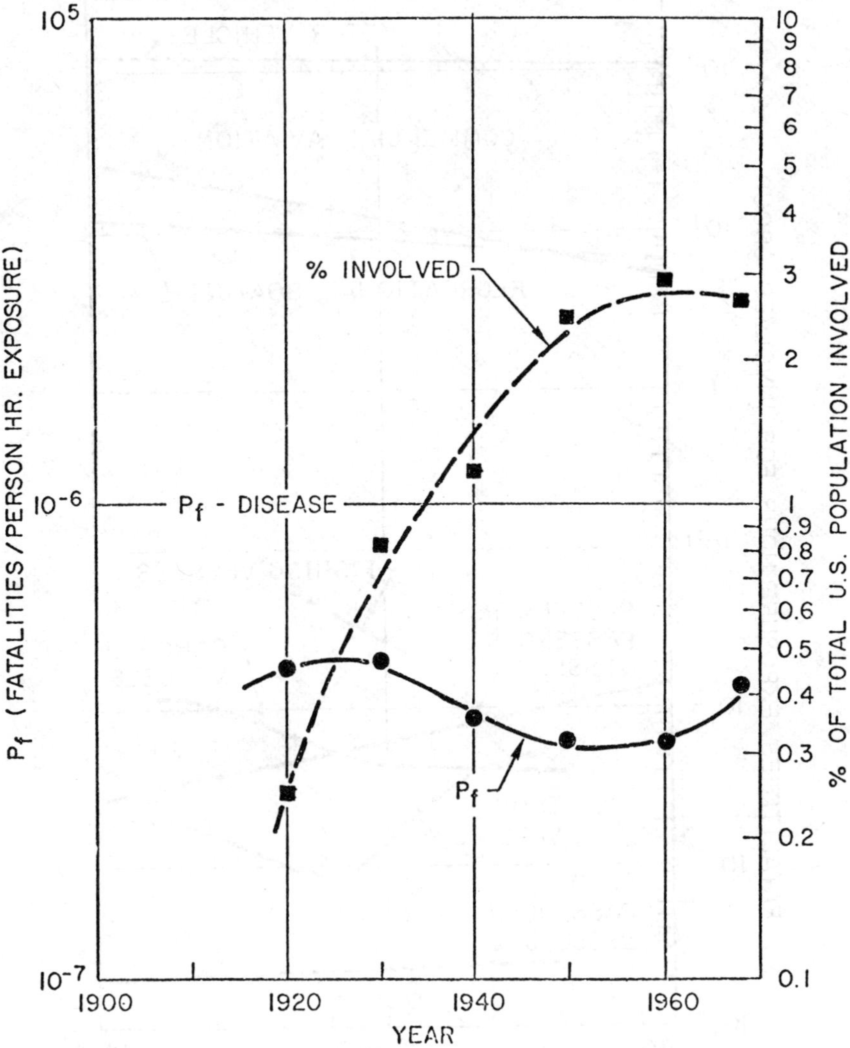

Fig. 12. Risk and participation trends for farm tractors

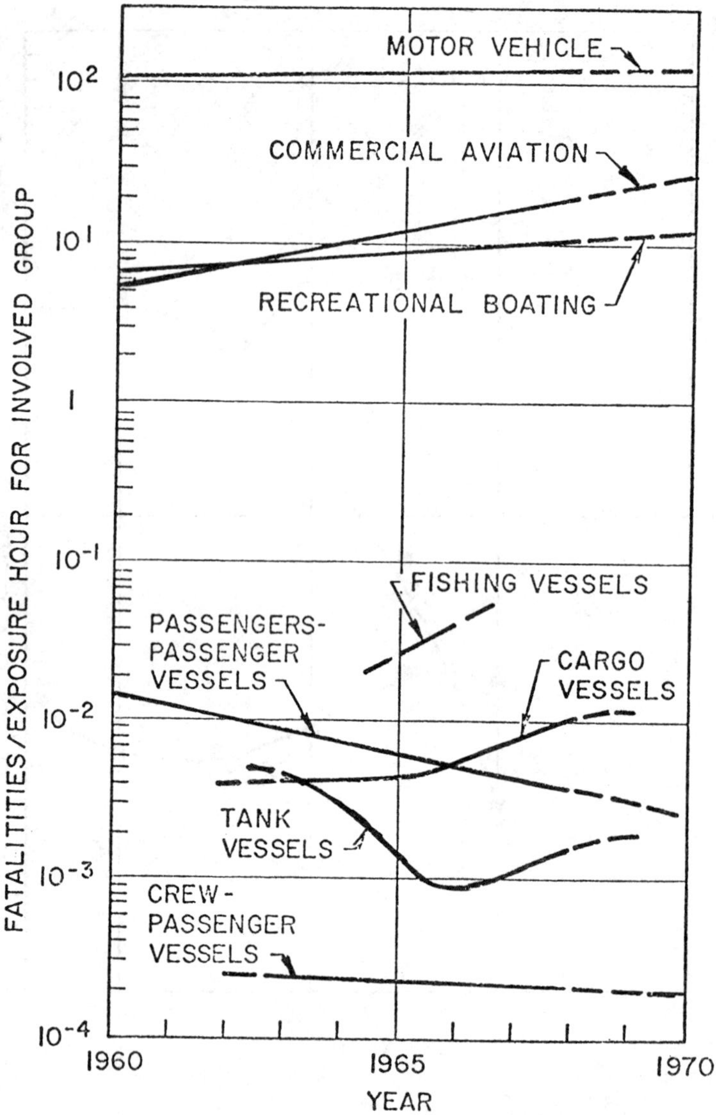

Fig. 13. Group risk vs year

The span of eight years (1960-68) is too short to show long term trends. However, the difference in risk between recreational boating and commercial vessels again shows the 1000 or more spread between voluntary and involuntary exposures accepted by the participants. The traditional risks of the professional seamen and fishermen appear modest as compared to that of the daily commuter driving to work.

PUBLIC AWARENESS

Finally, I attempted to relate these risk data to a crude measure of public awareness of the associated social benefits (see Fig. 14). The "benefit awareness" was arbitrarily defined as the product of the relative level of advertising, the square of the percentage of population involved in the activity, and the relative usefulness (or importance) of the activity to the individual. Perhaps these assumptions are too crude, but Fig. 14 does support the reasonable position that advertising the benefits of an activity increases public acceptance of a greater level of risk. This, of course, could subtly produce a fictitious benefit-risk ratio - as may be the case for smoking.

This study is only an exploratory application of the methodology of historical analysis for revealing social preferences, values, and benefit-cost relationships. On the specific illustrative question "How safe is safe enough?" it does reveal several interesting points. These may suggest an initial basis for judicious national decisions on involuntary risk associated with our socio-technical systems. The premises for such a societal policy may be summarized as follows:

1. Rate of death from disease is an upper guide in determining the acceptability of risk - somewhat less than 1 in 100 years.

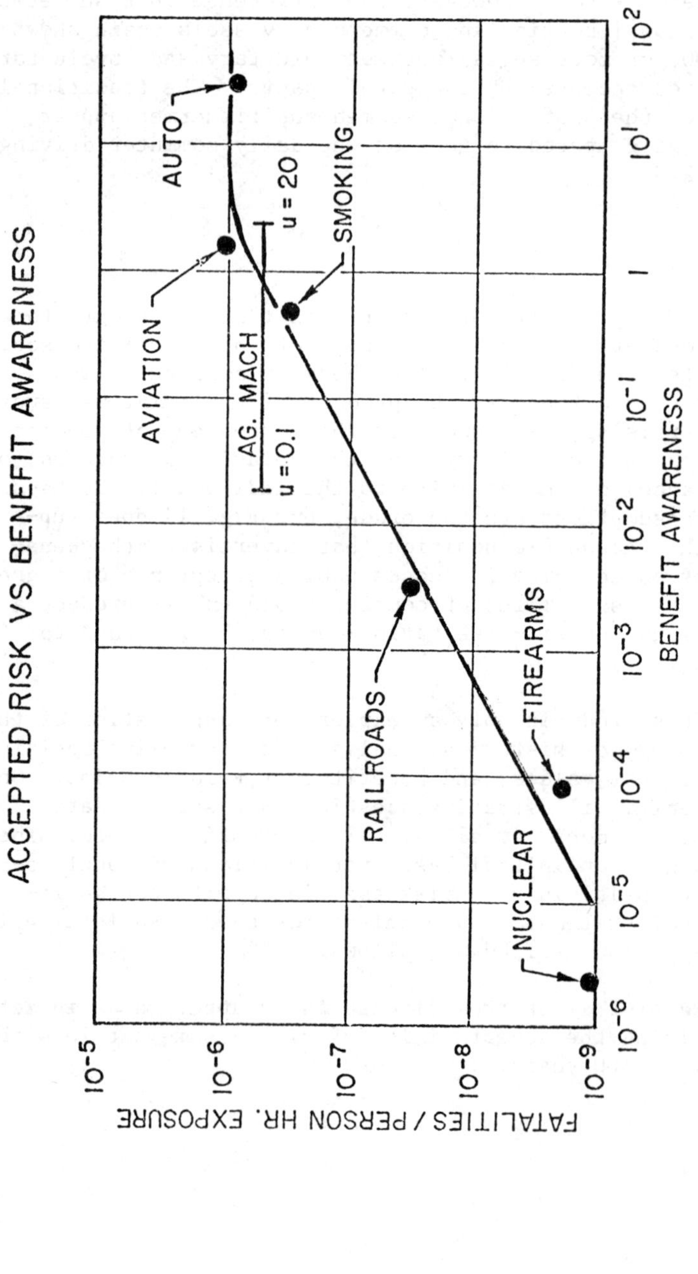

Fig. 14. Accepted risk vs benefit awareness

2. Natural disasters ("acts-of-God") tend to set a base guide for risk - somewhat more than 1 in a million years - similar to the intrinsic "noise" level of physical systems. Man-made risks at this level can be considered almost negligible, and can certainly be neglected if they are several magnitudes less.

3. As would be expected, societal acceptance of risk increases with the benefits to be derived from an activity. The relationship appears to be nonlinear, with this study suggesting that the acceptable level of risk is an exponential function of the benefits (real and imaginary).

4. The public appears willing to accept "voluntary" risks roughly 1000 times greater than "involuntary" exposure risks.

5. The quantitative risk analysis of a specific socio-technical system - such as atomic power - should take into account the continuous spectrum of accident frequency versus severity, both of the system under study and of comparative systems. The present study has not entered into the "fine-structure" of the problem. A more definitive and intensive analysis is needed for application to specific national policy guidelines.

CONCLUSION

Societal policy for the acceptability of public risks associated with socio-technical systems should be determined by the trade-off between social benefits and personal risk. As suggested above, the upper bound of such risk is the disease level and the lower bound is several magnitudes below the natural disaster level, as illustrated in Fig. 15. This provides a risk trade-off range of one million for social policy determination - from an individual risk of one fatal accident in a hundred years, to one in a hundred million years.

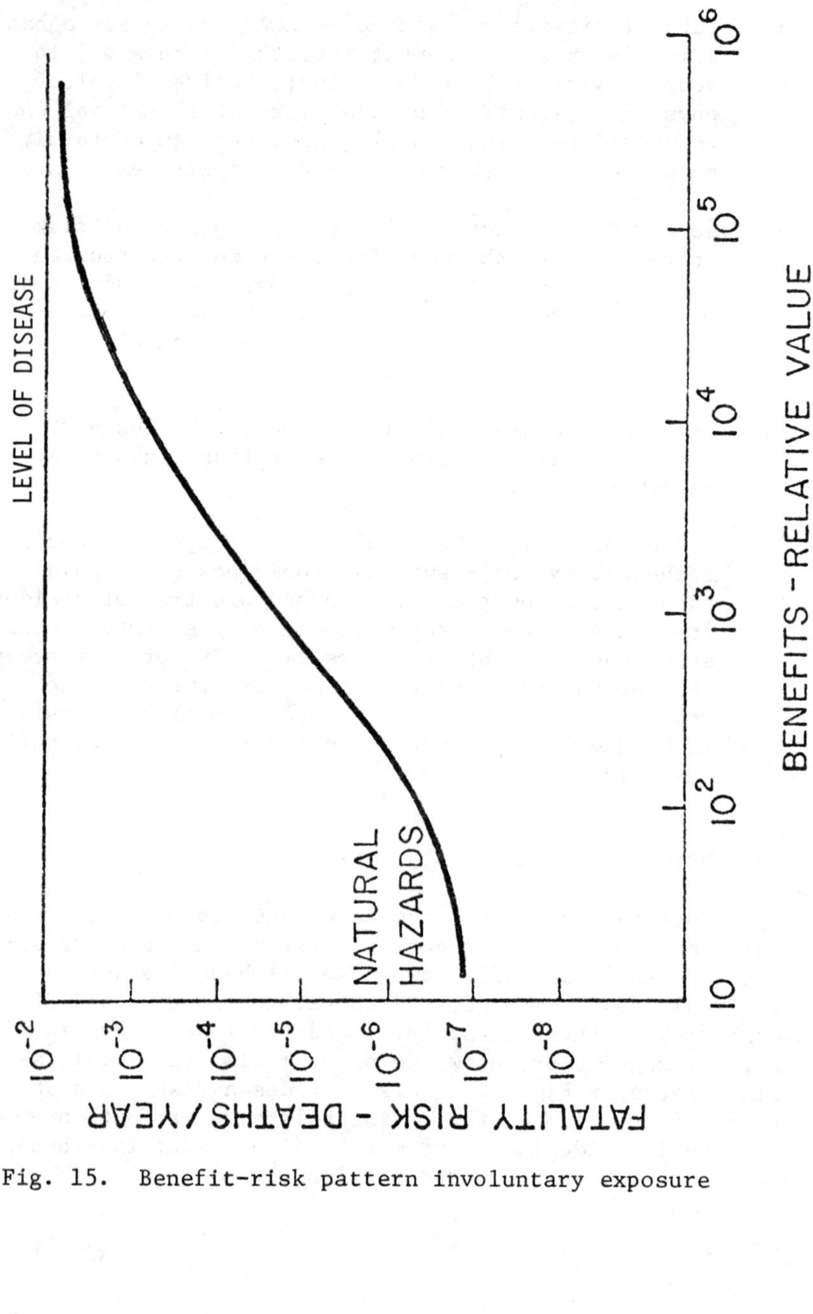

Fig. 15. Benefit-risk pattern involuntary exposure

It is evident that we need much more study of the methodology for evaluating social benefits and costs. The fatality measure of public risk is perhaps more advanced than most because of decades of data collection. Nevertheless, even the use of crude measures of both benefit and costs would assist in the development of the insight needed for national policy purposes. We should not be discouraged by the complexity of this problem - the answers are too important, if we want a rational society.

REFERENCES

1. OTWAY, H.J., "The Application of Risk Allocation to Reactor Siting and Design." 1969, *Ph. D. Thesis*, UCLA. Los Alamos Scientific Laboratory, University of California, LA-4316.

2. GIESEN, JAMES E., Product Safety Department, Deere & Company, December 3, 1969, Private communication.

3. BOSNAK, R.J., Commander, U.S. Coast Guard, December 24, 1969, Private communication.

TECHNOLOGICAL ALTERNATIVES FOR ENHANCED URBAN USES OF WATER AND RELATED LAND RESOURCES

MILLER B. SPANGLER

Director
Center for Techno-Economic Studies, National Planning Association, Washington, D.C., U.S.A.

THE ROLE OF THE RESEARCH AND DEVELOPMENT PLANNER

There are many kinds of technological alternatives in enhancing the urban use of water and related land resources and an even greater number of possibilities for technological change within each class of alternatives. To identify and discuss all of these alternatives even briefly would require several volumes. However, in serving the objectives of this study, which are directed to methodology, it will be sufficient to illustrate the broad range of technology assessment considerations that must be evaluated in deciding between these alternatives for only a few representative cases. These considerations include an examination of the potential impact of the R & D on social needs especially for urban areas, the cost of the R & D, its probability of success and the relation of institutional problems and measures to these interrelated decision variables. Unless the R & D planner broadens his assessment of these decision variables it will be increasingly difficult to generate the essential financial support to conduct the R & D.

For convenience of treatment the discussion of technological alternatives will be divided into two classes. The first of these classes constitutes those alternatives wherein systems performance objectives are dominant: (1) augmentation of water supply, (2) reduction of water requirements, (3) regulating water flows, (4) reducing costs of water supply, and (5) protecting or enhancing the quality of water and the related land environment. This class of objectives of R & D is principally concentrated on the resource itself, the technologies associated with the delivery of the resource to its users, and the engineering efficiency of resource use. In evaluating this class of technological alternatives water resources research planners will usually play a leading role. Moreover, the assessment of these alternatives will generally require an examination of resource alternatives within a given water basin or region.

The second class of alternatives includes those potential advances in science and technology related to the enhanced social uses of water and related land resources rather than the resources themselves. Examples include water transportation, recreation, fisheries, environmental beautification, agriculture, industrial products, power generation, household and commercial uses of water, and engineered structures or facilities on or underwater or on water-related land areas. The R & D planner responsible for evaluating these technological alternatives may have little if any competence in various aspects of water resources science and technology. Nevertheless, advances in this class of technological developments may hold important implications for water resources planning and most likely the benefits to be obtained from the prior class of technological alternatives more directly related to water resources.

Thus, if the evaluation of these two classes of technological alternatives requires overlapping analyses — albeit from a different vantage point — it is also true that institutional measures that facilitate or constrain the applications of these technologies likewise span the technology assessment functions of both classes of technology alternatives. Accordingly, the final section of this chapter will examine requirements for research in these related planning and institutional measures. This latter kind of analysis of research planning will relate heavily, though not exclusively, to the social and management sciences.

EVALUATING TECHNOLOGICAL ALTERNATIVES IN MEETING WATER SUPPLY PERFORMANCE OBJECTIVES

Augmenting Water Supply

There are a large number of technological alternatives for enlarging the water supply available to a given urban area. A fairly complete list would include:

1) Weather modification.
2) Desalination of seawater, brackish coastal waters, or inland saline groundwater.
3) Locating underground streams or unutilized groundwater aquifers.
4) Interbasin transfer of water.
5) Treatment of waste waters for reuse in desert areas.
6) Increasing water runoff rates from mountain catchment basins by altering the kind of vegetation to reduce evapotranspiration losses.
7) Melting snow or ice.

The latter alternative has already been employed in cold regions for limited sources of water supply by the simple application of heat. The natural melting of glaciers has long provided a number of rivers in Alaska and the Pacific Northwest with a substantial part of their annual flow. If there were a demonstrated need for additional fresh water supply in these areas or an interbasin transfer of water to adjacent areas, then R & D on technologies to cause a faster melting of these glaciers could serve to augment water supply. Some experimentation of this kind has been pursued in Russia using carbon particles to increase the absorption of energy from the sun. However, the limited U.S. requirements for this kind of augmentation in the involved river basins is a deterrent as well as the cost of interbasin transfer of such water augmentation, uncertainty over the cost effectiveness of the melting technology and various political complications. Another idea suggested by Professor John D. Isaacs of Scripps Institution of Oceanography is to tow icebergs from polar waters to water-short regions such as California by taking advantage of ocean currents to provide much of the transport energy [1]. One can envision certain problems in testing the feasibility of this alternative but it must be acknowledged, of course, that the very function of R & D is to provide solutions to engineering problems and to reduce the costs associated with the system technology to make it economically viable.

A set of technological alternatives that are receiving more serious attention is that of desalting seawater. Various processes being considered, or already applied, cost in the range of $0.50 to $1.00 per 1,000 gallons for plants with a capacity of 1 to 10 million gallons of fresh water per day exclusive of distribution costs [2]. One possibility for reducing desalination costs is to combine a large-scale desalination plant with a nuclear power plant in order to use the waste heat of the latter. A number of such potential projects have received study by the Atomic Energy Commission in cooperation with a number of other domestic foreign public agencies: Southern California (Camp Pendleton, Diablo Canyon, or the original Bolsa Island site at Huntington Beach); a Gulf of California site involving the States of Arizona, California, and the Mexican States of Baja California and Sonora; and agro-industrial complexes in the Middle East, India and Puerto Rico [3].

It would be instructive to examine some of the issues involved in performing a technology assessment of this R & D alternative. To do so it is necessary to examine the three basic components of the appraisal ratio of R & D (see Chapter 24 of this book): the net social [4] returns if the R & D is successful in achieving its objectives; the probability of attaining those objectives; and the cost of the R & D. As previously noted, these elemental considerations are importantly interrelated and cannot be evaluated in isolation of each other. For example, the appraisal of net social returns of the technological development involves an estimation of possible reduction in the prices of electricity and fresh water to the affected region versus alternative sources of supply; the outlook for protecting the environment or ecology against degradating effects of the technology; the outlook for utilizing the technological development as a stepping stone toward developing an even more efficient technology to benefit other regions or nations; and not unimportantly, the reduction of tensions between states competing for water rights in water-short areas or improving the quality of U.S. relations with other nations through cooperative programs of this kind including the prototype or demonstration project in question. Moreover, there is also a possibility of improving the U.S. balance-of-payments position through the potential increase in exports of the combination fresh water-electrical generating plants thus developed, especially if this promotes U.S. leadership in this application of technology.

While the technological feasibility of a dual-purpose development of this kind appears to be assured, there is a significant degree of risk that each of the above social objectives might not be attained. Thus, it is inappropriate to think in terms of a single probability coefficient for the appraisal ratio of R & D, but rather a separate probability estimate for each of the objectives involved in appraising net social returns. The magnitude of each kind of return and its probability coefficient will in turn be affected by the amount of R & D funds the sponsoring agencies are willing to commit (i.e., the cost estimate in the denominator of the appraisal ratio).

However, gaining agreement among the sponsoring agencies on the R & D objectives, the scale and design of the project (and hence its cost), and the cost-sharing formula is itself a highly uncertain and frustrating process [5]. Up to a point, if the prototype project is larger (and costlier) in scale, then the probability is greater for achieving substantial economic benefits in the form of lower-cost electricity and fresh water supply.

But the higher this cost the more difficult and uncertain it is that the essential public funds can be raised in view of other budgetary priorities. Also the higher this cost the greater the difficulty in negotiating cost-sharing formula between governments whether domestic or foreign. In fact the conflicts of purposes in this negotiating process would tend to discourage a full appraisal of the rather large variety of social benefits and costs noted above which are distributed unequally to the potential sponsors of the project. The obvious implication here is that if these net benefits were fully acknowledged by the sponsoring agencies who stand to receive them, then it could be argued that the distribution of costs should reflect the distribution of benefits.

Because of the considerable uncertainty over both the benefits and costs — as well as the opposing interests cited above — there is a tendency to understate or ignore them. For example, in the original proposal of the Bolsa Island dual purpose plant in Southern California potentially harmful environmental impacts were largely ignored. However, social measures have since resulted in appropriate studies such as one sponsored by the U.S. Office of Saline Water on "Environmental Impact of Brine Effluent on Gulf of California" [6] — a problem of major concern in the joint Mexican-U.S. project noted above. Likewise, there is inadequate attention to the aforementioned national benefits — social, economic, technological and political — that could result from prototype developments of these kinds. It is the opinion of this observer that if these potential benefits were taken more fully into account then a higher U.S. share of the project funding costs could be justified. Convincing relatively well-educated public administrators and legislators of these national benefits poses one kind of problem and persuading the public (who elect these officials) of the merits of such high-cost projects is quite another. Herein lies an important challenge for the "education and public information" mode of planning to lend needed support to the "capital projects" mode of planning involved in an R & D venture of this kind.

It is essential to appreciate that the evaluation of the appraisal ratio of R & D is not static over time. Continuing advances in various components of the system technology can reduce the cost of the prototype development as well as to enlarge the economic benefits. For example, the successful development of a fast breeder reactor or the fusion process for nuclear power generation would favorably improve the chances of a dual purpose project by reducing the charges for the by-product heat to be utilized in the desalination process [7]. Likewise on-going R & D to reduce the adverse ecological impacts of such a dual purpose facility — or to reduce the costs of avoiding unfavorable environmental impacts — could improve the appraisal ratio of R & D for the project in question. Moreover, growing scarcities of water may improve the appraisal ratio by increasing the cost savings over alternative sources of supply and by increasing the value of a technological alternative that will reduce political tensions between states and nations over shared water resources in the face of severe water shortages.

Weather modification, as a technological alternative in enlarging water supply, holds similar complexities of technology assessment. It is something of an understatement to say that experiments to increase precipitation by cloud seeding have not been uniformly successful. Overseeding is believed to result in an actual depression in the level of rainfall and it is rather difficult at present state-of-the-art technology to determine *a priori* in any set of existing weather circumstances if a given level of seeding will produce an increase, decrease, or no effect on precipitation [8]. Notwithstanding this, the National Academy of Sciences — National Research Council Report on Weather and Climate Modification came to the hopeful, if cautious, conclusion that "there is increasing but still somewhat ambiguous statistical evidence that precipitation from some types of cloud and storm systems can be modestly increased or redistributed by seeding techniques" [9].

Accordingly, as a technological alternative for increasing water supply through further R & D on weather modification, the probability of achieving very large social and economic benefits through this R & D does not at the moment seem particularly high. Moreover, once again there is the matter that emerges repeatedly in a broadly conceived technology assessment that those who pay for the costs of the R & D programs do not share equally in the benefits. Indeed, the seeding of clouds to increase precipitation in one area may reduce precipitation in adjacent down-wind areas, though overseeding may produce opposite effects.

Nevertheless, Myron Tribus, former Assistant Secretary of Commerce for Science and Technology, was hopeful that a plan of action would be developed in the not too distant future that would be approved by the Congress and the people [10]. However, he saw no Utopia ahead [11]:

> If the historical controversies in this field, when only the scientific community was involved, are any guide as to how people feel about weather modification, I wonder what will happen when everyone gets into the act and decides what kind of weather we wish to have.

Once again, the implication is clear for technology assessment: those who would seek to justify expenditures for one technological alternative over another must be as much concerned with "people planning" as with the more technical aspects of R & D planning. In this regard research on planning and related institutional measures to be discussed later in the chapter may serve a crucial role in influencing the choice of one technological alternative over another in augmenting the supply of water.

Reducing Water Requirements

Some industrial and household uses of water involve the intake of water and its return to watercourses without appreciable loss of quantity available to downstream users although changes in water quality may pose a major problem in meeting fresh water requirements. Yet other uses of water or malfunctionings of the distribution system can cause appreciable consumptive losses of water. Among the different kinds of non-productive water losses are the following: (1) evaporation from open reservoirs, canals, and watercourses; (2) evapotranspiration from plants; (3) irrigation losses of water not utilized by plants nor draining again into watercourses from which the water is subject to reuse; and (4) the leakage of water from mains, aqueducts, and other components of the distribution system.

Regarding the latter category, of the water being pumped through over 4,000 miles of water mains and feeders in the City of Chicago and the 66 suburban communities it serves, about 20 percent is lost through underground leakage [12]. This is believed comparable to other large cities with a significant segment of the distribution system having an appreciable age. For example, based on data from a number of cities it is estimated that savings of 10 to 20 percent of apparent water consumption, or 100 to 200 million gallons per day, could be achieved for New York City through an expanded leakage detection and remedial program applied to the City's water mains [13].

Accordingly, the support of R & D to improve the effectiveness of leakage detection technology could contribute significantly to the reduction of urban water requirements. Since the cost of this technology must enter into the choice of alternatives, it would appear that the value of this kind of R & D will rise as water becomes increasingly in shorter supply and more costly to develop new sources. Some leaks, of course, will be easier to locate than others. Thus a challenge is posed to develop instruments of considerable sensitivity to locate the more-difficult-to-find leaks. The lack of an easily developable market for such instruments would tend to discourage private sector R & D. However, the possible transfer of instrument technology from other fields such as military, space, medical or chemical industries would possibly entail little cost for the requisite R & D to modify or adapt the technology to this use.

In arid or semi-arid regions of the United States where water is used for irrigation there are numerous opportunities for reducing water requirements of irrigated crops. One such kind of R & D would be to improve the technologies affecting the manner of applying irrigation waters to avoid waste. In certain sandy or highly permeable soils much of the irrigation water if too rapidly or frequently applied may drain rapidly past the effective intake zone of the plant's root system. Likewise, in clayey or impermeable soils too rapid an application of irrigation waters may increase the rate of surface evaporation or runoff. Experimentation is under way leading to favorable results in reducing irrigation water requirements by determining the properties of the soil, periodic measurements of its temperature and moisture content, and the empirical results on plant growth relative to the application of irrigation waters by various kinds of drop or spray systems or other application techniques. There is little doubt that more R & D of this kind can bring additional benefits in the reduction of requirements for irrigation water.

Other gains can be achieved through genetic experiments to breed plants that utilize irrigation waters more efficiently and hence achieve greater crop yields per thousand gallons of water consumption. Thus far much of the experimentation in improving crop yields has been directed to increasing yields per acre of land rather than per unit of water consumption. However, focusing on increased yields per acre may, as a by-product, increase crop yields per gallon of water consumed. For example, in Thailand and India about 1,800 gallons of water were used per pound of milled rice, although Japan used only one-half as much and California about one-third this amount of water [14]. Land is a more valuable input than water in agricultural production and hence a higher priority would appear to be warranted for the R & D objective of increasing yields per acre than achieving greater yields per unit of water consumption. Nevertheless, in regions of growing water shortage the price of water may increase to the point where crop irrigation is no longer profitable (as is already happening in certain parts of the Phoenix region).

In such instances a higher priority for R & D that would reduce the consumption of water per dollar value of crop produced would tend to preserve a high level of agriculture production to the benefit of the urban economy which provides goods and services to the surrounding agriculture sector both within and outside the political boundaries of the metropolitan area. It is important that this statement of social benefits focuses attention on R & D to the dual objectives of reducing water inputs and increasing the dollar value of crop yields. The latter, of course, would divert research priority to high-value crops and possibly to those which already consume less water per pound of crop produced. For example, certain evidence suggests that potatoes and tomatoes require only about 10 percent as much water per pound as grain crops and oranges require about 40 percent as much [15].

Not surprisingly, Israel has been a leading nation in research to conserve water in the growing of crops [16]. However, the United States would stand to reap even larger economic benefits in the long run from such a research program. For example, the entire land area of Israel (prior to the Six-Day War) is only 7,993 square miles or 5.1 million acres, only a fraction of which is suitable for irrigation. On the other hand, the United States in 1968 had 8.4 million acres under irrigation with considerable potential for expansion [17]. Since about 80 percent of irrigation water is lost to the atmosphere by evapotranspiration from the plants to which it is applied, it would appear that considerable R & D attention is needed in the breeding of new strains of plants that would reduce these losses.

Research is also needed to develop low-cost linings for irrigation trenches that do not crack. Attempts to improve on the use of fairly expensive concreting of trenches in highly permeable soils have led to high maintenance costs as well as high losses of water due to cracks introduced by thermal expansion and contraction or by cattle or other heavy animals breaking less durable types of lining material. Still another kind of water loss that has received research attention is that due to evaporation of water from reservoirs, especially in dry, sunny climates. For instance, the volume of water evaporated from Lake Mead was 90 inches or 800,000 acre feet in 1957 and evaporation from other reservoirs on the Colorado River, from the river channel itself, and Arizona reservoirs is estimated at 500,000 acre-feet per year [18]. However, experimental attempts to use nonvolatile liquid films to cover reservoir surfaces as a means of reducing water evaporation have not been successful. Moreover, there is a question in using such techniques as to the possibility of adverse environmental or ecological consequences. Thus the outlook for pursuing R & D to promote this technological alternative does not appear especially bright at this moment.

Reducing the Costs of Water Supply

As a research objective, reducing the costs of water supply overlaps all other research objectives including those discussed above. In those cases where the development of a new technology is the object of R & D, the ability of the new technology to compete with existing technologies will chiefly be determined by its cost-effectiveness. While the effectiveness of the technology in serving social and political as well as economic needs must be taken into account, the cost element is always part of the consideration. The R & D alternatives to be discussed in this section are those for which the primary emphasis will be on cost reduction rather than augmenting water supply, reducing water requirements, or some other objective.

Very often alternatives of this kind are implicated in R & D efforts to bring about evolutionary or short-reach improvements in existing technologies. Generally R & D of this kind is performed by those firms or agencies that have had a history of involvement in the development of the technologies upon which cost improvements are sought. Moreover, many of the technological advances of this kind are the result of ideas contributed by personnel not directly employed in R & D functions of their respective agencies.

To examine technological alternatives for reducing the cost of water supply one needs to contemplate the various technological components involved: earth excavating equipment, dam and aqueduct construction, well-drilling equipment, pumps, pipes, motors, water treatment facilities, construction materials and methods, explosives, and aerial mapping, geological and engineering methods of determining the best cost-effective sites for locating dams and aqueducts. Normal profit incentives associated with an improved competitive position are usually sufficient to induce private enterprise to seek cost-saving advances in technologies of these kinds. Nevertheless, some of these areas of R & D are of sufficient strategic importance to the national economy to merit policy attention from the Federal Government.

One of these is the development of less costly earth excavating technologies which not only affect the cost of water supply in the construction of reservoirs, aqueducts and urban water mains but also excavations for sewers, subways, highway cuts and tunnels, basements for high-rise buildings, and the mining of various kinds of minerals. A wide variety of possibilities need to be evaluated for R & D: conventional and nuclear explosives, mechanical diggers and crushers, and materials transport equipment including trucks, carry-alls, conveyor belts, bucket conveyors, and hydraulic transport. Of the above areas the ones receiving most of the Federal support for R & D involve nuclear explosives.

Since 1957, the United States Atomic Energy Commission has conducted a research program (Plowshare) whose aim is to foster the development of peaceful nuclear explosive or excavation technologies. Its potential applications range from underground mining and stimulation of natural gas well production to the excavation of canals, dams, storage reservoirs, highway or railroad passes through mountaineous terrain, and deep water ports [19]. Although there have been impressive advances in reducing the hazards of nuclear radiation and improving the predictability of the application of the technology, there has been a general reluctance to utilize its full potentialities.

The feasibility of the application of nuclear explosives to water management in Arizona is under joint study by the AEC and the Department of the Interior in response to a proposal by Governor Williams [20]. Although the results of this study are not known, it does seem likely that further advances in nuclear excavation technology can be of considerable benefit in reducing the cost of water supply for certain regions of the United States, particularly where the interbasin transfer of water will involve large earth excavation projects. Nevertheless, uncertainty over the political acceptance of the technology not only by the region where it is applied but also by the Federal Government which must ensure that stringent requirements of the Nuclear Test Ban Treaty are met looms as a major risk factor in assessing the probability that the R & D will achieve its intended social benefits. Cost-sharing formula as between different levels of government or with private industry are also a controversial matter requiring a broad technology assessment of various benefits, costs and risks including their spatial, institutional or personal distribution.

Regulating Water Flows

There are several important reasons for desiring to achieve a more even flow of water in watercourses throughout the year: (1) to eliminate or reduce damage due to flooding, (2) to achieve a steady supply of water for urban, industrial or agricultural uses in periods of drought, (3) to sustain a uniform absorptive capacity for non-injurious levels of urban or industrial wastes including treated wastes that do not have all impurities removed, and (4) to maintain adequate water levels in certain watercourses for such uses as water transport, recreation, and to preserve the esthetic quality of beaches and shorelines in urban areas that may suffer from low water levels which expose smelly and unsightly bottom surfaces.

If the purposes of regulating water flows are diverse, so also are the present or future technological alternatives for achieving these objectives. These alternatives include the construction of dams for impounding water; channelization of watercourses (to reduce flooding); the use of canals or aqueducts for the interbasin transfer of water; forestation and soil conservation control programs; pumping excess water into selected groundwells for recharging aquifers; weather modification to increase rainfall in periods of drought; and long-range forecasting of precipitation in watersheds to facilitate improved management of water levels in reservoirs to meet the needs of various users. Some of these technologies have had a long history of development and are relatively mature, affording limited opportunity for radical advances. On the other hand, weather modification and reliable long-term weather forecasting afford a potential for marked advances in technology although there is considerable uncertainty as to when these will mature and their ultimate level of effectiveness.

Because there are so many alternative technologies that can be used singly or in combination to regulate water flows and in view of the wide variety of social benefits and costs, their evaluation in a probabilistic framework of R & D planning makes a most interesting and highly complex challenge for applying the methodologies of technological assessment. One such case is the proposed dam in the North Branch of the Potomac River that will displace 130 persons from their present homes in Shaw, West Virginia [21]. Among the primary benefits envisioned by the Army Corps of Engineers are additional water supplies to offset Washington Metropolitan area shortages in summer droughts and protection against flooding such as occurred in 1955 when two hurricanes arrived in successive weeks. Secondary benefits to the local area from the recreational potential of the lake created by the dam are of questionable value because of uncertainty as to how much the lake waters will be contaminated by coal mine acid drainage [22].

Also controversial to conservationists is the need for the dam to serve its primary purposes since it is claimed that the Washington Area could meet seasonal water shortages by charging users higher rates in summer to discourage lawn sprinkling or by developing the tidal area of the Potomac as a source of water [23]. The latter alternative would require R & D to improve water purification processes to eliminate viruses now found in sewage treatment effluents reaching the tidal area. Since it would take about three weeks for water to travel the 170 miles from the dam to the Washington Area, reliable long-term forecasting of drought conditions is required to ensure getting the water to the treatment plants when most needed. Clearly, technology assessment applied to so many issues would be a fairly costly exercise scarcely capable of producing highly precise and reliable estimates of various outcomes associated with the different alternatives.

Protecting or Enhancing the Quality of Water and Related Land Areas

Much of the literature on protecting or enhancing the quality of water and related land areas deals with the subject of pollution and its control or abatement. Unfortunately — or fortunately in the eyes of the environmentalist — pollution has become an emotionally charged word. The verb "pollute" has been defined: "to make physically impure or unclean" [24]. However, in view of the complex and adversary nature of social values associated with different states of environmental quality discussed in Chapter 24, it would seem almost desirable to shun the word pollution for the purposes of scientific analysis, especially if the word is interpreted according to its conventional definition given above. For example, virtually all natural watercourses that are unpolluted by activities of man would nonetheless contain impurities such as dissolved salts, suspended solids, bacteria and other biological substances, and dissolved gases. Without some of these impurities water for drinking purposes would have an unacceptably flat taste. Moreover, certain of these organic and inorganic impurities are valuable in sustaining various kinds of marine life.

Another form of natural pollution is the turbidity produced by suspended solids or the reddish-brown color of tannic acid that results from the decaying roots of trees and other vegetation. Since man prefers water to be sparkling clear for most kinds of water sports, discolorations of water produced by nature or man constitute a form of esthetic pollution with adverse impacts on these recreational pursuits and related land values. Yet sport fishing is sometimes improved by this turbidity which provides concealment for both small and large fish. An executive in a steel company expressed not a little exasperation over the "impossible" task of removing reddish-brown iron oxides from the plant's effluent waters to please the environmentalist whereas within full view of his office window sportfishermen visibly preferred to fish in the turbid discharge waters than in the clear lake beyond the breakwater for the company's docks.

Still another form of human intervention in environmental quality is the so-called "thermal pollution" related to increased water temperatures from thermal discharges of electric power plants or other industrial coolant waters. Since the temperature increases resulting in watercourses from such discharges do not of themselves introduce impurities or unclean elements into water, this intervention does not meet the above definitional criteria of "pollution." However, if a more socially operational definition of pollution were to include any intervention by man's activities that would produce *net* adverse social effects, a different interpretation would then be possible.

For example, thermal discharges from plants may reduce the level of dissolved oxygen that is essential to certain forms of marine life. Thermal discharges may also produce temperature levels that are not *preferred* by certain fishes or marine life or produce an ecological environment (e.g., rampant algal growth) that is not conducive to the livelihood of some forms of marine life. Thus, while it is certainly true that an average annual increase in temperature in certain portions of a watercourse will alter the ecology of the affected area, it is much less clear whether the *net* social effects will be good or bad — and this is a far more crucial test of "pollution" than whether the environment or the ecology has been altered.

For instance, at Marquette, Michigan far more persons were observed swimming in the middle of July in the heated effluent waters from a thermo-electric power plant than in any other place on the U.S. coast of Lake Superior visited on an extensive field trip. Moreover, the heated waters also attracted bass and other fish that prefer warmer to colder waters and hence was a popular spot as well for the fisherman. Even in the warmer waters of Lake Erie a similar boon to sport-fishing was observed at Dunkirk, New York in the warm effluent waters of an electrical power plant which promoted the additional benefit of providing an ice-free port for small boat owners during the winter months. Yet in other regions of the United States it is less clear that thermal discharges of these kinds can have what appears to be *net* positive social impacts. The Turkey Point power plant south of Miami has been most controversial in terms of the adverse ecological impacts that its expansion would create and the same has been true of other locations as well. Yet, Bernard I. Spinard feels that [25]:

> With regard to thermal pollution, the available evidence is that ecologists have overstated the the hazards to aquatic biota. There are relatively few cases where power plant effluents have affected lake or river ecology except over a very limited stretch of water which would seem reasonable to reserve for cooling use. Small lakes, used as cooling ponds by utilities, have normal and flourishing ecologies. The real question is whether any group — power companies, fishermen, swimmers or nature preservation enthusiasts — should have the exclusive say over water use. To me, it seems reasonable that, if different uses are in conflict, allocation of waters can be arrived at by public decision. As an enthusiastic swimmer, I would prefer that power plant effluent be

> directed to swimming areas of Lake Michigan rather than used to heat the air; and I resent being told that my reaction is less noble than that of fishermen, in whose behalf my money is already being spent to raise hordes of trout in hatcheries.

This is not to say that one might not find numerous cases of thermal and other forms of environmental modification that have net adverse social impacts. Rather it is to state that avoidance of environmental modification should not be made into a religion but should be the conscious decision of society after a full-blown technology assessment indicates whether the net social effects are likely to be positive or negative. The case is well stated by Dr. Richard L. Meier [26]:

> Tracing the conceptual origins of pollution illustrates how it springs from the violation of taboos. This basis has shaped both the public health and esthetic order approaches to environmental management. In its extreme forms, the present environmental stewardship approaches a religion. The apparently innocuous worship of Nature may have damaging consequences for the rest of society, through restricting freedom of choice and delaying the construction of needed public facilities. Planners and policymakers must be aware of the consequences of current conflicts over environmental values and seek strategies to postpone destructive confrontations.

An important development in the past decade is that concern for the degradation of environmental quality has developed into strident public issues at the local, state, national and international levels. Citizen action groups have come to wield such significant political clout in governmental policies and programs dealing with environmental issues that a new term, "*participatory technology assessment*," entered into the subject of discussion by several speakers at an August 1971 Conference on Technology Assessment [27]. The potential inputs to technology assessment by citizen action groups is reflected in the estimated 3,100 environmental organizations in the United States which include 2,500 local groups but exclude civic, church, or school groups and local chapters of national organizations [28]. There are also two important international organizations that deal with environmental matters, namely, The International Union for Conservation of Nature and Natural Resources (IUCN) and the recently established Special Committee on Problems of the Environment (SCOPE) of the International Council of Scientific Unions [29]. Moreover, several states — notably, Michigan and Illinois — have enacted laws giving citizens access to the courts to protect the environment [30].

However, if these citizen action groups are politically potent in protecting the natural environment and the social values inherent in preserving nature in a relatively pure state, it is also evident that a very high price is often extracted from the public especially in the form of prolonged delays in the construction of needed public service facilities. According to Meier [31]:

> In 1970, it was noted by engineers that every large public utility facility programmed for metropolitan New York was held up by litigation, federal investigation, or demonstrations by activists. The list included power plants, housing programs, jetports, harbor development, and freeway construction, all of them requiring two to five years to complete. This means increasing delays in air and ground travel, continuing scarcity of housing, and shortages of power. It could also mean a reduced ability of the metropolis to compete, an accelerated moving away of productive jobs, leading to the beginnings of a "depressed area" syndrome.

In part, the high price being paid by our society in pollution control and abatement efforts is due to the long years of neglect of environmental considerations and the inefficiency that often accompanies an atmosphere of crisis planning. Moreover, the inputs to the intellectual and political processes of participatory technological assessment by environmental extremist groups also contribute in an unbalanced (though democratic) way to the high costs of protecting or enhancing environmental quality by being too narrowly based on the special interests of a relatively limited segment of society. Hence their inputs constitute a highly restricted analysis and information base that yields maximum support for these special interests. Indeed, Dr. Walter W. Heller in a paper, "Coming to Terms with Growth" at a Forum on *Energy, Economic Growth, and the Environment* expressed concern over the elitism inherent in certain forms the environmentalist movement is taking [32]. Specifically, Heller called for taking the social planning for growth out of the one-dimensional context of the natural environment. In a broader context, he felt that environmental claims for a share of funds from the economy's growth proceeds should be directed toward:

Cleansing the physical environment of air, water, and land pollution and of urban congestion and sprawl;

Cleansing the social environment of the cancers of poverty, ignorance, malnutrition, and disease;

Cleansing the human environment of the urban ghetto and rural slum that blight not just the territory they occupy but the lives they hold in their vise;

Cleansing our personal environment of the fears of crime and violence. Meier also drew attention to the elitist nature of the environmental thrust [33]:

> Generally it is the poor who suffer the increased insecurity of employment in return for the benefits of better air, less noise, and more natural scenery that they did not ask for in the first place, or put far down on their priority list if they did. The intellectuals and the professional classes are in a position where they can usually evade the new inconveniences pollution control imposes upon them.

Thus, a responsible role for technology assessors and R & D planners in addressing the problem of protecting or enhancing the quality of water and related land areas is to take into consideration the broadest range of social impacts of technological alternatives. In particular, criteria for public sponsorship of R & D related to environmental measures of these kinds deserving high priority would include those projects which hold promise of: (1) reducing tensions between various groups in our society arising from environmental issues and present techniques utilized in dealing with them; and (2) reducing the high social and economic costs of pollution abatement measures. In addition to the high socio-economic costs of environmentalist-induced delays of constructing needed public services cited above for which no overall value has been inputed, it is estimated that $38 billion will be required during 1970-75 to meet the costs of water pollution abatement alone and $105 billion for the abatement of air and water pollution plus solid waste management [34].

Despite these high costs it is felt that our society will ultimately prove its willingness to spend the sums necessary to halt environmental degradation and repair its damages as witness the sacrifice of several hundred billion dollars of annual personal income by Americans in going from a work week averaging 70 hours in 1850 to the present 40-hour work week in order to receive the benefits of increased leisure [35]. Today the value of that leisure has been substantially impaired for an increasingly recreation-oriented society by the pollution of water and related land areas. Since many social tensions arise over the damage to recreation and esthetic values resulting from such pollution and the costs of repairing this damage are large, then advances in science and technology that will facilitate the reduction of these costs will also contribute to the reduction of social tensions. This social ferment is strongest in urban areas not only because this is where three-fourths of Americans live but also because industries located in urban areas and municipal waste disposal are major contributors to the degradation of the quality of water and related land resources.

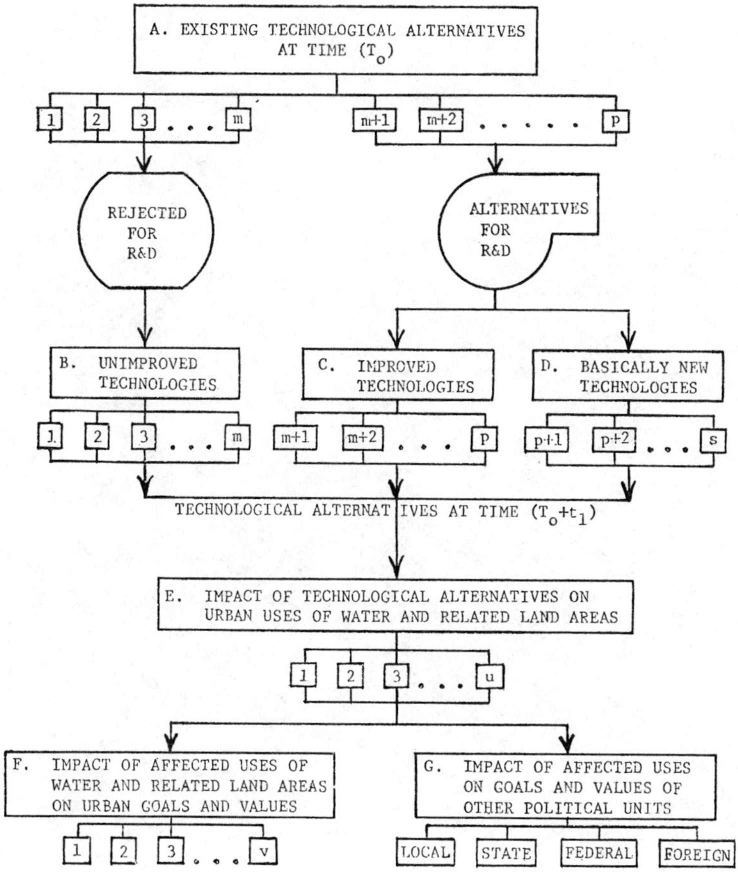

Fig. 1. A diagram relating technological alternatives considered in R & D planning decisions to impacts on water and related land uses and on urban goals and values.

TABLE 1. A Matrix for Evaluating the Impact Relationships Between Urban Uses of Water and Related Land Areas and Urban Goals and Values*

Urban Goals or Values / Urban Uses of Water and Related Land Areas	A. Protect or Enhance Environmental Quality						B. Protect or Finance Socio-Economic Values														
	1. natural environment			2. man-made environment			1. social values									2. economic development					
	a. Water, Air & Land Quality	b. Natural Land & Water Forms	c. Natural Life (Ecol.)	a. Patterns of Land Use	b. Buildings & Engr. Structures	c. Noise	a. Esthetic & Amenity Values	b. Community Identity & Cohesion	c. Health, Safety & Security	d. Recreation & Use of Leisure	e. Cultural & Spiritual Values	f. Social Justice	g. Freedom, Tranquility & Opportunity	h. Income Distribu'n & Econ. Stability	i. Population Growth & Distribu'n	a. Transportation	b. Housing	c. Public Services & Utilities	d. Commerce	e. Industry	f. Agriculture
1. Household & commercial uses of water																					
a. Drinking & food preparation							1-5	1-2	1-5			1-2			1-5		1-3		0-1	0-1	
b. Bath & laundry	1-5		1-5	1-3			1-5	1-3	1-2	1-5	0-1	1-2	1-4	1-2	1-3		1-2		0-2	0-2	
c. Sanitary purposes	1-5		1-5	1-3			1-2	1-2	1-5	1-5	0-1	1-2	1-4	1-2	1-3	0-1	1-3		0-2	0-2	
d. Lawns, gardens, pools, etc.	0-1		0-1	0-2	0-1	0-1	2-4	1-2	1-2	1-2	1-2		1-3		0-1	0-1	0-1	1-2			
2. Industrial uses of water																					
a. Consumptive uses	0-1		0-1							0-1								0-1		1-2	0-1
b. Cooling waters	1-5		1-5		0-1		0-1	1-3	0-1	1-3	0-1	0-1	1-2	0-1		0-1		1-2		1-2	
c. Cleansing & waste discharge	1-5	0-1	1-5	1-3			1-5	1-5	1-4	1-5	0-1	1-2	1-5	1-2	1-4				0-2	1-5	
3. Municipal uses of water																					
a. Fire fighting			0-1						1-3			0-1				0-1	0-1	0-1	0-1	0-1	0-1
b. Street cleaning	0-1		0-1				1-2	0-1			0-1	0-1							1-2		
c. Public greens & gardens	0-1		0-1	1-2	0-1	0-1	2-4	2-4	0-1	1-3	1-2	0-1	1-3		0-1		0-1	0-1	0-1	0-1	
d. Fountains, ponds & lakes		1-2	1-2	0-1			2-4	2-4	0-1	1-3	1-2	0-1	1-3		0-1	0-2	0-1	0-2	1-2	0-1	0-1
e. Public swimming pools				0-1			1-3	2-4	1-2	1-3	0-1	1-3	1-3				0-1		0-1	0-1	
f. Drinking water							1-5	1-2	1-5			0-1			0-1				0-1	0-1	
g. Sanitary facilities	1-3		1-3				1-2	1-2	1-3	1-3	0-1		1-3	1-2	0-1						
4. Recreational uses of watercourses																					
a. Swimming & water contact sports				0-1		0-1	1-2	1-3	1-2	2-4		1-3	1-3		1-2		0-1		0-1	0-1	
b. Boating	0-1	0-1	0-1	0-1	0-1	1-2	1-2	1-2	1-2	2-4	0-1	1-3	1-3		0-1		0-1		0-1	0-1	
c. Sportfishing					0-1	0-1	0-1	1-2	1-2	2-4	0-1	1-3	1-3	0-2	0-1				0-2	0-1	
5. Recreational uses of water-related land areas																					
a. Parks and trails		1-3	1-3	1-2	0-1		2-4	2-4	1-2	2-4	1-3	1-2	1-4		0-1		0-1		0-1	0-1	
b. Scenic vistas		1-2	0-1	2-4	2-4		2-4	2-4	0-1	1-2	1-3	0-1	1-2		0-1		0-1		0-1	0-1	
c. Golf courses & sports arenas	0-1	0-1		1-2	1-2	0-1	2-4	2-5	1-3	2-4	1-3	1-2	1-3		0-1	0-1	0-1	0-1	1-2	0-1	
d. Natural habitats or preserves	1-3	1-5	1-5	103		1-2	1-3	1-3	0-1	0-1	1-3	1-2	1-2								
6. Water transportation	0-3	0-3	0-1	0-3	0-1	0-1	0-2	0-1	0-1	0-3	0-1		0-1	0-2	0-1	0-3	0-1	0-1	0-2	0-4	0-2
7. Land transportation	1-2	1-2	1-2	3-5	1-3	2-3	1-2	2-5	1-4	2-4	1-2	2-5	2-5	2-4	2-5	3-5	2-3	1-2	2-4	2-4	1-2
8. Air transportation	0-1		1-2	0-2	0-1	1-4	1-2	1-2	0-1	1-2	1-2	1-4	1-3	0-2	0-2	0-3	1-2	0-1	0-2	0-3	0-1
9. Agricultural uses of water	0-3	0-2	0-3	0-1			0-1		1-2			0-1		0-1				0-1	0-2	0-1	0-5
10. Agric. uses of land	1-3	1-2	1-3	1-2		0-1	1-2	0-1	0-1	0-1	0-1	0-1	0-1	0-1	1-2				0-2	0-1	1-4
11. Construction on land	1-2	1-2	1-3	2-5	2-5	1-3	3-5	3-5	1-2	2-4	1-3	2-5	2-5	1-3	1-3	1-3	1-3	2-5	2-3	2-4	0-1
12. Construction in water & landfills	0-3	0-5	0-5	0-3	0-5	0-2	0-5	0-4	0-2	0-5	0-3	0-5	0-5	0-2	0-5	0-4	0-5	0-3	0-3	0-2	
13. Mining on land	0-5	0-5	0-3	0-1	0-2	0-1	0-5	0-1	0-2	0-3	0-2	0-2	0-2	0-2	0-1	0-2	0-1	0-3	0-3	0-3	0-1
14. Mining in water	0-3	0-2	0-4		0-2		0-5	0-1	0-2	0-1	0-1	0-2	0-4	0-2	0-1	0-2		0-3	0-3	0-3	

*Index ratings range from "0" (insignificant) to "5" (highly important). A vacant cell denotes an insignificant impact.

Because of the demonstrable increase of social concerns over environmental issues and the public laws and programs being implemented to deal with environmental problems, there has been an upsurge of interest in both the private and public sectors in sponsoring R & D in this field. Since there are a very large number of old and new technological alternatives in treating environmental quality problems, the evaluation of many of the possibilities for R & D is an exceedingly complex and difficult exercise. The efforts of R & D are directed toward improving existing technologies or creating basically new ones. Thus the value to society, and hence marketability, of the outputs of R & D alternatives involves a comparative evaluation of the state of technologies at some future time ($T_0 + t_1$) as between (1) basically new technologies, (2) improved technologies already in use, and (3) unimproved technologies (see Fig. 1). In this evaluation of alternatives two levels of impact analysis are involved. The first of these (category E in Fig. 1) includes the impact of the technological alternatives on urban uses of water and related land areas. The second involves an estimation of the positive or negative impacts of the affected uses of water and related land areas on urban goals and values (category F).

It is noted that the use-goal impacts inherent in the relationships between categories E and F have been tentatively evaluated (see Table 1). The latter evaluations represent the potential importance to urban goals of planned *changes* in water and land uses in various urban areas or *new* measures to *prevent* changes in these uses. Technological alternatives are among the key instruments of achieving desirable use-goal impacts, but a coordinated analysis of four basic planning modes in addition to the R & D planning mode is required. For the most part, advances in science and technology resulting from the exercise of the R & D planning mode can be expected to accelerate or enhance the use-goal impacts in the matrix. This would seem to be particularly true in the case of technological alternatives for protecting or enhancing the quality of water and related land areas if the R & D reduced the costs of the technological alternatives or made their application less politically controversial in other ways.

The potential application of present or future technologies to serve the goals and values of a particular urban area must also be based on analysis of their impact on the goals and values of other political units. These include other urban communities either upstream or downstream, non-urban countries, states, the federal government, and foreign nations (category G, Fig. 1). Federal or State laws have already been enacted to protect national or state interests against pollution damages including those of urban areas. For example, the State of Virginia's water quality agency has authorized filing a law suit to "force the Federal Government and the Maryland suburbs to act immediately to clean up the filth-laden Potomac River" [36]. Moreover, where national interests are involved positive as well as negative inducements may be implemented to encourage the utilization of desirable technological alternatives affecting the quality of water and related land uses in urban areas.

That the evaluation of R & D planning alternatives for protecting or enhancing water quality is no small task is seen in the legion of existing technological alternatives that have found application in treating different kinds of pollution problems. Kneese and Bower have provided an excellent review of many of these technological alternatives [37]. A partial list of technological measures derived from the treatment of Kneese and the treatment of Kneese and Bower include:

1. *Measuring the quality of waste discharges*. Biochemical oxygen demand (BOD), chemical oxygen demand (COD), aklalinity, acidity, color, hardness, taste, odor, turbidity, and electrical conductance.

2. *The treatment of water for various in-plant industrial uses*. Sedimentation, filtration (sand, diatomaceous earth, carbon), chlorination, ozonation, iron and manganese removal, silica removal, slime and algae control, coagulation, aeration and deaeration, dealkalization, demineralization, neutralization, and softening (lime-soda, sodium zeolite, hot-process phosphate).

3. *The treatment of water for domestic uses.* Sedimentation including the use of polyelectrolytes, filtration including the use of activated carbon, disinfection using ozone and various chloride compounds, and other methods under development.

4. *Methods for treating industrial and municipal wastes affecting water quality.* Screening, flocculation, chemical coagulation, flotation, sedimentation, centrifuging, filtration, stripping, neutralization, chemical oxidation, chemical reduction, wet oxidation, fermentation, emulsion breaking, evaporation, distillation, incineration, biological filtration, activated sludge, anaerobic digestion, stabilization lagoons, spray irrigation, disinfection, ion exchange, electrodialysis, reverse osmosis, and aerobic digestion.

5. *Discharge of thermal waste loads.* Various types of air or water cooling systems.

6. *Methods for reducing wastes generation to improve the quality of receiving waters.* Change in type of raw material inputs, change in production process, change in product outputs, and in-plant recirculation of water.

7. *Methods for reducing wastes after generation to improve the quality of receiving waters.* Materials recovery, by-product production, waste treatment (see above), and effluent reuse including groundwater recharge and waste water reclamation or renovation.

8. *Methods for increasing or making better use of assumilative capacity of the watercourse to improve the quality of receiving waters.* Addition of dilution water, multiple outlets from reservoirs, reservoir mixing, reaeration of streams, barriers against saltwater intrusion, and effluent redistribution including regulated discharge.

There is a profusion of literature that describes the advantages and disadvantages of these and other methods for protecting or enhancing the quality of water for urban uses and a review of these is beyond the scope of the present pilot study. The above methods of protecting or enhancing water quality are deliberative process procedures involved in water quality management. Another source of water pollution dealt with only incidentally by Kneese and Bower is the subject of *accidental* spills or discharges of wastes into watercourses. Much of the literature on this subject deals with the rather spectacular spills of oil in the ocean due to accidents involving tankers or offshore oil production. However, there are also numerous, if less dramatic, incidents of accidental discharges of contaminants in inland water bodies including estuaries, rivers, streams, lakes and groundwater aquifers.

Some of these involve equipment or human failures in the operations of industrial plants and others involve transportation accidents including marine vessels, railroad tank cars, truck tank cars, and pipeline facilities. A great deal of fervent effort is being made these days in developing new technologies and management procedures (a form of technology) for the prevention and control of accidental spills or discharges into watercourses whether inland or offshore in the oceans [38]. In general it is observed that technologies for the prevention of such accidental sources of water pollution and emergency containment or clean up measures are quite different, for the most part, from the technologies discussed above except perhaps for water quality monitoring, measuring, or warning systems. Accordingly, expertise in the technologies of accident prevention and emergency containment or clean up of spills is not usually in the province of the traditional skills of water resource specialists but rather those of industrial process or transportation engineers and a new breed of company or corporate department engaged in developing or marketing services and specialized products for preventing accidents and minimizing their adverse impacts. Included in the scope of application of such technologies is the detection, containment and removal of wastes deliberately dumped in a clandestine, and frequently unlawful, manner. The legal mode of planning is now being articulated in an effort to reduce clandestine discharges of wastes and accidental spills due to carelessness.

Within the urban area there are several other sources of water pollution that invite consideration of R & D alternatives. One of these is that many cities have combined storm drainage and sewage gathering lines that feed into the various municipal waste treatment plants. This technology works quite well until there is an unusually heavy storm whose water runoff surge exceeds the capacity of the waste treatment system. This requires raw sewage to be dumped into a river or lake with disastrous consequences for marine life and human safety in water contact recreation. The separation of storm drainage and sanitary sewage systems is a highly costly alternative running into the millions of dollars not counting the indirect costs due to disruptions to urban transportation and commerce during the period when streets are dug up to lay new lines. The Water Quality Act of 1965 established a 4-year program of R & D grants to demonstrate new or improved methods to eradicate the combined sewer problem leading to an appropriation by Congress of $20 million annually for the period FY 1966-69 [39]. Yet this amount is scarcely large in relation to the estimate of the American Public Works Association that it will cost between 15 to 48 billion dollars to remedy overflows from combined sewers [40].

Another urban pollution problem arises from airborne particulates and other contaminants that fall on roof tops, paved streets and sidewalks. During rainfall these are washed into storm sewers and may ultimately reach adjacent watercourses if urban sewage treatment systems do not remove them. Included are industrial pollutants, contaminants from automotive exhaust or engine leakages, and salts deposited on streets to melt snow or ice. The extent of the pollution of watercourses from sources of these kinds and their ecological and economic damages is not yet known. However, the fraction of urban areas covered by impermeable construction materials rather than green areas or soils is sufficiently great to suggest that these sources may contribute significantly to water pollution. This problem requires further R & D to determine its causes and effects and ways of reducing adverse impacts.

Soil erosion due to farming and highway and urban construction contributes more pollutants by weight than any other source. Although costs for erosion control on farms are not known, the Environmental Protection Agency (EPA) estimates highway erosion control costs to range from $130 million up to $7 billion annually [41]. The cost of controlling acid mine drainage was previously mentioned to cost $7 billion while the EPA estimates that $12 billion will be required for municipal waste treatment and related facilities [42]. The EPA also estimates that in 1970 the total costs of industrial waste treatment were $1.05 billion of which the following percentages were incurred by different industry groups: primary metals (23); chemicals (18); food and kindred products (15); paper (11); transportation equipment (10); petroleum (9); machinery (5); textiles (5); rubber and plastics (1); and all other manufacturing (3) [43]. An increase of 77 percent in these costs is estimated to be required to a level of $1.86 billion annually to meet 1974 water quality standards [44]. The largest relative increases will be required for the paper industry, rubber and plastics, and chemicals.

A different kind of problem is the need to protect or enhance the quality of water-related lands in urban areas. In part, the problem derives from the frequent deposition of solid wastes from urban and industrial sources in low-lying marshlands or in other areas where water runoff from precipitation carries pollutants from these solid waste dumps into watercourses. The problem is exaggerated because of the disappearance of suitable dumping sites such as abandoned quarries or mines. Landfills in coastal areas of lakes or estuaries also result from the desire of urban developers and local governments to obtain additional land in strategic areas of the city for a variety of uses including residential, commercial and industrial buildings, highways, airports, ports and marinas, and recreational and park facilities.

Despite the substantial social benefits that may derive from these landfills, these may also bring about a degradation in the quality of the environment for fish and other wildlife which depend on marshlands for breeding and nursery areas. Landfills, if improperly designed or located, may adversely affect the circulation pattern of estuaries and lakes creating stagnant conditions or interrupting the drift of sedimentary materials leading to starvation of beaches for sands or the unwanted build-up of sandbars that are hazardous to navigation. The use of spoilbank materials from harbor or channel dredgings for coastal landfills reduces the cost of the fills, but can lead to an undesirable source of water pollution unless these are properly bulkheaded prior to landfill operations. Research is needed to determine the adverse effects of different kinds of coastal landfills or solid waste disposal methods on the quality of adjacent water and related land areas as well as the development of technologies for reducing undesirable impacts.

The problem of dealing with the protection and enhancement of the quality of water-related land areas would appear to be far less complicated in terms of evaluating technological alternatives than the matter of water quality discussed above. The very large investments that are estimated to control various kinds of urban and industrial water pollution whether accidental, clandestine or process-induced may serve as a sizeable incentive to R & D since the large markets hold promise of substantial returns if the R & D is successful. On the negative side is that the prospective large markets and supporting pollution control legislation and enforcement programs have stimulated such intense competition to develop new technologies that the risk is rather formidable that many of the technologies being developed may not be the preferred ones. Moreover, the cost of the R & D to develop technologies for pollution abatement or control may also be relatively high for certain of these. Meaningful assessments of many of these technological alternatives must be coordinated in a systems analytic framework that includes both positive and negative impacts on various urban uses and goals. This will be dealt with in the final section of this chapter.

EVALUATING TECHNOLOGICAL ALTERNATIVES IN MEETING URBAN OBJECTIVES FOR WATER USES

Some of the urban uses of water are subject to potential technological developments that could alter substantially the requirements for water supply, its quality, and regularity of flow. A number of these are blue sky research possibilities that may be a long time, if ever, in their successful commercial development. Others already have been undergoing a measure of experimentation and application. However, their further successful development may have profound implications for urban water uses and requirements for water quality or permissable standards of waste discharges. A few of these potential innovations shall be mentioned only briefly [45].

Indications are that the technological revolution in agriculture is far from over. Genetic engineering may produce greater agricultural yields reducing agricultural requirements for water thus releasing water and related land areas for urban use. Experiments are in progress to manufacture synthetic proteins from petroleum using enzyme processes. This too could change the requirements of agriculture for land and water and reduce agricultural sources of water pollution. Moreover, the successful development of the hydrogen fusion process for generating low-cost electricity and related technological developments of batteries and fuel cells might usher in the electric automobile displacing requirements for petroleum to power automobiles. This would reduce water pollution from oil refinery operations or accidental oil spills. It might also reduce the need for coal mining and the related problems of pollution from acid mine drainage.

The continued development of new plastic materials could further replace requirements for steel and other metals that contribute more heavily to water pollution than do plastic manufacturers. The development of new technologies of mining and processing of metals from ocean sources (e.g., copper, nickel, cobalt and manganese from the so-called manganese nodules of the deep sea) could reduce requirements for the onland mining of minerals with reduced water requirements and pollution effects. Further advances in the manufacture of artificial turf might make this attractive for use in public greens and residential lawns thus reducing the requirements of water for lawn sprinkling. Revolutionary developments in laundry methods could likewise reduce water requirements or the amounts of detergents — or phosphates in detergents — which are a significant source of water pollution.

New forms of underwater sports are developing, or may develop, which would benefit from reduced turbidity in watercourses. Radical advances in marina construction and small boat storage could greatly increase recreational boating and hence pressure for cleaner water. Further advances in nuclear excavation technologies could reduce their cost and increase their utilization in excavating ports, canals and enlarged watercourses that would increase marine transport and water sports. The development of new structural designs for economic and highly functional floating cities, ports and airports as well as underwater habitats could stimulate increased activities in water and related land areas. Further developments in radical forms of marine transport such as air cushion vehicles, hydrofoils, LASH and others could stimulate marine transportation and passenger traffic.

Still another kind of development that would affect standards of water quality and waste treatment systems is that of aquaculture. Ryther and Bardach define aquaculture as "the rearing of aquatic organisms under controlled conditions using the techniques of agriculture and animal husbandry" [46]. The breeding of trout and salmon with superior qualities has been successfully performed by Dr. Lauren P. Donaldson of the University of Washington at Seattle. Some of these have already been transplanted in Lake Michigan with spectacular results for the growth of sportfishing and the related tourist industry. Further research in hybridization or genetic improvement in fish species might pursue one or more of the following objectives [47]:

To increase rate of growth at a given level of nutrition.

To improve resistance to disease, parasites, or pollutants.

To improve texture, flavor, color, or odor.

To increase survival rate in the face of harsh environmental conditions such as storms or the prevalence of predators.

To increase the tolerance of species among themselves and between species.

To increase procreative abilities in either natural or confined habitats.

To improve tolerance to waters of higher temperatures (e.g., coastal areas of thermal pollution from coolant waters of powerplants).

To improve the effectiveness of response to gating or coralling mechanisms, or to other means of containment.

To improve responsiveness to acoustic signals or other means of attraction to reduce the cost of harvesting.

Moreover, the addition of nutrients to increase aquaculture yields is a possibility for further R & D. Dr. John Ryther is conducting research at Woods Hole Oceanographic Institution with the ultimate objective of leading to a commercial system in which secondary-treated urban sewage would be used to fertilize algae which then will be fed to shellfish such as oysters producing both food and tertiary sewage treatment [48]. Being considered for future research of this kind are salmon, trout, haddock, and lobsters. Another research project is being conducted cooperatively between the National Marine Fisheries Service (NMFS) and the Ocean Systems, Inc. with the aim of investigating the feasibility of rearing and marketing pan-size salmon using aquaculture techniques [49]. The fish will be reared to smolt stage in freshwater ponds, heated to speed their growth during winter months, then transferred to floating saltwater pens in Puget Sound to reach marketable size only 14 months after hatching. At Humboldt State College in Northern California aquaculture experiments are underway to test the possibility of using warm water effluent from nuclear power plants for crab aquaculture and to use sewage as fertilization of fish ponds for rearing salmon and trout [50].

There are already a number of practices where thermal discharges are being used in aquaculture to accelerate the growth of fishery products. In Great Britain the White Fish Authority is using heat discharges from the Hunterston Nuclear Power Station in rearing sole, lobster, shrimp, crabs and oysters; in Japan mariculture is being conducted using thermal discharges for the growth and propagation of shrimp, yellowtail, eel, and whitefish since 1964; in Iceland geothermal water is mixed with cooler waters in raising salmon; and in the United States thermal discharges are being used in catfish farming [51]. At Louisiana State University experiments are underway to evaluate the feed conversion rates by post larval crustaceans such as penaeid shrimp of microcapsules developed from brewery mash and other food processing wastes with various kinds of pellet binders [52]. A somewhat different twist is a process developed by William B. Davis of Texas A & M University to convert industrial materials that pollute rivers and streams — especially those from paper pulp mills and petrochemical plants — into pellets that have been readily accepted by cattle and chickens [53].

Ralston Purina is establishing research facilities at Florida Power's Crystal River generating site to develop commercial production of certain salt water fish and shellfish using thermal discharges from the two fossil-fueled units and a nuclear plant now under construction [54]. In West Florida a nuclear power installation on Fisheating Creek is using cobalt irradiation to convert raw sewage to potable water in a pilot operation that destroys viruses and bacteria, decomposes pesticides, reduces phosphate and other undesirable nutrient and produces an effluent that is immediately suitable for reuse with no measurable radiation [55].

Many other examples could be found but these are sufficient to indicate that there is a vast potential for advances in science and technology that could impact the use-goal relations for urban areas affecting requirements for water supply and waste treatment. Of course, the time scale and probability of commercial success of many of the above examples is a highly debatable matter in many instances. This not only poses a problem of technological forecasting but the forecasting of the institutional measures and changes that may also be required to facilitate such developments assuming a technological assessment would show these to be socially desirable.

RESEARCH IN PLANNING AND INSTITUTIONAL MEASURES

Insofar as it is desirable that water supply performance and related use objectives be considered in a single system of interrelated decisions, it is seen in the above partial treatment that the assessment of relvant technological alternatives represents a complexity that tends to overwhelm the analyst. Not only is this a matter of frequently encountered information gaps and the normal hazards and uncertainties of technological and institutional forecasting. No less important is that the scope of needed expertise and intellectual effort exceeds the capacity of a single individual. In order that the collective intelligence of a diverse group of planners, forecasters and analysts can be applied in systems evaluations of these kinds it is essential that a conceptual analytical model be developed which facilitates the convergence of decision making on at least the issues and tradeoffs of the technological alternatives that are of major social importance.

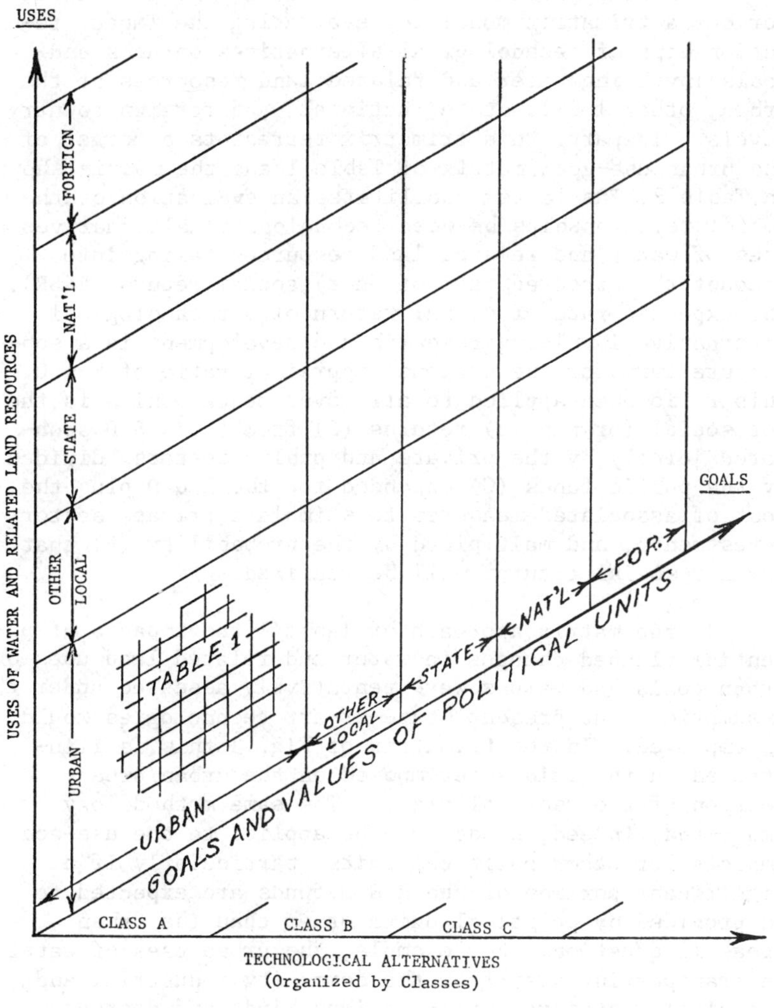

Fig. 2. A trimatrix model for evaluating the impact relationships of technological alternatives on uses and goals involving water and related land resources at the urban, other local, state, federal and international levels.

Such an approach is illustrated in Fig. 2 which describes a trimatrix model for evaluating the impact relationships of technological alternatives on uses and goals involving water and related land resources at the urban, other local, state, national, and foreign country levels. In part, this trimatrix represents a merger of the urban use-goal matrix of Table 1 and the matrix shown in Table 2. The latter facilitates an evaluation of the impact relationships between technological alternatives and uses of water and related land resources taking into account the expected rate of (net) social return (ERSR). The expected rate of social return of a technological alternative involving research and development is a substitute term for the national appraisal ratio of R & D. This ratio when applied to all governmental units is the net social (or public) returns (R) from the R & D sponsored jointly by the private and public sectors, divided by the public funds (C) expended for the R & D plus the cost of associated measures to stimulate private sector investments, and multiplied by the probability (P) that the net social returns will be realized.

In the matrix approach of Table 1 the impacts of potential planned changes in water and related land uses on urban goals and values were tentatively assessed under the assumption that present state-of-art technologies would be employed. In the trimatrix of Fig. 2 this is illustrated in the grid superimposed on the urban/urban section of the use-goal plane. The same methodology can — and, indeed, needs to — be applied to the use-goal impacts for other political units, particularly if a significant portion of the R & D funds are expected to be provided by political units other than the urban areas in question. For example, the urban uses of water in transporting away from the community industrial and municipal wastes receiving various kinds and degrees of treatment can impair water uses of other political units downcurrent as far as, and including, the ocean. Thus, R & D on behalf of pollution abatement measures may benefit persons or institutions in other political jurisdictions even more than certain urban areas which may bear more heavily the economic and social costs of implementing the pollution abatement technologies. In

TABLE 2. A Matrix for Evaluating the Impact Relationships Between Technological Alternatives and Uses of Water and Related Land Resources for Urban, Other Local, State, Federal and International Levels Taking into Account the Expected Rate of Social Return.

TECH'L ALTERNATIVES / USES OF WATER AND RELATED LAND AREAS	CLASS A																CLASS B															
	A_1^a				A_1^b				A_2^c				A_2^d				B_1^e				B_1^f				B_2^g				B_2^h			
	P	R	C	ERSR	P	R	C	ERSR	P	R	C	ERSR	P	R	C	ERSR	P	R		ERSR	P	R	C	ERSR	P	R	C	ERSR	P	R	C	ERSR
Urban Uses (see p. 124)																																
1. HOUSEHOLD																																
a.																																
d.	s	m	m	s	m	h	h	m																								
2. INDUSTRIAL																																
.	h	s	s	s	s	h	m	m	h	m	s	h	s	m	s	s	m	h	h	m	s	h	h	s	s	h	m	m	m	m	s	m
.																																
14. MINING																	m	h	m	m					s	m	m	s				
Uses by Other Local Units																																
1.																																
2.																																
.																																
.																																
L																																
State Uses																																
1.																																
2.																																
.																																
.																																
S																																
Nat'l Uses																																
1.																																
2.																																
.																																
.																																
N																																
Foreign Uses																																
1.																																
2.																																
.																																
.																																
F																																

Legend

For A_i^j

A = class of technology
i = specific technology within class
j = specific set of institutional measures to accompany A_i

P = probability that net social returns will be achieved if R&D is performed
R = net social returns if R&D is successful
C = cost of the R&D and other stimulative measures

ERSR = expected rate of (net) social return = $P\frac{R}{C}$
Symbols used in matrix cells:
s = small m = medium h = high

other words, the political feasibilities of raising funds for R & D and other supportive measures as well as the passage of laws and activities of other planning modes must be assessed from the standpoint of their social impacts on all affected jurisdictions.

Accordingly, by grouping the water uses and goals of various political units in distinct segments of the trimatrix it is possible to identify the potential voter interests of different kinds of constituencies. This in turn facilitates the essential intrusion of the political process in technological assessment with the result that greater social equity and justice is achievable as opposed to the unbalancing nature of participatory technology assessment where there may be a tendency to give undue weight to the more vocal of the special interest groups. The use of the trimatrix approach in technology assessment will also provide much useful information for the planning and management of water and related land resources as well as its intended use for appraising R & D alternatives vis a vis current state-of-art technologies.

Yet the high cost of technology assessments will probably for many years include a full-blown assessment of the kind envisioned in the trimatrix approach here presented. For example, Joseph F. Coates of the National Science Foundation estimated that a full-fledged technology assessment, not involving experimental work or the generation of new data, will cost between $150,000 and $250,000 [56]. Each of these would be a single line item (or slice of the assessment trimatrix of Fig. 2). One technology assessment conducted by the Stanford Research Institute under NSF sponsorship dealt with the prospective technological alternative of increasing water supply in the Colorado River Valley by the augmentation of snowfall through weather modification [57]. A technological assessment performed by the National Academy of Sciences at a cost of $300,000 involved the appraisal of a proposal to extend the Kennedy Jetport into Jamaica Bay. Indeed, the preparatory studies of the potential impacts of the Alaska Pipeline for filing the 102 environmental impact statement required by law cost around $8 to $10 million.

In view of the prohibitive cost of performing a full complement of technical assessments for the legion of R & D alternatives relating to water and land uses inherent in the system analytical framework of the trimatrix model it would appear desirable to develop preliminary assessments on an inexpensive basis using relative quick and crude judgments. This method would assist in identifying more profitable areas of the matrix that would merit more detailed analysis including experimentation and the generation of new information where present sources are inadequate.

The preliminary judgment approach in the use of the trimatrix model can be illustrated with a few examples. For instance, one class of alternatives might involve public support and encouragement for R & D that would reduce water requirements for urban use (Class A in Table 2). A specific technological alternative (A_1) within this class might be the stimulation of R & D in the private sector that would yield a low-cost, attractive artificial turf which, if adopted by a sufficient number of households, could bring about a substantial reduction of water consumed in lawn sprinkling. In line 1d of Table 2, two different sets of public measures are assumed to be employed to stimulate the R & D with each requiring a separate appraisal of the expected rate of social return (ERSR). In set (a) it is assumed that cost-sharing by the Federal Government of R & D performed by the private sector is offered as an inducement as well as tax abatement incentives, say, in the first years after the construction of the essential manufacturing facilities. Despite the medium costs of these stimulative measures, uncertainty over customer acceptance may afford only a medium estimate of net social returns (R). The latter estimate is derived by first evaluating the net social returns for each kind of goal or value affected by the R & D for each category of political unit in the trimatrix after which the net social returns are summed for all units [58]. Moreover, the estimated probablity of the success of the R & D in achieving its design objectives may be relatively small. Thus, taking all of these elemental estimates of the appraisal ratio of R & D into account, the expected rate of net social return may likewise be small.

Under an alternative set (b) of public measures, the outcome of the R & D might be somewhat enhanced. A major difference from the prior assumption would be the additional stimulation to the use of artificial turf by households by means of installing appreciably higher rates for water consumption during summer months or other periods of low precipitation [59]. Federal costs to persuade many municipal water suppliers to implement such a differential pricing policy as well as possible other measures would raise the public costs (C) to a higher level and at the same time increase the net social returns from the R & D by expanding the likely scale of market acceptance of the product. The latter in turn would tend to increase the probability of the R & D achieving its design objectives of a low-cost artificial turf by permitting greater economies in manufacturing technology from larger scale operations. The evaluations for comparing these two alternatives appearing in line 1d of Table 2 are very tentative and do not necessarily reflect what is meant by a quick and crude judgment. Instead of the few minutes of thought in arriving at the tentative judgments of Table 2, the recommended quick method would at least require several days or weeks of effort to talk with a sampling of private and public officials as well as potential household customers to ascertain their reactions and opinions concerning the potential product development, its desirable characteristics and prices, alternative sets of stimulative measures and other opinions concerning the various elements of the appraisal ratio of R & D.

Another class of technological alternatives affecting water use-goal relationships are those which are designed to contain or remove contaminants from accidental spills in mining or oil recovery in marine areas including protected bays or estuaries (line 14 in Table 2) [60]. Many competing technological alternatives under use, experimentation or development include [61]: (1) soaking up the contaminants at the edge of beaches with straw or other sorbant materials (e.g., urethane foam chips and urea formaldehyde; (2) a variety of floating oil skimming devices; (3) underwater tents to catch oil seepages; (4) a mechanically created vortex to concentrate oil for removal from the water; (5) controlled burning of heavy fuel oil spill with the use of porous glass beads; (6) consumption (or decomposition) of oil by release of biological agents; (7) a vessel mounted scrubber-wringer belt system; (8) chemical additives that jell the oil for removal; (9) chemical dispersants; (10) sink sand method; (11) magnetic polystyrene method; and (12) a wide variety of containment devices including air barriers and floating booms.

The cost-effectiveness of these technological alternatives varies in accordance with the size and nature of the oil spill; the chemical and physical properties of the oil (e.g., light or heavy crudes); the location of the spill and direction of surface currents and winds relative to environments requiring the greater protection; and the configuration of the shoreline environment or inlets. The social costs of oil spills that are not contained or removed will also vary from one area to another in accordance with the value of threatened recreational and marine and wildlife resources. Thus, it might be expected that among the contending technological alternatives some will prove to be complete losers and others will find some part of the spectrum of potential applications where they can compete successfully. In the hypothetical example in Table 2, case (B_1^e) is estimated to have a superior net social return (R) to that of technological alternative (B_2^g). Accordingly, recognition of this difference by responsible government officials may yield a higher priority for encouraging B_1 over B_2 and hence it might be expected that the set of institutional measures (e) designed to encourage B_1 would favor a higher probability (P) of successful development of B_1 than would the set of institutional measures (g) yield in regard to the probability estimate for B_2.

It is noted that in this hypothetical case technological alternatives B_1 and B_2 may also have application for industrial oil spills (line 2, Table 2) and oil tanker operations (not shown). In this event the evaluation of the expected rate of social return and the components of the appraisal ratios of R & D for these multiple uses in the various line items of Table 2 cannot be made independently of each other. Accordingly, the trimatrix approach is a useful device in identifying and coordinating the evaluation of interrelated technological alternatives as well as the utility of institutional alternatives that would encourage (or discourage) the pursuit of alternatives for R & D. Indeed, a considerable amount of research on the effectiveness of different kinds of institutional measures — singly and in combination — would facilitate the evaluations of the various cells in the trimatrix approach as well as the actual planning of R & D and measures of government-business cooperation to enhance the expected rate of social return.

SUMMARY

Although the urban sector has a sizeable stake in R & D affecting uses of water and related land resources, practically all of the relevant R & D is performed by R & D in the Federal Government and the private sector. There are a veritable legion of technological alternatives that need to be assessed in a broad analytical framework. A few representative alternatives of different classes of technology will suffice to illustrate the methodology and kinds of considerations that need to be assessed by the R & D planner. Among the latter is the potential impact of the R & D on social needs (especially those for urban areas), the cost of the R & D, its probability of success and the relation of institutional problems and measures to these interrelated decision variables.

There are two broad categories of technological alternatives to be appraised. The first of these constitutes those alternatives wherein systems performance objectives are dominant: (1) augmentation of water supply, (2) reduction of water requirements, (3) regulating water flows, (4) reducing costs of water supply, and (5) protecting or enhancing the quality of water and the related land environment. In evaluating this group of alternatives the water resources research planners will usually play the leading role. A second category of technological alternatives includes those advances in science and technology that would enhance the various uses of water and related land resources such as water transportation, recreation, fisheries, environmental beautification, agriculture, industrial products, and the like. The R & D planners responsible for evaluating these alternatives may have little or no competence in various aspects of water resources science and technology. Accordingly, a methodology of technology assessment is required that will span the interracting variables of both categories of R & D decisions especially as these relate to socially desirable or undesirable impacts of the technology alternatives.

An examination of the various appraisal considerations involved for representative kinds of technological alternatives led to a number of observations of possibly fundamental importance:

The elemental considerations of the appraisal ratio of R & D can be shown to have important interrelationships and therefore cannot be estimated independently of each other.

For technological alternatives having multipurpose objectives it is more appropriate to think in terms of a probability coefficient for attaining each of the objectives through the R & D than assessing a probability coefficient for all social returns taken collectively.

For those potential technological developments which have more than one kind of social benefit or social cost realized unevenly by differential jurisdictions which share in the cost of the R & D or consent to its applications, then the conflicts of purposes in the negotiating process may tend to discourage a full or unbiased appraisal of these costs and benefits.

The evaluation of the appraisal ratio of R & D is not static over time since continuing advances in various components of the system technology can reduce the cost of the prototype development as well as to enlarge the economic benefits.

Research in planning and related institutional measures may serve a crucial role in influencing the choice of one technological alternative or another.

Although cost considerations are part of all technology assessments, there is a particular need to give attention to those technological alternatives that would reduce the cost of adding new sources of water supply, improve the quality of water and related land resources, and ensure a reliable supply of water.

Full-blown and even-handed technological assessments of the net social effects of alternatives related to environmental modification or pollution control measures are especially needed as a corrective to the sometimes unbalanced "participatory technology assessment" inputs of various special-interest, citizen groups.

Criteria for public sponsorship of R & D on environmental quality measures deserving high priority would include those advances in science and technology which hold promise of: (1) reducing tensions between various sections in our society over environmental issues; and (2) reducing the high and economic costs of pollution abatement measures.

In regard to the latter it is estimated by the council on Environmental Quality (based on EPA data) that $38 billion will be required during 1970-75 to meet costs of water pollution abatement. Much of these expenditures will be required in urban areas for improved industrial and municipal waste treatment, to remedy overflows from combined storm and sanitary sewar systems and to curb soil erosion pollution resulting from highway and other construction activities. The value to society, and hence marketability, of the outputs of R & D alternatives involves a comparative evaluation of the state of technologies at some future time as between (1) basically new technologies, (2) improved technologies currently in use, (3) unimproved technologies currently in use. In the evaluation process two levels of impact analysis are involved: the impact of the technological alternatives on urban uses of water and related land areas; and the impacts of these affected resource uses on goals and values of urban areas and other political jurisdictions (other local, state, federal and foreign nations).

There are a number of blue sky research possibilities and other technological experiments whose applications appear more certain or closer at hand that would favorably enhance the uses of water resources. Included are such prospects as genetic engineering in agriculture and aquaculture; radical energy developments such as hydrogen fusion, fuel cells, etc.; development of synthetic foods from petroleum; increased substitution of plastic materials for metals and wood; substitution of ocean mining for land mining; new forms of water recreation and related structures and vehicles; emergence of revolutionary forms of water transport vehicles such as air cushion vehicles and hydrofoils; and a host of aquaculture developments that would increase the tolerance of marine species to waste and thermal discharges or to use processed waste by-products to cultivate commercial fishery products. The estimation of the probability of commercial success and the time-of-ripeness for innovations such as these poses difficult analytical problems of technological and institutional or political forecasting.

The latter problems are often compounded by information gaps and a scope of complexity and effort that exceeds the capacities of a single individual. In order that the collective intelligence and insight of a diverse group of planners, forecasters and analysts can be applied in systems evaluations of these kinds it is essential that a conceptual analytical model be developed. The major design feature of such a model is that it should facilitate convergence of decision making on at least the issues and tradeoffs between technological alternatives that are of major social importance. For this purpose a trimatrix model is proposed that provides a structure for evaluating the impact relationships of technological alternatives on uses and goals involving water and related land resources at the urban, other local, state, federal, and foreign country levels. The trimatrix represents a merger between the use-goal matrix and a matrix that relates technological alternatives to various use impacts. The technological alternatives are organized by classes of potential R & D activity. In each cell of the trimatrix the expected rate of (net) social return of the R & D alternative is to be evaluated in reference to: (1) specific uses of water and related land resources; (2) specific goals or values of various political jurisdictions; and (3) specific sets of assumed institutional measures that would encourage (or discourage) the development and application of the technological alternative in question. A considerable amount of research on the effectiveness of different kinds of institutional measures would facilitate the evaluations of the various cells in the trimatrix approach and, hence, the planning of R & D and measures of government-business cooperation to enhance the expected rate of social return from the most desirable of the technological alternatives.

NOTES AND REFERENCES

1. By personal communication.

2. SPANGLER, MILLER B., *New Technology and Marine Resource Developments*, New York: Praeger Publishers, Inc., 120-3, 1970.

3. Atomic Energy Commission, *Annual Report to the Atomic Energy Commission for 1968*, Washington, D.C.: Government Printing Office, 98-100, 1969.

4. Here the term "social" is used broadly to include social, economic, technological and political objectives.

5. For a revealing, though limited, discussion of some of these problems, see *AEC Authorizing Legislation, Fiscal Year 1972*; Hearings before the Joint Committee on Atomic Energy, 92nd Cong., 1 sess., Part 2, 542-552 and 1173-1189, March 4, 1971.

6. *Ibid.*, 1180.

7. Indeed, if agriculture uses of desalted water are to be justified, Young estimates that a price of water in the vicinity of 20¢ per thousand gallons will be required. See, YOUNG, GALE, "Dry Lands and Desalted Water," *Science*, *340*, January 23, 1970.

8. NEIBURGER, M., "Rainmaking: Claims and Questions," *Science*, 1256, December 5, 1969.

9. *Ibid.*

10. TRIBUS, MYRON, "Everybody Talks About the Weather, But...," *Science*, *1341*, June 20, 1969.

11. *Ibid.*

12. *The Comprehensive Plan of Chicago: Analysis of City Systems*, *1*, 116, Chicago: Dept. of Development and Planning, October 1967.

13. HIRSHLEIFER, JACK, DeHAVEN, JAMES C. and MILLIMAN, JEROME W., *Water Supply: Economics, Technology and Policy*, 260, Chicago: The University of Chicago Press, 1960.

14. YOUNG, *op. cit.*, 340.

15. *Ibid.*

16. GOLDBERG, D., GORNAT, B. and SHMUELI, M., "Advances in Irrigation in Israel's Agriculture," a paper presented at the 1st World Congress of Engineers and Architects, Israel, 1967.

17. *Statistical Abstract of the United States, 1970*, 591, Washington, D.C.: U.S. Government Printing Office, 1970.

18. HARSHBERGER, J.W., *et. al.*, *Arizona Waters*, Geological Survey Water-Supply Paper 1648, 5, Washington, D.C.: U.S. Government Printing Office, 1966.

19. *Annual Report to Congress of the Atomic Energy Commission for 1970*, 201, Washington, D.C.: U.S. Government Printing Office, January 1971.

20. *The Nuclear Industry, 1968*, prepared by the Division of Industrial Participation of the U.S. Atomic Energy Commission for delivery to Atomic Industrial Forum's International Conference, Washington, D.C., 193, Washington, D.C.: U.S. Government Printing Office, November 14, 1968.

21. "D.C. Water to Bury Town," *Washington Post*, C-1, Sept. 6, 1971.

22. The FWPCA estimated in 1969 that a 95% reduction of acid pollution from active and inactive mines throughout the United States will require $7 billion. "States Making Headway on Mine Drainage," *Environmental Science and Technology*, 1237, December 1969.

23. "D.C. Water to Bury Town," *loc. cit.*

24. *Webster's Seventh New Collegiate Dictionary.*

25. SPINRAD, BERNARD I., "America's Energy Crisis: Reality or Hysteria?," *Science and Public Affairs, Bulletin of the Atomic Scientists*, 5, September 1971.

26. MEIER, RICHARD L., "Interpretation: Insights Into Pollution," *Journal of the American Institute of Planners*, 211, July 1971.

27. Sponsored by the Engineering Foundation at Proctor Academy, Andover, New Hampshire, August 30-September 3, 1971.

28. *Environmental Quality*, The Second Annual Report of the Council on Environmental Quality, 89-90, Washington, D.C.: U.S. Government Printing Office, August 1971.

29. *Ibid.*, 33.

30. *Ibid.*, 56.

31. MEIER, *op. cit.*, 215.

32. Sponsored by the Resources for the Future, Washington, D.C., April 20, 1971.

33. MEIER, *op. cit.*, 215.

34. *Environmental Quality*, *op. cit.*, 111.

35. Testimony by Miller B. Spangler, *The Environmental Decade*, Hearings before a Subcommittee of the Committee on Government Operations, House of Representatives, 91 Cong., 1 sess. 4, February 2, 1970, Washington, D.C.: U.S. Government Printing Office, 1970.

36. "Va. to Sue on Pollution of the Potomac," *Washington Post*, A-1, Sept. 12, 1971.

37. KNEESE, ALLEN V. and BOWER, BLAIR T., *Managing Water Quality: Economics, Technology, Institutions*, Baltimore: Published for Resources for the Future, Inc. by The Johns Hopkins Press, 1968.

38. See especially, *Prevention and Control of Oil Spills*, Proceedings of a Joint Conference sponsored by the American Petroleum Institute, Environmental Protection Agency and the United States Coast Guard, Washington, D.C.: The American Petroleum Institute, June 15-17, 1971.

39. BIGLER, ALEXANDER B., *The Role of Marine Sciences in the Multiple Uses of the Coastal Zone of Lake Erie and Lake Superior*, A study by the National Planning Association for the Marine Sciences Council, Report No. PB-185163, 147, Springfield, Va.: National Technical Information Service, June 1969.

40. *Environmental Quality, op. cit.*, 145.

41. *Ibid.*, 148.

42. *Ibid.*, 145.

43. *Ibid.*, 148.

44. *Ibid.*

45. Some of these are mentioned in Kahn and Wiener's list of one hundred technical innovations likely before 2000 A.D. See, KAHN, HERMAN and WIENER, ANTHONY J., *The Year 2000*, 51-55, New York: The Macmillan Co., 1967.

46. RYTHER, JOHN H. and BARDACH, JOHN E., *The Status and Potential of Aquaculture, I*, Pt. I, No. PB 177767, 1, Springfield, Va.: National Technical Information Service, 1968.

47. SPANGLER, *New Technology and Marine Resource Developments, op. cit.*, 412.

48. "Fine Mess to Set Before So Many Hungry Shellfish," *Oceanology International*, 26, March 1971.

49. "Wanted: Pan Size Salmon," *Oceanology International*, 17, January 1971.

50. "Aquaculture: An Overview," *Sea Grant 70's*, 3 and 5, May 1971.

51. DIAMOND, HENRY L., "Beneficial Uses of Thermal Discharges," *Catalyst for Environment Quality, 1*, 4, 17-20, 1971.

52. "Aquaculture: An Overview," *op. cit.*, 4.

53. *Ibid.*, 3.

54. "Use That Heat," *Oceanology International*, 26, May 1971.

56. COATES, JOSEPH F., "Interdisciplinary Considerations in Sponsoring Technology Assessments," a paper presented at an Engineering Foundation Conference on Technology Assessment, *20*, August 30-September 3, 1971.

57. *Ibid.*

58. To be sure, some of the social values affected positively or negatively will prove to resist measurement or quantification so that "summed" must be interpreted in an aggregative judgmental sense.

59. In many urban areas this could require water meters to be installed for each household and may be expected to produce voter antagonism unless public information services demonstrated why the new measures could prove in the long run to be in the affected parties' best interests.

60. The threat of oil pollution in estuaries or other coastal areas of concern in fresh water planning may also arise from oil spill accidents in the ocean if the oil should drift shoreward to tidal areas of bays and inlets.

61. See especially, "Pollution Reports," Parts 1-4 in *Ocean Industry*, June, July, August and September 1970 issues.

Note: This chapter is an edited version of Chapter 6 in, Miller B. Spangler, *The Role of Research in Water Resources Planning and Management for Achieving Urban Goals*, a contract study prepared by the National Planning Association for the U.S. Office of Water Resources Research (Springfield, Virginia: National Technical Information Service, May 1972), 169-222. The work upon which this chapter is based was supported in part by funds provided by the United States Department of the Interior as authorized under the Water Resources Research Act of 1964, as amended.

FOOD PRODUCTION WASTES - SUMMARY OF A CASE STUDY OF A SYSTEMS APPROACH TO RESEARCH PROGRAM PLANNING*

THEODORE B. TAYLOR

President
International Research and Technology Corporation
Washington, D.C., U.S.A.

1. INTRODUCTION

This paper briefly summarizes a systems approach to research program planning that was developed for the National Science Foundation (NSF) by International Research and Technology Corporation (IR&T), Washington, D.C., and the Futures Group, Glastonbury, Connecticut, between May and September, 1971. Work was performed under a contract in which the prime contractor was IR&T, with the Futures Group as subcontractor. As a test of the utility of the method, it was applied to the specific problem area associated with the production of wastes by the U.S, food production system. IR&T assumed primary responsibility for parts of the study related to food plants, and the Futures Group for those related to food animals. Only those parts of the study related to food *plant* wastes are summarized in this paper.

The case study was a follow-on to a brief joint effort by IR&T and the Institute for the Future, carried out for NSF in December 1970 and January 1971, that led to the preliminary formulation of a systems approach to

*Summary of a paper presented at the Technology Assessment Symposium, Stockholm, Sweden. October 11-12, 1971.

formulating actions to alleviate present and future national problems. A schematic diagram of the principal steps in this approach, is shown in Fig. 1. The method starts with the identification of a set of major areas of national concern, such as the threat of nuclear war, environmental pollution, deterioration of central cities, prospective electric power shortages, etc., and prescribes steps to be taken for detailed evaluation of each of the problem areas, specification of relative priorities that the problem areas should be given (in terms of their present and future impact on society), formulation of action programs to alleviate the high priority problems, selection of specific action programs to be implemented, and, finally, the monitoring of the effectiveness of new programs as they are implemented. Following a review of this over-all approach to problem oriented planning, NSF decided to carry out a study in which the focus would be narrowed to a specific problem area (not comparing it to others) and to the formulation of a specific research program in that area. This decision led to the study that is summarized here.

2. THE METHOD

A. *Over-all Description*

The method is illustrated in Fig. 2. A preliminary description of the system, in terms of inputs, conversion processes, inventories, effluents, and outputs, is made in such a way as to exhibit all major likely categories of potential problem sources in the broad problem area under study (1)*. For this purpose, the system products or services, consumers, processes, and problem-causing attributes (wastes, in the present case study) are subdivided into progressively more distinct categories. This is done for three reasons:

i. To facilitate the subsequent analysis

ii. To help assure completeness at whatever level of detail is compatible with the resources available for the analysis

---

*() Numbers refer to steps depicted in Fig. 2.

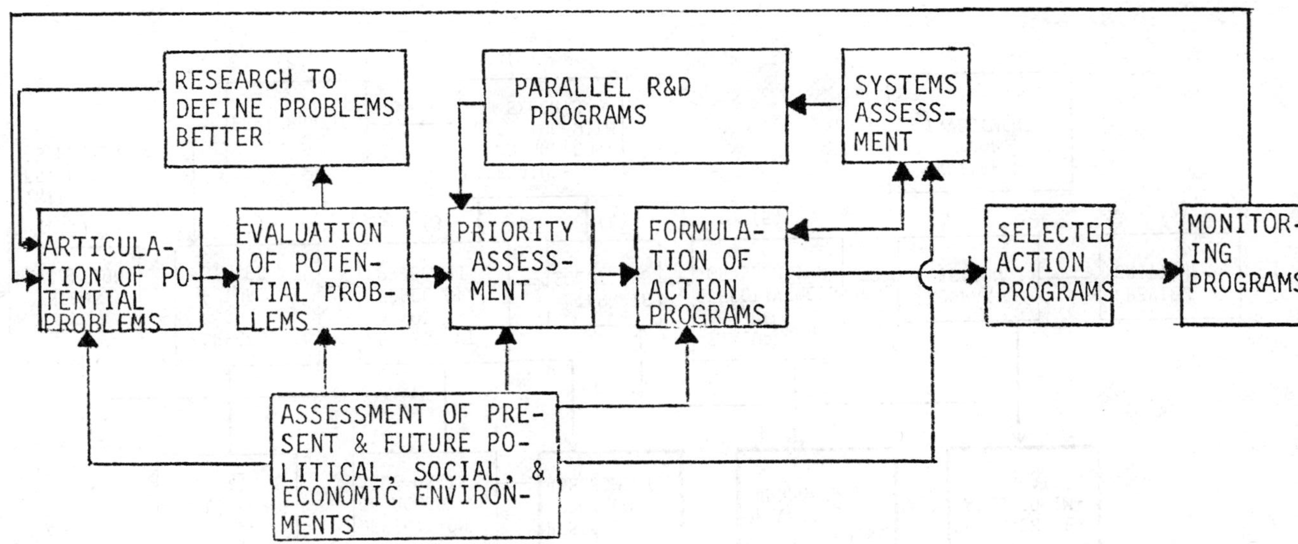

Fig. 1. Flow chart of actions to alleviate present and future national problems

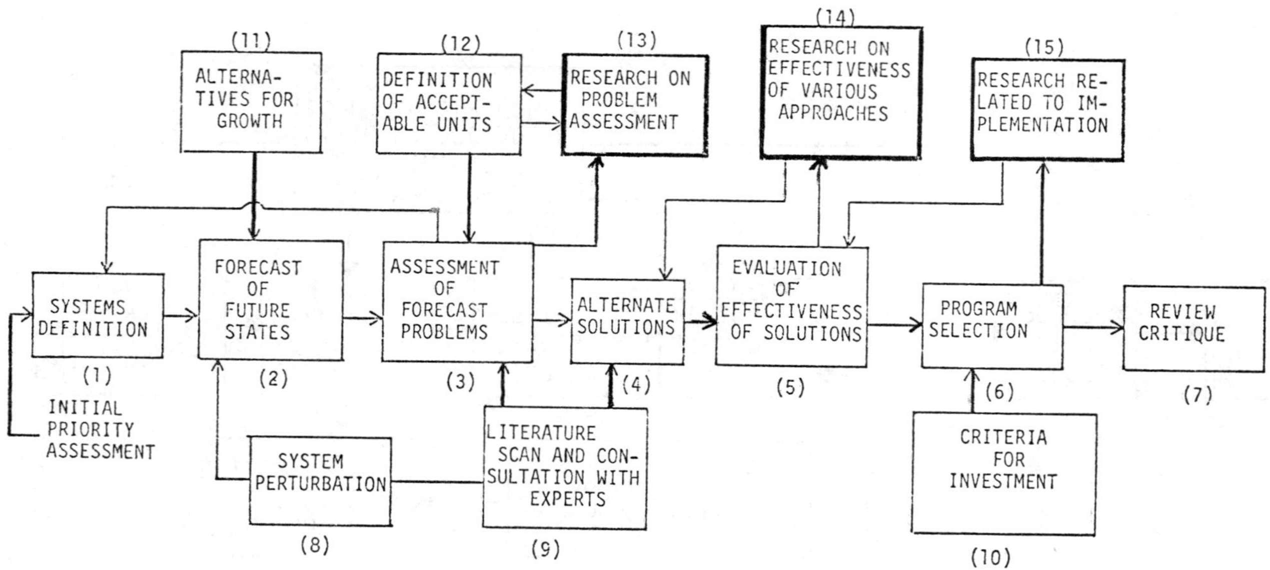

Fig. 2. Flow chart for formulation of research programs

iii. To help exhibit possible problem solution strategies that will simultaneously alleviate many different types of problems.

Using "surprise-free" models of future change (1), the future states of the system elements are described (2) by determining how each major process, output, and potential cause of problems changes with time. With preliminary information about the present system, and prospective changes in hand, possible future problems are identified, described, and assessed (3). Futher information about the system is then obtained to help determine which present problems are most severe and when the measures of intensity of future problems may exceed acceptable limits (12.) The establishment of such limits depends heavily on the state of knowledge about what constitutes a danger; hence, attempts at the definition of acceptable limits often results in the identification of need "problem-defintion" research (13). Further assessments are then made of the relative severity and confidence in understanding of present problems, evident or possible, as well as future problems (3).

Alternative solutions to the problems are identified (4) by systematically examining all major categories of approaches suggested by a hierarchical categorization of types of solutions. A literature scan (9) is important to provide tests for omissions in such a hierarchy and to help identify possible solutions. The approaches to the solutions are evaluated (5) to determine not only their likely effectiveness in mitigating the problems but also any major harmful or beneficial side effects of implementing the solutions. While this analysis can sometimes be rigorous, often it must rely on judgment. Thus, it gives rise to the identification of a class of research programs designed to test the likely effectiveness of proposed solutions to the problem (14). Program selection (6) from among the most promising candidates can be made on the basis of explicit criteria (10), which may include estimated program costs, both fiscal and social as well as the level of benefits expected. This evaluation process sometimes identifies a third category of needed research: research required to develop solution strategies and test and evaluate their consequences.

Up to this point the analysis has been based on the present and expected changes in the system; to be more general, it is necessary to examine possible problems resulting from changes that are not predicted by using recent trends. The literature scan (9) is used to define several more or less extreme states (8) of the system under study and the environment in which it operates. The entire process can then be repeated under these new boundary assumptions to determine the need for additional research. Although a number of possible deviations from "surprise-free" projections were identified in this study, this last step was not carried out for lack of time.

B. *Description of the System*

Any system for providing goods or services can be described in terms of three primary elements: the products (goods or services), the primary consumers, and the processes used for providing the products. The components of each of these primary elements can be organized into hierarchies that subdivide each category into subcomponents corresponding to greater and greater disaggregation or degree of detail. For illustration, examples of components of hierarchies of agricultural products and their consumers, at various levels of disaggregation, are shown in Fig. 3. A hierarchy of the types of processes used to provide the products is illustrated in Fig. 4.

Associated with the processes that make up any system are four major items: inputs, inventories, outputs, and effluents, the latter being any items released by the systems that are not directly incorporated into the products. Each of these four categories of items can also be disaggregated into a hierarchy to whatever level of detail is desired. The inventories, outputs, and products of any system can contain wastes (energy or materials) whose value is either zero, or negative, in the sense of being undesirable product contaminants, interfering with the system processes, or being environmental pollutants. Since the subject of our case study was problems associated with various types of agricultural wastes, these were also organized into a hierarchy, the components of which are illustrated in Fig. 5.

FIG. 3. Hierarchy of Products and Consumers

Level	Products	Primary Consumers
I	ALL AGRICULTURAL PRODUCTS	ALL CONSUMERS
II	FOOD, TEXTILES...	DOMESTIC, FOREIGN
III	WILD, CULTIVATED	ANIMALS, HUMANS
IV	LAND, FRESH WATER, MARINE	FOOD ANIMALS, OTHER ANIMALS
V	GRAINS, FRUITS, LIVESTOCK....	LAND, FRESH WATER, MARINE
VI	WHEAT, APPLES, BEEF CATTLE....	LIVESTOCK, HOUSE PETS, WORK ANIMALS
VII	DWARF MEXICAN WHEAT, BLACK ANGUS CATTLE....	BEEF, CATTLE, DOGS, HORSES....
VIII		BLACK ANGUS CATTLE, ST. BERNARDS, PALO-MINOS....

After a set of hierarchies has been constructed that apply to a particular system and the problem area to be considered, the system is then quantitatively described by preparing lists or matrices that exhibit the values of suitable quantitative indicators associated with elements, or combinations of elements, of each hirarchy. This permits the quantitative aspects of the system to be easily viewed at a wide variety of levels of detail without requiring the construction of many large and unwieldy flow charts of the types that are characteristic of many types of systems analysis. An example of such a data display, that is a partial description of the U.S. food production system, is shown in Fig. 6.

For the case study under discussion, emphasis was placed on displays of indicators directly related to the rates of accumulation and release of waste products from the major categories of processes associated with the U.S.

FIG. 4. Hierarchy of Food Related Processes

Level	Processes (Food Plants)
I	ALL PROCESSES
II	PRODUCTION, PROCESSING, DISTRIBUTION, CONSUMPTION
III	SOIL PREPARATION, GROWING, HARVESTING, CANNING....
IV	PHYSICAL CONDITIONING OF SOIL, PLANT FEEDING, PEST CONTROL, CROP WASTE MANAGEMENT....
V	PLOWING, APPLICATION OF CHEMICAL FERTILIZERS, INSECT CONTROL, SEPARATION OF CROP WASTES....
VI	NITROGEN APPLICATION, CONTROL OF PLANT-EATING INSECTS, CONBUSTION OF CROP WASTES....
VII	APPLICATION OF GASEOUS AMMONIA, CONTROL OF SPECIFIC SPECIES OF INSECTS, OPEN FIELD COMBUSTION OF CROP WASTES....
VIII	SPECIFIC PROCESS FOR APPLICATION OF AMMONIA, CONTROL OF INSECTS, OPEN FIELD COMBUSTION OF CROP WASTES....

food production system, both as it now operates and as it might operate in the future. Although not carried out in any detail in this study, several alternative possible future states of the system could be described based on various assumed major future changes that would affect the U.S. food production system. A few examples of such changes are given in Fig. 7.

C. *Problem Assessment*

After a hierarchically organized, specific problem-oriented description of the system under study has been prepared, the system is examined to identify possible

FIG. 5. Hierarchy of Food Production Wastes

Level	Food Plants
I	ALL WASTES
II	EFFLUENTS, ACCUMULATED WASTES, PRODUCT CONTAMINANTS
III	AIRBORNE, WATERBORNE EFFLUENTS, WASTES ACCUMULATED IN SOIL
IV	NUTRIENT-RELATED PEST CONTROL AGENTS....
V	MAJOR NUTRIENTS, TRACE NUTRIENTS, INSECT CONTROL AGENTS....
VI	FIXED NUTRIENTS, NUTRIENT BY PRODUCTS, CHEMICAL INSECTICIDES....
VII	NITROGEN, PHOSPHORUS, ORGANIC INSECTICIDES....
VIII	NITRATES, NITRITES, ORGANO-PHOSPHATE INSECTICIDES....
IX	NITRATE IONS, AMMONIA GAS, SODIUM NITRATE, PARATHION....

specific problem areas. In the present study, lists of possible problems associated with each of the major categories of wastes were prepared by members of the study team in consultation with various outside experts. Once these lists were prepared, a specific procedure was used to screen the specific problem areas and place them in decreasing order of priority for abatement action. For this purpose members of the study team were asked to provide their assessed values of a "screening factor" associated with each problem area, defined as follows:

$$F \text{ (screening factor)} = \frac{1}{.5 + C(1-C)},$$

where 1 is a measure of impact (on a scale from 0 to 10) of the problem on society, and C is a measure of the

FIG. 6. Yield, Acreage, and Farm Value of U.S. Cultivated Crops (1968)

Crops	Total Yield ($X10^6$ tons)	Acreage ($X10^6$ acres)	Farm Value ($X10^6$ \$)	Farm Value per harvested acre (S/A)
Grains	246	155	8,678	55.9
Wheat	47	55	1,920	34.7
Rye	0.7	1.0	23	23.1
Rice	5.2	2.4	526	223
Corn	131	56	4,601	82.5
Oats	28	17	557	32.0
Barley	12	9.6	364	38.1
Sorghums	22	14	687	49.2
Oilseeds	38	53	3,225	59.3
Cottonseed	4.6	10.2	233	23.0
Soybeans	32	41	2,617	64.4
Peanuts	1.3	1.4	299	208.7
All other	<0.1	<0.1	76	36.0
Vegetables	36	4.4	2,265	491.2
Potatoes	14.6	1.37	609	442.3
Snap Beans	0.83	0.37	107	294.4
Cabbage	1.19	0.1	65	593.9
Carrots	0.94	0.07	81	1,127.4
Celery	0.79	0.034	72	2,131.9
Sweet Corn	3.1	0.7	128	181.8
Cucumbers	0.8	0.2	87	436.8
Lettuce	2.2	0.22	200	900.2
Onions	1.5	0.1	100	940.6
Green Peppers	0.24	0.05	51	993.0
Tomatoes	8	0.5	512	999.5
Other	2.4	0.74	253	293.4

FIG. 6. (Continued)

	Total Yield ($X10^6$ tons)	Acreage ($X10^6$ acres)	Farm Value ($X10^6$ \$)	Farm Value per harvested acre (S/A)
Fruits & Nuts	19.4	3.5	2,068	668.8
Citrus Fruit	8.3	1.0	606	571.7
Apples	2.7	0.4	332	
Grapes	3.5	0.6	237	
Peaches	1.8	0.3	186	
Other Fruits	1.2	0.2	251	
Nuts	0.28	0.29	182	
Berries	0.08	0.34	134	1,615.3
Melons	1.38	0.4	139	337.4
Hay & Silage	222	65.7		
Alfalfa	73.4	27.1		
Timothy & Clover	23.7	13.4		
Corn for Silage	90.4	7.7		
Other Hay	20	13.2		
Sorghum Silage	14.6	3.4		
Sugar	50	2.0	490	245
Sugar Cane	25	0.58	126	218
Sugar Beets	25.4	1.42	364	257
Other Crops	2.1	4.5		
Seeds	0.33	2.0		
Dry Beans & Peas	0.8	1.5	165	97
Other	~1	~1		
Totals	614	288	16,892	

level of confidence of the analyst in his assessment. I was thought of as measured on a logarithmic scale, a value of 10 to be associated with impact on the level of a major national catastrophy. The specific form of F that was chosen was rather arbitrary and was designed to emphasize the effect of the estimated impact level (I) more than the

FIG. 7. Possible Major Changes Affectung U.S. Food Production System

1. THE UNITED STATES PLAYS A MAJOR ROLE IN ALLEVIATING WORLD SHORTAGES BY EXPORTING FOOD.
2. CULTIVATED ANIMAL FEEDS NEARLY COMPLETELY REPLACE WILD GRASS AS FOOR FOR LIVESTOCK.
3. MOST LIVESTOCK SPEND THEIR LIVES IN DENSELY POPULATED ENCLOSURES.
4. HYDROPONIC GREENHOUSE AGRICULTURE PRODUCES MOST VEGETABLES AND SIGNIFICANT AMOUNTS OF FRUITS AND ANIMAL FEED.
5. HYDROPONICALLY CULTURED ALGAE, PROCESSED ORGANIC WASTE PRODUCTS, AND OTHER UNCONVENTIONAL FEEDS LARGELY REPLACE WILD AND CULTIVATED GRAINS AND GRASSES AS FOOD FOR LIVESTOCK.

confidence level (C), while also tending to indicate the importance of problems that are known with high assurance to be severe as well as the importance of problems about which relatively little is known. Final choises of the screening factors were made by the project managers, following discussions with members of the study team and outside experts. An alternative method, not used for the study, could have been to construct group averages of the response of a selected panel of experts.

D. *Formulation and Assessment of Solutions to Problems*

Once the specific problem areas have been rank ordered, in terms of the values of the screening factor, a search is made of ways to alleviate the problems. For this purpose, two types of hierarchies of kinds of action that could be taken to alleviate waste problems were constructed, one that describes technological actions and one that shows institutional actions. These are shown in Figs. 8 and 9. The primary purposes in constructing such hierarchies were to help address attention at types of solutions that might apply to large numbers of different types of problems simultaneously and to help assure completeness in the search for solutions. Examination of

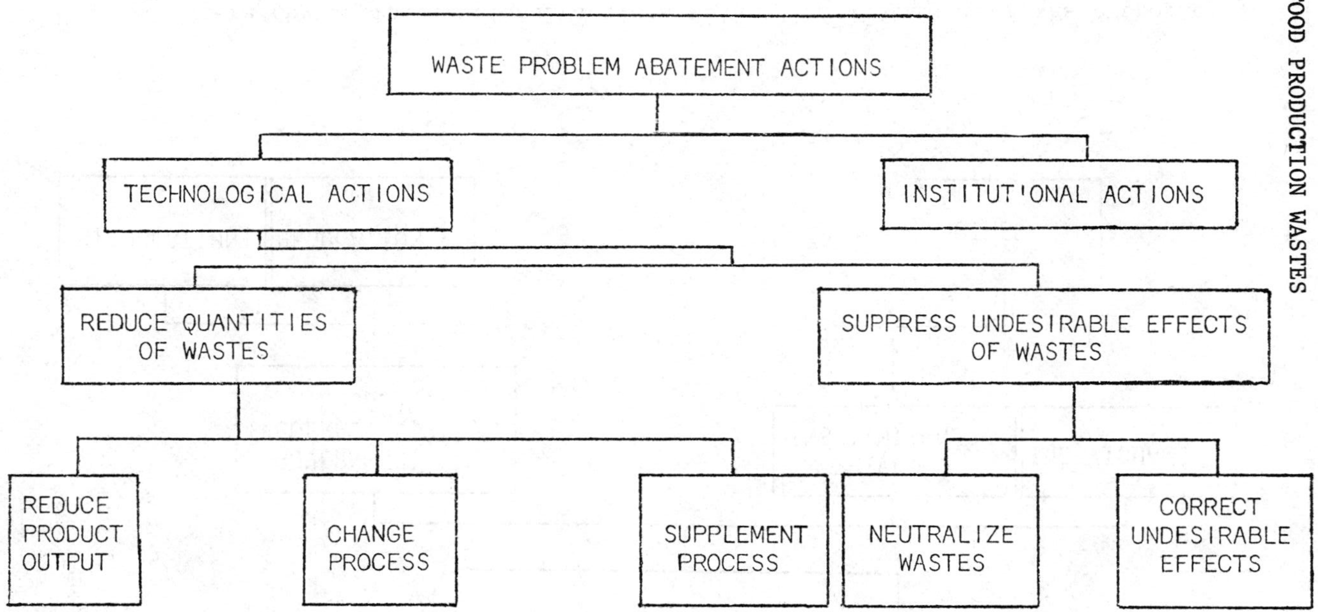

Fig. 8. Hierarchy of technological options for abatement of waste problems

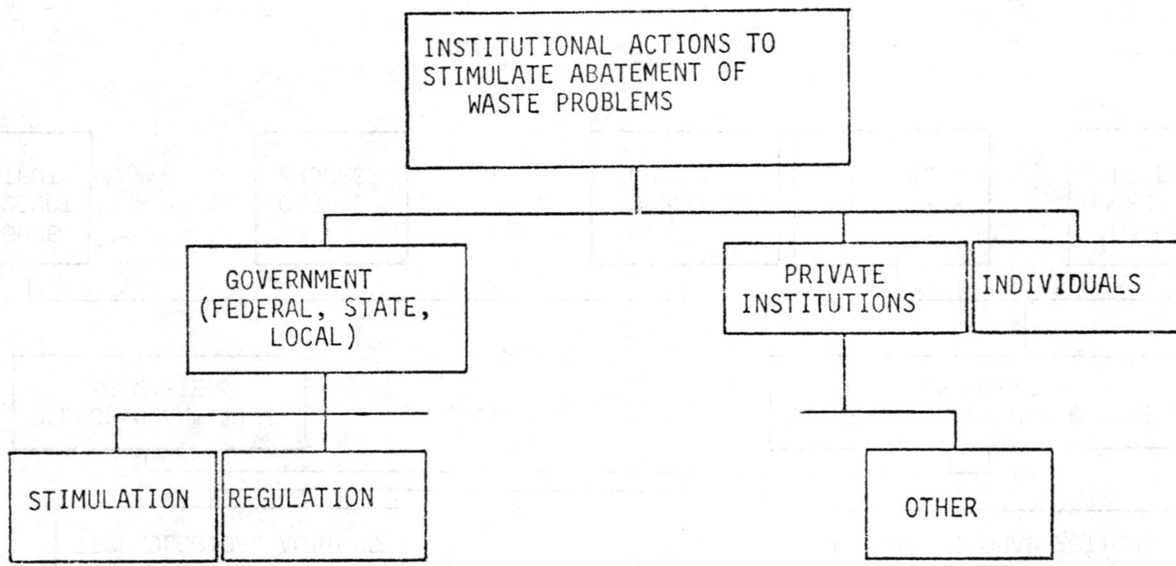

Fig. 9. Hierarchy of institutional actions related to abatement of waste problems

these hierarchies suggested various specific problem abatement actions. These were recorded and their likely effect on each of the problem areas was assessed in terms of the expected change in the impact on society of each problem. An over-all measure of attractiveness of each suggested abatement action was then derived by summing all the expected changes in impact (ΔI) on problems and multiplying the result by the confidence that the action would be technically and economically feasible by 1985 ($C_{\dagger,e}$), and the confidence that it would be politically and socially acceptable by 1985 ($C_{p,s}$). Thus,

$$\text{Attractiveness (A)} = (C_{\dagger,e} C_{p,s}) \sum_j \Delta I_j ,$$

where ΔI_j is the expected change in the impact of the jth problem to be expected by implementing the abatement action.

D. *Formulation of Needed Research Programs*

Once the assessments of specific problems and possible problem solution options have been completed, the research program related to further assessment of problems or of problem solutions can be formulated. Problems judged to have low impact, but with low confidence, may require additional problem assessment research if the problems are judged to be significantly likely to be more severe than indicated in the initial assessment. Such research would presumably carry a lower priority than problem assessment research on problems initially judged to have high impact. Research on problem solution options for which the attractiveness (A) is high tend to carry high priorities.

3. SUMMARY OF RESULTS OF CASE STUDY

A. *Problem Assessment*

The six highest priority problems associated with wastes produced by the production of food plants are listed in Fig. 10, along with the assessed values of impact, confidence in understanding of the problem, and the values of the screening factor F. Although no new problems

FIG. 10. Problem Assessment Summary

Problem	Impact (I)	Confidence (C)	Screening Factor (F)
1) ENVIRONMENTAL DISRUPTION BY PERSISTENT CHLORINATED HYDROCARBON PESTICIDES	8	.1	12.8
2) SEDIMENTATION IN RIVERS AND STREAMS AND TONS OF	8	.6	10.6
SOIL CAUSED BY WATER EROSION	8	.6	
3) CONTAMINATION OF FOOD BY PESTICIDES	5	.1	8.7
4) ACCUMULATION AND REPRODUCTION OF PLANTS, PESTS IN CROP RESIDUES THAT ARE PLOWED UNDER	4	.8	6.2
5) POLLUTION OF SURFACE WATER BY WASTES FROM FOOD PROCESSING PLANTS	4	.8	6.2
6) AIR POLLUTION BY AIRBORNE SOIL	4	.7	6.1

not already recognized by at least some experts were identified, we believe that this has been the first effort to rank order and assess, relative to each other, the seriousness of such problems. Some of them, such as problems associated with the contamination of food by pesticides, are not *evidently* major hazards to society but, in the judgment of the study team, *might* be and therefore should be thoroughly investigated.

B. *Problem Solutions*

A preliminary assessment of alternative ways to alleviate the high priority problems yielded several especially high leverage approaches, in the sense that they either

FIG. 11. High Leverage Options for Abatement Problems Associated with Food Plant Wastes

	"Attractiveness" A
1. CONTROLLED ENVIRONMENT ENCLOSURES FOR CULTIVATING ALL MAJOR FOOD PLANTS	19
2. REPLACEMENT OF CONVENTIONAL LIVESTOCK FEEDS WITH HIGH NUTRITIVE VALUE CULTIVATED FOODS GROWN AT HIGH DENSITY	10
3. INTERMIXING OF CROPS TO PRODUCE ECOLOGICAL STABILITY	9
4. BIOLOGICAL CONTROL OF ALL MAJOR PLANT PESTS	5.6
5. LOCAL CONTAINMENT AND RECYCLE OF SOIL FROM SURFACE WATER FLOWING FROM CROPLANDS	3.2

contributed toward solving many problems simultaneously or had major impact on the most urgent problems. These are listed in Fig. 11.

Controlled environment enclosures (e.g. inflated plastic greenhouse systems) for food crops rank especially high because they could effectively contain all important pollutants. Greenhouse agriculture can now yield extra high quality vegetables at prices that are competitive with those of vegetables grown in the open. There is a distinct possibility that the costs of very large plastic greenhouse systems for containing the growth of grain crops could be reduced sufficiently to make it economically practical to grow all major foods, for both human and domestic animal consumption, in enclosures. Reported yields of various types of food plants, including such grains as rice, are considerably higher in greenhouses that contain atmospheres enriched in carbon dioxide than for those grown in the open--typically by factors between four and ten. Large scale use of greenhouse agriculture would also dramatically reduce the land areas under cultivation.

The production of food for livestock now accounts for about three quarters of the cultivated land in the United States and for more than 90% of all land used for growing food, including natural grazing lands. It is for this reason that replacement of conventional livestock feeds with cultivated plants grown at very high density could have a major impact on many problems associated with agricultural wastes.

Trends toward cultivating very large areas with a single crop have brought with them requirements for massive use of pesticides to control pests in an ecologically unstable region. Intermixing of different crops, though increasing planting and harvesting costs, can restore ecological balance to the agricultural process.

METHODOLOGY FOR TECHNOLOGY ASSESSMENT

Case Study Experience in the United States

GABOR STRASSER*

Director of Planning
Columbus Laboratories, Battelle Memorial Institute,
Columbus, Ohio, U.S.A.

I. WHAT IS TECHNOLOGY ASSESSMENT?

"Technology Assessment" may be defined as a systematic planning and forecasting process that delineates options and costs, encompassing economic as well as environmental and social considerations, that are both external and internal to the program or product in question, with special focus on technology-related "bad" as well as "good" effects.

If you agree with this definition, which in substance is reasonably widely accepted, then you may conclude that "technology assessment" as a label is too limiting.

The purpose of this paper, however, is not to argue semantics, but to address the development of substantive methodologies that are needed for the practice of the process that we have labeled "technology assessment".

The concept of technology assessment is not new, not even in its broader interpretation. What *is* new, is that

*The author has done most of his work relating to technology assessment while affiliated with the White House's Office of Science and Technology in Washington as Technical Assistant to the Director, during 1970 and 1971.

the problems which today are in need of some form of technology assessment have become more numerous, more severe, and more complex. Also, people's awareness of these problems has become more acute, and their insistence, that something be done about them, more articulate.

In essence, technology assessment is a natural followup to what we have been calling systems engineering with few differences. These few differences, however, do cause real problems; the most important ones being, (1) our program objectives are more diffused, since they are not primarily economics and technical hardware or systems oriented, (2) it is more difficult to orchestrate the contributing disciplines to solutions, as they are more numerous as well as more diverse, (3) there is out-of-context emphasis on deleterious side-effects per se. Out-of-context, because these side-effects whatever their form are in fact part of the overall "cost" and should be looked upon in this context. For example, the overall "cost", as defined here, includes such items as environmental pollution, the "cost" of which is borne by the public by virtue of being exposed to it.

It should not be surprising, therefore, that adequate methodologies, required to address these types of problems, would turn out to be more complex than what we have been using in the past. Even more important, such methodologies have not been developed adequately as yet.

II. WHAT HAS BEEN GOING ON IN TECHNOLOGY ASSESSMENT?

Most of the Washington-based technology-assessment activities have centered on the six-year effort of Congressman Emilio Q. Daddario to develop the concept and mechanism through his Subcommittee on Science, Research, and Development [3, 8, 9]. Meanwhile, less visible efforts of Government agencies and other organizations took place.

Some of these other efforts were commissioned by the Daddario Subcommittee, such as those by the National Academy of Engineering [5] and National Academy of Sciences [4].

As of 1969/1970, by far the majority of technology-assessment-oriented literature dealt with three things:

(1) The proliferation and ever-increasing severity of the undesirable side-effects of many of our actions.

(2) The difficulties associated with coping with these problems; realizing that our current, conventional approaches and techniques do not measure up to this need.

(3) Possible organizational arrangements as to where the technology assessment function should be assigned.

What has received very little attention is the development of the substantive methodologies for technology assessment, without which only the dialogue can continue but without much progress.

OST's recognition of this "methodology-gap" led to the issuance of a contract to the MITRE Corporation in 1970 [2]. The objective of the effort was simply to make a start to bridge this gap.

How OST and MITRE went about doing this is discussed in Section IV, after delineating a framework for the approach in Section III.

III. A FRAMEWORK FOR TECHNOLOGY ASSESSMENT METHODOLOGY DEVELOPMENT

We are used to talking about a resource or a *Program Cost* (A) being committed to the achievement of a *Program Objective* (B) (See Fig. 1). These may be called *Direct Costs* (F) and *Direct Benefits* (E), in that they represent an intended set of cause and effect relationship.

In the process we can have two side effects. These may be *Desirable Side Effects* (C) or *Undesirable Side Effects* (D). By virtue of these being "side effects" they may be termed "Indirect" (G), and (H). The *Desirable Side Effects* (C) together with *Program Objectives* (B), sum up into *Total Benefits* (I), whereas *Program Cost* (A) and *Undesirable Side Effects* (D) sum up into *Total Costs* (J).

Fig. 1. *Total* program costs vs. *Total* program benefits

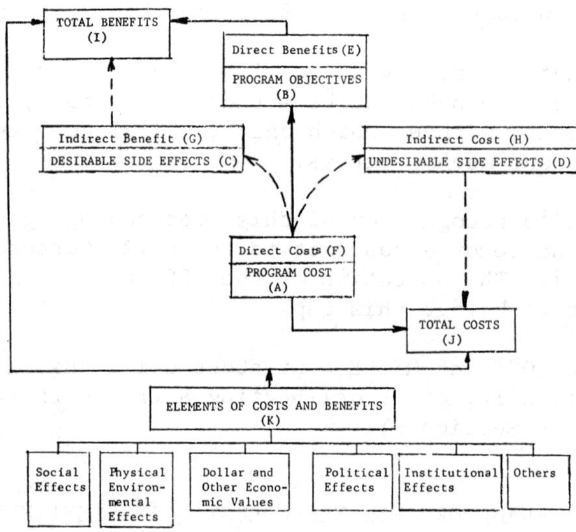

By the terminology of the economist, what is defined here as "Indirect" may be termed "external costs and benefits."

These *Total Benefits* (I) and *Total Costs* (J) can be expressed in such terms as: Social Effects, Physical Environmental Effects, Dollar and Other Economic Values, Political Effects, Institutional Effects, and various others (K). These terms are incommensurable and call for the establishment of some sort of social, environmental or urban indicators, with three characteristics: (1) relevance, (2) utility, and (3) public acceptance. We do not have such indicators today.

Since there is also a time scale involved, we cannot talk only about what is happening now, but must include in our deliberations discounted future costs as well as benefits. It is when we do this, as we must, that the problem of uncertainty comes into sharp focus, calling for some popularized treatment of risks and expected values, which is difficult enough when only dollar "costs" are involved alone.

In times past the primary concern has been to go from Program Cost (A) to Program Objective (B) while spilling into our social and physical environments some of the costs which have been "external" to the primary route from (A) to (B). What has happened is that both the "social sink" as well as the "physical environmental sink" are becoming saturated and there are demands to turn off the "faucet" that permits the spillover from (A) to (D). But such faucets cannot be turned off. The spillover can merely be rerouted back into the original program cost from (D) to (A). For example, if a paper pulp manufacturer who has been polluting a river (which has been his "physical environmental sink") is prevented from doing so in the future, then he must install devices which can control and eliminate his pollutants from being discharged into the environment. This, in fact, adds cost to his production (A). Since many such companies have been only marginally profitable, it would not be surprising if a massive effort to force industries to internalize their external costs would turn the ink on most of their profit-and-loss statements from black to red. Many similar examples could be cited. In total they convey the message that many of our industrial products are profitable only because some of the costs have never been absorbed by the producer himself.

The real challenge is two fold. One, how to balance in an equitable manner internal and external costs and benefits? Two, how to devise an expeditious and fair transition without causing unacceptable disruptions in our economic, social, and political system?

Our old, reliable models of the free market economy have been giving us some surprises lately. Perhaps the reason is, that most of these models exclude (or insufficiently treat) what has been referred to here as "external costs", about which an increasingly vocal public is becoming more and more concerned. Because of such public concerns, external costs came to represent some of the most relevant potential inputs into any model, aiming at depicting the process of doing business. Any model that insufficiently treats such relevant inputs cannot be expected to yield correct results.

We need a supplementary accounting system which addresses not only the path between Program Costs (A) and Program Objectives (B) but also takes into account the external cost (Undesirable Side Effects (D)) as well as demands credit for external benefits (Desirable Side Effects (C)) which, incidentally, has not been the practice either.

The understanding of some rational interrelationships among these many considerations is essential, before we can begin to address in a meaningful and practical way the many problems facing us.

The need for some such overall scheme should be self-evident. A "loose" approach (that some have called "muddling through" or "hit and miss") can only work when available resources substantially exceed the demands that we place upon them. As our resources and demands for such resources come into balance, or turn negative, the necessity for efficiency, overall understanding, and better management becomes obvious. This is the problem we are facing today. When crises occur intermittently with long periods of lulls in between, it is not imperative that we look at our problems in their totality. But, when crises begin to overlap (e.g., intermittent sporadic wars throughout the globe, international monetary crises, social unrest, urban blight, environmental pollution, and many others), it behooves us to sort out these problems and their interrelationships and ask ourselves questions like the following: How can we use better what we have for those things that we hold most dear, and then list them in some decreasing order of importance? If we should find that we cannot meet all or even most of our aspirations, where and how can such aspirations be compromised? Can we afford to keep addressing individual problems, one at a time, rather than in some broader, traceable rational context?

Therefore, the task before us is to seek some way by which all of us will better understand the interrelationships and trade-offs of the many different objectives that we seek, and the many different costs that we must incur, in a language that permits dialogues among policy-and decisionmakers, their constituencies, including the intellectual community, industry, and analysts, or in short, all of those who are involved in one way or another in our collective actions as a society.

It is within this framework that OST/MITRE approached the development of technology assessment methodologies.

IV. THE OFFICE OF SCIENCE AND TECHNOLOGY (EXECUTIVE OFFICE OF THE PRESIDENT)/MITRE APPROACH TO METHODOLOGY DEVELOPMENT

This approach is really a combination of two approaches.

One of these analyzes and assesses a given *real* problem. It focuses on how to resolve specific issues as these relate to the problem itself. (How should we go about deciding what we should do about DDT, or the SST, or automobile emission standards?)

The second approach concerns the extension of the state of the art of existing methodologies to encompass technology assessment in a generic sense. For example, develop new modeling and simulation techniques — industrial or urban dynamics (?); or, develop broader costing/financing techniques; or, better define interfaces between long-range and short-range planning; or, perform more realistic trade-offs between qualitative versus quantitative, direct versus indirect, or current versus deferred costs and benefits; or, better orchestrate hard-to-relate enabling mechanisms, as represented by such disciplines as technology, economics, sociology, etc.

The former of these approaches may be termed the "case history" approach, which lends relevance, practicality and reality to our efforts. When used alone, however, as a rule it suffers from the lack of transferability of findings to other situations.

The second approach, on the other hand, has the advantage of yielding transferable results, but these are usually too abstract and academic, and are often transferable only in a theoretical sense. They are difficult to apply to complex, real situations.

When the two approaches are used in combination, in an iterative fashion, the first approach lends practicality and relevance, whereas the second one provides needed generality and transferability to the resulting methods.

Such combination approach calls for patience, understanding and a lot of compromise on the part of all of the participants.

The case historians must understand that their sole function is to contribute to the development of generic methodologies, and not to the advancement of the state of the art of any discipline, or to the final resoltuion of any existing conflict within their areas of interest. Rather, they must place emphasis on how to look at problems; on what are the important things to consider; on what are subjective versus objective considerations; on what is relevant and what is not (often contrary to initial appearances); on what decisions are highly value-judgment sensitive; on how to determine whose value judgment it should be; on how to help him make the value judgment; and on how to incorporate considerations of risks and uncertainties into the analysis among many other things. In short, the emphasis must be placed on the many elusive aspects of "how to go about solving problems", rather than actually solving them.

The model builders, on the other hand, who like orderly, traceable processes, must realize that many things are not orderly and some are not even traceable. What must go into models is what is *relevant*, however elusive, rather than what is quantifiable and manipulable. In short, pragmatism must be stressed via practicality and relevance, even at the expense of quantifiability and elegance.

In this spirit OST/MITRE selected five case histories, called "pilot studies", and proceeded with the identification and development of a methodology for technology assessment.

The resulting seven major generic steps in making a technology assessment are listed in Fig. 2.*

A brief description of each of the five pilot studies is given in Figs. 3 through 7.

Each of the seven steps is illustrated in Figs. 8 through 14 by various pilot studies.

For the full treatment of the subject, see Reference 2.

Now what has Reference 2 accomplished? First let's see what it has *not* done:

*Figures 2 through 14 are based on Reference 2.

(1) Reference 2 has *not* resolved any substantive issues related to any of the five pilot studies that it treated.

(2) Reference 2 has *not* provided a comprehensive, final "cook-book" for making technology assessments.

However, Reference 2 *has*:

(1) *Delineated* seven generic steps covering what we believe to be the entire spectrum of making technology assessments.

(2) *Illustrated* how such steps may be applied.

(3) *Developed* a generic framework within which technology assessments in general can be performed, provided

(a) the assessor first reviews the volumes of Reference 2

(b) the assessor studies Volume I, especially Appendix B

(c) the assessor is already familiar with the technical substance relevant to the subject of his assessment

(d) the assessor realizes that we are at the beginning, rather than at the end of methodology development; and, that his effort will probably contribute to the state of the art of methodology as much as it will benefit from what is already available in Reference 2.

In short, Reference 2 allows a knowledgeable individual to make a first stab at undertaking a technology assessment of his area of interest, where he would not even have known where or how to start before.

STEP 1 DEFINE THE ASSESSMENT TASK

Discuss relevant issues and any major problems
Establish scope (breadth and depth) of inquiry
Develop project ground rules

STEP 2 DESCRIBE RELEVANT TECHNOLOGIES

Describe major technology being assessed
Describe other technologies supporting the major technology
Describe technologies competitive to the major and supporting technologies

STEP 3 DEVELOP STATE-OF-SOCIETY ASSUMPTIONS

Identify and describe major nontechnological factors influencing the application of the relevant technologies

STEP 4 IDENTIFY IMPACT AREAS

Ascertain those societal characteristics that will be most influenced by the application of the assessed technology

STEP 5 MAKE PRELIMINARY IMPACT ANALYSIS

Trace and integrate the process by which the assessed technology makes its societal influence felt

STEP 6 IDENTIFY POSSIBLE ACTION OPTIONS

Develop and analyze various programs for obtaining maximum public advantage from the assessed technologies

STEP 7 COMPLETE IMPACT ANALYSIS

Analyze the degree to which each action option would alter the specific societal impacts of the assessed technology discussed in Step 5

Fig. 2. Seven major steps in making a technology assessment.

This study analyzes the automobile's role in causing air pollution from the standpoints of the kinds of pollutants emitted, the amounts emitted, and the damages inflicted as a result of these emissions. The study reviews automotive emission control technology which would help abate automotive-generated pollution, including devices which modify emissions produced by the current internal combustion engine; alternative power sources (e.g., electric power plants); and institutional factors which would reduce automotive-generated air pollution (e.g., driving bans within central business districts).

The study identifies and defines state-of-society conditions which are directly and indirectly related to automotive emission control technology. Among the topics covered are: present and future emissions control technology; regulation and testing of automotive emissions; emissions criteria; air quality standards; air pollution monitoring; automotive use and demographic trends; urban mass transit technology and patronage; physiological effects and property damage from air pollution; and emission control costs. The time period covered is 1965-1985.

This study concludes that large societal impacts may result from seemingly small changes in technology or in emissions control actions. The study provides an analytical framework for relating the costs incurred in reducing automotive-generated air pollution with the costs saved in the resultant improvements in air quality, health, and national welfare.

Fig. 3. Automotive emission controls.

This study provides an approach for integrating and evaluating the exploding volume of knowledge about computers and communications. The components of a computer-communications network include general-purpose computers, data-communication facilities, and minicomputers or sophisticated terminals and data banks. The study emphasizes man-machine interaction, especially for such fields as medicine, law, and education, including decision making and policy/systems analysis.

Projections concerning the major technology and complementary technologies (communications, microform) are made to the 1980-85 time period. It is anticipated that the primary mode of operation for this period would be distributed computer-based information networks and data banks.

Action options which would monitor, control and/or obviate the impacts of this technology are discussed. The problem of security/privacy is used to compare the effects of designated action options upon specific possible adverse impacts.

This report recommends research and analysis in both the technological and policy areas. These recommendations include: the formation of a permanent nucleus planning and analysis group for the technology assessment of computers and their complex impacts; directed research to achieve a practical and economically viable solution to the security/privacy problem; a research, development and monitoring program; and legislative and educational programs aimed at guiding development of computer technology and channeling their impacts for the benefit of society.

Fig. 4. Computers-communications networks.

The enzyme industry is still in its infancy — scientists have crystalized about 100 pure enzymes, have described the characteristics of about 1,000, and believe that there may be as many as 100,000. Due to the relative newness of this technology, this study gives major emphasis to understanding the enzyme process. Topics covered include: the direct and indirect uses of enzymes in products; uses of enzymes secreted by microorganisms; and production and technological innovations. Forecasts of trends and developments in this technology are made to the 1980 and 1990 time periods.

Some of the applications of enzyme technology are considered to be replacement of high temperature and pressure processes, lower-cost sugars, and new and improved pharmecutical products. This study suggests the possibility that some of these applications could have adverse consequences. For example, an enzyme process which would displace cane and beet sugar as the primary source of sweeteners could have demographic and international political consequences in terms of displacing workers and changing our protectionist policy toward foreign producers of cane and beet sugar.

Some of these applications could have large impacts on a complex variety of phenomena, such as energy consumed, pollution, capital equipment purchased, and the spectrum of viable products. This study provides a framework for the analysis of such impacts and discusses various control options which would minimize the adverse impacts of enzyme technology.

Fig. 5. Enzymes (industrial).

This pilot study addresses the project applications of mariculture in abating the malnourishment problem through economic stimulation. This assessment anticipates that the main avenue through which mariculture would exert its influence is in building up the export posture of 67 selected developing world countries. As a result of more prosperous economies, the malnutrition crisis would be lessened. In this context, the primary purpose of mariculture development would not be food production for internal consumption, but rather for external consumption which, in turn, would provide an increased GNP for the exporting country.

Due to overriding institutional factors, mariculture technology in this study is confined to coastal and brackish waters. The present status of mariculture in the developed world is described and the applications and impacts on the developing world are projected for the 1970-1990 time period.

Action options which would enhance the beneficial impacts of mariculture are discussed, such as paying some wages and salaries in food products. This option could ultimately result in an annual reduction of 4.6 million cases of malnutrition and 1.2 million cases of infant mortality for the developing countries by 1985. Additional action options dealing with certain disbenefits, such as water pollution caused by mariculture effluents and required dredging, are also discussed.

This assessment provides the key parameters for a cost/effectiveness study to determine the proper framework in which mariculture should be applied in a developing country. Specifically, it shows that mariculture application can increase a country's economic development and, concomitantly, provide additional animal protein, both by increasing income and decreasing food prices.

Fig. 6. Mariculture (sea farming).

The pollution of our rivers and lakes has created myriad, complex problems. For example, water pollution has reduced the availability of drinking water supplies, detracted from recreational areas, imbalanced certain ecological systems, caused health hazards and aesthetic nuisances, contaminated shellfish fields in coastal regions, etc. The inadequate treatment of sewage before it is returned to the water resource cycle is one source of water pollution.

Approximately two-thirds of the United States' population is served by centralized sewer systems. The remaining population utilizes individual home septic tanks and associated soil disposal systems. The nature of the soil in most of the United States makes these systems inadequate. It is this area, the individual home or local community which does not utilize a central sewage disposal system, that this study addresses.

The major areas examined are the collection, treatment, and disposal technologies relating to waterborne domestic wastes. Other subjects discussed are domestic water use patterns and the effects of pollution from domestic wastes. Organizational considerations, water quality standards, and administrative, financial, and community development considerations are assessed.

This study provides the first step toward construction of a dynamic interactive model which interrelates the complex factors that affect individual domestic waste systems. Several examples of these factors are: the time interval between discovery and initial application of a new technology; the rate of its widespread application; the economic life of each type of system; and its total cost. An illustrative quantitative support model is exercised in various ways (such as increasing the speed at which the technology is adopted) to show potential impacts. Action options are also applied to this support model, and their possible effects are examined.

Fig. 7. Water pollution: Domestic wastes.

STEP 1

Fig. 8. Water: Scope of study

Breadth of Study	Depth of coverage			Comment
	Major	Minor	None	
Range of Technologies				
Supporting	x			Soil system technology
Competitive		x		Conventional centralized waste treatment, advanced treatment technology
			x	Package plant technology
Range of Topics				
Technology forecasts	x			Parametric rates and magnitudes of technology diffusion
State-of-Society conditions		x		Utilized in a dynamic interactive conceptual model
Action options	x			70 action options associated with 16 interest groups
Groups affected				
Beneficiaries	x			16 interest groups affected
Sponsors	x			
Third parties		x		
Time period analyzed				1970-1989
Extent Retrospective		x		Prior experience with individual system technology
Extent futuristic	x			

Fig. 8. contd:

Breadth of Study	Depth of coverage			Comment
	Major	Minor	None	
Types of impacts				
Economic	x			Economic costs of sewage service
Social		x		Community development considerations
Environment	x			Quality and quantity of surface water and groundwater
Political			x	
Legal		x		Water quality standards
Institutional		x		Federal financial assistance programs
Other		x		Organization of waste treatment services, technology transfer mechanisms
Levels of Impacts				
Primary	x			Five primary impact areas
Secondary		x		Tertiary impacts in all five areas
Tertiary		x		
Impact Measurements				
Qualitative Approximate or Precise	x			Economic impacts
Quantitative Approximate or Precise	x			Selected environmental impacts
Uncertainty Analysis		x		Parametric analysis

STEP 2

Fig. 9. Computers: Technology description background statement

MATTERS ADDRESSED	COVERAGE
1. Physical and Functional Description	Computers and communications supported by such scientific disciplines as operations research, systems/policy analysis, mathematics; EDP and communications industries heavily involved; computer professionals, mathematicians, engineers, systems analysts; wide variety of products affected.
2. Current State of the Art	Highly advanced with projected relevant interaction between computers and communications; modeling simulation advanced but needs significant and further research and development and significant data gathering.
3. Influencing Factors	Breakthroughs and directed research required: voice input at least to a limited extent; relatively inexpensive sophisticated terminals/minicomputers, increased software certifiability and verifiability; increased acceptance on the psychological and political level of the most innovative technology; a practical and economically viable solution to the security/privacy problem; performance standards and measures — hardware and software.

Fig. 9. contd:

MATTERS ADDRESSED	COVERAGE
4. Related Technologies	Communications, microform, TV.
5. Future State of the Art	Initial applications, presently operational; and projected applications, operational by the 1980 to 1985 time frame.
6. Uses and Applications	Extremely wide spectrum including decision making/decision maker; policy sciences/analysis and systems analysis; professions, medical, legal, educational.

STEP 3

Fig. 10. Mariculture: A classification of major state of society attributes. (With an example from the Mariculture Pilot Study)

CLASSIFICATION	MARICULTURE EXAMPLE
1. Values & Goals	The extent of mariculture application will depend on the tendencies of developing countries to accept assistance from other countries in the development of mariculture application for food and economic growth, also the extent to which money and talent are devoted to mariculture research.
2. Demography	The magnitude of impact of mariculture will depend on the total population, and its rate of growth, and public tastes.
3. Environment	The extent of mariculture application will depend on the availability of water areas and the degree of pollution.
4. Economics	The extent of mariculture application will depend on the amount of foreign aid assistance, foreign capital investment, balance of trade, and rate of economic growth.
5. Social Factors	The magnitude of impact of mariculture application of nonluxury products will depend on the social state of developing countries especially in relation to malnutrition and its many effects.
6. Institutional Factors	The extent of mariculture application will depend on the government policies relative to economic development, use of public waters, outside capital investment, government assistance in education, and government policies on food distribution.

STEP 4

Fig. 11. Automotive: Selected impacts of the automobile (1895 to present)

VALUES

- Geographic mobility.
- Expansion of personal freedom.
- Prestige and material status derived from automobile ownership.
- Overevaluation of automobile as an extension of the self — an identity machine.
- Privacy — insulates from both environment and human contact.
- Consideration of automobile ownership as an essential part of normal living (household goods).
- Development of automobile cultists (group identification symbolized by type of automobile owned).

ENVIRONMENT

- Noise pollution.
- Automobile junkyards.
- Roadside litter.
- Land usage for highways — takes away from recreation, housing, etc.
- Land erosion from highway construction.
- Water pollution (oil in streams from road run-off.)
- Unsightly billboards.
- Air pollution — lead, asbestos, HC, CO, NO_x, SO_x.

DEMOGRAPHY

- Population movement to suburbs.
- Shifts in geographic sites of principal U.S. manufacturers.
- Displacement of agricultural workers from rural to urban areas.
- Movement of business and industry to suburbs.
- Increased geographic mobility.

ECONOMIC

- Mainstay and prime mover of American economy in 20th century.

Fig. 11. contd:

ECONOMIC contd:

- Large number of the jobs directly related to automobile industry (one out of every six).
- Automobile industry the lifeblood of many other major industries.
- Rise of small businesses such as service stations and tourist accommodations.
- Suburban real estate boom.
- Drastic decline of horse, carriage, and wagon businesses.
- Depletion of fuel reserves.
- Stimulus to exploration for and drilling of new oil fields and development of new refining techniques, resulting in cheaper and more sophisticated methods.
- Increased expenditures for road expansion and improvement.
- Increased Federal, state, and local revenues through automobile and gasoline sales taxes.
- Decline of railroads (both passengers and freight).

SOCIAL

- Changes in patterns of courtship, socialization and training of children, work habits, use of leisure time, and family patterns.
- Created broad American middle class and reduced class differences.
- Created new class of semiskilled industrial workers.
- Substitution of automobile for mass transit.
- Ready conversion of the heavy industrial capability of automobile factories during World War II to make weapons.
- Many impacts on crime.
- Increased tourism.
- Changes in education through bussing (consolidated school versus "one room country schoolhouse").
- Medical care and other emergency services more rapidly available.
- Traffic congestion.
- Annual loss of life from automobile accidents about 60,000.

Fig. 11. contd:

SOCIAL contd:

- Increased incidence of respiratory ailments, heart disease, and cancer.
- Older, poorer neighborhood displacement through urban freeway construction.

INSTITUTIONAL

- Automotive labor union activity set many precedents.
- Decentralized, multidivisional structure of the modern industrial corporation evident throughout the auto industry.
- Modern management techniques.
- Consumer installment credit.
- Unparalleled standard of living.
- Emergence of U.S. as foremost commercial and military power in world.
- Expansion of field of insurance.
- Rise of entrepreneurship.
- Basis for an oligopolistic model for other sectors of the economy.
- Federal regulation of interstate highways and commerce as a pattern for other fields.
- Highway lobby — its powerful influence.

STEP 5

Fig. 12. Enzymes: Impact analysis for enzyme-based cane and beet sugar substitute

	TECHNOLOGY
Development	Process might replace some of the cane and beet sugar crop by an enzyme process making fructose from potato or corn starch.
Application	This could reduce the cost of sugar and shift sugar production to a potato-or corn-based agricultural crop.

	SOCIAL
Social Impact	The first level of impact here might be to reduce the number of people working in the cultivation and harvesting of cane and beet sugar.

	IMPACT CHARACTERISTICS
Affected Group	Agricultural workers, farmers, and sugar refiners would be most affected.
How Affected	There would be a trend away from sugar crops toward corn and potatoes leading to large-scale aggregation of farms. Corn and potato growers could benefit, cane and beet growers would be harmed. There would be an enhanced flow from farms to the cities.
Likelihood	Estimated 25% chance.
Timing	Initial impact, 1975; widespread effect, 1980.
Magnitude	Assuming a transfer of 25% of the industry to the new method, this would amount to about $200 million/year.
Duration	The displacement would take 5-10 years, with some benefiting and some harmed.
Diffusion	The breadth would be widespread and non-uniform having demographic implications both in regional terms and also affecting cities.
Source	The source is the chemical processing industry.
Controllability	A public program could dampen bad impacts.

STEP 6

Fig. 13. Computers: Action option evaluation criteria (Strong national program versus local options or ordinary market place actions RE security/privacy)

ACTION OPTION / CRITERIA	STRONG FEDERAL PROGRAM	MARKET PLACE AND LOCAL PROGRAMS
(1) Controllability	There is a relatively high probability that the problem of violation of security/privacy can be significantly altered if adequate resources (planning, R & D and legislation all at the national level) are allocated.	Although some think that the operational market place or a local control could handle the problem, nonetheless, this does not seem to be very realistic.
(2) Worth	The advocates for a strong Federal program have considerable support from the information presently available to indicate that action should be taken.	The advocates for the minimal control position either feel that a minimal alteration is sufficient or that perhaps no outside effort is required.
(3) Priority	This would be one of the highest priority items ranking close to and perhaps at the top.	Relatively low priority.

Fig. 13. contd:

ACTION OPTION / CRITERIA	STRONG FEDERAL PROGRAM	MARKET PLACE AND LOCAL PROGRAMS
(4) Effectiveness	This control option should have a major impact on reducing the security/privacy problem.	There might be some reduction in the security/privacy pr problem but it would probably remain as a major liability.
(5) Cost (Sponsor)	The derivation of the costs would be one of the objectives of the R&D program described later in this chapter.	This again would be the subject of a R&D effort. Although the direct cost may be minimal, the indirect costs of various types of damages could be quite large.
(6) Non-financial	Again, the subject of R&D program, but there may be additional administrative or overhead problems.	Again, this requires additional R&D because, although on the surface this option may have minimal indirect problems, still the impact of the security/privacy issue itself could be enormous.
(7) Institutional obstacles	Requires new legislation, probably a new administrative system or at least a modification of the interaction of some of the government agencies.	Minimal institutional obstacles — indeed, institutional obstacles may very well be the major force backing this option.
(8) Uncertainty	Considerable uncertainty with respect to the details of solutions, and therefore, a major objective in the R&D program plan.	Considerable uncertainty with respect to the claims that minimal effort is adequate.

STEP 7

Fig. 14. Enzymes: Key impact comparison — enzymes and consumer exposure with and without action options. Action option: Establish laboratory for toxicity testing and tolerance establishment for a wide variety of substances

TECHNOLOGY	
Development Application	Use of enzymes in consumer products. Detergents for laundry and dishes and also spot removers and drain cleaners.

SOCIAL	
Social Impact	May be harmful health impacts to some users such as allergic reactions

Impact Characteristic	Impact	
	Without Action Option	With Action Option
Affected Group	Manufacturers and consumers.	Manufacturers and consumers.
How Affected	Consumers will be mass testers of toxicity.	Toxicity will be tested before mass marketing.
Likelihood	Low-medium likelihood of problems.	Much lower.
Timing	May have problems in 2-3 years.	Delay things by 3-4 years.
Magnitude	Allergic reactions could be widespread.	Much lower probability at injury.
Duration	Potentially long lasting.	Smaller potential Bad impact.
Diffusion	Affects most people in society. May Harm large numbers.	Affects fewer people.
Source	Industry.	Industry and Government.
Controllability	Very little government control.	Stronger control.

V. A RELATED NEED AND A RELATED EFFORT

The Need

One of the most relevant related needs is for some methodology to help implement Section 102 of Public Law 91-190 which established the Council on Environmental Quality (CEQ) in the U.S.

Section 102 calls on all Federal Departments and Agencies to perform environmental impact analyses and prepare environmental impact statements on all of their proposed actions which might impinge in some manner on the environment. CEQ in the Executive Office of the President is to review such documents.

It is safe to say that at present neither are the Departments and Agencies sure how to carry out such environmental impact analyses and prepare impact statements, nor is CEQ (together with the Environmental Protection Agency — EPA) quite sure how to review such documents if and when they get them.

The Effort

Perhaps the most interesting potential related effort is a Bill to establish an Office of Technology Assessment (OTA) within the U.S. Congress. Most of the work leading up to the submission of the Bill was done under Congressman Daddario's Chairmanship of the Subcommittee on Science Research and Development, and was introduced by the current Chairman, Mr. Davis of Georgia. The House already passed the Bill, (256 to 118).

What will be the ultimate fate of this Bill is difficult to tell. But, if it should pass, it could cause considerable stir within the Executive Branch. Assuming that OTA would be set up according to the particulars of the Bill as introduced, and assuming that it would be headed by the right man, we would be talking about nothing short of Congress, the "questioner", becoming better informed on a host of issues than many of the "expert testifiers" would be from the Departments and Agencies of the Executive Branch.

Nothing would give a greater impetus to the practice of the principle of "Technology Assessment" in the Executive Branch, than the establishment of an Office of Technology Assessment (OTA) in Congress.

VI. CONCLUDING REMARKS

The history of mankind is full of sporadic, isolated efforts undertaken in the name of progress or to redress past wrongs. In retrospect, these efforts often turned out to be counterproductive. At times they were downright disastrous. After all, from the beginning of recorded history war has been man's primary instrument to "fix things". What has been lacking throughout the ages, and what is still lacking today, is some integrated, "surprise-free" approach to our affairs as individuals, as groups, as nations, and even as the world. Myriads of efforts have come and gone; few have taken root. Is technology assessment going to be just another passing fad? It could be, and in all likelihood it will be unless we broaden it and better understand its potentials as well as its limitations.

First, we must realize that what we do or don't do is really up to our socio-political system, not to our scientific-technical establishment, however extensively many of our industrial products and Government programs may depend on science and technology. Furthermore, science and technology represent only one set of the many "enabling mechanisms" that help us attain our objectives. Others lie in such areas as economics, management, labor, political science, institutional arrangements, etc. It is the integrated use of these enabling mechanisms, under the direction of our socio-political system, that can make the difference between success and failure. It is a mistake to look primarily at technology when something has gone wrong. It is also wrong to search primarily for technological solutions, since the best solutions generally involve a combination of technology and other means, or even a combination of non-technological means without any new technology. Hence, what we need is a more encompassing process than what we have labeled "Technology Assessment".

Second, this new process cannot be superimposed on our current way of doing things as if by afterthought; it must be woven into the very fabric of our system. It must become an integrated part of our lives, our way of planning, our way of doing business.

All too often, when we want to get something done but lack faith in the existing institutional mechanisms, we set up new mechanisms rather than ascertaining what is wrong with the existing system and fixing it. This institutional proliferation results in an ever-decreasing effectiveness of the organizations that have been "end-run". Despite the ineffectiveness, the organizations continue to exist and, to a certain extent, become overhead items. The newly created mechanisms may be superbly efficient (until they themselves get "end-run") but the overall efficiency of our system progressively declines.

All too often, progress is stifled because many of our institutions, established to cater to past needs, cannot respond to the challenges of today and of tomorrow. Our system must be nudged to evolve, to remain current and effective vis-a-vis the ever-changing nature of the problems to which they must respond. The technology assessment concept represents one such necessary nudge.

Whereas our system may indeed be in need of an evolutionary overhaul, anyone who believes that the future power-center of our policy and decision-making process can be anywhere but in the socio-political system is fooling himself.

We must think of the concept of technology assessment not as something in lieu of, or separate from the political system, but as a well-integrated part of it; as one of the means to help the political system be more responsive to the people, and become more efficient. Only in this manner will the socio-political system accept the technology assessment concept; only in this manner can the technology assessment concept contribute to the upgrading of our socio-political system; only in this manner can the concept of technology assessment take root and make badly needed contributions to our way of life.

NOTES AND REFERENCES

1. STRASSER, GABOR:

a. Methodology for Technology Assessment — Experience in the United States — Technology Assessment Seminar-Organization for Economic Cooperation and Development (OECD) Paris January 1972.

b. Solving the Environmental Problem — Too Little Too Slow or Too Much Too Fast? In: *Environmental Quality, Now or Never*, Michigan State University Continuing Education Service, pp. 281-293, 1972.

c. Technology Assessment: What's in a Name? *The Futurist*, 5, 240, No. 6, December, 1971.

d. Impediments to Societal Problem Solving — What must Happen Before we can Succeed? *IEEE Spectrum*, July, 1971.

e. Technology Assessment: What is New? *Aeronautics and Astronautics*, pp. 11 and 12, November, 1970.

f. Essential Changes in Engineering to Respond to the Needs of the 1970's — Implications for Engineering Education, *National Academy of Engineering*, Washington, D.C., Workshop on: Social Directions for Technology, July, 1970.

g. Developing a "Technology Assessment" Capability — New Analysis and Planning Methods with a Scope Much Broader than Technology, *Technology Assessment Hearings before the Subcommittee on Science Research and Development, U.S. 91st Congress, 13*, 503-510, November, December, 1969.

2. Technology Assessment, (6 Volume + Summary Report-MTR-6009), by the MITRE Corporation, Washington Operations, in cooperation with and for the Office of Science and Technology (OST), Executive Office of the President, June, 1971. MITRE Project Leader and Principal Author: Dr. Martin V. Jones; OST Project Officer: Gabor Strasser. Now available through: U.S. Department of Commerce (NTIS), Washington, D.C. Reports: PB 202778 — 01 through 07.

3. Technology Assessment Seminar, Subcommittee on Science Research and Development of the Committee on Science and Astronautics, U.S. House of Representatives, Ninetieth Congress, September 21 and 22, 1967.

4. Technology: Process of Assessment and Choice, National Academy of Sciences, July, 1969.

5. A Study of Technology Assessment, National Academy of Engineering, July, 1969.

6. The Behavioral and Social Sciences: Outlook and Needs, National Academy of Sciences, October, 1969.

7. Knowledge into Action: Improving the Nation's use of the Social Sciences, National Science Foundation, May, 1969.

8. Technical Information for Congress, Report to the Subcommittee on Science Research and Development of the Committee on Science and Astronautics, U.S. House of Representatives, Ninety-first Congress. Prepared by the Science Policy Research Division, of the Legislative Reference Service, Library of Congress, April 25, 1969.

9. Technology Assessment, Hearings before the Science Research and Development Subcommittee of the Committee on Science and Astronautics, U.S. House of Representatives, Ninety-first Congress, November 18, 24 - December 2, 3, 4, 8 and 12, 1969.

10. Institutions for Effective Management of the Environment, National Academy of Sciences/National Academy of Engineering, January, 1970.

PART IV

THE FUTURE OF TECHNOLOGY ASSESSMENT

TECHNOLOGICAL FORECASTING IN TECHNOLOGY ASSESSMENT

DAVID M. KIEFER

Senior Editor, Chemical & Engineering News, Washington, D.C., U.S.A.

The link between technological forecasting and technology assessment is both clear and strong. The tie is so close, in fact, that as the concept of technology assessment wins wider acceptance, technological forecasting and planning studies are not infrequently equated with assessments, despite the latters' focus on secondary and indirect impacts of technology on the environment and society at large as well as on technology itself. Yet much as the formulation of economic programs would be hamstrung without some understanding of likely future economic trends, so the development of policy for controlling technological change — and that is what technology assessment is all about — would be much more difficult lacking a sense of future trends in technology.

Forecasting is important not only to provide insight into new technologies that are emerging, to outline the probable future course (or courses) of their development, and to sketch the range of potential costs and benefits, but also to indicate when and where they might first appear. It provides a guide to possible future achievement against which technological progress may be judged. It is needed, no less, as a means for identifying technologies which may be useful in the future but remain undetected or undercultivated in the present. "Technological forecasting," as former U.S. Congressman Emilio Q. Daddario has pointed out, "means predicting the results of the competition among technically feasible goals for political acceptance."

Yet the bulk of the rapidly expanding literature dealing with technology assessment to date seems often to focus,in a philosophical manner, more on how the process might be institutionalized, where it should be undertaken, by whom, and for whose use, rather than on what tools and techniques (of which technological forecasting is but one) will be required to successfully carry out the assessment process. One of the deficiencies in such limited attempts at technology assessment as have been made so far is that they generally have been retrospective and restricted in scope. And for the most part they have been undertaken in response to a specific problem created by the introduction of new technology into society, rather than in anticipation of innovation. Ours tends to be a crisis-oriented society in a world where policy that is initiated only as a reaction to real and present threats is proving increasingly inadequate for coping with those threats. Assessment in the past has often been on a trial-and-error, hit-or-miss basis, with little perspective beyond short-term hazards, opportunities, and alternatives. It has viewed the future narrowly — if at all — as no more than an extension of the immediate past. Technological forecasting must come into play, therefore, to lengthen the time horizon in which technology assessment and planning can take place.

It is possible for an established or emerging technology to be assessed, in a limited manner at least, without a forecast. Much of the information needed for assessment may be already in existance; it remains only to be uncovered by a diligent search of the pertinent scientific and technical literature, reinforced perhaps by a limited amount of original, independent laboratory or field research and testing. This is especially the case when the assessment is restricted to a relatively narrow innovation or advance, such as a specific new drug, pesticide, or food additive or a modification of an established device or technique.

Until assessment becomes more accepted and widely used, therefore, many proponents of the concept see it based, to a considerable extent, on a careful and thorough documentation and evaluation (as to relevancy, authenticity, acceptability, and like criteria) of data presently available. The intent will not be just to supply a greater amount of information to policy-makers, to be sure, but rather to improve the quality, pertinence, and completeness of the information on which they must base their decisions. In many cases, a painstaking analysis of existing information sources would be adequate for spelling out incipient dangers and drawbacks posed by emerging technological developments.

But it is nevertheless also clear that if technology assessment is to probe the future in order to fulfill many of the wider objectives that its backers envision — if it is to act not just as a screening device for eliminating potential hazards but also as an early warning system and as a means for systematically evaluating secondary and indirect consequences and allocating limited technological resources with minimum waste — then the act of assessment demands more than just a documentation of existing information dealing only with present technology. A study is required also, of where emerging technological developments are likely to lead and of their possible impact on other areas of technology, as well as on society as a whole and its environment. Hence the forecasting of technology is essential, both to uncover otherwise unexpected or unforeseen problems and to disclose unappreciated or unexpected opportunities, options, and alternatives. This is especially true because assessment, if it is to be effective, should be triggered as early as possible, before a new technology has become well entrenched or developed a momentum of its own.

But can technological forecasting meet this challenge? Economic forecasting is largely a post-World War I development. Over the past 50 years it has evolved a rather formidible methodology involving econometric models, input-output analysis, the study of leading indicators, and the like. Despite these refinements, its shortcomings are clear. It remains an imprecise and uncertain, if nonetheless unavoidable, tool for policymakers in business and government.

In contrast, technological forecasting, although much in vogue now, is still in its infancy. Its antecedents lie in the writings of science-fiction authors and the speculations of science popularizers during the first half of the 20th century. But its development as a formal endeavor stems largely from the need, following World War II, for organizing and planning huge military and aerospace research programs. It has come into its own only within the past decade. Its reliability is still largely untested and its worth must still be accepted with a considerable degree of faith.

Yet even within this relatively short time many interesting techniques have been developed for predicting technological change in an organized, logical, and systematic manner. The literature on the subject has burgeoned. The current state of the art is well surveyed in several recent books [1].

In general, there are two broad approaches to technological forecasting. One of these, often termed *exploratory* forecasting, seeks to hang the future on the framework of the past and present. It attempts to delineate what alternative opportunities for the future lie within the capabilities of the present. Its rationale rests on the assumption that the technology that will be put in use tomorrow is foreshadowed by the science of today or is a direct outgrowth of current technological knowledge. Hence an important aspect of forecasting is a careful monitoring of the present.

One of the earliest — and still one of the most widely used — methods for forecasting consists of extrapolating historic data into the future, a technique which also has long been used for economic, business, and demographic forecasting. Many series of data (the maximum speed of aircraft, for example), when plotted as a function of time, form a pattern to which a curve can be fitted and then projected. A refinement of curve-fitting techniques involves the use of an envelope curve to superimpose a single general line to a number of more specific ones. Thus a single attribute (such as the maximum speed of transportation) can be forecast without stipulating the means (horse and buggy, automobile, jet plane) by which it is attained.

Correlation analysis, based on precursor events, can be used to forecast related technologies when the attainment of one specific characteristic (such as the speed of commercial aircraft) is found to normally lag in time behind another (the speed of military aircraft). Somewhat similar is the use of analogy: Technological development may be analogous to biological growth rates, for instance, or generalizations regarding the appearance of technological events of the past may be used to suggest what might happen under similar circumstances in the future.

Quantitative forecasts are also possible through the use of network analysis and decision tree techniques, although these probably are more strictly applicable to planning than for forecasting as such. A decision or relevance tree essentially is a graphical array of possible pathways toward a desired objective which displays (in a quantitative manner insofar as possible) the relationships between actions and events. It is designed to provide an orderly way for looking at the manner in which one event influences or leads to another, with emphasis placed on the points of branching where decisions regarding alternative courses must be made.

Cross-impact analysis, developed at the Institute for the Future, and its related techniques make use of a matrix which arrays and evaluates, in terms of probabilities, the influences and interactions within a set of possible parameters or future events and attempts to identify chains of causal linkages. Cross impacts stemming from one possible development or alternative course of action may change the probability of occurrence for other events within the set. Statistical analysis of changes in the likelihood of occurrence of some events as a result of changes for others permits the most important chains of events to be singled out.

Intuitive forecasting methods also have gained prominence. One such technique is scenario writing, popularized by Herman Kahn and the Hudson Institute but used more widely for sociological and political forecasting. A scenario attempts to describe, in systematic but hypothetical and largely qualitative terms, the future sequence of events that would appear logically to evolve, step by step and through cause-and-effect relationships, from any given set of conditions or recognized trends. Emphasis is placed on those critical decision points from which alternative chains of events might arise and on the simultaneous interactions between events and their environment. A single set of assumed initial circumstances can generate an entire family of related scenarios (or alternative futures), any one of which may be plausible.

The widely publicized Delphi technique, originated at Rand Corp., is designed to apply the collective expertise and intuition of a panel of anonymous experts by developing a consensus through several steps of systematic questioning and polling about future events. The polling process is carefully organized so as to minimize the biases that might otherwise arise from interacting personalities or other psychological influences within the expert panel. A feedback system is used to sharpen or narrow the forecast through successive rounds of polling which are designed to call the attention of the panelists to factors that they might have initially overlooked or dismissed and to force them to rethink or defend responses which differ markedly from the panel's overall views. Personal factors presumably cancel out one another in the process. The outcome typically is an approximate schedule of future occurrences. Delphi exercises can be particularly useful for exploring the future when adequate historic data are unavailable or when future developments are likely to be strongly influenced by such nontechnical considerations as changing social values or political feasibility.

In contrast to all these methods of forecasting from the past and present into the future, *normative* forecasting reverses the process. It starts with some future need, mission, or objective and attempts to work backward in time toward present capabilities so as to define the technological pathways and means by which a goal might be reached and to identify the technological barriers which must be overcome in the process. The aim is less to prophesy than to "invent" the future, with the focus not on that which might happen but on that which should happen. Since such goal-oriented forecasting by intent is designed to be self-fulfilling, normative techniques tend to be more suitable for planning than for forecasting alone. (Of course, any forecast will tend to be self-fulfilling, when it is optimistic, or self-defeating, when it is pessimistic, if it is sufficiently credible to encourage steps toward attaining a promised benefit or toward avoiding a disagreeable or unwanted outcome.)

Many other techniques for technological forecasting have been formulated. As a whole, however, they are mostly modifications, combinations, expansions, or refinements of those summarized above. In their attempts to map the future, forecasters in a few short years have constructed a surprisingly complex (and often jargon-ridden) methodology.

It would be unreasonable to single out any one or more of these techniques as most appropriate for technology assessment. Each assessment must be designed to fit the specific requirements and characteristics of the particular problem under investigation.

At first glance, normative approaches might seem best in the light of technology assessment's fundamental concern with goals and priorities. Unfortunately, the establishment of national goals that are concrete and well-defined, rather than merely statements of broad generalities or platitudes, has met with scant success. And even more difficult than the identification of publicly acceptable goals is the definition of explicit priorities for attaining those goals.

Perhaps the best approach would be an eclectic one integrating several techniques of forecasting in order to sharpen the reliability of the overall forecast while minimizing the shortcomings to be found in any one method. Intuitive Delphi forecasting, for example, might be used in conjunction with more quantitative methods, such as curve fitting, that would provide a statistical link with the past. Or a carefully prepared scenario could be used to provide a broad perspective for the deliberations of a Delphi panel. The use of cross-impact matrices has increasingly proved useful in conjunction with Delphi exercises.

The difficulty in combining techniques, of course, lies in the cost and complexity that ensues. Forecasting is but one step in the overall assessment process, and most often a relatively early one at that. A forecast that requires several months to prepare would only unduly postpone the completion of the final assessment. It is possible to conceive of very elegant and exhaustive forecasting procedures. But in practical terms, decisions must be made within a time frame that does not allow unlimited delay in order to generate new information or sharpen techniques. Assessments must be timely, therefore, if they are to serve as early warning devices and be of practical rather than merely academic interest. Of course, the earlier they are done the fewer hard facts and figures will be available. Nevertheless, while forecasting is designed to reduce uncertainty in the assessment process, it cannot be expected to totally eliminate it.

For all its apparent sophistication, moreover, technological forecasting is still in its formative years and must be viewed with skepticism and caution. It is easy — perhaps all too easy — to be overly critical of what has been accomplished so far, however.

Quantitative techniques for fitting curves or correlating related trends can be criticized for being naive or simplistic. (One of the advantages of such methods, on the other hand, is their intrinsic simplicity and the ease with which they can be applied, given an adequate array of historic data, and understood.) Extrapolations and correlations are meaningful only if it is clear that past trends will continue unchanged and that present patterns are likely to repeat themselves. These methods are most suitable for relatively short-term forecasts dealing with rather limited technologies. They cannot be used to forecast beyond the range of what is currently known. They tend, too, to focus on developments that can be described statistically to the exclusion of that which can be stated only qualitatively, although the latter may be no less significant or real.

Analogy, meanwhile, can only suggest possible courses for future occurrences. Cross-impact techniques are limited to interactions between pairs of events or dichotomous situations and are dependent on the ability to assign reasonable probabilities to future events.

The Delphi method hinges heavily on how well the panel of experts is organized and how well informed it is, as well as on the formulation of the questions put to the panel. Emphasis on consensus tends to stifle minority opinions, no matter how intelligent, for the sake of general agreement. And the Delphi procedure also is likely to be cumbersome, time consuming, and comparatively expensive. Wide-ranging scenarios may also be time consuming, and their validity is difficult to evaluate.

No technique of technological forecasting, moreover, can really envision the flashes of innovation or the unpredictable discoveries which bring about much technological change. Nor can the social, political, economic, and other external pressures and constraints which significantly influence technological development readily be taken into account. Forecasting that is purely technological tends to see innovation as something that will happen regardless of social change and policies and to assume a steady-state society and a steady-growth-rate economy. Even when social factors are brought into play, they generally are done so in terms of the impact of technology on society rather than the other way around. Yet is is clear that social values can affect technological priorities or the allocation of resources; technological innovation often results as a response to social or economic demand and takes place only if a willingness exists on the part of society to put a new technology to use. It is rather ironic the aerospace industry, which probably has done more than any other to apply technological forecasting to practical needs, has had the greatest problems in coping with changes in technological priorities and political policy during the past couple of years. The fundamental difficulty in foretelling social and political change — or of even devising meaningful social indicators for measuring such changes statistically — remains a serious obstacle not only for technological forecasting but for technology assessment as well.

Any forecast for use in technology assessment, of course, probably must include a large component of subjective judgment, for much of the insight needed can be acquired in no other way. Quantitative statements about the future may appear more scientific (especially if they are presented in the form of a computer readout), but qualitative forecasts are not necessarily less realistic or objective. Quantitative measures should be included where they can. But judgment is often the only source of information concerning future social, political, and environmental effects. To exclude it from the forecast would only unduly restrict the forecaster's perspective.

It is important, too, that the forecast be presented in terms of alternative futures. By identifying possible alternatives or options, together with the probability of their attainment (especially in the light of reinforcing or conflicting events), the forecaster provides an array of possible actions from which the decision-maker can select a single course of action.

Questions of reliability can be resolved, at least to some degree, once it is recognized that any forecast of more than the near-term future must be continuously reviewed and revised if it is to remain useful. Any technological forecast has a limited shelf-life, especially when it embraces areas where technological change is most rapid. All long-range forecasts are certain to become obsolete before their terminal data is reached. Hence, both forecasting and assessment must be ongoing processes subject to periodic updating and modification. And any forecast should have continuity designed into it in the form of interim milestones that permit its reliability to be tested periodically prior to its completion.

In any event, to judge the worth of a forecast by stressing its "accuracy" (which can only be verified by hindsight) is a mistake, although a common one. A technological forecast, after all, is less a prediction of what is likely to happen than a statement of potential technological capabilities and opportunities which will be realized only after rational choice has been made to implement the forecast.

Forecasters must not lose sight of the present while they engage in the intellectually stimulating task of weighing the future. Their task usually lies less in foretelling what will be at some time to come than in providing a frame of reference for current policy making, in reducing at least some of the uncertainty that befogs the critical decisions that must be reached here and now, and in minimizing the technological surprises that thwart systematic planning. The greatest value of a forecast may well be its ability to stretch the imagination or catalyze fresh thinking. It is worthwhile, as well, if it just serves to pose previously unconsidered questions, leads to a searching review of established assumptions and interpretations, or points to unexplored opportunities. Raising new questions may be more important in the long run that merely devising better answers to questions that have gone stale. The very act of forecasting, if it leads to a disciplined, logical evaluation of current conditions and spotlights problems and deficiencies, is in itself of value.

Nevertheless, the user of a forecast must be wary that the forecast, whatever its intended purpose, is not biased or self-serving or nothing more than a projection of the forecaster's own personal beliefs, commitments, and prejudices. Forecasts have been made merely to enhance the prestige of the forecaster or to win for him an audience. They may be undertaken, also, to gain support for a pet project or program or acceptance for a decision already reached. Thus it is important to know not just how a forecast was made, but why it has done as well, in evaluating its worth.

The point is that if a forecast is to prove useful, either for technology assessment or for any other purpose, it must be credible. And it must be communicated to decision-makers in terms that they can relate and react to. Plausibility is more important than accuracy or sophisticated techniques and tools. Only when a forecast is believable and captures attention will anyone take serious action on it. But despite the increasing vogue of futures research, the acceptance of technological forecasting for policy making and long-term planning has been far from unanimous.

In part, the blame may rest with the forecasters themselves. Their focus on the mechanics and methodology of forecasting and their interest in turning what had heretofore been a rather arcane and obscure art form into something more of a scientific discipline is readily understandable. Yet they seem to be doing all too little to win converts or credibility. The rapidly expanding literature of technological forecasting and long-range planning is especially opaque, if not impenetrable, to the unitiated, muddied as it is by an unfortunate tendency toward coining obscure words and acronyms.

Forecasters frequently seem more enthralled with the entertaining tasks of model building, manipulating and massaging series of data, and imposing some sort of formal, stylized structure on the seemingly random process of scientific discovery and technological innovation than they are with the more mundane chore of explaining to the world outside what their studies and speculations are all about or how they might find practical application. Increasingly sophisticated and complex methodology may appear designed, as a result, less to make forecasting more reliable and rational than to conceal its shortcomings and veil its relevance to the world at large. At the same time, excess concern with methodology may only stultify the imagination and intuition that must be an important part of even the most quantitatively based technique. Technological forecasting, in short, is in need of plainer language and of writing unencumbered by dense and tangled jargon.

The importance of forecasting and futures research is clear if we are to set meaningful national goals and priorities, to allocate our finite resources of money and technology over an increasingly widening range of national needs, and to assess the direct and indirect consequences and benefits of accelerating technological change. The advent of technology assessment as a formal, institutionalized endeavor presents forecasting with both a challenge and perhaps its greatest opportunity.

Until we learn to forecast better, progress in technology assessment will be slow and uncertain. Nevertheless, we must apply whatever methodological tools are now available as best we can, meanwhile honing them through wider use.

Economic forecasting has advanced largely because of a growing government commitment in the past quarter of a century to promoting economic growth and dampening the fluctuations of business cycles. Technological forecasting should benefit similarly once technology assessment becomes an accepted process for guiding technological growth.

REFERENCES

1. JANTSCH, ERICH, *Technological Forecasting in Perspective,* Paris, Organization for Economic Cooperation and Development, 1967.

2. BRIGHT, JAMES R., editor, *Technological Forecasting for Industry and Government,* Englewood Cliffs, N.J., Prentice-Hall, Inc. 1968.

3. CETRON, MARVIN J., *Technological Forecasting: A Practical Approach,* New York, Gordon & Breach, 1969.

4. AYRES, ROBERT U., *Technological Forecasting and Long-Range Planning,* New York, McGraw-Hill Book Company, 1969.

5. CETRON, MARVIN J. and CHRISTINE A. RALPH, editors, *Industrial Applications of Technological Forecasting,* New York, John Wiley, 1971.

FOUR AMERICAN FUTURES: REFLECTIONS ON THE ROLE OF PLANNING*

HAROLD A. LINSTONE

Director, Systems Science Institute, Portland State University, Portland, Oregon, U.S.A.

ABSTRACT

Long range planning is considered in its relation to four alternative futures for the United States: surprise free, internal reconstruction or reform (maximum planning), a new society (minimum planning), and authoritarianism. Our ignorance in understanding complex systems frustrates meaningful holistic planning at a critical time. This fact, combined with the uncertain future decision making environment in this country, suggests that highest priority now be placed on three "good risk" elements: system cognition, creation (design), and communications. Several steps in these directions are outlined. They involve basic systems research, modeling and gaming of complex systems, use of the computer as a high level communicator, and introduction of two new types of organization. By focusing on these three elements a gain is possible in the soundness of decisions which have long range impact in each of the environments.

In the United States the air is rent today with the pronouncements of future oriented thinkers. There are optimists, like Charles Reich and Buckminster Fuller, and pessimists, like Donald Michael and Paul Ehrlich. We are deluged with new labels — future shock, consciousness III, the technological man, world dynamics, Spaceship Earth, and the Japanese Century.

The basis for today's profusion of ideas and slogans is clear enough. In the technologically most advanced country an oppressive sense of malaise has intensified the probing of the winds of change.

I. THE PRESENT

The rapid evolution of technology in this century has introduced grave imbalances across the spectrum of social structures. Today the family is undergoing unprecedented stresses. Mobility splits apart family clans and children from parents. The difference in the learning environment for young and old reduces the authority of the latter. Communications tie in the young to the outside world directly, with the parents no longer acting as filter. The value systems of young and old indicate widening differences.

Outside his home the individual has been transformed into the organization man of the corporate state, a highly structured, technology centered organism. Its cells form a bureaucracy which has now grown to a size and power that make it extremely difficult to control. Organization charts change in detail but not in concept. Meritocracy demands that the individual fit into the "system" if he wants to prosper. His own views may differ radically from those of his organization, but they must remain submerged. There is no room for a General Gavin in the military or a Secretary Hickel in the Cabinet. Loners such as Ralph Nader and Robert McNamara are the Don Quixotes of this generation; they are certainly short term nuisances, but their long term impact remains to be demonstrated.

A liberal like Lyndon Johnson becomes more conservative as he enters the White House and a conservative like Richard Nixon becomes more liberal. The chief must adjust himself to the system; the gigantic system cannot adjust itself to an individual. Technology has altered the character of the power held by the President: the "system" reduces his decision space but increases his communication space. The vaunted democratic voting concepts lose much of their significance in the process and the bumper sticker advice on the preference between candidates — "vote no" — is more meaningful each year.

The same situation prevails in today's large corporation. As one writer has put it, the executive suite is empty — there is no one there to lead. The chief is in reality a coordinator and monitor. The technologists bring forth the new product ideas and their success or failure can make or break the corporation, no matter who sits in the front office. Not surprisingly, the actual corporate objectives (e.g., technical leadership, employment stability) often differ from the stated objectives (e.g., profitability). The stockholder meeting is as much an annual ritual as the board chairman's Christman speech to his employees.

The effectiveness of the corporate state is clearly a matter of serious concern. In the wealthiest and technologically most advanced nation the huge hierarchical government structure is today failing to fulfill those functions which constitute its major concern. It is unable to (1) protect its citizens against destruction from external sources (nuclear attack), (2) efficiently wage combat against a much smaller and poorer people, (3) fight a successful "War on Poverty" at home, (4) provide a physically healthy and safe environment (traffic safety, clean air and water, infant mortality rate, expected life span of its men) and (5) plan the evolution of its society beyond the industrial stage.

This concern is reflected in the decline of the proportion of Americans having high confidence in the Federal Government — in 1964 62%, in 1970 37% [1].

Let us now turn to the global stresses. The precipitous growth of population and technology (e.g., communications, transportation, energy consumption) in the last fifty years has shrunk the earth to a "global village." In one lifetime we have suddenly become aware that we must consider the earth as a spaceship, a closed system of very limited size. Environmental protection and natural resources can no longer be viewed as national problems; they have become global. We observe that the undernourished and undereducated two-thirds of the world is mulitplying much faster than the wealthy one-third: it will constitute three-quarters by the year 2000. This means that the average education and intelligence are declining. Global education via satellite transmission is feasible, but televised programs which give evidence of the tremendous material gap between the world's rich and poor may exacerbate urban unrest.

If the state has failed to perform its functions, the modern supranational organizations have, until very recently, exhibited an even worse record. The League of Nations, the Organization of American States, and the United Nations have been largely ineffective in solving regional or global problems. No effective global control of the environment, population, conflict, use of natural resources, or capital investment has evolved. Only the European Economic Community is developing into a success story and its unifying impact may extend beyond the economic sphere.

The current approach to long range state planning (ten years and beyond) is perhaps best illustrated by the modus operandi of that government branch which has undertaken the largest planning effort — the defense establishment.

Figure 1, adapted from Jantsch [2], shows the mechanism for change. Initially the focal point is the dialogue between the planner in one branch, say, the tank area, and the technological community (blocks 10 and 11). Normative thinking is minimal. There are, of course, alternative future scenarios (block 2) available to the tank planner and he studies them. If a scenario indicates that any future war will be nuclear, he considers how to use tanks in nuclear operations. If a scenario assumes that any future war will be a guerrilla action, he determines how to use tanks in guerrilla

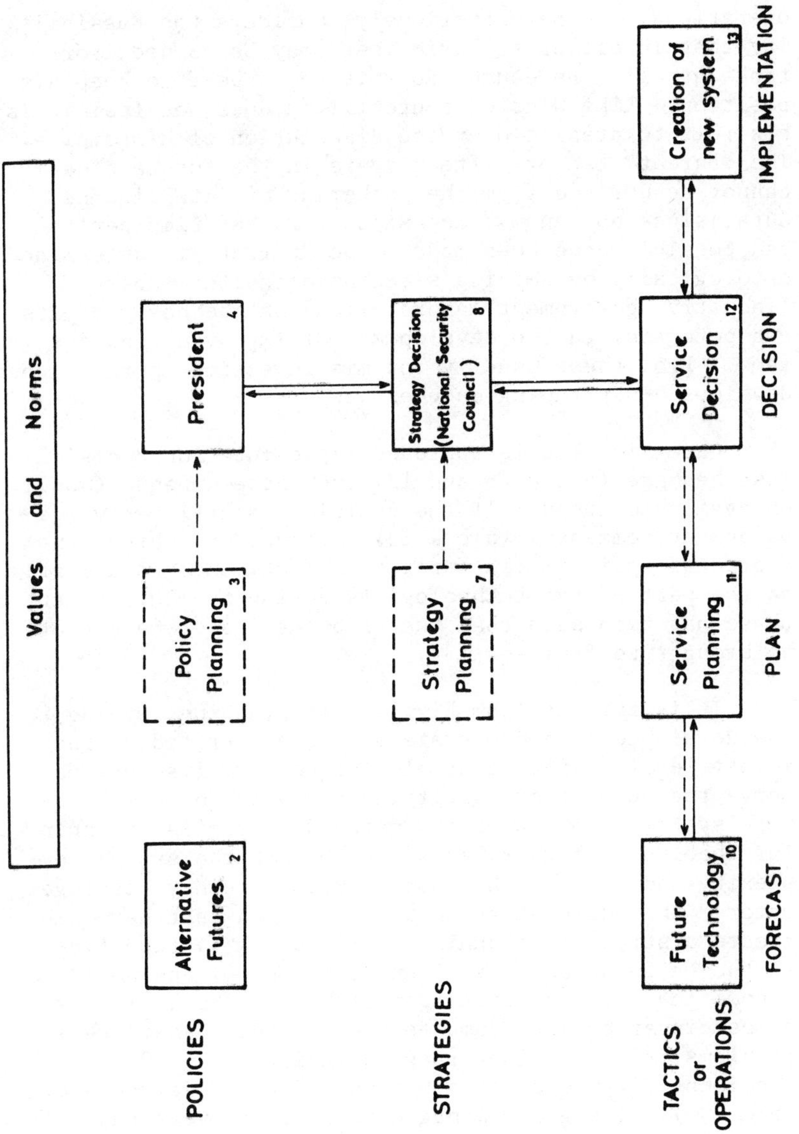

Fig. 1. Planning the realistic setting.

operations. He never seriously considers the possibility that in either scenario there may be no need for tanks at all. He cannot do this and expect to keep his position. Like Winston Churchill he does not feel it is his role to preside over the dissolution of his empire. The "threat" ten or fifteen years in the future clearly cannot be derived from the gathering of intelligence data, since no "enemy" decisions for that time period can possibly have been made. The "threat" is determined ontologically by America's technological community (industry, government laboratories) on the basis of its own potential in the development of improved tank systems. From these usually optimistic estimates the planner develops his planning document.

Once his plan is approved implementation decisions must be made (blocks 8 and 12) and these depend strongly on available funds. If the decision is affirmative development commences (block 13). Often, of course, the degree of optimism (in terms of effectiveness and cost) on the part of the technologists during the development phase has been such that the stipulated systems cannot be brought to fruition [3].

It is evident from Fig. 1 that planning leading to the development of new systems is concentrated at the lowest level. If used at all, higher level scenarios serve the purpose of justifying the need for new hardware systems. It is rarely matched by similar planning for changes at higher levels where both the system complexity and the time horizon increase. Henry Kissinger entered the Administration with the idea that he would create a strategic planning staff (block 7) reporting to the National Security Council (block 8) and would not become involved in short term problems. Such efforts almost invariably succumb to the law that "short term problems drive out long range planning." In 1969, President Nixon created a National Goals Research Staff (block 3). The group managed to last one year and quietly died after publication of its first (and only) report. The report did not develop alternative goals or design new policies so as to permit public debate; rather, it consisted of expansive generalities and truisms such as the following:

> "The central ingredient in the development of a growth policy will be for the American people to decide just what sort of country they want this to be ... We will have to develop better institutional arrangements for the people to relate to the leadership and better mechanisms of policy analysis to serve all parties" [4].

Earlier national goals studies, such as that done in the Eisenhower Administration [5], similarly accomplished little besides salving the executive conscience.

There is no mechanism for the design of new systems (institutions, instrumentalities) at the higher levels. There is little feedback to change norms and values or accept new ones. Interaction between blocks at the different levels is sporadic at best. The most common driving force for change resides in technology at the lowest level (block 10). The result is today's imbalance between continuously changing hardware systems and unchanging or slowly changing policies and strategies. It took over twenty-five years to change America's China policy and fifteen years (and four presidents) to change its Vietnam strategy. We still have the Rural Electrification Administration of the 1930's with us.

To provide a perspective Fig. 2 depicts the ideal controlled change process — a dynamic feedback system within government and between government and people. All blocks are strongly interconnected so that a change at any point reverberates throughout the system. It should be noted that arrows lead into the box "values and norms" as well as out of it. This signifies that values often change in consequence of actions, in contrast to the popular view that values should inform the planning process. In this approach to planning technology is no longer an overpowering driving force. Control and self-regulation are feasible and change is homeostatic.

We may further characterize system levels as shown in Table 1.

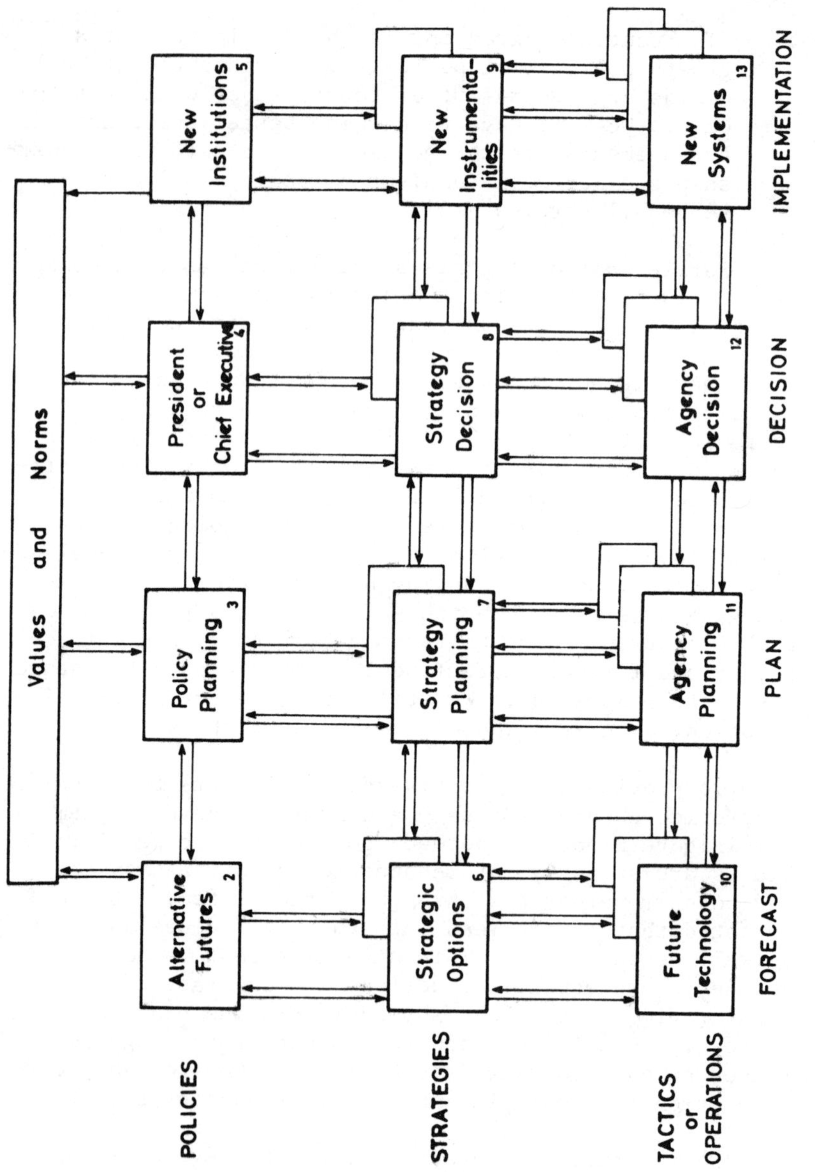

Fig. 2. Planning the ideal setting.

TABLE 1. System Levels (Based on [6]).

Level	System description		
Policy	Transdisciplinary	Multigoal	Multilevel
Strategy	Interdisciplinary	Multigoal	Two level
Tactics	Subsystems and Technological Systems	Single goal	One level

Recent studies on large scale systems exhibit several vital characteristics (Table 2):

TABLE 2. System Characteristics.

high degree of interdependence between parameters
non-linearity of relationships
system stability in the face of most changes in control, i.e., small perturbation
discontinuous (catastrophic) changes in behavior with certain continuous (gradual) changes in control

The overall impact is the occurrence of unexpected or counterintuitive consequences in response to intervention in the system. Jay Forrester's system dynamics approach applied to urban areas shows that an intuitively obvious step, building housing for the poor in the central city, has an undesirable effect: accelerated deterioration of the urban environment [7]. This conclusion has been dramatically supported in St. Louis, where well-intentioned high rise, low rent housing projects built in the 1950's have created a dismal "new" ghetto and are now being torn down [8]. On a global scale an intuitive action, birth control, merely has the effect of shifting crises — from an overpopulation to a pollution disaster. Industrialization of underdeveloped countries and world reduction in the usage rate of natural resources are similar non-solutions.

The work of John B. Calhoun [9] with a population of mice in a closed environment also yields counterintuitive results. Starting with a few mice, the population does not reach the expected level and then stabilize. It begins to decline after reaching a point considerably below the calculated limit and then dies out — despite adequate food and physical space.

Holling and Goldberg observe remarkable similarities between ecological and complex urban systems [10]. Application of insecticides to control malaria in Borneo and cotton damage in Peruvian valleys leads to the realization that ecological systems function as interdependent systems, exhibit historical linkages as well as spatial linkages, and have a non-linear structure. Turning to modern cities they find the same characteristics. They note that all policies tried have

> "unexpected consequences ... The basic trap is that people conceive of a small fragment of a whole and the very best policy for that fragment can produce the reverse effect through interaction with other parts of the system ... radically different hypotheses can have little effect on the output of the model. The effect is absorbed by other parts of the model ... There are, however, strategic points in the model that have a very sensitive effect on the output" [11].

With this brief background, let us now consider four alternative futures for the next generation and focus on the role of planning in this context.

II. THE SURPRISE FREE FUTURE

There is ample justification for extrapolating the recent "technological era" into the next quarter century. Figure 3 (dashed border) shows the essence of the governmental change process in this case. The bureaucracy remains most massive at the lowest levels; it constitutes a gigantic machine which functions autonomously without

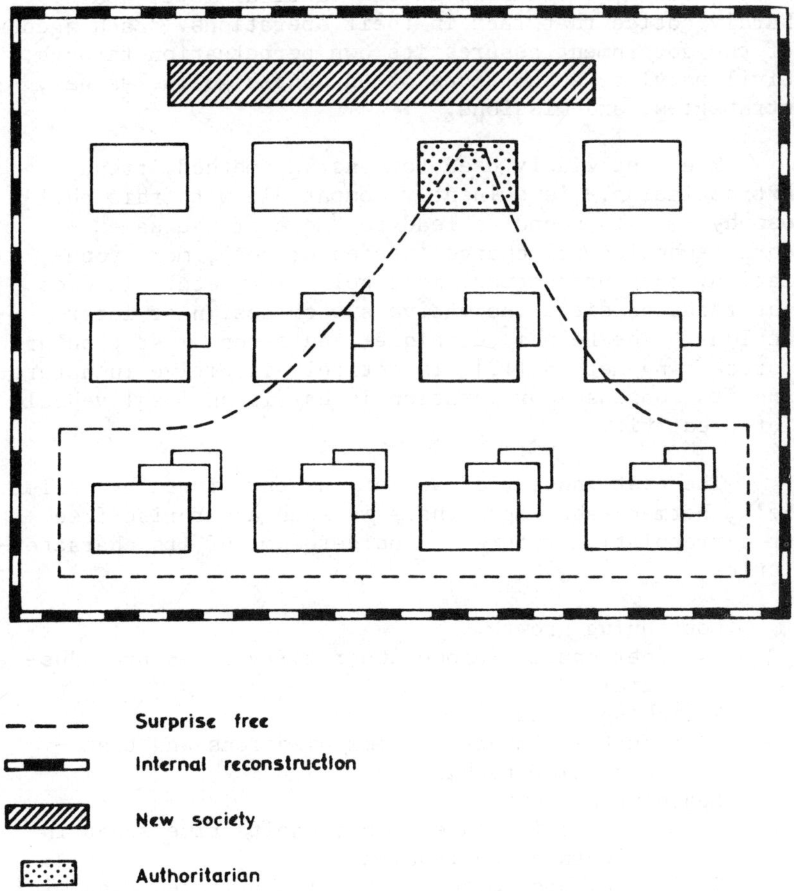

Fig. 3. Focus of planning activities. (See Fig. 2 for connecting lines).

much of the feedback and control shown in Fig. 2. The most effective planning involves one goal, one level systems (Table 1). If there were no president in the White House the planners at the tactical level would hardly notice that fact in their operations. Each agency of the government assures its own perpetuation through civil service, assumptions of unchanged goals, values, strategies, and missions.

The most widely used forecasting method, trend extrapolation,* is eminently compatible with this philosophy — it is bound to lead to "more of the same" — more technological change (needed or not), more population, more production, more pollution, etc. It does not focus on either normative system design or interactions. The Delphi technique, the second most popular forecasting method [12], is entirely intuitive in nature and its consensus orientation is hardly an ideal vehicle for creativity.

The post-industrial society as envisioned by Daniel Bell, Herman Kahn, and others is such a surprise-free or extrapolative future. Consider some of its characteristics:

- Continuing growth
  - per capita income about fifty times preindustial
- Shrinking world
  - further impact of communications and transportation technology
- Learning society
  - further increase in per capita time spent in a learning environment
- Shift from production to service oriented economic activities
  - this shift is already in progress

---

*Cetron reports that 90% of the input provided to the Navy Technological Forecast project consisted of trend extrapolations and correlations [13].

Business firms no longer the major source of innovation
with government a multibillion dollar source of R & D funding
since 1940 this trend is already in full swing (e.g., the first all-electronic computer, the ENIAC, was built for the U.S. Army; the Boeing 707 was based on R & D performed previously for a U.S. Air Force tanker)
Widespread cybernation
Computers — and hence automation — have constituted one of the most dynamic growth technologies of the post-World War II period

As W.I. Thompson observes,

> "The difficulty with the liberal imagination is that it is so owned by the myth of progress that it cannot think of the future in any other terms except more of the same" [14].

Our scenario for a surprise-free U.S. may be summarized as follows:

"Responsible center" middle-of-the-road administration
Continuation of present trends (government follows more than it leads)
Maintenance of current institutions, few steps to eliminate obsolete ones
Dissident elements alternately accommodated and restrained
Continuation of current level of military spending and reduction of manned space programs
Some recognition of a shift in the threat to survival from external military to non-military external and internal problems
Some increase in social and environmental control programs (necessarily superficial approach in view of existing institutions and vested interests)
Large scale action only when crisis or disaster strikes

Progress measured in terms of GNP (total and per capital)
Lip service paid to long range planning, little motivation for anticipatory design of new systems

The surprise-free future is, in fact, one to be expected of a people attuned to the optimistic conviction that "growth is good." Continuation of high military spending is based not on a "threat" but on the thesis that there is no surrogate for the non-military functions (socio-economic, behavioral) of a defense economy in the United States [15]. The recent sequence of events may be cited in its support: decline in defense spending, massive unemployment, inability to shift to large new technological programs in the civilian sector, and finally reversion to higher defense spending.

In this future long range planning is itself an activity characterized by "more of the same." Detailed planning is carried out in the dominant low level bureaucratic structures. Sporadic activities, control and feedback, occur at higher levels. The Department of Commerce planners concentrate on steps to assist industrial growth, the Department of the Interior planners forecast natural resource usage rates; each tends to ignore the cross-impact. Lip service is paid to dynamism in long range norms and values and to alternative futures design. Full emphasis is placed on short range policy and strategy objectives.

The justifications are strong:

a. The political horizon of the decision maker

The electorate has a short memory and interest span.* Long range benefits often entail short term sacrifices — with resulting unpopularity. The short term pay-off has a visible impact while the long term pay-off is drastically discounted. New housing for the

---

*This is typified by the voter's proverbial comment: "But what have you done for me lately?"

urban poor, increases in welfare payments, and support of large failing business concerns are acceptable; incentives to stop migration of the poor to decaying central cities, to eliminate obsolete industries, and to limit energy consumption are not.

There will, of course, continue to be exceptions, as there have been in the past. Fear of survival, the most potent basis for large scale public funding support, leads to occasional programs with long term payoff. The popular fear induced by the Sputnik permitted a shift from the minor Vanguard to the major Apollo program. A specific goal, a manned lunar landing, was proclaimed in 1961 to be attained by 1970. More often the long term payoff is unplanned (e.g., the impact of the KC-135 tanker program on commercial jet aircraft.)

b. Resistance to change

The government bureaucracy reflects the individual's search for security. Rapid technological change is assumed to continue and even accelerate "future shock." Pressure to keep some "fixed points" in this environment increases. People who have not experienced a previous disaster "interpret signs of danger as signs of familiar normal events until it is too late to take effective precautions" [16]. A perfect example is the inability to "sell" a civil defense system in a cold war environment. Perhaps the strongest argument for a surprise-free future was given by Hamlet — we would "rather bear those ills we have than fly to others that we know not of."

c. System complexity and the state of the art of planning

Ozbekhan writes on planning:

> "We have neither a general theory nor special theories of it, we have no widely accepted language that communicates its logic, no structure that organizes its predictive or explicatory propositions into an operational model, we do not even have such propositions. In sum, whatever it is that we call *planning* lacks that solid foundation

> from which it should be possible to strengthen and flesh out, to direct and evolve, to dispute and adjust the *corpus* of the practice itself" [17].

Dror sees the need for a scientific revolution to develop new paradigms to place policy sciences on a sound footing [18]. And Michael declares [19]:

> "Our technological ability to change our world will exceed our ability to anticipate whether we are using it wisely.
>
> "We are almost certain to face disaster if we don't plan ... but we are also almost certain to be in deep trouble even with planning because our best plans will be developed and fostered by limited human beings picking and choosing among limited knowledge, very often ignorant of the extent of their own ignorance."

If we are only now taking the first tentative steps toward a systematic basis of planning [20], it is a reasonable assumption that we will not have reached such heights of success in the next decade as to give the decision maker the confidence to base strategy and policy decisions (involving interdisciplinary and transdisciplinary systems) on planning.

The approach in the surprise free future has the obvious advantage that the future is not constrained by a firm plan, and the equally significant disadvantage that major crises are not anticipated and thereby averted.

III. TWO ALTERNATIVES — MAXIMUM AND MINIMUM PLANNING

The scope of alternative U.S. futures considered for the next generation ranges from effective full scale planning at all levels to its complete rejection. Both possibilities are currently being expounded by intellectuals in the United States.

A. Internal Reconstruction or Reform

This future is based on the conviction that the surprise-free extension of the present is doomed to disaster and that the needed changes in the system can be understood, planned, and implemented without dismantling the basic framework. This philosophy is expressed in the Bellagio Declaration:

> "We ... feel compelled to put forward our views on the potentials of planning as a method of approach to solving many contemporary problems ...
>
> "Having discussed the present state of the art of planning and the diversity of its new approaches we believe that its possibilities including the appreciation of human values transcend mere technocratic objectives. Scientific attack on those problems of complexity and interdependence is a matter of the utmost urgency ... we are ... convinced that a corpus of knowledge already exists capable of immediate exploitation, and that there is expectation of further and fruitful development ...
>
> "The need is to plan systems as a whole ...
>
> "Planning must ... be undertaken simultaneously at different levels and must be integrated across these levels ...
>
> "We take it upon ourselves therefore to issue this collective warning that social and technological developments already clearly foreseen can exacerbate matters beyond any hope of peaceful relief. In doing so we express the belief that a basis of remedy already exists to help man to define and create his own future" [21].

The focus is on systems involving nature, man, society, and technology and the aim is bold: creation of a society with a high quality of life in dynamic equilibrium with its environment, offering a new range of opportunities for individual fulfillment and homeostatic societal evolution.

We give the Internal Reconstruction scenario the following characteristics:

Inward rather than outward political orientation (i.e., reduced role in world affairs)
Major shift in values and priorities to address critical internal problems
Belief that changes can be effected within the "system"
Militarily a return to the Eisenhower era, i.e., reliance on strategic offense/defense and greatly reduced "general purpose" forces
Creation of new institutions, but also proliferation of obsolete ones
Internal stability sought through combination of strong, modernized police forces and a large scale attack on underlying social and ecological problems
Technical task forces reminiscent of the Manhattan project formed to work on selected crisis problems
Serious effort to implement integrative and interactive planning involving all levels with the objective of purposeful continual self-organization and self-regulation*

There is recognition that survival of the nation depends more strongly on internal reconstruction than on defense against external threats. The need for adaptive and inventive systemic planning is accepted and the ability to activate most, if not all, of the feedback shown in Fig. 2 is considered essential (Fig. 3). Homeostasis during change is of utmost concern and loss of control is viewed as the true threat.

One visualizes not only massive support of social systems analysis and synthesis, but development of action

---

*The term used by Jantsch for the most advanced level of system planning.

oriented long range planning staffs at all levels, interactive mechanisms (e.g., policy research group — legislative advisor dialog) and line executives with responsibility solely for long range planning. Dennis Gabor's theoretical concept of planning consciously to maximize the degree of freedom of the next generation of planners is further developed [22]. One consequence of the application of this idea may be a drive to reduce population (not merely population growth).

Business, at times more successful in long range planning than government (particularly at the strategic level), is also harnessed to these tasks. It is interesting that in its own visions of the future, management does not see profits and sales, but total control and centralized direction of the world's resources by large organizations under professional management to achieve the appointed goals of national societies [23].

It is inevitable in this environment that bureaucratic structures from the State Department to the local school system must undergo drastic changes. More fluid, ad hoc teams and fewer permanent, static organizations are probable. Potential crises must be vividly presented to the public in ample time to gain the necessary support for major corrective actions.

This alternative may be condemned as a planning utopia. Its weaknesses correspond to the strengths of the surprise-free future: the political nature of the decision maker, resistance to change, and the state of the art of effecting homeostatic change in complex systems. Total system planning affects all subsystems, hence subsystem bureaucracies and vested interests can be expected to oppose it. The Bellagio Declaration itself is hardly reassuring in its wording. A "basis of remedy" is not a remedy; an "expectation of further and fruitful development" is not a realization.

The use of communications technology to date also offers little comfort. An average television news program contains less than 2% of the news wordage as the same day's New York Times. Vested interests manipulate telephone feedback systems. One minute spot announcements replace the Lincoln-Douglas debates in presidential campaigns [24].

Recent advances in system dynamics provide no assurance that ten years' study will result in a model which includes all critical parameters, exhibits sources of instability, and correctly simulates all major interactions in the system. And even if the model is successful it is still extrapolative and not necessarily a creative tool in the design of new systems.

One must even be prepared for the possibility that successful planning will prove totally incompatible with a democratic or participative philosophy of government. At the very least, a new elite, a "planning bureaucracy," may result. At worst, failure and disillusionment resulting from implementation of decisions based on apparent "in depth" planning may have catastrophic consequences.

B. A New Society

A prospect receiving increased expression is a more revolutionary (not necessarily violent) change. It is based on the belief that the change in the governmental system (blocks 2 to 13 in Fig. 2) from within, as suggested in the Reconstruction alternative, is a false hope and cannot succeed. Charles Reich voices this idea in his popular book, "The Greening of America."* His concept involves a drastic change in values (block 1, Fig. 2) on the part of a majority of the *individuals* in the nation. Control is concentrated at this point and the government (blocks 2 to 13) is forced to adapt to changes originating there (Fig. 3). Furthermore, value changes are in the direction of non-materialistic desires. Its advocates insist that man has been subordinated to technology and that he must reassert his dominance of, and control over, technology.

---

*In Reich's terminology the "New Society" corresponds to Consciousness III [25].

This is, in many respects, a romantic anti-establishment future. There is skepticism of rational analysis and planning. The failures of the 1960's — Vietnam, social legislation — are blamed in part on the expanded use of rational tools. The emphasis is on the individual rather than on group efforts, on compatibility rather than competition, on freedom rather than group discipline, on the present rather than the future.

> "(Consciousness III) does not try to reduce or simplify man's complexity, or the complexity of nature. It values the present, not the past, the future, or some abstract doctrine of mythical heaven."*

The scenario can be described as follows:

- Most profound change — a second American Revolution
- Belief that the "system" and its institutions are unworkable and must be restructured
- Drive to eliminate obsolete institutions and rewrite the Constitution
- Emphasis on experimentation involving new social arrangements and groupings, encouraging diversity
- Decentralization of decision making, increased autonomy of regional and local governing bodies**
- Minimal formal long range planning
- Massive decline in funding of federal programs (e.g., national security except strategic retaliatory systems, space)
- Shift of financial control from federal to regional and local organizations
- Decline in influence of business, military

There is a strong classical utopian flavor to this alternative. It denies the negative aspects of man —

---

* Cf. [25], p. 426.

**This direction may, in fact, also be a consequence of the full scale planning approach in the Reconstruction future (cf. [11]).

his aggressiveness which finds an outlet in competition, his desire for security which is reflected in the permanent bureaucratic structure, his penchant to follow rather than lead. A fundamental problem is that no one has yet developed an alternative to bureaucracy, although Toffler's "adhocracy" is perhaps a beginning. One cannot suppress the fear that nihilism and anarchy will be the long first step in the creation of this new culture.

Analysis is suicidal to any revolutionist. Logic must be replaced by fanaticism when the chance of success is small (as it nearly always appears a priori) and action is nevertheless desired. Even after success is achieved the philosophy of this alternative assigns a small role to formal planning. It may be described as trial and error. Experimentation with new solutions determines the changes to be implemented. Increased field experimentation has been recognized as a desideratum to improve planning even in more orthodox futures [26]. While this process accentuates innovation and creativity, it may prove impractical in the case of very large systems (e.g., cities, states). Small scale experimentation frequently cannot represent such systems and large scale, or total system, experimentation entails a tremendous financial burden. Decentralization may help in some areas but often results in the abandonment of the holistic view. The obvious fluidity of such an environment is also likely to create massive "future shock." Finally, an experimental approach is slow and the crises may descend upon us too rapidly to permit the luxury of time.

IV. ANOTHER ALTERNATIVE — AUTHORITARIANISM

Failure in any of the three preceding futures can easily drive a state toward an authoritarian or totalitarian regime, i.e., counterrevolution. Both Napoleon and Hitler were the harvest of new or revolutionary experiments (i.e., French revolution, Weimar republic).

Intellectuals have succumbed to LSD in their search for new doctrines and experiences as easily as their predecessors turned to the charisma of Hitler. With today's communications media a new personality can be packaged with unprecedented ease. This future may be described as the residue of failure.

The scenario may be characterized as follows:

Reactionary illiberal regime based on military and populist support
Intolerance toward, and suppression of, dissent and minorities
Use of Communist or other "threat" to motivate and unite population
Aggressive foreign policy (e.g., leadership of a crusade to prevent the poor two-thirds of the world from engulfing the rich)
Military and police needs given top priority
Control of communications media and universities
Rigidly controlled planning

In this case control of the system is exercised by the Chief Executive (Fig. 3). Technology is not the driving force but is exploited to further the stipulated normative values, goals and policies.

In an authoritarian setting voter pressure is not a basic concern and long range strategic and policy planning appears at first glance feasible. Napoleon and Hitler had long range plans for Europe (more correctly described as visions). Stalin created the COMINTERN and Mao is trying to create a new society in China. In the late 1950's the Soviet State Economic Council developed a plan for the 1961 to 1980 period and this plan was incorporated in the Party Program and presented at the 22nd Party Congress in 1961. However, since then long range forecasting "seems to have lost its 'home'" [27]. Romania has established formal planning organizations but its efforts are characterized by rigid acceptance of norms and policies with superficial efforts at linear extrapolation [28].

The suspicion persists that an authoritarian approach may successfully raise a feudal society to a high industrial level in a span of fifty years but may well prove totally incapable of moving the society beyond the industrial stage.

On historical evidence the probability of successful and beneficial authoritarian long range planning must be rated as small. The chances of a demagogue are high and those of a wise statesman low. There are few men of the stature of Moses and Solon and many like Napoleon, Mussolini, and Hitler. It is axiomatic that power corrupts. Typical failures are dilletantism, rigidity of plans, imposition of arbitrary values and undesirable goals (e.g., personal glory, military conquest).

In today's nuclear age the result can be fatal to the survival of the nation. The possibility of a "crazy state" — irrational, criminal — is a real one [29]. In the "global village" such a state poses a threat of genocide to others. If it does not prove fatal, the regime may precipitate the decline of intellectual achievements and the Kafkaesque atmosphere seen in many authoritarian societies from Sparta to the Soviet Union.*

We have examined three alternatives to the surprise free future (A). From the planner's viewpoint these futures are summarized in Table 3. None promises a "Golden Age."

Future B can lead to analytic blunders which may accelerate or intensify the catastrophic events anticipated in future A. Grave errors of omission or commission in the models may be exposed at a point in time when it is too late to avert disastrous consequences. Future C's non-planning may bring forth a society of cosmic dissipation which drains away the reservoir of human capability. Future D presents the historical last resort, the authoritarian regime. It may be the outcome of failure of any of the other alternatives. Even if initiated with the best intentions it can readily deteriorate into an oppressive and aggressive police state with the loss of that most precious of commodities, freedom.

---

*Cf. Andrei Amalrik on the collapse of the Soviet Union [30].

TABLE 3. Planning in Four Futures: A Summary.

Future	Description	Emphasis: Exploratory	Emphasis: Normative
A	Surprise free "continue as is" bureaucratic subsystem planning, primarily at operational level minimal long range integrative planning		
B	Reconstruction "must and can plan systematically" formal dynamic integrative planning, striving for flexibility, self-regulation emphasis on large scale social systems analysis		
C	New Society "any plan is worse than no plan" planning by experimentation minimum of formal planning at any level, maximum freedom, romanticism		
D	Authoritarianism "any plan is better than no plan" centralized planning exercised by leadership minimum freedom and flexibility		

Obviously, many other scenarios are possible. Paradoxically, the most probable event is the occurrence of completely unanticipated events — political, economic, behavioral, and technological — which will have major impact. The American posture during the next generation may well be decided by external events: refusal of most nations to recognize the need to reverse the population growth pattern or to reduce pollution, nuclear weapons in the hands of poorer nations, an unforeseen ecological disaster, a world-wide economic collapse, or a fanatic new ideological movement may provide irresistible intrusions. And in the words of William Butler Yeats,

> "Things fall apart; the centre cannot hold;
> Mere anarchy is loosed upon the world;
> The blood-dimmed tide is loosed, and everywhere
> The ceremony of innocence is drowned;
> The best lack all conviction, while the worst
> Are full of passionate intensity."

V. ANOTHER PERSPECTIVE

Two considerations brighten the picture:

(a) Long range forecasts are characteristically pessimistic (and short term forecasts optimistic). In the near term there is invariably wishful thinking on how easily the clearly possible can be implemented; in the long term lack of imagination obscures new alternatives and solutions [31].

(b) Man has shown himself to be remarkably adaptable in his brief existence on earth (e.g., expanding his conceptual space when exhausting his physical space) [9] and this trait offers a chance of survival even in a rapidly changing, occasionally catastrophic, environment.

Our concern in this section is of a more activist nature, however. The analysts can hardly continue as is and they have no real "solution." Is there a reasonable direction toward a solution — even if the solution itself now eludes us? Can we at least avoid disastrous developments? In other words, is there a "good risk" approach which increases the likelihood of a satisfactory future in view of the uncertainties?

I believe that answers to these questions emerge through a process of refocusing, a shift from the familiar comprehensive Forecasting/Planning agenda to three of its elements — *system cognition, creation* and *communications*. The objective is increased insight on the broadest possible scale, i.e., holistic system understanding and design capability effectively communicated to a large audience. At first glance, this appears to be a harmless and useless platitude. Yet its implications are far from trivial.

(a) The emphasis is shifted from the problem solving view of planning to that of the problematique. This is not to denigrate the desirability of planning but to recognize its variable status in the alternative futures considered in this discussion. The shift is based on the proposition that the emphasized activities — cognition, creation, and communications — lead to improved action decisions in any of the alternative futures. Decision makers in each case benefit, no matter what the degree of planning sophistication or structural authority. The stress is placed on expanding the user's horizon, on comprehension of relationships and of old and new alternatives and their implications. Analytic reduction of choices (or selection of an "optimum solution") and derivation of a matching plan are assigned lesser significance. The formulation of better questions is considered more valuable than the statement of answers.

(b) In view of the nonlinear feedback characteristic of most such systems, the forecasting tools upon which overwhelming reliance is placed today — trend extrapolation and Delphi — are totally inadequate. Improvements in forecasting are strongly linked to system cognition.

(c) A change in the attitude toward, and use of, communications is indicated. The potential of the "computer age technology" in this context needs to be plumbed in depth. Traditionally, most scientists and analysts have not concerned themselves seriously with the communication of their work except to their peers. It is also evident that most long range planning studies in industry and government have been academic exercises, not used in the decision making process. Virtually no curricula in management science or systems science require courses in communications.

(d) Cognition, or knowledge, must share the stage with other elements of experience which are necessary for social system creation. These include, among others, drives, desires, values, motives, aesthetics, ethics, conflict, and anthropomorphic modes of expression (e.g., symbolism) [32]. A creative architectural system design is by no means only a product of knowledge. Long range planning itself requires a cultural appreciation which has been lacking in our society and has not successfully appeared as a product of knowledge.

Carrying these thoughts a step further, we envision a major thrust in several areas in this decade:

(1) Basic systems research.

The mathematical foundations of complex systems clearly require development. The characteristics shown in Table 2 suggest the existence of regions of stability and instability. The complexity of the system makes the identification of their boundaries difficult. The superposition of technology on a natural system in effect constricts the areas of stability. Encounter of instabilities thus becomes ever more likely as technology advances.

Biological systems are known to exhibit a behavior unlike that of most classical physical systems. It should not be surprising that the corpus of mathematical analysis which evolved, after all, in conjunction with the physical sciences (e.g., differential equations for mechanics), is not suitable for biological and social systems. The non-divergent behavior of the former (e.g., differentiability) is replaced by divergence in the latter. Marginal changes at a crucial instant lead to major changes subsequently.

Conflicting influences result in a bifurcation of the dynamic system and the presence of discontinuities should be predictable. Topology is currently providing the stimulus for a new calculus which may be appropriate for such systems (e.g., the cusp catastrophe concept).* The aim is not to optimize, but to lessen the counter-intuitive aspect of system behavior, to recognize threats to homeostasis, to disclose options which increase system resilience and thereby avert catastrophic occurrences. Here, then, is a worthy challenge for the academic research community.

(2) Modeling and gaming of complex systems.**

Examples of techniques are the system dynamics concept of Forrester, the urban and regional modeling under way at institutions such as the University of British Columbia [11], the simulation of legislative programs undertaken at the Institute for the Future [35], and Buckminster Fuller's World Game. Much more extensive and intensive work is needed at the interdisciplinary level.

The Holling and Goldberg game approach at the Resource Science Center at British Columbia suggests a step-like process which is particularly promising. Expanding on their experience, sequential workshops might proceed as follows:

(a) Overall conceptualization of the simulation game, evolution of rules, and crude model development.

---

* A new mathematical theory, Structural Stability, is being derived from Qualitative Dynamics and Differential Topology. See Thom [33].

**Cf. D. Gabor's comment in [34].

(b) Improvement and detailing of model and submodels; visual output emphasized in model development.

(c) Enlargement of participation to bridge institutional and social gaps (e.g., academicians and bureaucrats); emphasis on man-computer interaction.

(d) Enlargement of participation to include constituencies (e.g., local and regional government, public); larger system view and extensive "playing" with computer model, i.e., visually aided dialog.

(e) Design of more normatively oriented model and interaction with existing model to assess longer range alternatives.

(f) Communication with wider spectrum of "users" to increase insight.

The continuing involvement and interaction with non-academic individuals of widely varying backgrounds and interests is critical: no computer model can hope to yield adequate insight into complex processes in which man constitutes a significant element. In particular, values and norms cannot be specified a priori, but must be supplied externally to the model by each "user" or "player" in the context of his own system view.

(3) Use of the computer as a high level communicator.

In dealing with complex systems the computer has two obvious roles: (a) storage of, and rapid access to, vast quantities of data, and (b) ability to efficiently perform the numerous computations characteristic of the model of such systems. As already implied in the previous paragraph, however, its role as a communications device is equally important.

The computer can give the individual a new ability to play with very complex systems, to get the "feel" or experience which he relies upon to comprehend complex systems. We refer to an informal, intimate dialog, the concretization of *Gedankenexperimente*. The key is the use of multiple sensory modalities in interacting with the computer (manual-visual, audio-visual) [36, 37].

The computer can break down traditional social and organizational compartmentalization. Individuals from diverse backgrounds must use the same language in communicating with the computer and thus find it easier to communicate with each other. Information nets and conferences can be created to bypass structural, space, and time constraints using the computer together with remote terminals. The computer based Delphi conferencing is a recent example [38].

Iterative telephone-computer feedback systems are being developed to permit massive interaction between the public and the expert or the government on policy questions [39]. Umpleby reports that an elementary version of a computerized participative exploration of alternative futures is already in operation at the University of Illinois [40] and audiovisual taped scenarios may be the forerunner of three dimensional multiple motor sensory environmental simulations. If unscrupulous manipulations of public communications systems can be prevented, the value of such tools in terms of improved insight communicated to planners or decision makers and bridging of the cognitive gap between the societal elements and levels is inestimable.

(4) New organizations.

(a) Systems Study Centers. In the past few years "interdisciplinary" units have been formed out of existing organizations in governments, universities and other institutions. Almost always it is a case of old wine in new bottles. Resistance to change and the traditions of problem isolation, compartmentalization of knowledge, disciplinarity and specialization (typified by the doctoral dissertation) prove too strong.

Entirely new systems study groups must be formed if these constricting ties to the past are to be severed. Each group should be project oriented and characterized by:

(a) Focus on the problematique of a significant

system involving nature, man, society, and technology.*

(b) Holistic approach to system design (including design of new instrumentalities and institutions).

(c) Stress on the conjunction of creativity and cognition (synthesis and analysis).

(d) Openness of the group and inclusion of individuals with different fields of knowledge and experience but a common interest in the particular system area.

(e) Recognition of the importance of communications to bridge social gaps.

A number of such groups can be physically collocated** and subsumed under a loose umbrella organization. As one group ends its work another may be created with personnel shift occurring as in any industrial project. Key personnel in such project groups who desire a degree of security and have no other attachment can be made regular members of the umbrella organization. Such a center can provide library, computer, communications (e.g., audio-visual production) and housekeeping facilities for use by all groups. Seminars may be desirable to provide a means of communication among the different groups. Students interested in interdisciplinary experience may also become attached to this center. They participate in the group activity and augment their learning process through seminars.

---

* The systems studied initially should preferably be local, rather than global, in scope.

**This should not imply that a group performs its entire work in one location; considerable field in real life settings may be necessary.

(b) Alerting activities.

Another kind of organization is needed for the scanning and alert function. Congressional consideration has already been given to the formation of Technology Assessment Centers for the study of the impact of technological (sub)systems on their ecological and social environment, i.e., on higher level systems. New technology is obviously not the only source of instability. Early warning of human and social sources of potential crisis is equally significant. On the positive side new ideas and alternatives may be disclosed in the scanning operation. Following Calhoun [43] a hierarchy of interconnected centers (e.g., local, regional, and national) might be considered.*

Reliance cannot be placed solely on the systems study groups described in (a). The latter probe selected systems in greater depth whereas the emphasis of the alert activities is breadth. There is considerable complementarity between the two: in (a) critical new problems and ideas may be uncovered which are peripheral to the area under study, in (b) priorities for new systems studies should be determined and unanticipated interactions and alternatives applicable to a system under study in (a) may be uncovered.

Sponsorship of the new organizations will vary with the particular future. Federal or local government, international organizations, business, foundations, and citizens' groups may provide the impetus. We already see a few clues today to the growing diversity of support (e.g., the Club of Rome, the Ford Foundation sponsorship of the Resource Science Center at the University of British Columbia, Duhl's invisible college [43]).

---

*Calhoun conceives of a global network of alerting subsystems covering all societal levels with ideas and concepts flowing continuously through its elements.

VI. A FINAL WORD

Our ignorance in complex systems and uncertainty of the American socio-political environment move us to focus in this decade on *system cognition, creation*, and *communications*, rather than on formal, full scale planning. This shift to more modest dimensions is not an escape from responsibility. It constitutes a recognition of the overwhelming need for holistic awareness and system insight to open new paths and for avoidance of grandiose "scientific" planning based on naive suboptimizations and/or an unreal, utopian decision-making environment.

Upon reflection we are also led toward a possible redefinition of the "post-industrial society." The industrial society was based on an understanding of physical and chemical processes. Dissemination and application of this knowledge initiated and advanced our present society. Widespread understanding and creative design of complex systems encompassing nature, man, society, and technology may provide the true foundation of the post-industrial society.

REFERENCES

1. Center for Political Studies, University of Michigan, poll quoted in Time, November 15, 1971.

2. JANTSCH, E., "Technological Planning and Social Futures," Chapter 2, Cassell/ABP London 1972.

3. BROWNE, S.H., "Cost Growth in Military Development and Production Programs," December 1971 (unpublished).

4. "Toward Balanced Growth: Quantity with Quality," Report by the National Goals Research Staff, The White House, Washington, D.C., July 4, 1970.

5. "Goals for Americans," Report of the President's Commission on National Goals, Prentice-Hall, Inc., Englewood Cliffs, New Jersey, 1960.

6. JANTSCH, E., "Inter- and Transdisciplinary University: A Systems Approach to Education and Innovation," *Policy Sciences, 1*, 403-428, 1970.

7. FORRESTER, J., "Urban Dynamics," M.I.T. Press, Cambridge, Mass., 1969, and "World Dynamics," Wright-Allen Press, Cambridge, Mass., 1971.

8. *Time*, 38, Dec. 27, 1971.

9. CALHOUN, J.B., "R_x Evolution, Tribalism, and the Cheshire Cat — Three Paths from Now," Unit for Research on Behavioral Systems, National Institute of Mental Health.

10. HOLLING, C.S. and GOLDBERG, M.A., "Ecology and Planning," *Journal of the American Institute of Planners, 37*, 4, July 1971.

11. HOLLING, C.S. and GOLDBERG, M.A., "Resource Science Workshop," 20-21 (unpublished), Sept. 1971.

12. CETRON, M.J. and RALPH, C.A., *Industrial Applications of Technology Assessment*, 209, John Wiley & Sons, New York, 1971.

13. CETRON, M.J. and DICK, D.N., "Producing the First Navy Technological Forecast," *Technological Forecasting, 1*, No. 2, 192, 1969.

14. THOMPSON, W.I., "At the Edge of History," Harper and Row, New York, N.Y., 1971.

15. "Report from Iron Mountain on the Possibility and Desirability of Peace," with Introductory Material by LEWIN, L.C. (pseudonym), The Dial Press, New York, N.Y., 1967.

16. FRITZ, C.E., "Disaster" in MERTON, R.K. and NISBIT, R.A. (eds.), "Contemporary Social Problems," 1st ed., Harcourt, Brace & World, 665, 1961, quoted by van Court Hare, Jr., "Systems Analysis: A Diagnostic Approach," Harcourt, Brace & World, 423, 1967.

17. OZBEKHAN, H., "Toward a General Theory of Planning," in JANTSCH, E. (ed.), "Perspectives of Planning," OECD, Paris, 48, 1969.

18. DROR, Y., "Design for Policy Sciences," American Elsevier Publishing Co., New York, N.Y., 1971.

19. MICHAEL, D.N., "The Unprepared Society: Planning for a Precarious Future," Basic Books, New York, 1968.

20. LENK, H., "Prolegomena zur Wissenschaftstheorie der Planung," in "Kolloquium über Forschungsplanung," Franz Steiner Verlag, Wiesbaden, 1971.

21. JANTSCH, E. (ed.), "Perspectives of Planning," OECD, 7-9, 1969.

22. GABOR, D., "Open-ended Planning," in E. Jantsch (ed.), *ibid.*

23. DRUCKER, P. (ed.), "Preparing Tomorrow's Business Leaders Today," Prentice-Hall, Inc., Englewood Cliffs, Jew Jersey, 1969.

24. McGINNISS, J., "The Selling of the President 1968," Trident Press, 1969.

25. REICH, C.A., "The Greening of America," Bantam Books, Inc., New York, N.Y., 1971.

26. LINSTONE, H., "When is a Need a Need?", *Technological Forecasting, 1*, No. 1, 68, 1969.

27. JANTSCH, E., "The Organization of Technological Forecasting in the Soviet Union: Notes from a Brief Visit," *Technological Forecasting, 1*, No. 1, 83, 1969.

28. JANTSCH, E., "The Organization of Forecasting in Romania: Notes from a Brief Visit," *Technological Forecasting and Social Change, 4*, No. 1, 1972.

29. DROR, Y., "Crazy States," D.C. Heath & Co., Lexington, Mass., 1971.

30. AMALRIK, A., "The Fall of the Soviet Empire," Atlas, February 1970.

31. BUSCHMANN, R., "Balanced Grand-Scale Forecasting," *Technological Forecasting, 1*, No. 3, 221, 1970.

32. JANTSCH, E., "Education for Design," Presentation to the International Symposium, "The Universitas Project-Institutions for a Post-Technological Society," Museum of Modern Art, New York, N.Y., 8-9 January, 1972.

33. THOM, R., "Topological Models in Biology," *Topology, 8*, 313-335, Pergamon Press, 1969.

34. GOLDFARB, A., "Technological Forecasting: What and Whither?", *Technological Forecasting and Social Change, 3*, No. 1, 55, 1971.

35. GORDON, T.J., ENZER, S. and ROCHBERG, R., "An Experiement in Simulation Gaming for Social Policy Studies," *Technological Forecasting, 1*, No. 3, 241, 1970.

36. JOHNSON, A., "Information Tools that Decision-Makers Can Really Talk With," Innovation No. 10, 1970.

37. JOHNSON, A., "Organization, Perception, and Control in Living Systems," *Industrial Management Review*, Winter 1969.

38. TUROFF, M., "Delphi Conferencing: Computer Based Conferencing with Anonymity," *Technological Forecasting and Social Change, 3*, No. 2, 1972.

39. RUDINGER, G. and KRAUCH, H., "Ein Rückkoppelungs-System zwischen Rundfunkanstalt und Zuhörerschaft," Studiengruppe für Systemforschung, Diskussions-papier 1970/4, Heidelberg, 1970.

40. UMPLEBY, S., "Fourth Generation Electronic Mass Communications Media," *Technological Forecasting and Social Change*, *4*, No. 1, 1972.

41. LINSTONE, H., "A University for the Post-Industrial Society," *Technological Forecasting*, *1*, No. 3, 263, 1970.

42. WEIZSÄCKER, E.v., "Baukasten gegen Systemazwänge: Der Weizsäcker Hochschulplan," Pieper Verlag, Munich, 1971.

43. CALHOUN, J.B., "Promotion of Man," in ATTINGER, E.O. (ed.), "Global System Dynamics," Karger, Basel/Munich/New York, 36-58, 1970.

---

*From *Technological Forecasting and Social Change*, *4* No. 1, to be published Sept. 1972, (preprinted by permission of the author and publisher)

RELATING THE IMPACTS OF TECHNOLOGY TO THE ROLES OF AGENCIES

DOUGLAS G. BAILEY with ANDREW R. STRAUSS

Director, Intendes Research Foundation, Toronto, Canada

One of the most important long-run issues currently confronting both developing and post-industrial nations today is national identity. With rapid advances in the technologies of movement of goods, people, and information, territory is rapidly fading as the measure of a society's boundaries. Similarly, performance and efficiency measures are no longer considered the meaningful criteria for evaluating technological change.

No agency operates in isolation. The delivery of its intentions must match more expectations than its own. However, each agency maintains itself in an environment about which it has incomplete information. The purpose of this paper is to examine phenomenological viewpoints concerning the consequences of technological change in the environment of an agency.

The *role of* any agency may be defined in terms of three broad sets of factors: (1) intent, (2) conduct, and (3) consequences (changes) which relate to intent. The intent of an agency consists of its desired consequences: its purposes, goals, mandate, and so forth. Intent is a construction of expectations built on the foundation of an anticipated reality. The policies, plans and programs of organizations are statements of expectation based on the anticipated environment in which they are to act.

The conduct of an agency is its behavior, regardless of intent and/or consequences. Conduct, or action, has no meaning unless it is related to the role of the agency. Serendipity in scientific conduct is a consequence of conduct rather than the intent of conduct. Similarly, many changes occur in the environments of agencies without specific intent, conduct, and/or consequences relative to those agencies. Waste is a behavior associated with all production; pollution is a consequence which makes waste meaningful.

A consequence is a change which is meaningful to the identity of an agency, whether or not it is intended. Statements about pollution are commitments on tolerable concentrations of waste as opposed to statements about waste itself (or the condition of waste itself). The commitment to do something about pollution is directly related to the intent of conduct. We have the following model of the role of an agency:

INTENT

CONDUCT
(PHENOMENA IN ENVIRONMENT)

CONSEQUENCE

All technology assessment activities, as related to this model, must treat conduct in relation to the intent of agencies and consequences relative to agencies. Criteria for technology assessment activities are drawn from this model of "role."

Further development of this model, in order to detail criteria for the assessment of impacts of technological change, involves the analysis of three key sets of concepts:

1) Time and change
2) "Clients" of change
3) Choice

Time measures conduct, regardless of intent or consequence. As such, calender time has little relationship to the assessment of alternative applications of technologies, or to the evaluation of the consequences of change relative to roles.

Considerable confusion surrounds the use of calender time to measure intent and/or consequences of change. Consider, for example, the concepts of short-, mid-, and long-term as employed by various disciplines:

Time concept	Small business	Large Business	Economists	City planners	"Futurists"
"LONG"	1 ± yrs	5 ± yrs	10 + yrs	20 + yrs	30 + yrs
"MID"	1-6 mo	1 year	5 years	10-15 yrs	20 + yrs
"SHORT"	1 mo	1 mo	1 mo to 1 year	1-5 yrs	10 yrs

In most larger agencies, all of these attitudes about short-, mid-, and long-range or term are operative. The "long range" impacts of a new transportation system are very different depending upon to whom one is speaking. Frequently, the different concepts of time when applied to intent and/or consequences generates issues (contentions of viewpoints) based more on the time idea than the meaning. A major issue in many public agencies is, "What is our goal?"

One measure of intent is investment, or risk. Three rather arbitrary degrees or levels of risk are: (1) probability, (2) feasibility, and (3) possibility. Probability may be defined as the ratio between desired consequences of conduct and the number of alternative consequences which could occur. Feasibility is the expected capability of the achievment of the desired consequences. Possibility is all consequences that may or may not occur.

These notions of time as a measure of risk are of course not new in themselves; however, when applied to the construction of expectations built on anticipated reality they generate some interesting concepts:

"Time"	Expectations	Anticipations
Possibility	Myth	Belief
Feasibility	Postulate	Predilection
Probability	Supposition	Prediction
Certainty	Proof	Truth

This argument differentiates between expectations and anticipations. Expectation is extrapolation requiring internal consistancy and coherence. Anticipation, on the other hand, requires a reasonable correspondence with knowledge about the phenomenal environment (and conduct in it).

What do these differentiations mean to the agency manager or planner? A *plan* or *program* is an expectation, while a forecast is anticipation. In this set of definitions, "exploratory forecasting" is the development of a plan or program, while "normative forecasting" examines the relationships of decisions made to anticipated reality. Such techniques as cross impact analysis employ time as a means to measure the relationship between decisions and anticipated realities.

Further definitions and arguments, drawn from field experience with public agencies, could be posed for this application of notions of risk as a measure of intent. For the moment this discussion has been adequate to build a case for using risk, a measure of time, as a criteria for technology assessment. The next concern is the development of measures of time as a consequence.

In this case, relative distance rather than risk has been developed as a useful analogy. Distance relates to perception of phenomena as opposed to expressions about it. Any change is relatively "near" or "far" from an identity in terms of the latter's perceived commitments. Some arbitrary degrees of distance are: (1) range, (2) function, and (3) domain. Range is a perceived ratio between a particular consequence and other consequences. The range of a consequence establishes how much change can occur before changing the identity of an agency. The domain of a consequence is the field of consequences which could affect the agency. A function might be considered the relationship between the limits of consequences — range — and the field of consequences perceived — domain.

Again, these measures of distance of consequences are not new, and provide interesting concepts only when applied to the questions: "What is perceived?" "How is it perceived?"

Time	What	How
Domain	Alternatives	Qualities
Function	Preferences	Values
Range	Adjustments	Quantities

This model of perceived consequences must correspond to the commitments of agencies. A consequence which is an alternative quality of identity is much more drastic than a consequence which is an adjustment in the limits of an agency's identity. When an agency loses or gains new qualities of identity, whole staffs may be added or relieved; adjustments to a budget have far less drastic consequences.

The role of an agency model is now significantly expanded, in terms of measures of intent and consequence. The next step in this argument is to predicate conduct, or phenomena in the environment. The environment of an agency naturally includes itself, but for simplicity in this analysis, consider the agency a separate actor, and the client the "not-agency," or phenomena, in the environment.

Most public and private agencies maintain both expectations and anticipations of "clientele" without consideration of perceptual influences on phenomena. They maintain a myth of, or belief in, certain clients, such as "the general public" or the "free market economy." These myths need analysis when consequences are perceived which don't correspond to the agency's intent.

For example, in studies carried out in three Federal agencies of the Canadian Government, organizations which claimed to serve "the public" were found to serve (design their programs for) three primary phenomena: the relevant Minister in Cabinet; the status of particular key government employees; and the financial community of Canada. It appears that the more experience an agency has had, the greater the diversity of myths and beliefs as alternative criteria for the assessment of intent and consequences.

Where myth is the primary criterion for the design of the conduct of an organization, it loses its relationship to the more subtle consequences and responds only to consequences in the domain. In other words, the more dependent the role of an organization becomes on its own myth structure, the more drastic consequences of conduct or phenomena in the environment may become.

All clientele also maintain intents and perceptions of conduct's consequences. Where an agency's intent/consequence structure tends to move from the corresponding structure of the clientele in a direction of higher risk and greater distance, one may term that agency's *conduct* as portraying confidence. Where a "client's" intent/consequence tends to move from an agency, in a direction of higher risk and a greater distance, the agency's attempts to adjust or correspond may be termed as the *conduct* of relevance. Thus any agency may be portrayed as a structure of intent/consequence relative to the intent/consequence perceptions of a client in terms of a position somewhere on the continuum:

Relevance ________________________ Confidence

Some examples are in order. The Treasury organizations of the United States and Canadian Federal Governments are, by mandate, to direct the domain of the economy. As such they may be considered to conduct their activities so as to inspire confidence. When massive strikes "halt" the economy, they may act directly or indirectly to maintain a belief in continued economic activity, by directing alternative perceptions of consequence.

Welfare organizations and programs are typically intended to adjust their activities by "true" client needs. As such, they may be considered to conduct their activities so as to insure relevance. When a particular problem, such as the illness of a client, arises, they may act directly or indirectly to change the range of their services.

It should be clear that the examples posed above are gross generalizations about the real output of applying this theoretical material to agency rules. In fact, all agencies must be both relevant and confident in their conduct. The consequence of the theoretical material discussed above is: is the generation of (1) criteria for the assessment of change relative to roles; and (2) guidelines for the design of process-oriented agencies which take such criteria into explicit account.

The criteria proposed are as follows:

A. Relevance

1) Postures regarding issues (contending interests)
2) Postures regarding problems

B. Conduct

1) Efficiency
2) Performance
3) Internal harmony
4) External harmony

C. Confidence

1) Consistency of consequence
2) Correspondence to expectations

Each of these criteria must be related in turn to the intent and consequence factors of an agency's role. That is, issues created by change or intended change must be evaluated considering their significance in relation to (1) expectations and (2) anticipations of agency intent, and in relation to perceptions of consequence.

In general, it could be said that relevance criteria are "short-run" — that is, problems and issues in the environment of an agency tend to change relatively rapidly compared to the criteria of confidence. The "mapping" of the relationship of any agency's role to current problems and issues may be an inventory of problems and issues with consideration of (1) intended attitudes toward them, and (2) the alternative consequences which changes in them would have on the agency's intent.

The consistency, or repetition of consequences, creates confidence — whether or not the methods employed to produce the consequences are redundant. An agency which generates different quantities and qualities of consequence each time it acts cannot hope to inspire public confidence. The implementation of a new welfare system by a responsible agency must provide consistent consequences — not only over the time of the system's implementation, but also in relation to the past behavior of the agency.

There must be a correspondence between the environment's expectations and the agency's intent. Interests and institutions in an agency's environment *expect* the agency to react in specific ways. For example, if the agency was developing parks, the institutions would expect it to purchase land. There must be some correspondence between the agency's intent and those sets of expectations if the agency is to maintain confidence.

The conduct (or behavioral criteria) is more familiar to most public and private agencies. Efficiency refers to the optimal use of resources in achieving desired objectives. A housing program is most efficient if it accrues the greatest possible net benefit to the agency and the "publics" it serves. Traditionally, efficiency measures employ cost/benefit models in comparing the use of resources in alternative programs.

Performance refers to the relationship between intent (plans and programs) and actual conduct. A program is working well if it is operating according to plan.

Internal harmony refers to the degree to which different sets of actions in a program are mutually beneficial to the program as a whole. When the achievement of sub-objectives in a particular program tend to aid or abet the achievement of other sub-objectives, the program can be considered to be internally harmonious. When some activities of a program strategy act as obstacles to the intent of other activities, there are problems in internal harmony.

External harmony is seldom considered in the programs of a government or large corporation. Frequently the achievement of some objectives necessary to the intent of one program may act as obstacles to another program. Both internal and external harmony are questions of degree as opposed to a desire to achieve absolute harmony of action. That is, the degree to which the programs of one government department are to complement the programs of another, is a matter of policy decision on the part of each of the agencies.

When each of these criteria is applied to any given technological change, and that change is assessed in relation to intent and consequence over the kinds of "time" suggested above, even the most reduced analysis begins to provide information on which role decisions may be made. The use of one criterion alone (for example, the traditional efficiency criterion) applied only to one element of an agency's role (for example, consequence) would leave to chance the other bank of decisions to be made. In public agencies this is far too frequently the case.

One means of insuring that all of these criteria are explicitly considered in decision-making processes is their explicit inclusion in process-oriented organizational designs. Some practical examples of these designs, as they have been developed in Federal agencies in Canada, may give greater clarity to the discussion. One of these examples is currently in use, while the other two are recommended structures based on analysis of current agency problems.

In 1970, the Canadian Federal agency responsible for housing and related affairs maintained over 30 separate task force assignments on policy and planning issues, varying in subject matter from urban assistance programs to sewerage hardware policies. The agency required a new division which would (1) relate the impact of changes in the Canadian environment to its own role, and (2) assess the potential impacts of alternative agency conducts. The new division was to take over program evaluation and policy-making activities, in addition to the work of the task groups.

The division was to report to senior management and to the Minister of State for Housing and Urban Affairs. Four operational groups were established in a structure as follows:

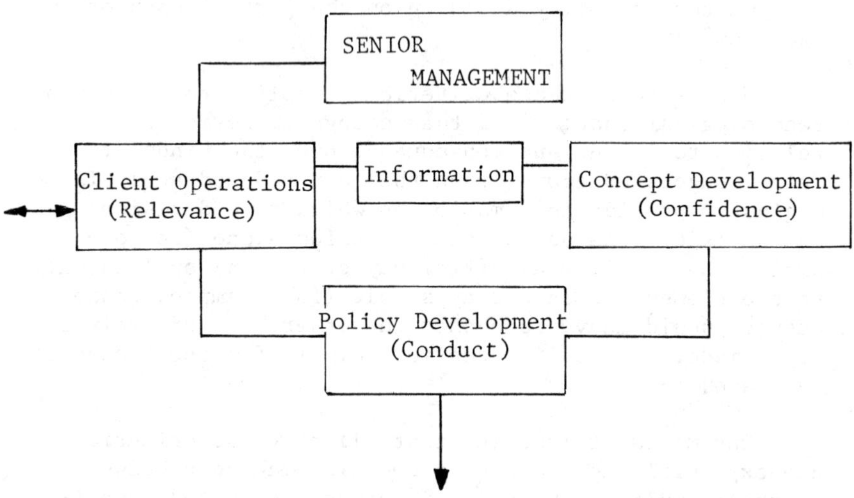

The responsibilities of the Client Operations Group were to gather information concerning current issues and problems "in-field," including the expectations of component interests about actions of the agency. In addition, a sub-group within this arm of the division was to represent the long-run (feasible and possible) expectations and anticipations of the "clients" of the agency.

Four types of client expectations were identified:

Client	Consequences and Expectations/anticipations
Political interests	Authority gains
Private interests	Personal gains
Professional interests	Influence qualities of programs
Popular interests	Sufficiency of programs and services

These longer-run "clients" acted as judges of the overall significance of each problem and issue encountered by the division. A primary function of this group was to insure relevance and optimal solutions/resolutions for the problems and issues.

The Policy Development Group was established to conduct workshops composed of field personnel from the five regional offices of the agency. The Group was structured to manage both forecasting and planning problems, employing the time concepts discussed earlier. Workshop managers represent "near," "mid," and "far" range perspectives of intent and consequence in their running or monitoring of workshops. In policy exploration/development problems, the "time range" managers monitor workshop activities to insure that adequate consideration is being given to each time range, while "planning managers" run the workshops:

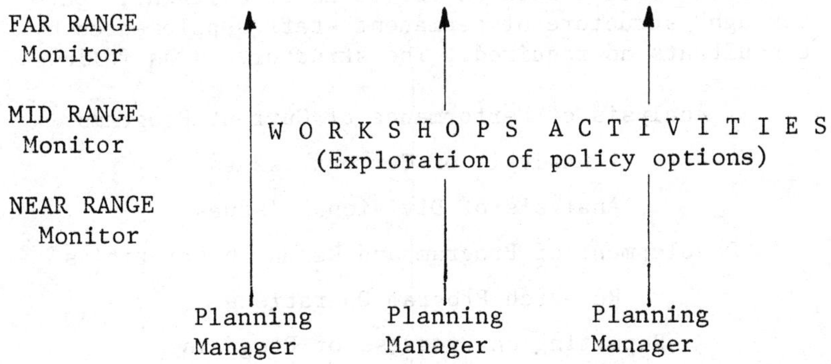

Once policy options have been developed, they are evaluated by the Client Operations Group (relevance criteria) and presented to Senior Management for decision. Once decisions have been made, they return to the Client Operations Group for a second evaluation, and are again the basis of workshop activities. In program development problems, however, the "time range" managers operate workshops while the planning managers evaluate content and product of the workshops:

FAR RANGE MANAGER	Planning Workshop on Contingency Environments →
MID RANGE MANAGER	Planning Workshop on Alternative Objectives →
	Planning Workshop on Operational Program →
NEAR RANGE MANAGER	Planning Monitor — Planning Monitor — Planning Monitor

The function of the Concept Development Group is to develop the "strategic" objectives for the agency's role, including research priorities, program priorities, agency initiatives, evaluation activities, and improvements in forecasting and planning methodologies. This Group was established on the basis of a linear, "put-through" structure of permanent staff supplemented by consultants as required. The structure is as follows:

Analysis of Performance of Current Programs
↓
Analysis of Current Issues
↓
Analysis of Divisional Issues
↓
Development of Program and Research Priorities
↓
Research Program Operations
↓
Reporting on Progress of Programs
↓

Turn-around time within this structure has averaged two weeks, while turn-around for policy development and planning operations runs approximately four weeks, including all reporting and presentations.

The information Group was established as the "memory" device of the organization, storing and distributing information from both the external and internal environments. Less emphasis has been placed on advanced technology than on appropriate information presentation, since, as is often common in government operations, there is too much data and too little comprehensive understanding of the utility of this data in information packages.

As an actual operation, this Policy Planning Division of Central Mortgage and Housing Corporation has extensive advantages over the more traditional planning operations in other agencies. In the first place, it examines all policy and planning problems from the points of view of different kinds of clients — something that most politicians do intuitively anyway. But in this operation, viewpoints concerning social costs and benefits are constructed on a systematic, issue-related basis. This "soft" analysis is backed up by all of the traditional assessment methods, such as cost/benefit and performance analysis, and the application of the "confidence" criteria discussed above in the theoretical materials.

In the Policy Planning Division example given above, the structure of the organization corresponded directly to the suggested criteria for technology assessment activities. Another potential structure for operations would differentiate organizational components on the basis of "role component analysis," as follows:

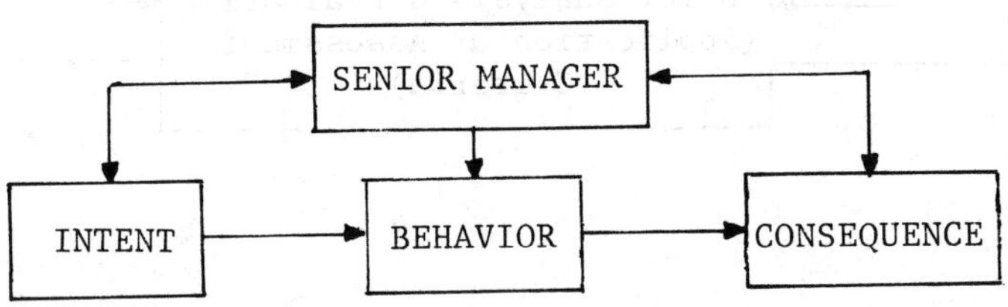

In this structure, an intent Division is organized to undertake exploratory and normative forecasting and planning. It is divided into sub-groups concerned with each of the time ranges discussed above.

Expectations	Anticipations
Myth Analysis	Analysis of Goals of Interest Groups
Contingency Forecasts & Program Plans	Contingency "Client" Reactions
Development of Role Postures in Relation to Current Issues	Analysis of Current Problems in the Field

The function of the Behavior Division is the operation of programs. Organization is largely dependent upon the types of programs; nonetheless, there must be a strong relationship to the criteria discussed in the theory portion of this paper. A suggested structure for Behavioral operations which would match all of the assessment criteria noted within a framework based upon the passage of social interests follows:

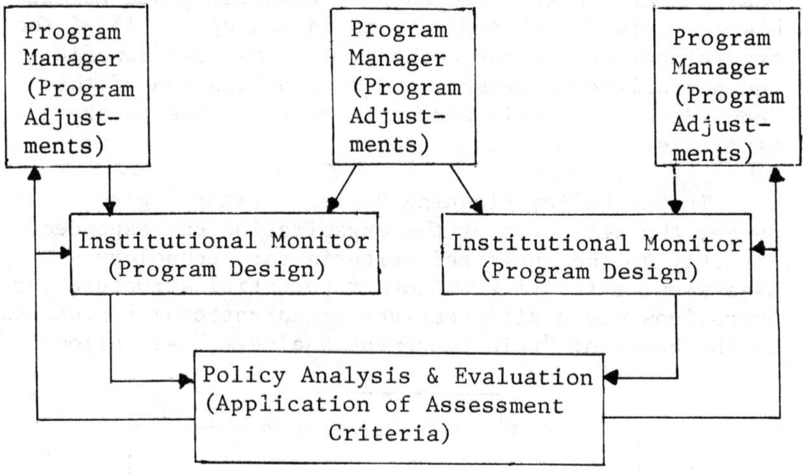

A Consequence Division concerns itself with the impacts of behavior and the relationships of these impacts to the agency's role. The Division's organization may also correspond to the theory. Separate groups in such a division might examine: (1) alternative environments which could have an impact on the agency; (2) the relative importance of these alternative environments to the role of the agency; and (3) changes in the agency's intent necessary to continue its current role.

Material in the form of forecasts, program options, and perceived role performance might be presented to Senior Management for decision, and subsequent implementation in the Behavior Division. Beyond particular elements of this structure, it amounts to a homeostatic model of the environment.

This policy-planning and implementation structure would be most appropriate to agencies which act in a functional capacity as opposed to those which act in a purely advisory capacity.

A third structure might be developed which is relevant to agencies which do not in fact implement their recommendations; organizations which are concerned with the delivery of choices as opposed to actual objectives and programs. Organizations which develop science and economic policy options are an excellent example of situations and conditions for which a mixed criteria and role/component design would be applicable.

An example of such a model follows:

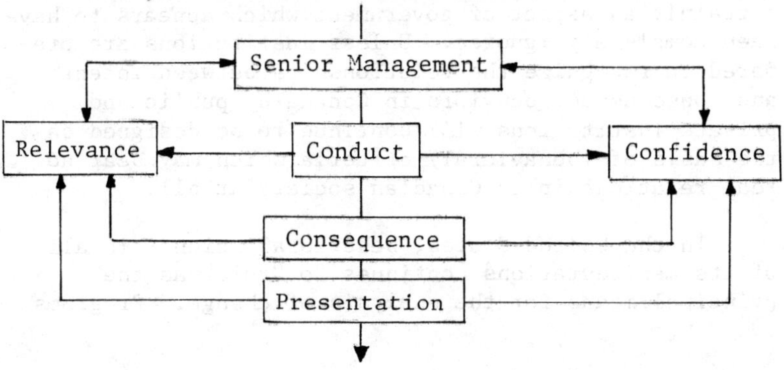

In this case, each of the criteria divisions carry on continuous analytical activities in applying specific measures to an environment perceived by the role components. A presentation division's outputs are options, recommendations, and indications of "client" interests, Senior Management considers current conditions using the assessment criteria and indicates the correct intent in light of the conditions.

These three models are presentations of organizations to effect operating styles for the assessment of technological change relative to the role of agencies. One of these models (the first one presented) has been implemented, and its operations have proven successful, in the Canadian context. Applications of this analysis of roles and criteria for the assessment of technological change relative to the identity of an agency will depend upon its context, but this theory goes a great deal further to complete analysis of consequences than most models that have been presented to-date.

We may go further in applying these concepts to questions of national identity. National identity in the United States and Canada has become a public issue with both institutional and social implications. This analysis suggests that national identity might be approached in terms of a rational construction which relates current interests and issues to expectations and myth. The myth of a nation need not correspond to an extant reality, but it must be supported institutionally in terms of criteria used to judge the consequences of international action. In Canada, this is certainly an aspect of government which appears to have been completely ignored. Unless institutions are prepared to recognize the relationships between intent and consequence, behviors in Canada by public and private institutions will continue to be designed on the basis of (behavioral) criteria which may bear no real relationship to Canadian society at all.

In the United States, behavioral science in all of its manifestations continues to "rule" as the principal means for the analysis of change. Programs and institutions, technologies and objectives, are evaluated and assessed in terms of such measures of performance, efficiency and internal and external harmony. Little if any attention has been afforded the intent and consequence of alternative technologies. If technology assessment is to be anything more than rationalization of past mistakes, such criteria must be employed in an approach, both in theory and in operation, which accepts actions of all kinds as phenomena and treats them as such, without a necessary relationship to "causes" or "effects".

THE FUTURE OF TECHNOLOGY ASSESSMENT IN POLICY FORMULATION

WALTER A. HAHN

Senior Specialist in Science and Technology, Congressional Research Service, Library of Congress, Washington, D.C., U.S.A.

Those who have read this far into this volume will have an up-to-date view of the state-of-the-art and practice of technology assessment. Though still inadequate, we can perceive the growing strength and utility of a variety of tools and concepts. There also are increasing numbers of cases in which the preassessment of the consequences of proposed or actual technological initiatives have been undertaken. This chapter, in parallel with the others in Part V on the Future of Technology Assessment, looks at one aspect of the question, "Where do we go from here?" In particular, it will attempt to illuminate the role of technology assessment in policy formulation.

There have been almost as many meanings given to the words technology assessment as there are authors in this volume. Central to all of them, however, is the notion of it as a *process*. It is a process concerned with the basic *management* problems of a technological society. Four elements of the process stand out as essential: awareness, thought (analysis/synthesis), resolve or commitment, and action.

These are elements in common with the policy process. The writer will resist the temptation to launch into definitions and conceptual schemes for wrestling with the slippery topic of policy. Suffice it to say that, "Policy means ... intelligently directed action toward consciously determined goals — as distinct from aimless drift and blind faith"[1].

Man's concern for control over his increasingly pervasive technology is not new. One thing that is new is the extraordinary increase in the rate of change — things are happening more suddenly and we expect the trend to continue. Also, today's events are more complex. A speech or article without some variation of the phrase "everything is related to everything else" marks its producer as out of tune with the times. Complicating man's incomprehension of the full set of interactions of events is that many phenomena are also beyond man's five natural senses and even his senses extended through knowledge and instrumentation. Today, the matter is more urgent with the realization that some current (and past) actions are irreversible and some may be catastrophic.

Like many other terms, technology assessment has come to have a variety of meanings as its popularity and number of advocates expand. The problem of defining "TA" was forcefully demonstrated during the August 1971 Engineering Research Foundation Conference at Andover, New Hampshire. The subject was "Technology Assessment: Management, Manpower, and Methodologies." It was attended by almost 100 of the key people involved in technology assessment from government, industry and academia. The variety of terms and topics included: problem-oriented TA, resource allocation TA, Daddario-like TA, mini and macro TA, TA methodology, management of TA, net TA, real (physical science and technology oriented) TA, and broad-gauge TA (like "real TA" but including social technology, e.g., credit cards). There was another class of terms involving the institutional label of the performer of the TA: academic TA, industrial TA, etc. Still another approach used terms indicating the degree to which the assessment was carried, e.g., first-order TA, and second (and nth)-order TA.

No attempt was made during the conference to develop a set of coherent terminology but there was consensus that clarification and some degree of standardization was highly desirable. Since the introduction of still more terms can only slightly increase the degree of verbal pollution, one more approach is suggested in this chapter on TA in policy formulation. A simple formula describes the scheme.

$$TA = \Sigma\ TA_p + TA_i + TA_m + TA_t.$$

TA is the symbol for the overall and generalized concept of technology assessment. All of the terms used at the conference and referred to above are included in TA. There are four sub-elements of this overall TA.

TA_p is *policy*-oriented TA. It is in harmony with the Daddario definitions and those in most of the proposed legislation that looks to impact assessments as an explicit input to policy, legislative and strategic processes. Engineering, environmental, economic, and social phenomena are all included. Public legislative, executive, regulatory and possibly judicial institutions are the prime sponsors and users. Private corporations similarly can be sponsors and users. The key concept is the use of the analysis/synthesis for *policy* purposes.

TA_i is *issue*-oriented TA. It refers to a problem that needs solving,be it environmental, social, economic, marketing, etc. Social or engineering "technology" is suspected (or known) to be one element of the problem but this aspect of it may be vague at the start of a TA_i.

TA_m is the label for TA *methodology*. The major focus here is how to do it. It includes all disciplinary, modelling, and forecasting techniques and approaches. It also includes the management and organizational (or team) arrangements to perform the assessment or impact study.

TA_t is *technology*-oriented TA. As the term technology assessment was first used, this would be a redundant term. Since the original term "technology" when paired with "assessment" has become so broad, it appears necessary to refocus on the simple question, what are the impacts of this specific technological capability?

Any institution can be a sponsor, doer, or user of any of these categories of TA but some roles will stand out. Congress, state, and local legislatures will almost exclusively be concerned with TA_p. They will often sponsor specific TA_i or TA_t and will probably have the least direct interest in TA_m. Universities and not-for-profits might be expected to have a reverse interest profile. Industrial firms in their product market planning can be expected to be heavily involved in TA_t. Universities, not-for-profits, industrial and executive branch organizations all can be doers of TA_i and TA_t and contributors to TA_m. There may be an important role for professional societies in establishing norms and "best practice" criteria for management and quality control in TA.

Throughout the remainder of this chapter the words technology assessment will be used to mean the overall TA unless noted otherwise.

TECHNOLOGY AND POLICY

Technology, either directly or indirectly, is a critical element of almost all policy decisions. This is true for both the public and private policy maker. What does technology assessment have to offer him? How can the policy maker incorporate its contribution into his decision making? What is the prognosis for his doing so?

Earlier, technology was cited as *a* critical element in almost all policy decisions. While technology may often be identified as a causal factor, it usually is not the controlling decision element. Has technology for some reason been singled out to be the whipping boy? Or is it just handy as a relatively more tractable and explicit feature to manipulate in complex policy level decisions? In the language of the operations researcher, are we suboptimizing on technology? Why don't we hear the phrases economic assessment, social assessment, environmental assessment, total goals assessment, or assessment? Except for the semantic identity, of course, we do hear these other assessments called for. And that is the point, to remind us that we are dealing with a large multi-faceted socio-economic-environmental system involving technological features. Unless we have equal capability to think about, measure, and manipulate the non-technological aspects of problems, we will not have adequate "technology assessment" as the term is now in use.

Technology assessment appears as an "obvious" necessity to technologists, anti-technologists, systems and policy analysts, and many other groups in our society. Yet it has taken a long time coming into its own. It is far from being a continuing feature in national policy making. For all of the talk about technology that some of us perceive, technology *per se* does not show up in the *expressed* concerns of the public. In the dramatic results of the Cantrill and Roll survey, *Hopes and Fears of the American People* [2], technology is not even mentioned. For example, the highlights of the findings (quoted in their entirety pp. 11-12) are:

> 1. The American people believe their nation is in trouble. They feel that the United States has slid backward over the past five years.
>
> 2. Public anxiety over the state of the nation focuses particularly on tensions and divisions among its citizens. Almost one in every two Americans sees these tensions as serious enough to 'lead to a real breakdown in this country.'
>
> 3. The public does not dismiss this national unrest as simply the work of radicals and troublemakers. Sizable numbers sense systemic causes related to the quality of leadership

and the performance of institutions.

4. The hopes and fears of the American people about their own lives reflect general satisfaction. There is lessened reference to many traditional material aspirations — a higher standard of living and owning a home, for example — which may indicate overall contentment and assumption of fulfillment of much of the traditional American Dream.

At the same time, there is recognition of a new body of concerns — among them threats to political stability, world peace, and the environment, as well as drugs and crime — demanding social rather than individual solutions.

5. Public opinion toward the Vietnam War has moved dramatically toward disengagement. A clear majority now wants the war ended, even at the risk of an eventual communist takeover of South Vietnam. This sentiment has risen markedly in the last three years and represents a shift since 1968 on the part of one American in four.

6. Public attitude toward existing and future American commitments abroad is one of caution and selectivity. Although the feeling has grown that the nation should concentrate more on domestic than on international problems, it would be an over-simplification to label this mood "isolationism." Rather, the public is critical and discriminating in assessing foreign commitments; its mood appears to range between noninterventionism and selective interventionism.

In another survey, Congressman Roth of Delaware tabulated the concerns of 24,000 voters in his state. Like those in the national sample, Delawareans did not mention technology. In rank order, they rated the following problems as the most important ones facing our country:

Southeast Asia
Crime

Inflation
Drug Use
Pollution
Militancy
Population
Welfare Costs
Unemployment
Racial Discord
Defense Spending
Transportation [3]

The argument here is not that technology is uninfluential or unimportant — we *know* it is both. We know that technology is a basic cause of much that is good and some that is bad in our society, our economy and our ecosystem. It is argued that the public and their elected, appointed and hired policy makers do not necessarily see the issue as one of *technology* assessment. It is possible that these people, who in the main are not technologists, see the simplistic truth that technology is neutral, it is people who are not.

In the preceding chapters a variety of activities has been identified as integral to the technology assessment process. Elsewhere Franklin Huddle [4], describes four major elements of "formal" technology assessment as: technological forecasting, social indicators, technology assessment per se, and national goods. This writer would add three more explicit elements, economic and environmental impact analysis and the new effort called "futures research." Most readers would also add or substitute their own favorite activities. By using the term "formal" Huddle can properly exclude the informal unstructured "people's technology assessment" identified in the National Goals Report [5] but it too is an essential part of the total process. Whatever the set of elements, there is consensus that a wide range of substantive and methodological skills are involved. Clearly "TA" is multidisciplinary, interdisciplinary or transdisciplinary, according to which word is fashionable. But often we forget that more than "disciplines" are involved if the process is to be meaningful.

THE POLICY MAKER'S VIEWPOINT

A policy maker finds it useful to take a different viewpoint and to identify the *roles* played in assessing the impacts of technology. Four roles seem to stand out as most meaningful: alerters, constituents, analysts, and decision makers. On some occasions one person may play multiple roles. But, as in the traditional view, the point is not the number or labels on the several roles — it is to perceive the TA process in terms of its actors rather than its substantive or methodological features. The "alerters" identify the problem, need or opportunity to perform a technological assessment; they are the causal or motivational factor. The "constituents" include the partisans for selected values, approaches and outcomes. Most importantly, they include — or should include — the unrepresented, unorganized affected parties; the ones who feel the direct and indirect outcomes of that which is being assessed. The "analysts" are the entire group of substantive and methodologically-oriented persons discussed above. The "decision makers" include policy, program and sometimes enforcement personnel — the ones who take some kind of action following the intellectual, analytical, debating aspects of the assessment process.

The policy maker's view of the problems in technology assessment varies in emphasis and, in part, in content from those of the analyst. Naturally he seeks more powerful methodology, longer range and more accurate forecasts, and more comprehensive, reliable information. He looks to the analysts, forecasters and information technologists for these improvements.

The following paragraphs will sketch some of the problems of TA which are of more immediate concern to the policy or "decision maker" in the role nomenclature previously used.

Complexity

This is the "everything is related to everything else" problem that involves all role players in the TA process. More effective models and techniques will help. So will experience with a variety of actual assessments. At present, that scarce, but critical, quality with the old-fashioned name wisdom, must be the mainstay to cope with complexity.

Who's Values?

Little has been said about the criteria and/or weights that shall be assigned the different value scales that inevitably must be taken into account in each TA. Majority over minority? Present versus future? Economic-social-environmental-technological? Or really tough ones, like ours or theirs? In view of the rapidly changing and widely diverging life styles and values of the day, this becomes both a more important and more difficult consideration.

Rationalism

Anyone reading this book can be expected to agree that the real problem is not rationalism but the absence of it. How does one deal with the irrational or anti-rational participant? How can technology assessments accommodate the growing mood and felt right of self interest and individual choice? And what about the everpresent factors of ignorance and apathy?

In a different context there is the problem of what is perceived versus what really is. Rational objective processes can determine a situation to be one way. But prejudice, ignorance, emotionalism and the like can cause that same situation to be perceived quite differently by other beholders. Bridging this gap can range from extraordinarily difficult to impossible. Yet such situations are certain to arise and must be dealt with effectively.

Motivation

Under what conditions will the policy maker convene a technology assessment? Gross problems, strong pressures or obstacles to achievement of the policy maker's goals can all be expected to produce appropriate assessments. But what about weaker stimuli? How does one select that set of technological initiatives to assess from the limitless set that will always be before us? Another aspect of motivation concerns the will to act upon the findings of the assessment. We can all think of cases where the undesirable, uneconomical, foul and even dangerous consequences of technologically-based activities are known and yet the actions proceed. What is it that distinguishes the "acceptable" from the unacceptable negative consequences? Even an approximate answer could save effort and expense (and embarrassment) on assessments which will result in no action. And thus we have backed into that unaskable question — which things shall we not assess since we don't want to know the answer?

Communications

The policy maker, be he a public or private executive or a legislator, is most often a layman in technological and analytical affairs. The excellent publication, "Technical Information for Congress," deals with the problem at length [6]. The criteria for information needed in a TA listed in that report include: authoritative, accurate, objective, technically sound and complete. Not mentioned but implicit are the additional characteristics: authentic, relevant, clear and understandable. This problem of communication is an old one; it is mentioned frequently but it is still far from being solved. For viable assessment of technology there must be good communication among the parties.

The reverse side of the communication problem is equally thorny — how can the results of these highly technical, complicated, and often highly political processes be made known to the voter or the public in general in a clear and simple fashion?

Measurement and Feedback

The title here is given in TA-language. The policy maker asks, how are we doing? What are the results of our decisions or actions? What do people think? How do we know that...? These trigger a wide range of additional problems. Surveillance and monitoring systems. Productivity measures in a service economy. How much environmental quality? Standards and norms. Social indicators and polls. Timeliness and accuracy. These aspects of TA are relatively more important to the policy maker than the analyst for developing policy is an incremental dynamic process requiring continuing feedback. In reading the literature of TA there is a tendency to think of even the largest assessments as events with beginnings and ends. This may be convenient and correct for the analyst, and while necessary for the policy maker, it is insufficient.

Benefits/Risks Allocation

Actions based on technological assessments will result in rewards and gains for some, and restrictions and losses for others. What kinds of incentive-constraint mechanisms shall be built into the technology assessment process? Can risk spreading arrangements be created before the assessment is made? Who pays (if anyone) compensation for damages and missed opportunity gains? These and related effects questions must be dealt with if policy makers are to embrace the TA process as a positive instrument rather than a means of getting into situations with unknown routes out.

SOME INSTITUTIONAL CONSIDERATIONS

To be an effective instrument in the policy process, technology assessment must occur in a favorable institutional setting. Policy, institution, and TA are all such large concepts that it appears certain that there is no single institutional form that will meaningfully relate all three. Assuming consensus that TA will occur in public and private, in large and small, and in executive, legislative and academic settings, what are some of the key considerations for designing the appropriate setting?

Prior to designing the home for TA activities it might be helpful to look at the criteria for convening a technology assessment event. Five are posed:

1. Recognition of a problem, opportunity or need.
2. The availability of substantive knowledge: technological, economic, social, environmental, ...
3. The availability of concepts and methods for analysis and synthesis; process knowledge.
4. The possibility (belief?) that something can be done with the results of the assessment. (This is discussed in the material to follow.)
5. The provision of adequate human and financial resources and the institutional means to conduct the assessment.

The first criterion implies a knowledge of the desired goal, objective, or end-state that is different from the present situation or that which may obtain without assessment. One would like to meet the requirements of the operations researcher who, in his own jargon, asks that the objective function be clearly, explicitly and unambiguously stated. But in the real world this can seldom be done. It is particularly difficult in complex social situations and this is what technology assessment is all about — what are the social (among other) impacts?

Many persons and groups have advanced arguments that it is neither possible nor desirable to treat all goals explicitly. They point to the increased possibilities of reaching consensus on means and on next steps if the ends are left implicit or at least a little vague. Many governmental approaches to goals have proceeded on this assumption. The National Goals Research Staff Report of July 4, 1970, deliberately took this stance in saying, "This report is designed not as a listing of specific goals to be sought, but as a springboard for discussion and an aid to decision. It does not presume to say *what* our choices should be. Rather it defines the questions, analyzes the debates and examines the alternative sets of consequences" [7].

The second and third criteria are relatively self evident. There must be sufficient information as to *what* the assessment concerns and there must be knowledge as to *how* to use it and how to proceed.

The fourth criterion of the possibility that something can be done with the results of an assessment can probably safely assume that there will be no bars to anyone's assessing anything he wants. The reverse question as to who has *responsibility* to assess given classes or levels of things is the difficult one. This question also includes the so-called problem of "who shall assess the assessors?" i.e., the review and adjudication issue. The National Environmental Policy Act may give us some guidance. Section 102(2)(c) requires Federal agencies proposing legislation or planning to undertake an action "significantly affecting the quality of the human environment" to file a statement as to the nature and degree of that impact. These statements are reviewed by the Council on Environmental Quality and are available to the public. If experience shows this to be a workable arrangement for environmental impacts, it might be desirable to explore the possibilities of the Council of Economic Advisors reviewing "economic impact statements." If the Council of Social Advisors which has occasionally been proposed were created, social impact review might similarly be possible. But whether there are one, two or x number of executive branch councils there still remains the overall or balanced impact assessment problem. The only remedy here appears to be some instrument of the legislative branch. A variety of proposals are under active consideration and therefore it is too early to prophesize the specific form it will take.

Regardless of the specific form, there is a compelling requirement for government to be involved in technology assessment. Only government can be expected to have sufficient objectivity and has the power to enforce action in the best interests of all of the people. In cases involving security or proprietary information, only government can be assured access. In other cases, only government can arrange for processes that do not violate antitrust laws. Lastly, government is uniquely positioned to deal with many of the intergovernmental and international aspects of assessment activities.

There is consensus that no single organization, team, or other array of skills, ever will be equipped to perform the wide range of assessments that will come to a legislative or even a high-level executive branch component. Additionally, there is strong evidence indicating that a major portion of the resources applied to each assessment must be assembled on an ad hoc basis especially for that assessment. A centralized institution like, for example, the Office of Technology Assessment proposed by Senator Jordan et al. in S.2302, must therefore incorporate two major capabilities. One concerns the ability to design analyses and to perform them with temporary task teams or to contract for them. A complementing continuing capability must exist to advise in the selection of tasks, to review the results, and to present the findings in a form suitable for policy decision and action. One would hope that the continuing core staff would also (1) have the ability to accumulate knowledge (substantive and procedural) about the TA process, and (2) to act as, or to arrange for, a central clearinghouse for knowledge of on-going activities and as a repository of the results of completed ones.

Managing the TA process or team is a critical issue just now coming to serious attention. Should the leader of the assessment effort be one highly skilled in the key technological area? A representative of the affected parties? An analyst skilled in methodology? A professional manager? At the previously mentioned 1971 Andover conference the only qualification for a TA team leader on which all could agree was that he be able "to walk on water!"

A major feature of technology assessment is, or should be, participation. All of the role players described earlier need an explicit and meaningful way to get involved and to execute their roles. All participants will not be involved simultaneously but neither is the process strictly a sequential one with little direct interaction. The hardest part of designing a participative mechanism concerns incorporating the views of the affected parties, the public, and those of valid but minority interest groups. Who shall speak for the public? How can the unorganized organize? Can one conceive of priorities for the number, intensity and duration of invited and self-invited marginal participants? And who is marginal?

SOME ALTERNATIVE FUTURES

The most unlikely future scenario is that we will continue to drift along without a formalized function that at least today would be called technology assessment. Whether that name will endure is less certain than that there will be something to name. TA in some form and under a variety of labels is old stuff. The degree of formality, sophistication of method, and self-consciousness are the main changes. The TA movement has developed sufficient momentum and adherents and its essence is embodied in law (at least in part) through the National Environmental Quality Act of 1970.

There is insufficient information upon which to predict the directions more formal and institutionalized technology assessment will involve. One can however resort to the "surprise free prediction" technique which allows one to explore a number of alternatives. The surprise free term means that if developments are not along one or a combination of the directions cited, then it will be a surprise.

One class of technology assessments that have a high probability of increasing and staying with us are the reactive ones. These are analyses initiated in response to large, well recognized, sometimes dramatic events. Thalidomide, ADX-2 and the SST are recent examples. Occasionally, long recognized problems spring to prominence and enter this reactive category. Mercury in fish and miner's lung are examples here. The term "reactive" tends to have a negative connotation but with forethought a more positive strategy can evolve. Resources are always scarcer than needs and without focusing events — such as impact crises of varying kinds — interest and support for TA will decline. One deliberate and open strategy could be to capitalize on dramatic events to move the entire process forward and in turn to decrease the number and degree of surprises, and severity of future events.

Another approach is the incremental one where some level of effort for TA activity is assigned through normal priority setting and budgeting mechanisms. Within these constraints the assessors would "optimize" by selecting for analysis those areas with largest suspected negative impacts and those deemed to have the highest payoff from knowledge of the effects.

It will be a surprise if the legislature does not continue to be the leader and prime mover for the establishment of some form of institutionalized technology assessment. From its own perspective the Congress cannot afford to lose control over so vital and dynamic an area of society's concerns. Also, Congress is the only institution sufficiently broad and detached from specific motivations to deal with the issues presented. Most importantly, because of its legislative powers, it has the capacity to act innovatively and directly on behalf of and in the balanced interests of *all* the people. In the vernacular of the day, TA is Congress' bag.

Technology assessment will most likely flourish in executive branch components. Some activities will be motivated directly by law as in the case of environmental quality. Other assessments will be ancillary to each agency's mission and performed in the course of competent and responsive public administration. Intending no pejorative tones, other assessments will be performed offensively or defensively in the normal competition among bureaucracies and between the executive and legislative establishments.

Private technology assessments are likely to increase and least likely to be identified by such terminology. Corporate reports, advertising, product statements, etc. will all contain evidence of some credible attempts to assess impacts. Unfortunately, it will be difficult to ascertain which are based on high-quality objective efforts and which on more flimsy or prejudiced grounds. Unless the mood of citizens toward business changes rather markedly, all will be suspect. Volunteered academic and not-for-profit assessments are likely to be more plentiful. Many of these will be based on the highest motives of public service and some will be incidental to other activities. They are more likely to be characterized by novelty or sophistication of method (TA_m) with substance playing a lesser role. This is as it should be and all assessment activities will benefit from the results.

In a slightly different type of forward look,one can focus more on the technology assessment process. In the previously cited report "Technical Information for the Congress" an argument is made for a sequential approach. The point here is that the technical (and scientific) arguments should be resolved first and in the normal manner by which these are settled among scientific and technological peers. *Then*, the "political" issues are introduced and resolved (or not resolved) through extant political mechanisms. This approach may be desirable and efficient from the analytical standpoint but it raises the emotionally-laden issues of secrecy and elitism. Unless the public, its representatives, and particularly the affected parties feel at a minimum they have the right to observe or participate in the technical arguments, additional barriers to effective assessment are raised.

An alternative to this sequential approach is the systems approach. Complex models — systems — are constructed including all types of variables: scientific, economic, social, ecological, etc. Currently, some computer models and expert-team efforts approach problems in this systems fashion with growing insight and finesse.

Much has been said and written about the adversary process and technology assessment. The arguments as to whether or not adversarial actions should be used in TA are spurious — it is inevitable that they will. The key problems will in focusing the arguments (cases?) on the real issues and not on those bringing more power than clarity to the outcome.

It seems most likely that the sequential, systems, adversary, and "best efforts" (i.e. all of the above and more) approaches will all proceed in parallel for the foreseeable future. Conceptual and methodological advances coupled with improved skill and experience will help this field to grow, probably in spurts, like many similar complex endeavors.

Earlier in this chapter one of the role players in the technology assessment process was identified as the "alerter". In the opinion of this writer this subject has not received sufficient attention by students and practitioners of TA. But it is far from a new idea as demonstrated by this 1937 comment: "Though the influence of invention may be so great as to be immeasurable, as in the case of gunpowder or the printing press, there is usually opportunity to anticipate its impact upon society *since it never comes instantaneously without signals* "[8]. [Italics in the original.]

There are at least three aspects of this alerting function that requires attention. The first concerns the invention and innovation of a new technological capability. How do we "alert" society that this new thing exists and how do we perceive what its impacts might be? This is an anticipatory activity involving a now extant technological capability but with no knowledge as to its impacts as yet.

Theoretically, one could argue for prior impact assessment of possible technological advances as a guide (positive and negative) to research and development. In time this may come in cases where huge R & D expenditures are anticipated and in these circumstances it may be warranted. In the light of present resources and a limited TA capability, it would seem prudent to give existing and about-to-be-introduced technologies priority attention over R and D.

The second and third aspects of alerting are complementary functions of early warning. One type of monitoring must involve tracking known technologies and substances to detect the nature and extent of their direct and indirect impacts on society, the economy and the ecosystem. The everpresent impossibility of doing this for all things reminds us of the need for a selection scheme.

The other side of the monitoring for early warning function is to observe society, the economy and the ecosystem for undesirable effects and track down their source, which may or may not be technologically caused.

As has been suggested in several places, a key element in both the alerting and doing areas is the selection of what to monitor and what to assess. A monitoring and alerting scheme in which society could place reasonable confidence could go a long way toward easing the second problem of setting TA priorities.

TECHNOLOGY ASSESSMENT IN PERSPECTIVE

For the most part, the present policy-making machinery in this country does not yet include a viable technology assessment function. Like all generalizations there are notable exceptions. Some technological assessments have been mounted in response to crisis and others in response to the actions of specific pressure groups. A major set of TA activities varying in quality and intensity are underway in response to new laws — principally the National Environmental Policy Act and food and drug statutes. There is a growing literature; substantive, methodological and exhortative. Most significantly, there is increasing awareness of the need for the potentiality of assessing the impacts of our actions before they overwhelm us.

Technology assessment (as distinct from assessment of the impact of other activities) is in the forefront for a variety of reasons. To say that ours is a technologically-based society has become trite but no less true. Congressional leadership and focusing attention on a concept with a handy label has created a movement now beyond party and bureaucratic identification.

Not so happily, TA is out in front because of a persistent though disorganized anti-technology anti-rationalism mood pocketed around our country and the globe in a variety of forms. That this is a self-defeating approach is a rational thought and, by definition, uncommunicable, and unacceptable to adherents of those views. Negativism, ignorance and apathy are, and will continue to be, troublesome aspects of TA and policy making and are areas we are ill-equipped to deal with.

Government leadership is both required and appears to be forthcoming, albeit, more slowly than many desire. Technology assessment cannot exist as a separate something — it must be an integral part of the total social, economic, environmental policy making system. Therefore, the highest focal point for its development and execution must be in the legislature. TA also will be an important aspect of policy making, planning, and action throughout the executive branch, in the regulatory agencies and in the private sector.

Presently technology assessment with respect to environmental impact is the area of greatest activity and of advance. Alarmists tell us survival is at stake and they may be right. While preservation of the ecosystem is absolutely essential, only a few inhabitants of space ship earth are willing to return to the noble savage condition however harmonious was his relationship to his environment. Problems of balance, timing, who pays, and spotting critical irreversible areas for priority actions offer meaningful challenges to the technology assessor.

Social and economic survival is also at stake. At the time of writing of this chapter, the Nation is in the grip of freezes, tariffs, and other controls designed to gain mastery of its economy and to preserve its position as a world power. And all this in times of unprecedented plenty and growth. Even the President has been moved to remark that we are a nation with more than ever before but are enjoying it less. If technology assessment can achieve only some of the claims made for it, the conditions are certainly right for its use.

But a condition precedent for assessing the impacts of technology, or anything else for that matter, upon society is a system of measures or indicators as to what is happening. We appear to be moving very slowly and haphazardly toward the development of meaningful social indicators and accounts. Additionally, there is only a loose coupling between the TA_t technology assessors and the "soft science" TA_i assessors. In addition to lack of contact and poor communications when contact is made, one finds evidences of hostility and snobbery among the disciplines that somehow must be brought together if TA_p and TA in general are to be successful. Who shall be the integrator and who shall be the referee?

The reader will properly observe that this chapter has raised more questions than it has answered. Developing answers to those and many more questions is the future of technology assessment. There is much to learn. But enough is known and the pressures are strong enough to act now. We can learn much more by doing. There is room for and necessity for a multiplicity of institutional approaches. The Congress is positioning itself for action and can be expected to supply the major missing ingredient of centralized leadership and inspiration. Technology assessment is here, and it is here to stay. It has a bright future as a key element in national public and private policy formulation and program implementation.

NOTES AND REFERENCES

1. "Resources for Freedom," A Report to the President by the President's Materials Policy Commission, June 1952, p. 17; (The "Paley Commission").

2. A Potomac Associates book: Universal Press, N.Y., 1971.

3. *Congressional Record*, pp. S 13823-4, August 6, 1971.

4. "The Social Function of Technology Assessment." Paper presented by Dr. Franklin P. Huddle, Science Policy Research Division, Legislative Reference Service, Lib. of Congress, at Seminar Series: Processes of Technology Assessment, Program of Policy Studies in Science and Technology, The George Washington Univ., January 22, 1970.

5. *Toward Balanced Growth: Quantity With Quality*; Washington, D.C.; Government Printing Office; 124, July 4, 1970.

6. House Committee on Science and Astronautics, pp. 515 ff, April 1969.

7. *Op. Cit.*, 23.

8. "Technological Trends and National Policy — Including the Social Implications of New Inventions," A Report to President Franklin Roosevelt, June 1937.

SOME MODEST CONCLUSIONS IN TECHNOLOGY ASSESSMENT

MARVIN J. CETRON

President
Forecasting International Ltd., Arlington, Va., U.S.A.

The technology assessment movement marked a turning point in the public's attitude toward science and technology. It wasn't just its jargon, but rather it heralded the beginning of a new and more enlightened phase in appreciating how technology could be used to improve our quality of life. Technology assessment should not be misinterpreted as a barrier to the advancement of technology and its applications, but as a means to promote technology which will on the whole help society and at the same time deter technology which may have detrimental effects. Technology assessment is in its infancy as a newly emerging management science.

WHAT IS REQUIRED?

The overall task of the technology assessment process is to provide explicit information that will assist managers and politicians in arriving at more rational decisions and which will promote the development and application of new technologies to bring about solutions to today's societal problems. Technology should be supported that:

(a) Can solve existing or potential problems caused by the application of earlier technologies

(b) The application of which does not produce greater, new problems including the secondary and tertiary order consequences.

(c) Can lead to the betterment of the quality of life for Mankind.

The other side of the coin — the prevention of the development and application of existing or new technologies which tend to produce negative side effects — must also be addressed. There are several criteria — some in the form of "catalysts" — that must be present in order for a viable technology assessment program to work. Among others, these include:

(a) *A list of qualitative or quantitative goals, needs or areas of concern.* These problem areas or goals should consist not only of present goals but also time-oriented future goals or potential future problem areas.

(b) *A set of operational societal indicators*, (i.e., social, technological, economic, legal, political, demographic, etc.). Hopefully, quantitative subjective weights can be assigned to these indicators. This quantification would be "nice to have," but is not considered to be a firm requirement.

(c) A set of analytical skills in the form of *tools, methods and techniques* that can be brought to bear on various aspects of the technology assessment problem. This includes such specialized "tools" as: the structuring and projection of plausible total future environments ("New Worlds") in scenario form; the development of Models of both individual and group behavior that will reflect changes — in selected indicators — brought about by the introduction of technology into the socio-political environment; and, finally, a technique for measurement of — within reasonable limits — the total impact that technological innovations will have on every aspect of the environment. The advantages and disadvantages of the different methods should be determined and shown, case studies should be available, and the relationships between various indicators should be explained and modeled. This would make it possible for others to develop better instruments for the systematic integration of various aspects of the problem and to balance the short-and long-term benefit/cost, including social benefits.

(d) *Organizational implementation patterns and procedures* to provide the institutional framework that must be present in making operational decisions. One must make the information on technology assessment studies available in a manner that is easy for the decision-maker to assimilate and use (including available options) in a realistic manner.

WHO SHOULD PERFORM THE ASSESSMENTS?

Another problem arises in technology assessment which comes under the heading "Should technology assessment be accomplished strictly by various government agencies or should the private sector participate?" There are proponents on both sides — those who contend that the government as the "guardian of the public interest" should be the focal point; and, of course, certain individuals in the private sector feel that technology assessment will never pay for itself. This latter viewpoint is about as short-sighted as saying that R & D, market research, advertising, education, and training do not pay for themselves. In the short run they don't; however, in the long run I think we all understand that there are the real payoffs. It is ever so easy to become imbued with the corporate philosophy which says that the private segment has no memory for work that took place over a year ago.

I think a second item should be considered in the same category and this concerns the so-called "inherent untrustworthiness" of industry. Those who are against industry participating in the area say that industry can't be trusted; they can't perform a technology assessment for essentially ultruistic purposes, since they are either selfishly motivated, or have an axe to grind. However, technology assessment is a very definite component of the rampant social consciousness much touted these days and explained in more detail in my chapter "The Social Responsibility of Business or the Changing Role of Goals in Corporate and R & D Planning." As far as the contract research or "think tanks" are concerned, it is a welcome new source of direct business. But it is not strictly a windfall. Technology assessment is viewed by the research labs with mixed blessing since it may give some of their departments extra work, it may also cut off research that may appear to have some negative effects, or it may lead inexorably to more stringent regulations. Nevertheless, it is my opinion that technology assessment should be performed by profit-making organizations that can

(a) explore and expose exploitable technological opportunities and

(b) which can also be used by corporations and industries to anticipate legal and political restraints.

Technology assessment is still more art than science but we hope that this volume will help to put more of the concepts in clearer focus.